A Field Guide to the Birds of Borneo, Sumatra, Java, and Bali

THE AUTHORS

John MacKinnon is currently Director of the Asian Bureau for Conservation. He is a professional conservationist who has worked in Asia for over 20 years and lived for seven years in the region covered by this book. He is the author of several other books on natural history including *In search of the red ape*, *The ape within us*, *Animals of Asia*, *Borneo*, and the *Field guide to the birds of Java and Bali*. He has spent several years developing field methods for assessing bird species richness in forests and computerized species databases and monitoring systems.

Karen Phillipps was born and brought up in Borneo and lived in the region for many years. She is a professional artist and birdwatcher. She has illustrated a number of other field guides including *A field guide to the mammals of Borneo*, *A new guide to the birds of Hong Kong*, *A guide to the fruit of south-east Asia*, and *Common birds of Malaysia*. She currently lives in Hong Kong.

A FIELD GUIDE TO

The Birds of Borneo, Sumatra, Java, and Bali

THE GREATER SUNDA ISLANDS

JOHN MACKINNON and KAREN PHILLIPPS
In collaboration with
PAUL ANDREW and FRANK ROZENDAAL

Illustrations by
KAREN PHILLIPPS

OXFORD
UNIVERSITY PRESS

Great Clarendon Street, Oxford OX2 6DP

Oxford University Press is a department of the University of Oxford.
It furthers the University's objective of excellence in research, scholarship,
and education by publishing worldwide in

Oxford New York

Athens Auckland Bangkok Bogotá Buenos Aires Calcutta
Cape Town Chennai Dar es Salaam Delhi Florence Hong Kong Istanbul
Karachi Kuala Lumpur Madrid Melbourne Mexico City Mumbai
Nairobi Paris São Paulo Singapore Taipei Tokyo Toronto Warsaw
with associated companies in Berlin Ibadan

Oxford is a registered trade mark of Oxford University Press
in the UK and in certain other countries

Published in the United States
by Oxford University Press Inc., New York

© John Mackinnon and Karen Phillipps

The moral rights of the author have been asserted
Database right Oxford University Press (maker)

First published 1993
Reprinted 1994, 1995, 1997, 1999

All rights reserved. No part of this publication may be reproduced,
stored in a retrieval system, or transmitted, in any form or by any means,
without the prior permission in writing of Oxford University Press,
or as expressly permitted by law, or under terms agreed with the appropriate
reprographics rights organization. Enquiries concerning reproduction
outside the scope of the above should be sent to the Rights Department,
Oxford University Press, at the address above

You must not circulate this book in any other binding or cover
and you must impose this same condition on any acquirer

A catalogue record for this book is available from the British Library

Library of Congress Cataloging in Publication Data
MacKinnon, John Ramsay.
A field guide to the birds of Borneo, Sumatra, Java, and Bali, the
Greater Sunda Islands / John MacKinnon and Karen Phillipps, in
collaboration with Paul Andrew, Frank Rozendaal; illustrations by
Karen Phillipps.
Includes bibliographical references and index.
1. Birds—Greater Sunda Islands. I. Phillipps, Karen.
II. Title.
QL691.I5M33 1993 598.29598—dc20 92-30340
ISBN 0 19 854035 3 (Hbk) ISBN 0 19 854034 5 (Pbk)

Printed in Hong Kong

*For Mary
who put one and one together
and made three.*

Acknowledgements

WE WOULD like to thank a large number of people who have helped in the compilation of data, preparation, reviewing, and editing of text, making materials available for artwork, and commenting on plates and text. Three collaborators deserve special thanks. Paul Andrew revised and edited the entries on species distributions and status in line with his new Checklist for Indonesia (Andrew 1992). Frank Rozendaal did a detailed editing and technical checking of the text and plates. Bas van Balen made a thorough review and update of one author's previous work, *Field guide to the birds of Java and Bali*, from which the present book is partly drawn.

Other experts who reviewed sections of the text and gave valuable comments include Derek Scott, David Melville, Ken Searle, Verity Picken, and Phil Round. Thanks are also due to many people who have given advice, comments, and information to help with the preparation of the book or earlier lists and data on which it is based, namely David Wells, Derek Holmes, Murray Bruce, Charles Francis, Edward Dickinson, Chris Hails, Hugh Buck, John Schmitt, and Tim Inskipp.

We thank other colleagues who have provided unpublished notes, species lists, tape recordings, and other privately held information: Frank Rozendaal, Paul Andrew, Ben King, David Wells, Johan Iskandar, Sugarjito, Pak Soemadikarta, Randy Milton, Glynn Davies, Jan Wind, Rob de Wulf, John Ash, George Ledec, Nico van Strien, Alan Robinson, Ken Scriven, James MacKinnon, Derek Holmes, Bill Harvey, Steven and Anne Nash, David Gibbs, David Bishop, Jared Diamond, Joe Marshall, Wim Verheugt, Marcel Silvius, Bas van Helvoort, Harti Amman, Arthur Mitchell, Mark Leighton, Tim Laman, Paul Gittins, David Pearson, and Jesper Madsen.

We would like to thank the staffs of the various regional museums for their assistance in examining birds in their collections: in Bogor Museum, Ibu Sudarianti, Pak Nurjito, Pak Boeardi, Pak Su. Mohd. Amir, and Director Pak Sutikno; in the National University of Singapore, Mrs Yang Chang-man and staff; in the Muzium Sabah Conservation Building, Mr Raymond Goh and staff; and in the Sarawak Museum, Director Lucas Chin and zoologist Charles Leh.

Particular thanks are due to all those who have helped in the physical preparation of the book—in the typing, editing, and other tedious labours of love: Jani Budiman, who typed the original drafts; Kathy MacKinnon, who nursed the *Birds of Java* book to press and whose secretary Katarina Panji transferred the text to word processor; and Florence Lai, who helped type the final draft.

We would like to thank those who have been of particular help by providing hospitality, logistical support, and other encouragement: Mary Ketterer, Ken Searle, Liz Bennett, Eric Wong, and Vicky Melville.

Hong Kong J.M.
March 1992 K.P.

Contents

The plates are to be found between pp. 44 and 45

List of colour plates	x
Anatomy and plumage of a bird	xii
Glossary and abbreviations	xiii

BACKGROUND

Introduction to this book	1
Introduction to the region	5
Biogeography	14
Conservation	20
Field techniques for birdwatching	26
When and where to see birds	35

FAMILY AND SPECIES DESCRIPTIONS 45

APPENDICES

Appendix 1 Endemic and threatened and endangered species in main reserves	410
Appendix 2 Endangered and threatened species by island	417
Appendix 3 Land birds found on offshore island groups	421
Appendix 4 Bornean montane birds by mountain group	434
Appendix 5 Annotated list of birds of the Malay Peninsula not described in the text	438
Appendix 6 Sonosketches of characteristic bird calls	442
Appendix 7 Regional ornithological clubs, journals, and museums	446
Bibliography	448
Index	459

Plates

1. Petrels, Shearwaters, and Storm-Petrels
2. Tropicbirds and Boobies
3. Pelicans and Cormorants
4. Frigatebirds
5. Herons and Egrets
6. Smaller Herons and Bitterns
7. Storks, Ibises, and Spoonbills
8. Ducks
9. Osprey, Hawks, Kites, and Eagles
10. Eagles, Harriers, and Hawks
11. Eagles, Buzzards, and Hawk-Eagles
12. Falcons
13. Hawks and Falcons in flight
14. Megapodes and Partridges
15. Partridges, Quail, Buttonquails, and Pheasants
16. Fireback Pheasants
17. Pheasants
18. Rails
19. Rails, Finfoot, and Jacanas
20. Large Plovers
21. Small Plovers
22. Curlews, Godwits, Tattler, Turnstone, and Dowitcher
23. Sandpipers
24. Painted Snipe, Snipes, Woodcocks, and Ruff
25. Knots, Stints, and Sandpipers
26. Grebes, Stilts, Phalaropes, Thick-knee, and Pratincoles
27. Jaegers and Gulls
28. Terns
29. Terns
30. Crested-Terns and Noddies
31. Green-Pigeons
32. Green-Pigeons and Fruit-Doves
33. Pigeons
34. Pigeons
35. Parrots
36. Large Cuckoos
37. Small Cuckoos
38. Malkohas, Coucals, and Ground-Cuckoo
39. Barn Owls and Owls
40. Scops-Owls
41. Frogmouths and Nightjars
42. Swifts and Treeswifts
43. Trogons
44. Kingfishers
45. Kingfishers
46. Bee-eaters, Roller, and Hoopoe
47. Hornbills
48. Barbets
49. Barbets and Honeyguide
50. Woodpeckers
51. Woodpeckers
52. Broadbills
53. Pittas
54. Larks and Swallows
55. Cuckoo-Shrikes and Fruit-Hunter
56. Minivets
57. Leafbirds
58. Bulbuls
59. Bulbuls
60. Bulbuls
61. Drongos
62. Orioles and Fairy-Bluebird
63. Crows and Bristlehead
64. Tits, Nuthatches, Shortwings, and Robins
65. Jungle-Babblers

66 Scimitar-Babbler and Wren-Babblers
67 Tree-Babblers
68 Tit-Babblers, Fulvettas, Sibias, Yuhinas, and Rail-Babbler
69 Laughingthrushes, Mesia, and Shrike-Babblers
70 Shamas and Forktails
71 Cochoas, Chats, Rock-Thrush, and Whistling-Thrushes
72 Thrushes
73 Flyeater, Leaf-Warblers, and Reed-Warblers
74 Warblers, Cisticolas, Tesia, and Bush-Warblers
75 Tailorbirds and Prinias
76 Jungle-Flycatchers and Flycatchers
77 Flycatchers
78 Flycatchers
79 Paradise Flycatchers, Fantails, and Monarchs
80 Whistlers, Wood-Swallow, and Shrikes
81 Wagtails and Pipits
82 Starlings
83 Sunbirds
84 Spiderhunters and Honeyeater
85 Flowerpeckers
86 White-eyes
87 Finches and Munias
88 Sparrow, Weavers, and Buntings

Anatomy and plumage of a bird

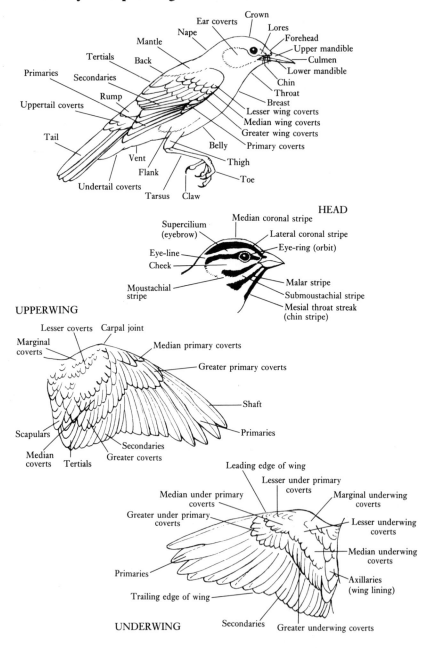

Glossary and abbreviations

Glossary

accidental: a stray.
adult: birds mature enough for breeding.
apical: terminal, outer end of something.
aquatic: water-living.
arboreal: tree-living.
axillaries: the feathers in the axil or wing lining.
basal: refers to the base.
cap: usually used to denote a larger area than the crown.
casque: an enlargement of the upper part of the bill.
cere: bare, wax-like or fleshy structure at the base of the upper beak containing the nostrils.
cock: male galliform bird. Used as verb to describe jerky raising of head or tail.
collar: a band or bar of contrasting colour passing around either the front or back of the neck.
coronal stripe: a streak on the crown running front to back.
cosmopolitan: a species which is widely distributed, having a near worldwide distribution.
crepuscular: active in twilight and just before dawn.
crest: tuft of elongate feathers usually on the head which, in some species, may be raised or lowered.
crown: top of a bird's head or top branches and foliage of a tree.
cryptic: having protective colouring or camouflage with associated behaviour.
deciduous: of a tree (or forest) which is leafless for part of the year.
diagnostic: character of sufficient distinctiveness to allow identification to be made.
dimorphic: existing in two well-defined, genetically determined plumage types.
dipterocarp: a tree belonging to the family Dipterocarpaceae.
distal: terminal; refers to the outer end; opposite of proximal or basal.
diurnal: active in daytime.
duetting: male or female of a pair singing together, in response to each other.
echolocation: emission of high-frequency sounds to locate objects.
eclipse plumage: post-breeding plumage in which distinctive breeding features are obscured; found in ducks, sunbirds, etc.
endemic: indigenous; restricted to a particular area.
face: the lores, orbital area, cheeks, and malar area combined.

feral: domesticated species released and living wild.
ferruginous: rusty brown colour with orange tinge.
fledgling: a young bird partly or wholly feathered, flightless or partly flighted, but before flight.
flight feathers: the primaries, secondaries, and tail feathers which give buoyancy in flight.
flush: frighten out of concealing cover.
foot: claws, toes, and tarsus combined.
foreneck: the lower part of throat.
frontal shield: bare, horny or fleshy skin on forehead, which extends down to the base of the upper bill.
fulvous: brownish yellow.
gait: manner of walking.
gape: the fleshy interior of the bill.
gliding: straight, level flight with wing outstretched or slightly swept back, without flapping.
gorget: a necklace or distinctively coloured patch on the throat or upper breast.
grace note: soft introductory note given just before main song.
gregarious: living in groups.
gular pouch: an expandable patch of bare skin on the throat of pelicans, cormorants, etc.
Gunung = *Gn.*: Malay name for mountain.
hackles: long, slender feathers found on neck of some birds.
head: forehead, crown, nape, and sides of head combined. Does not include chin and throat.
hepatic: usually used as a label for the brown colour morph found in some cuckoos.
Holarctic: Palaearctic and Nearctic regions combined.
hood: a dark-coloured head and (usually) throat.
immature: used to denote all plumage phases except the adult plumage, i.e. includes juvenile and subadult.
juvenile: fledging to free-flying birds, with the feathers which first replaced the natal down.
leading edge: the forward edge of the wing.
lobe: a fleshy, rounded protuberance (usually on the feet, as an aid to swimming).
local: discontinuous or uneven in distribution; found in only certain areas.
malar area: the area bounded by the base of the bill, the throat, and the orbit.
mantle: the back, upper wing coverts, and scapulars combined.
median: pertaining to the middle.
melanistic: a blackish morph.
mesial: median; running down the middle, usually of throat.
migratory: relating to regular geographical movement.
morph: a distinct, genetically determined colour form.

nocturnal: active at night.
non-passerine: refers to those species having a different foot structure from the passerines.
notch: an indentation in outline of feather, wing, tail, etc.
occiput: the rear part of the crown.
ochraceous: dark brownish yellow.
Old World: the Palaearctic, Afrotropical, and Oriental zoogeographical regions.
Oriental: referring to a zoogeographical region, from Himalayas, India, SE Asia east to Wallacea.
padang: Malay/Indonesian word for open space.
paddy: irrigated rice fields.
Palaearctic: North Africa, Greenland, and Eurasia.
pan-tropical: distributed around the world's tropics.
passerine: refers to the families of 'perching birds'. They differ in foot structure from the non-passerines with three toes always forward and one always back.
pelagic: ocean-living.
pied: patterned in black and white.
piratic: stealing food from other birds or species.
'pishing': making a harsh squeaking noise in imitation of bird alarm calls to attract attention of birds.
plume: a greatly elongated feather, usually used in display.
primary forest: original or virgin forest.
proximal: the near, or basal portion.
race: a colloquial term for subspecies.
raptor: a bird of prey.
rictal bristles: bare feather shafts arising around the base of the bill.
roots: a resting or sleeping place/perch for birds.
secondary forest: new forest growing where the original or primary forest has been removed.
shaft streak: a contrasting line of pigment along the midline of a feather which forms a pale or dark streak.
shank: bare part of leg.
skin: a study skin or specimen.
skulk: action of creeping or flitting about in unobtrusive and furtive manner close to the ground.
soaring: rising flight, riding updraughts to gain height without flapping.
spangles: shimmering points in a bird's plumage.
speculum: iridescent dorsal patch on a duck's wing which contrasts with the rest of the wing.
storey: a level of the forest.
streamers: greatly elongated, ribbon-like tail feather or projections of the tail feathers.

subterminal: near the end.
subadult: a later immature plumage phase.
sub-montane: lower elevations and foothills of mountains.
subspecies: a population which is morphologically distinguishable from other populations of the same species.
terminal: at the end.
terrestrial: ground-living.
trailing edge: the hind edge of the wing.
undergrowth: herbs, saplings, and bushes in a forest.
underparts: undersurface of body from throat to undertail coverts.
underwing: the entire undersurface of the wing, including coverts and flight feathers.
upperparts: upper surface of the body.
vagrant: rare and irregular in occurrence.
vent: the area surrounding the cloaca, sometimes used as shorthand for undertail coverts.
Wallacea: zoogeographic region between Oriental and Australasian regions showing relations to both.
washed: suffused with a particular colour.
wattle: a patch of bare skin, often brightly coloured, usually hanging from some part of the head or neck.
web: a fold of skin stretched between two toes; or a vane of a feather.
wing bars: bands formed on the wings when the tips of wing coverts are a different colour from their bases.
wing coverts: a term used for all or part of the lesser, median, and greater coverts of the upperwing or underwing.
wing lining: a term loosely applied to the underwing coverts.

Abbreviations used in the text

AG	(see p. 3 for explanation)	N	North
App.	Appendix	Pl.	Plate
C	Central	P.R.	Phil Round
C.H.	Chris Hails	S	South
D.A.H.	Derek Holmes	Tg.	*Tanjung* (= cape)
E	East	T.H.	Tom Harrisson
Gn.	*Gunung* (= mountain)	W	West
M.M.N.	Mary M. Norman	>	(see p. 3 for explanation)
M. & W.	Medway and Wells (see Bibliography)		

BACKGROUND

Introduction to this book

THE Greater Sunda Islands—Borneo, Sumatra, Java, and Bali—are biologically one of the richest and most complex areas of our planet. Their complexity led to a large extent the Victorian naturalist Alfred Russel Wallace to identify the process of evolution by natural selection, which then prompted Charles Darwin to publish his own conclusions on the subject. These form the foundations of modern biology. Yet 150 years later, our knowledge of the precise relationships of the bird faunas of these important islands remains only a little better than it was when Wallace was collecting in northern Borneo. Worse, the forests are now being destroyed, and if we do not document quickly which species occur where, the information content of the region may be lost for ever.

One handicap in gathering a complete picture of the distribution of the birds of the region has been the paucity of regional field guides. For Sumatra there is a *Checklist* (Marle and Voous 1988) but no illustrated field guide. Most Sumatran endemic species have never been figured before. For Borneo there is Smythies (1960) which is rather outdated and too heavy to be regarded as a modern field guide, although the Malayan Nature Society's re-publication of the annotated plates is a partial solution. For Java there is a field guide (MacKinnon 1990) which is a limited Indonesian printing and has rather poor plates. The only general book of the whole region is the 1947 *Birds of Malaysia* by Delacour which has no colour illustrations and uses a greatly outdated taxonomy. The excellent book *A field guide to the birds of South-east Asia* by King, Woodcock, and Dickinson (1975) only covers SE Asia as far as the Malay Peninsula and Singapore. The present book is the next logical geographical field guide.

This book gives descriptions of 820 species living or reported in the region. These are illustrated in 88 colour plates and a few additional black

2 BACKGROUND

and white illustrations, including most of the insular forms, sexual variants, and immature forms of polymorphic species. Each plate is accompanied by a page of notes, drawing attention to the most important and diagnostic characters. The letters S, B, and J after species names indicate their occurrence on Sumatra, Borneo, and Java/Bali.

The appendices give additional useful information such as distribution details of the birds on the smaller islands, distributions of Borneo's montane birds, and where to find the region's rare and endemic birds. There is a list of species found in the Malay Peninsula which are not described in the main text.

In 1990 Sibley and Monroe published their *Distribution and taxonomy of birds of the world*. This work is the most comprehensive revision since the 15-volume treatment of Peters (1931–86). The volume is based on DNA hybridization studies which provide a far more objective measure of the relationships of genera and families than mere morphology. As a result the authors have produced a rather revised ordering of families. The most dramatic changes involve inclusion of whistlers, wood-swallows, orioles, fantails, drongos, monarchs, leafbirds, and helmet-shrikes as tribes under an enlarged crow family Corvidae; and subjugation of flowerpeckers under the sunbirds.

Over the next few years these revisions are likely to become accepted as the new standard. I have mentioned these realignments in the relevant family and species descriptions in the text but continue to use the more traditional family and species ordering of King *et al.* (1975) which was also followed by Marle and Voous (1988) in the British Ornithologists' Union (BOU) check-list of Sumatra. This is because most birdwatchers in the Sunda region are now familiar with this ordering and will find their way around this field guide more easily if it is retained. I have, however, followed the nomenclature of Sibley and Monroe (1990) as closely as possible in an effort to limit the number of current names in common usage.

The process of revision of scientific names will continue to follow well-established rules. The process of lumping and splitting forms on the basis of new or improved information will continue. Tracing these revisions is a laborious but straightforward task.

There is usually a correct scientific name and all others are incorrect. Recent scientific names invalidated by new revisions and corrections are listed in the index which will refer the reader to the correct species. Where acceptable alternative names remain valid these are discussed in taxonomic notes at the end of species entries.

It is a much more difficult task to keep control of the common names in regular usage. A bird watcher may not readily know whether the Common Flameback of SE Asia is the same as the Goldenbacked Threetoed Woodpecker of India. I have therefore listed most currently-used common names on the second line of species entries using the following conventions. Alternatives are separated by a diagonal slash and the common general name is given only for the last name of a list. The symbol > indicates that names following are applicable to only part of the whole species. AG indicates that the alternative general name so introduced can be applied to all preceding names sharing another general name. Thus an entry (Russet/Mountain Bush-Warbler > Javan/Timor Bush-Warbler, AG: Scrub-Warbler) indicates that Russet Bush-Warbler and Mountain Bush-Warbler are the same species, that Javan and Timor forms are included in this species, and that the general name Bush-Warbler can be replaced by Scrub-Warbler. All names are also traceable through the index. Each family is introduced with a short description followed by entries for each species.

Species entries cover nomenclature; description of plumage and soft parts and length from the tip of the bill to the tip of the tail; voice; range (global); distribution and status (within region); and habits, including diet only where this differs from the general dietary habits of the group which are covered in the family introduction.

For the sake of brevity the directions north, east, etc. are abbreviated to capitals without full stops. Thus SE Asia means South-east Asia. N Borneo and S Sumatra refer to northern Borneo and southern Sumatra and do not mean the administrative division North Borneo (old name for Sabah) or South Sumatra (one province of S Sumatra). N and S are also used for directions as in 'migrates S in winter' only where brevity is essential. Species reported or expected from the region but still lacking confirmation are described in the text but placed within square brackets [ ].

In addition I have occasionally used abbreviations in the text when descriptions of voice or other details are based on secondary sources that I have not been able to check. A list of these abbreviations is given after the Glossary.

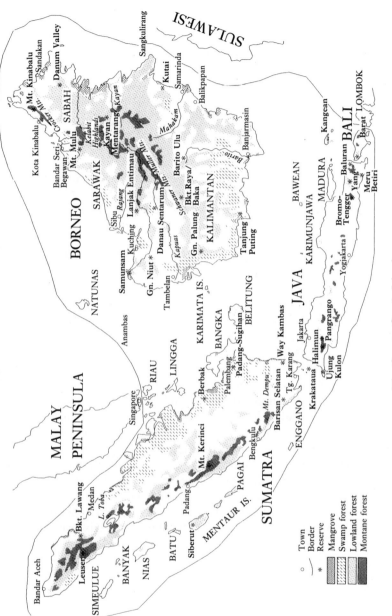

The Greater Sunda Islands.

Introduction to the region

Geographical limits

THIS field guide includes all birds recorded from the Greater Sunda Islands—Sumatra, Borneo, Java, and Bali together with their offshore satellites (the Mentaur Islands off western Sumatra, Riau and Lingga archipelagos, Bangka and Belitung off eastern Sumatra, Natunas, Anambas, Tambelan, Karimata, Karimunjawa, Madura, Bawean, Kangean, Nusa Penida, and the islands off the north and east coasts of Borneo). The book does not include Palawan and Andaman which, although faunistically related to the Greater Sundas, are generally not included under that term and are ornithologically included in guides for India and Philippines respectively. Bali, which some geographers include in the Lesser Sundas, is included in this book because it is faunistically and geologically an offshore island of Java. The endpaper map (also shown opposite) shows the region covered. This area consists of the following political territories: the whole of East Malaysia including Sarawak and Sabah, the whole of Brunei and large areas of Indonesia including the eight provinces of Sumatra, three provinces of Java, the province of Bali, and the four provinces of Kalimantan.

This geographical area totals almost 1.4 million square kilometres, broken down as follows:

Sumatra	km^2	Borneo	km^2	Java and Bali	km^2
Aceh	55 390	Sarawak	124 968	W Java	48 690
N Sumatra	70 790	Sabah	76 115	C Java	37 370
W Sumatra	49 780	Brunei	5 765	E Java	47 920
Riau	94 560	Kalimantan		Bali	5 560
Jambi	44 920	West	146 769		
S Sumatra	103 690	Central	152 620		
Bengkulu	21 170	South	37 660		
Lampung	33 310	East	202 440		
Sub-totals	473 610		746 337		139 540
Grand total	1 359 487				

Physical description

The Greater Sunda Islands are relatively recent in geological terms, having risen from the sea about 15–30 million years ago as a result of high tectonic and volcanic activities.

Sumatra and Java lie along a line of volcanoes. In Sumatra these have given rise to the Barisan mountain range which runs the entire length of the island, split in some areas by rift valleys and graben lakes. The highest peak is the active volcano of Mt. Kerinci (3805 m), while forests reach their highest limits on the relatively dormant Mt. Leuser (3419 m). The peaks are less high in the south. To the west of the Barisan chain, the land falls steeply to the coast and the deep Indian Ocean. There is a rather narrow coastal plain. By contrast, to the east, the land falls gradually to the much shallower Sunda Sea over the Sunda continental shelf. Here there is a much broader coastal plain and alluvia form extensive swamps. Faulting and rifting have severed fragments from the west coast of Sumatra resulting in several groups of islands collectively known as the Mentaur Islands. To the east of Sumatra are a series of small islands of the Riau and Lingga archipelagos and two large islands, Bangka and Belitung.

In Java, the volcanoes do not form a continuous mountain chain. Instead, scattered isolated volcanoes stand out from the surrounding plains and, being more active, give the Javan soils greater fertility than on Sumatra. The islands of Madura and Bali have only recently broken from Java. Several peaks on Java and Bali reach over 3000 m, the tallest being being the active Mt. Semeru (3676 m), while forests reach their greater altitude on the dormant Mt. Pangrango (3019 m). As in Sumatra, the alluvial plains lying on the Sunda shelf (north) are broader than those facing the deep ocean (south). The south coastline of Java is rugged and rocky with occasional sandy bays.

Borneo has no active volcanoes but its main mountain ranges are also igneous. Mt. Kinabalu in the north east of the island (Sabah) is, at 4101 m, the highest peak in the whole of South-east Asia. It consists of a granitic plug forced up by volcanic pressures and is still rising. Few other peaks exceed 3000 m, and indeed only 6% of the whole island is higher than 1000 m above sea level. The main mountain backbone of Borneo runs from the north-east of the island in an arc inside the northern coastline with a southern branch, the Mueller range, extending down to the Bukit Raya mountains; a second southern branch extends to the Meratus mountains in the south-east. Between these ranges lie the major river basins and extensive

INTRODUCTION TO THE REGION

Borneo mountains.

coastal plains in the south and west, with narrow plains to the north and east.

Climate

The climate is moist equatorial with year-round rainfall and few dry months (when rainfall is less than 100 mm). The northern winter monsoon (November to April) is generally wetter than the southern summer monsoon. Some coastal areas have a bimodal rainfall distribution and some regions such as northern Aceh in Sumatra, the east coast of Borneo, and the eastern end of Java including Bali have a more seasonal and drier climate than the rest of the region. In inland hilly parts of Borneo, Sumatra, and West Java, rainfall is between 2000 and 4000 mm per year. But in the north of Bali rainfall is below 1000 mm.

Natural vegetation

Most of the region was originally covered in lush tropical rainforests, but deciduous monsoon forests occur in the drier areas of East Java and Bali. A variety of distinct forest types can be recognized, and support different communities of birds.

Coastal scrub and beach forest. These are light, simple forests or scrub growing on sandy, fast-draining soils. They are characterized by *Casuarina* groves on sand bars, *Barringtonia asiatica*, *Morinda citrifolia*, *Pandanus* spp., *Heritiera littoralis*, and coconuts. The total area of such forests is small and they contain only a few rather common bird species.

Mangroves. These are low tangled forests growing between the high tide zone and the muddier beaches. They contain a few tree species such as *Rhizophora*, *Avicennia*, and *Brugieria* but are rich in fish and crustaceans, and support high densities of waterbirds and a few common forest birds. Mangroves are very extensive on the east coast of Sumatra and the west and south coasts of Borneo and occasional along the west coast of Sumatra and north and east coasts of Borneo. They used to be common along the northern coast of Java, but only a few patches now remain there.

Peat swamp forest. These are stunted forests of relatively low tree diversity growing on poor drainage areas where lenses of peat have accumulated. This creates rather acid conditions which only specialized trees can withstand. Such forests occur extensively in eastern Sumatra and in the coastal plains of Borneo. They support a specialized subset of the lowland forest bird-fauna plus some important wetland birds.

Fresh water swamp forest. These are tall forests growing in low-lying areas of regular submersion by flowing stream water. The understorey is relatively bare and dominated by palms. Such forests are extensive in Sumatra and Borneo and rare in Java. They support a rich and interesting bird fauna, though less rich than the dry land forests.

Heath forest. These are rather low, specialized forests on areas of poor, sandy soil, leached of nutrients, and characterized by a shallow accumulation of peat. They contain many mistletoes and other flowering bushes and support an impoverished but distinctive bird fauna.

Limestone forest. Parts of northern and eastern Borneo, southern Java, and the central highlands of the Barisan chain in Sumatra are characterized by spectacular karst limestone scenery. These areas are riddled with caves and underground water passages, so they tend to be fast-draining and have distinct renzina soils. They support rather dense but less tall forests of specialized trees, and have a slightly impoverished but distinctive bird fauna.

Monsoon forest. These are small rather open forests with a rather park-like aspect and with grass in the understorey. They are restricted to east Java and Bali and support a rich bird fauna, though less diverse than that found in evergreen rainforests.

Ironwood forest. The Ulin or Belian tree *Eusideroxylon zwageri* is an immensely durable, heavy wood which grows in moist low-lying forests of Borneo and Sumatra. It is often just one of a diverse species mix in lowland rainforest, but in a few places in Borneo and in quite large areas of Sumatra this species becomes dominant, forming near mono-species stands. These forests are less rich for birds than the more diverse mixed forests and have now largely been destroyed for their valuable wood.

Lowland dipterocarp rainforest. These are the tallest and lushest rainforests of the region as described and depicted in so many texts and films. Trees of the dipterocarp family dominate the main canopy, but legumes such as *Koompassia* and *Intsia* tower even higher as emergents. Huge branch-free boles are supported by flared buttresses, and the whole complex is draped in lianas, epiphytes, and abundant strangling figs. Great rivers wind through the plains with smaller streams trickling down the valley-sides. Such forests are highly prized for logging. They are being reduced fast in extent in Borneo, have almost been cleared in Sumatra, and have virtually been eliminated from Java. They support the greatest diversity of plant species and an immensely rich bird fauna. They are of great conservation significance.

Hill dipterocarp forest. In hilly areas, different species of dipterocarps dominate the ridges and the steep hillsides are clothed in mixed forests that contain a wealth of bird niches. Frequent landslides maintain a mosaic of seral succession. These are some of the richest habitats for bird diversity but it can be difficult to walk through them or see what is flying through the canopy. Such forests are still extensive on Borneo and Sumatra but have become rare in Java where they are thus a serious conservation concern.

Submontane rainforest. At about 1000–1200 m the forest composition changes. Dipterocarps become scarce and forests are dominated by laurels, oaks, and more temperate genera. The bird fauna is still rather rich and many interesting endemics are found in this zone.

Upper montane rainforest. The highest forests are stunted and increasingly covered in lichens and mosses, with an abundance of epiphytic orchids and ferns. They are sometimes called elfin forests. They support a less diverse bird fauna but include several rare endemic species.

Montane dry forest. These drier montane formations are found on the mountains of East Java and Bali. They are usually dominated by *Casuarina* trees with a fire-prone grassy ground cover. They contain few bird species but some of those are rare and interesting for the birdwatcher.

Alpine scrub. These scrublands are found above the tree line on the highest mountains of the region. Such scrub is usually dominated by ericaceous plants such as *Vaccinium* with *Schima*, *Potentilla*, and *Hypericum*. In total area they are very small and they support only a few bird species. However, such mountain tops are highly isolated and the bird fauna shows high levels of endemism so that some of the region's rarest and most exciting birds are found in this zone.

Human population and current land use

Java and Bali is one of the most densely populated parts of our planet, with over 96 million people—a density of nearly 800 persons per sqare kilometre. Moreover, population is still rising at almost 2% per year. It is perhaps not so surprising, therefore, that little forest survives and pressure on bird habitats is very high. Less than 10% of the land area is still covered in natural forests and much of that has been affected by man. These forests are found on the higher mountain slopes or infertile and remote places. Only small patches of lowland forest survive in nature reserves and national parks. Cleared lands are occupied by irrigated paddy in the plains, dry land farms on uplands, or plantations of teak, pine, and *Agathis* on forestry lands. Most villages have extensive fruit groves of durian, rambutan, mango, *Areca*, and bamboos, which provide some cover for countryside birds.

The population density in Sumatra is not so high, at nearly 80 persons per square kilometre. However, the soils of Sumatra are less fertile than those of Java, and the Sumatran population have cleared on average six times as much land as their Javan counterparts. In addition, population growth in Sumatra is far higher, at about 3.3% per year as a result of high birthrate combined with both sponsored and spontaneous transmigration from neighbouring Java. As a result the forests of Sumatra are disappearing very fast. The main chain of the Barisan range remains forested but is being encroached by coffee farmers to over 1000 m altitude. There are still extensive swamp forest and mangrove along the eastern seaboard but these are being targeted for conversion to agriculture and transmigration areas and are also rapidly diminishing. Lowland dry land forests have almost all gone unless specifically protected in reserves. Even here they face illegal exploitation. Many have been replaced by plantations of rubber and oil palm which provide poor habitats for forest birds, or have deteriorated into secondary scrub or *Imperata* grasslands.

Population density in Borneo is much less, at only 12 persons per square kilometre, reflecting much lower soil fertility. However, the very economically valuable lowland forests of Borneo have proved most profitable and very large areas of original forest have been selectively cut over. Unlike Sumatra these are not usually converted to agriculture but regrow to form new forests, less rich than the original but certainly better habitat for birds than agricultural land or plantations. About 60% of the island remains under some sort of natural forest cover, compared with only 30% in Sumatra and less than 10% in Java. The area under permanent agriculture in Borneo remains small, but large areas that were cleared for shifting cultivation have now become *Imperata* grasslands. In the more fertile areas there are some plantations of peppers, cocoa and oil palm.

Birds in the local economy and culture

Swiftlets are a valuable industry in the region. Two species' nests are regularly harvested—*Collocalia fuciphaga* which makes the most valuable white nests of almost pure saliva, and *Collocalia maxima* which makes the so-called 'black' nests which are somewhat less valuable because of the many blackish feathers embedded in the saliva, and the elaborate cleaning required.

These birds nest in caves and rock crevices, particularly in limestone regions. Main harvesting areas include the Nusa Barung off the south coast of Java, the Sangkulirang area of E Kalimantan, and the caves of Sabah (Tapadong, Madai, and Gomantong) and Sarawak (Niah, Mulu, and Baram district). Some of the collecting methods defy belief, with collectors risking their lives to climb flimsy bamboo poles over 30 m high to scrape the nests from the cave gallery roofs in almost total darkness.

A few districts have developed adequate harvesting controls to ensure sustainability of yields but many other sites are over collected, with the result that populations are dwindling. However, in Java the nests are farmed in special houses.

Singapore is the main clearing house for the trade before the valuable gourmet items are redirected to the restaurants of Hong Kong, Taiwan, USA, and Canada. Imports to Hong Kong alone in 1990 were declared at a value of US$ 360 million.

The trade in live cage birds is also a cause for grave concern. A network of collection channels funnels an estimated one million birds a year through Jakarta and Singapore—cockatoos, parrots, starlings, munias, bulbuls, chats, white-eyes, shamas, doves, and jungle-fowl.

Cage birds are also very popular house pets in Indonesia and Malaysia, and as many birds are kept to meet domestic trade as are exported. Some species have almost vanished as a result such as—Straw-headed Bulbul, White-rumped Shama, and Peaceful Dove in W Java—and it is noteworthy to see that local bird markets are now stocked with a high proportion of imported birds from China. This may seem to indicate that domestic sources are no longer able to match local demand.

One bird formerly targeted for persecution is the Helmeted Hornbill whose ivory-fronted casque was avidly sought after by the Chinese craftsmen for carving intricate pieces of *ho-ting* artwork. The heads still fetch a high price but the bird is protected throughout its range, and all such trade is illegal.

In the uplands of Borneo, birds are used as a farmers' calendar. Fields are prepared and sown with the arrival and passing of migrant wagtails (see the Avian Year below). The Iban and other Dyak tribes of Borneo have taken the art of using bird indicators a whole lot further. They have developed a complex bird-omen science similar to the augury practised in ancient Greece and Rome. Singalung Burong, in the form of the Brahminy Kite, is recognized as a great overseer of man's affairs and the omen birds—Crested Jay, Banded Kingfisher, Scarlet-rumped Trogon and others—are his special messengers. These birds visit man to guide his actions. Special behaviour or calls of these omen birds indicate the need for caution or the signal to proceed boldly with one's endeavours. Important events such as house building, clearing a new field, or setting out on a hunting raid must be preceded by favourable signals from the birds

Head-hunting and war raids were preceded by even more elaborate 'gawai' ceremonies in which the ancestral spirits were invoked in the image of hornbill symbols. These spirits proceeded in advance of the warriors to destroy the enemy's guardian spirits before the real battle was started.

Fishermen respect the frigatebirds and flocks of terns in a highly pragmatic way as guides to find the shoals of mackerel and tuna on which their livelihood depends. Fishermen leave their nesting sites alone, but the huge eggs of the Scrubfowl enjoy little respect and are harvested avidly wherever they are found.

Other beneficial birds of the region are the egrets and rice field warblers which eat a great many insect pests and the raptors which control rodents in this rice-dependent society.

On the other side of the coin are the large flocks of munias and finches that jeopardize the rice harvest. Many elaborate and ingenious bird scaring devices are employed to protect the ripening crops such as wind-driven clappers, flags, scarecrows, and small boys, hidden in tiny huts, controlling long strings attached to rattles all over the fields. It is a battle that man is winning. The level of crop predation by birds is down dramatically compared to thirty years ago, and some species that used to arrive in hundreds of thousands, such as the parrot-finches and Scarlet Finch, are now rather rare and even locally extinct.

Biogeography

The avifauna of the Greater Sundas

THE Greater Sundas lie in one of the world's most interesting zoogeographical areas, the Malay–Indonesian archipelago, an arc of some 17 000 islands straddling the Equator and extending for 5000 km between mainland Asia and the continent of Australia.

The archipelago can be divided into three distinct faunal sub-regions: the Australo–Papuan subregion, which consists of all the islands lying on the Sahul or Australian continental shelf such as Aru, New Guinea, and New Britain; the Sundaic subregion, which includes all the islands lying on the Sunda or Asiatic continental shelf such as Borneo, Sumatra, and Java plus the Malay Peninsula—which, although not an island, is faunistically more similar to the other Sundaic areas than to the rest of the Asian mainland; and finally, the Wallacean subregion, which consists of all the islands that lie between the two continental shelves such as Philippines, Sulawesi, the Moluccas, and the Lesser Sunda Islands.

During the ice ages of the Pleistocene epoch, between 3 million and 8000 years ago, the climate of our planet was very unstable and swung several times from warm, wet 'pluvial' periods to cold, dry 'interpluvials'. Interpluvials correspond with the 'glacial' periods of temperate regions, and at such times so much of the northern and southern latitudes was covered in ice that sea levels were reduced by as much as 100 m. In the Greater Sundas area this resulted in the exposure of much of the land now submerged by the Sunda and Java Seas. Borneo, Sumatra, and Java were thus all connected to the Asian mainland by dry land and could be easily colonized by Asiatic fauna and flora. The last connection was only 10 000 years ago.

The pattern and duration of land connections has had a profound effect on the present distribution of species and occurrence of localized endemic forms. The islands have been differently affected. Java, for instance, is a smaller and more isolated land mass than either Sumatra and Borneo. It has been connected to Asia less often and for shorter periods than the other two. As a result it has retained a less rich fauna but a higher proportion of island endemics.

On the other hand, Java has a drier climate. Hence, its climatic conditions are more similar to those prevailing at times of maximal land connection. Java has thus been able to retain 30 monsoon forest birds from Asia that are no longer found on Sumatra and Borneo such as Green Peafowl, Lineated Barbet, Small Minivet, and many more.

Borneo is the largest land mass but is far more remote from the mainland than Sumatra. Its overall bird fauna is slightly less rich than that of Sumatra but it has more endemic species. Most of these island endemics are montane. This is because the mountains remained as evergreen islands in 'interpluvial' periods when drier lowlands were periodically recolonized from Asia. Whatever progress towards speciation had been made by lowland species during periods of isolation would have been swamped or diluted through such intermixing between the islands.

The map on p. 16 shows the region and those areas of shallow sea that were exposed during the periods of lowest sea-level. It shows clearly why there is such a marked change in bird fauna when one crosses the comparatively short distance from Borneo to Sulawesi; and also why Palawan has a bird fauna as closely related to Borneo's as to the Philippines'; and why there are some curious similarities in faunas between Java and the SE corner of Borneo. A deep trench separating Sumatra from the Mentawai islands helps explain why the bird fauna of Mentawai is more distinctive than on the other Mentaur Islands that were connected to Sumatra at various times. The endemism of Mentawai mammals, which cannot fly and were therefore even more isolated than birds is even more striking, with no less than 10 endemic species being found there.

We can see the relationships between the different islands best if we limit our analysis to resident land birds (omitting migrants, seabirds, waterbirds, waders, and wide-ranging raptors). This leaves us with 541 species distributed as follows: 171 (34%) are found on all three major islands, and most of these (164) are also found on the Malay Peninsula. 135 are endemic to the Greater Sundas (there are an additional 6 endemics in the families excluded).

The Javan fauna is comparatively impoverished, with only 289 species of the sample. Of these 57% are the 164 species shared by all islands. 176 (61%) are shared with Borneo compared with 215 (74%) shared with Sumatra. 49 (17%) are shared with other non-Sundaic islands and 30 (10%) are endemic. Of particular interest on Java are some 30 species shared with SE Asia but not found elsewhere in the Greater Sundas, and many not even in the Malay Peninsula. These are mostly monsoon forest and scrubland

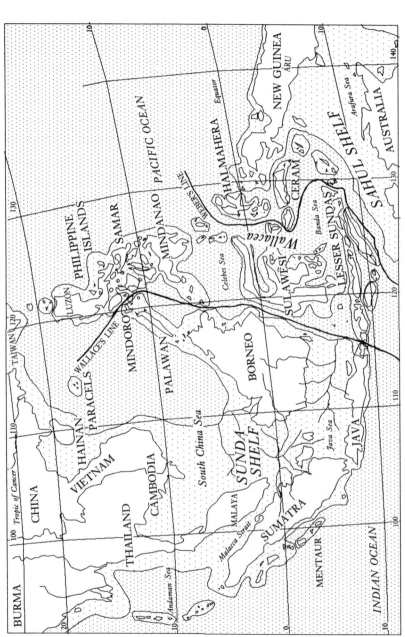

The Sundaic and Wallacean region showing 100 fathom depth contours. At times of lowest sea level, all area above 100 fathoms (light stippling) was connected by dry land.

birds such as Green Peacock, Brown Prinia, and Common Tailorbird, reflecting E Java's generally much drier climate and more extensive savanna and monsoon habitat. About 28% of Javan land birds are montane, a slightly higher proportion than on the other two islands. Bali is clearly seen as a further impoverished subset of the Javan fauna, with 97% of the Bali species being shared with Java, compared with only 54% shared with the Lesser Sundas.

Borneo has 358 species, or 66% of the resident land birds of the region. The 164 species found on all islands constitute 46% of this total. 306 (85%) are shared with Sumatra, and a similar number (297 or 83%) with the Malay Peninsula. Only 177 species (49%) are shared with Java and 42 (12%) with non-Sundaic islands. Some 37 species (10%) are endemic, mostly montane, but there are some lowland endemics such as Bulwer's Pheasant and the Bornean Bristlehead. Overall, 24% of Borneo's land birds are montane, slightly less than on Sumatra, and perhaps reflecting that Borneo has proportionally less montane habitat (6%).

Sumatra benefits from its closer links to the Asian mainland in supporting 397 species out of the total of 541, but being less isolated it has less endemics—22 (6%), including some small island endemics. None of the mainland endemics is a lowland form. Sumatra shares 306 species (77%) with Borneo and 345 species (87%) with the Malay Peninsula. 211 species (53%) are shared with Java.

The close relationships between the birds of the Malay Peninsula and those of the Greater Sundas show that the Malay Peninsula belongs in the Sundaic avifaunal province rather than in SE Asia. There are only 46 species found on the Malay Peninsula that are not found in the Greater Sundas. Ten of these are migrants, vagrants, or only marginally found within the peninsula. Only three species are endemic and the remaining 33 are SE Asian species that extend onto the peninsula. A list of these species is given in Appendix 5.

The island of Palawan is of particular interest. Unlike the rest of the Philippines, which is separated from Borneo by a deep trench, Palawan lies on an extension of the Sunda shelf and has at various times enjoyed land connections with Borneo. As a result its fauna is a mixture of Philippine/Wallacean species (77 out of a total of 118 species) and Bornean species (80), with 14 species (12%) endemic to the island.

The map on p. 18 shows these inter-island relationships. The size of circles is proportional to the number of resident birds on each island, the thickness of connecting lines indicates the absolute number of shared

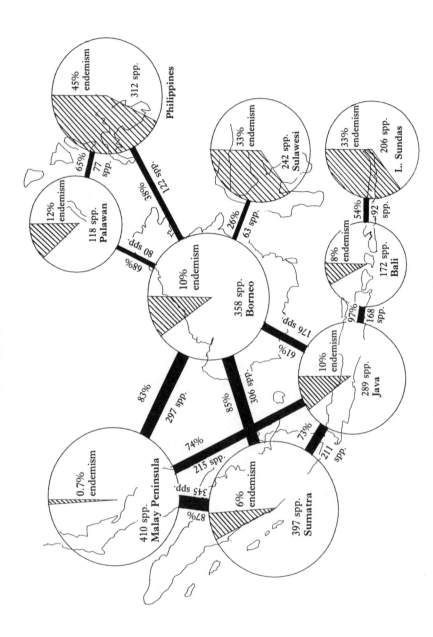

Relationships between resident land bird faunas of the Sundaic region.

species, and the percentage shown indicates the proportion of the smaller of the two faunas which is shared with the larger.

Bird faunas of smaller islands

Families and species show very different abilities to disperse to and survive on small islands. Appendix 3 shows the distribution of land birds found on the small islands in the region. Data for some species are rather incomplete but distributions generally match the predictions of island biogeography theory, namely that smaller islands and more distant islands will have fewer birds than large ones or islands closer to the mainland.

Some families such as pigeons show good ability to colonize small islands while others, such as woodpeckers and babblers, show very poor ability. Only Bangka, which is both large and very close to Sumatra, shows a good assemblage of these latter families. Some species, such as Nicobar Pigeon, several Imperial-Pigeons, and Scrubfowl, are almost exclusively limited to small islands. The small isolated islands have several endemic species and many endemic races.

It is interesting to note that, in several instances, Bangka and Belitung harbour Bornean rather than Sumatran forms of some species. This probably reflects the fact that these islands are the only part of Sumatra to contain extensive heath forest, a habitat prevalent on Borneo. Moreover, these islands lie on the bridge that would have linked Borneo with Sumatra in the periods of lowest sealevel during the Pleistocene 'interpluvials'. A similar linkage is reflected in the occurrence of several birds shared between Java and Sulawesi in the SE corner of Borneo that are absent or rare in other parts of Borneo. Examples include Great Tit, Yellow White-eye, Moustached Parakeet, Sunda Teal, Black Moorhen, and Comb-crested Jacana.

Conservation

WITH such a large human population, and increased pressures to exploit all economic resources that the land can offer, it is inevitable that nature is in a state of retreat. Forests are pushed back to the highest peaks or most unfarmable lands, and the birds themselves are persecuted for food, sport, or sale. How are they coping in such a changing world? What conservation measures are being taken to protect them? Java and Bali are clearly the most affected islands of the Greater Sundas, with 68% of the region's population living on only 10% of its land surface. Only 10% of the area of Java and Bali remains forested, mostly in mountainous regions. The lowland forests are almost gone and less than 3% cover remains, only half of which lies inside protected areas. It is thus not surprising that some birds seem to have completely disappeared from Java, that others have not been seen for many years, and that many more are rare now. Appendix 1 lists the main species of conservation concern.

The loss of bird species on Java is of serious concern and an indicator of what can be expected in the other Greater Sundas as forests are cut and fragmented. It is worthy of closer examination. Studies of the distribution of forest birds in remaining patches of all sizes on Java (van Balen 1987*b*) show several interesting patterns. For example, it is noteworthy that Java shows an atypical profile of bird altitudinal distributions. The bar charts opposite show the proportions of the resident fauna living at different altitudinal levels on Java compared with the proportions on Borneo and Sumatra. The Borneo and Sumatra graphs are typical of those found for other tropical regions such as New Guinea, with maximum bird richness in the lowland dry land forests and a gradual reduction in species number with increasing altitude which is only slightly compensated by the gradual increase in numbers of montane species appearing at different altitudes.

The Javan graph is peculiar in having a dip in the hill zone between 300 and 1500 m. This is interpreted as the result of long-term deforestation of lowland forests starting well before this century. Lowland species have failed to maintain themselves in hill zones where they could be expected to live in an unbroken forest profile, because they have been cut off from the

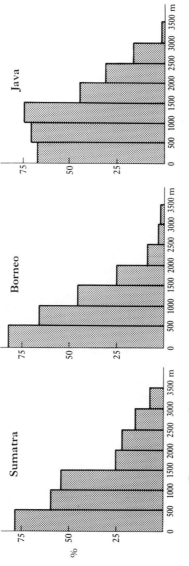

Proportions of bird species at different altitudes on Sumatra, Borneo, and Java.

lower altitude populations on which depend the regular recolonization and restocking of middle altitudes. We can see a small but well-documented example in the case of the Bogor Botanic Gardens. In 1947 Hoogerwerf wrote a book on the birds of these gardens listing 142 species. Some of these were only occasional visitors, but over 100 were regular. Today there are only about 40 regularly seen species. Only 81 of Hoogerwerf's species have been seen in the last 20 years, representing a loss of richness of 46%. Other species that were common are now rare, yet the habitat is no less good today and the garden still seems like an oasis of birds. What has changed is that there is no longer any neighbouring forest in the vicinity of Bogor, and we are seeing the effects of forest fragmentation and isolation. On a larger scale this has been happening throughout the lowlands of Java for a long time.

Families of larger birds have lost proportionally more species than families of small birds. Thus while 5 out of 7 (70%) of Sumatran *Diceaum* flowerpeckers persist on Java, only 2 out of 6 *Phaenicophaeus* malkohas, 1 out of 5 *Batrachostomus* frogmouths, 1 out of 5 *Malacopteron* babblers, 1 out of 5 *Harpactes* trogons, and 2 out of 7 *Cyornis* flycatchers survive across the Sunda Strait. Another pattern of interest is that Java has lost more 'real' forest lowland birds than species that are able to live in secondary forest, forest edge, or open habitats.

In relation to habitat patch size, van Balen's findings show clearly that small forest patches lose far more species than larger blocks. Extinction rates of forests of only 10–40 ha are as high as 80% compared with rates of only 25% for areas over 10 000 ha. These results have an important bearing on the design of nature reserves. More than 100 nature reserves have been established in Java and Bali, but most of these are very small and clearly inadequate to protect complete bird communities. There are only 12 reserves larger than 100 km^2 though more could be established, especially along the south coast.

Clearance of coastal wetlands, mangroves, and swamp forests along Java's north coast are also threatening many of Java's more interesting birds. The Javanese Lapwing is extinct, the Sunda Coucal very rare, the White-winged Duck probably extinct, Mangrove and Lemon-bellied White-eyes very rare, the Milky Stork down to one breeding colony, and the Royal Spoonbill is no longer a resident. In addition the trapping of hundreds of thousands of migrant waders along the north coast each year for sale as food in local markets is placing a disastrous strain on those species, and their numbers are also dwindling.

In West Bali National Park, there is an ICBP (International Council for

Bird Preservation) project to save the Bali Myna. Studies and habitat management have failed to stem the decline of this species from a few hundred individuals in 1970 to a few tens of individuals today. The problem appears to be uncontrolled capture of fledglings for sale in the bird markets, and a general scarcity of suitable nesting trees as a result of fires and woodcutting in the species' limited habitat area. Efforts to release captive-bred birds to boost the local population have also not been very successful. Another captive-breeding rescue operation is being planned to save the declining Green Peafowl on Java.

Birds in the Javan countryside are depressingly scarce in both numbers and diversity, owing to a combination of pesticide use, loss of habitat and cover and direct persecution with air rifles and slingshots. In Sumatra the situation is only a little better. There is still 20 to 30% remaining forest cover but this too is disappearing very fast, and human population is rising faster on Sumatra than on the other islands.

Forest loss in Sumatra is not spread evenly among all types. Ironwood forests have totally disappeared, lowland dipterocarp forests survive only in nature reserves, and peat swamp forests are being cleared and burned at a frightening pace. Only the high montane forests are still in good condition, by fortune the home of all Sumatra's endemic species, but even these face hunting pressure and slash-and-burn agriculture. Thus several of Sumatra's lowland and mid-level species have almost totally vanished. A few have not been seen for many years, and other species that could be expected to be common are rare. Appendix 1 lists 37 species regarded as endangered. On the east coast of Sumatra are some very important breeding areas for wetland birds and also some important staging areas for migrating shorebirds. These are significant wetland sites in urgent need of protection but they are largely not included in the nature reserve system.

Sumatra's reserve system is, however, much more comprehensive than that on Java. Leuser and Kerinci National Parks are both very large and between them protect all known hill and montane species of Sumatra. The other reserves fill most gaps in habitat coverage and in total 30 reserves cover 45 000 km^2, or 10% of the island. This would be adequate to protect almost all avian species of Sumatra if these were inviolate reserves, but sadly they are far from secure. Timber concessions have operated for years on the eastern and southern flanks of Leuser and on the west side of Kerinci. Farmers and coffee growers continue to clear fields further and further inside both parks. Kerumatan reserve was totally logged and burnt. Way Kambas reserve was partly logged and burnt. Berbak reserve is threatened

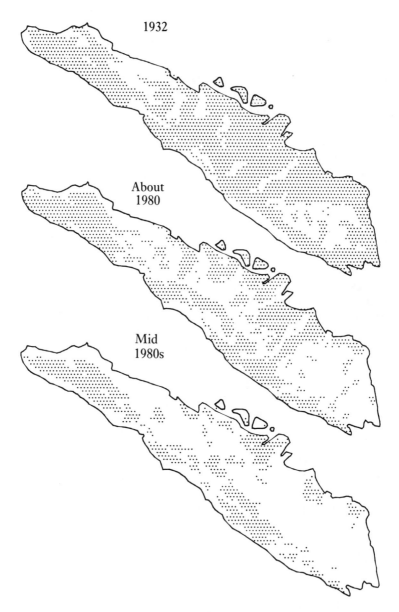

Loss of primary forest in Sumatra in recent years.

by Bugis settlers clearing and farming the coastal zone. Padang Sugihan reserve has been severely damaged by fires. Timber concessions have logged large areas inside Taitaibatti and Barisan Selatan reserves. All these reserves face heavy pressure from collection of firewood and from poaching. The problems are simply enormous, and the nature protection department is still too weak to control either the farmers or those powerful commercial interests that threaten the reserves. The map opposite shows the startling recent shrinkage of Sumatra's forest cover as revealed from satellite photos.

The situation in Borneo appears a little better. Almost 60% of the island is still forested, and 37 main reserves cover a total of 35 000 km^2, or 5% of the land area, with much larger areas proposed as reserves or with status of protection forests. There are areas such as West and South Kalimantan or the coastal parts of Sarawak and Sabah, where forest loss is as bad as in Sumatra. However, Borneo is very large, so there are still huge areas of forest in the south and centre of the island. Almost all the mountains, where the bulk of endemics occur, are protected as either nature reserve or protection forest.

This does not mean all is well in Borneo. There are many problems. There are very few lowland reserves and these are always threatened. Kutai has been largely logged, burnt, and invaded by farmers. The limestone forests of East Kalimantan remain unprotected and have been damaged by fires. The peat swamp forests of East Kalimantan were totally burnt in the great fires of 1982 when a total of 3 million hectares were destroyed by fire. Large areas burnt again in 1991. Lowland areas of Kinabalu Park were excised and cleared, other parts damaged by copper mining. Gunung Palung, Tanjung Puting, and Bukit Raya are all encroached upon by loggers and settlers. Danau Sentarum is threatened by the activities of hundreds of seasonal fishermen around the lakes. These problems are all serious for local wildlife populations but, as yet, there are only a few bird species endangered on Borneo. Appendix 4 lists only 16 species as endangered.

Field techniques for birdwatching

BIRDWATCHING is an absorbing pastime, and the broader the birdwatchers' experience, the better and more acute their observations become. Birdwatchers are the scientists' eyes and ears for monitoring the state of our planet. Their bird lists and observations are very important to scientists in indicating any deterioration of the environment. This book is designed to help birdwatchers make their observations more accurate, meaningful and, indeed, useful when submitted to the relevant body.

Hints on birdwatching in forest

Watching birds in tall forests is not easy. You may walk for an hour without seeing anything then suddenly be surrounded by so many twittering birds you cannot focus on any. A bird may be so high up and so obscured by foliage that you cannot get a good view. In the rain, water on your lenses may blur your vision. The newcomer to the forest should first get to know the common forest-edge species by walking along roads and wide trails through forest. Observation conditions are better, and the birds appear more numerous because the light falling sideways on the forest edge creates a very rich feeding zone. Some of the true forest-canopy birds can be seen along roads, and where forest roads cut through mountains you may have excellent side views into the canopy that you may never get when walking on narrow trails.

Leeches are an accepted irritation to the hardened birdwatcher, but may be quite distressing to the newcomer to the forest. There are several ways to minimize damage, such as wearing two pairs of fine mesh socks or spraying inside boots with insect repellents (Shelltox or benzyl benzoate).

Rainforest humidity causes problems with binoculars, cameras, and spectacles. Make sure your kitbag is fairly waterproof, keep equipment in polythene bags and wrapped with dry absorbent cloth, paper, or desiccant packets. If you wear spectacles in rain, wipe them dry with a small piece of soft (chamois) leather whenever you are watching something. If you get moisture inside your binoculars, dry them out as quickly as possible. Quick ways to remove moisture include laying them in the sun with the lenses facing directly at the sun, putting them overnight in a bag of desiccant, or sleeping with the binoculars held close for warmth. More drastic measures such as warming them near a campfire will work but shorten the life of your binoculars considerably!

When buying binoculars, select good waterproof varieties that will last for many years. Cheap binoculars are not an economy. It is also important to select a relatively small magnification with a wide aperture for use in forest. Powerful but narrow-angle glasses or use of 'spotting scopes' (telescopes) are not good for forests. Most of the very small, lightweight binoculars are quite useless.

Most birdwatchers walk very slowly through the forest trying not to miss anything. This works quite well for canopy birds but you will miss a lot of ground birds—pheasants, pittas, and thrushes that are very wary, hear you coming, and have already slipped away before you arrive. Moving quickly but silently will bring more success with these shy birds. There are also great rewards for sitting quietly and patiently on a log. You may have to wait for some time but you can get views of birds you would never see otherwise. Ideally you should alternate between bouts of slow careful searching, faster silent walking, and resting periods of silent waiting. Particularly rewarding places for waiting are fruiting fig-trees, red-flowering trees, large mistletoe clumps, or by pools and streams. Limit conversation to a minimum, and if you have guides or porters persuade them to keep quiet or remain at a distance. Wear drab clothes and never white.

Three methods are used increasingly by birdwatchers to call birds into view. The first, 'pishing', involves making sibilant, squeaking, or rasping sounds. This can have the effect of exciting small birds, especially skulking babblers, to call back or even emerge from the understorey to investigate the source of sound. A similar effect can be achieved by imitating the call of the Collared Owlet, or other small raptors, tempting small birds to come to mob the threat. The third method is the use of playback of tape-recorded songs which can cause a territorial reaction, and attract birds. In the depths of the forest all three methods are harmless, but problems can arise in areas regularly visited by birdwatchers such as the main trails of national parks. Here it may be necessary to ban such techniques which can disturb the territorial and nesting behaviour of local bird life, and reduce the natural alarm responses of wild birds. This is particularly relevant with taped calls.

Birds are most active in the early morning and this is the best time to see them. Their activity drops off towards midday. The afternoon activity peak is never as energetic as in the morning, though there is sometimes a 'false dawn' peak of activity after prolonged rain. Midday and late afternoon are, however, quite good times to wait for birds to visit water sources in the drier times of year.

Keeping field notes

The bird fauna of the Greater Sundas is still rather poorly known. Bird lists by locality are still incomplete, lists of offshore islands are very inadequate, and most habitat and feeding generalizations are based on very few records. Migration patterns and breeding seasonality are also poorly documented. There are few ornithologists in the region, which is geographically extensive and complicated. You may watch birds for your own amusement but your observations are nevertheless important. You may be the only source of information from a given locality in months or years. Keeping good notes and sharing your observations with other birdwatchers, or through clubs and journals, will help fill some of the gaps in our knowledge. A few pointers are pertinent.

- Always carry and use a notebook. Keep it in a waterproof bag.
- Use a pencil or ballpoint, not ink nor felt-tip (which run when wet).
- Indicate clearly the locality and date of all notes or lists recorded.
- Record all birds seen at a locality.
- Make especially detailed notes of any rare or unusual bird sightings and send these to a bird club or journal later.
- Note the dates of migrant species seen, and also the condition of their plumage—breeding, eclipse, non-breeding etc.
- Keep notes of any breeding records.
- Record any unusual feeding or behaviour.

It is important to take notes at the time of observation, and not wait till you get home. Without them, the observer will soon forget which birds were seen in a given area or on what dates (important in building a picture of breeding season and migration periods). Moreover, notes are essential for later identification of birds not recognized immediately in the field.

Identification hints: what to look for

Recognition is based on a combination of several characters, including a bird's appearance, voice, and behaviour. Check as many parts of the bird as possible, especially diagnostic features when known. The most conspicuous character, such as a white bar on the tail, may be remembered vividly but other features are often overlooked. On checking in the field guide later, the observer may find two similar birds with white tail bars and cannot remember if the bird seen had a brown or grey head.

With a new or unfamiliar bird, it is best to make a sketch in your notebook, as shown below. This does not need to be a work of art, simply enough to record details such as size; shape; length of bill; presence of crest or other features; colour of plumage; length of wings and tail; colour of any bare facial skin; colour of bill, eye, and feet; any other unusual features. Additional notes about the call, behaviour, and locality all help in later

Example of field sketch from notebook (Dollarbird, *Eurystomus orientalis*).

identification. If an unknown bird resembles a known species, list the variant features. For example, a note entry might read: 'small pinkish dove, similar to spot-necked dove but no white spots on black neck patch, greyer head, and more uniform red brown on back.' Such a description can be readily identified later, in this case as the Island Collared-Dove. It is useful to have a field guide close at hand for quick reference so that you can recheck features overlooked on first encounter.

Some species are difficult to identify in the field, and others are virtually impossible without close examination and measurement. With experience you will learn which are the 'difficult' birds and which distinguishing characters to look for. These diagnostic features are given in the individual species descriptions in this book, and in the accompanying notes facing each plate.

Making bird lists

Many birdwatchers keep lists of the birds they see—day lists of all birds seen on a given day, locality lists of all birds ever seen at a given locality or reserve, island lists, trip lists, year lists or life lists, and so on. Such lists are fun to compile and give an added sense of purpose and achievement to the activity of birdwatching. They are also valuable sources of scientific information both for the study of birds for their own sake and also for the conservationists and wildlife managers who have to select, protect, and manage reserves and other areas to ensure the survival of important and valuable wild species. Obviously, to be useful such lists must be accurate. Do not dilute valuable accurate information by the inclusion of doubtful records. By all means make a note of uncertain species, but separate these from species positively identified. Even experts do not expect to identify every bird they see. Ensure that all lists are clearly identified with the location and dates or period in which birds were observed.

The most common lists made by birdwatchers are probably day, locality, or trip lists. The drawback of these is their incompleteness. It takes a long time to locate and identify all resident species in a given area, and new visitors can be added to the list year after year. A partial list for an area does not give a full picture; moreover, it does not indicate the relative abundance of different species, nor the richness of the locality. A short list may indicate a rather impoverished fauna or may only reflect the shortness of the visit or adverse weather conditions. Bird lists become far more useful if they indicate the abundance of species and the intensity of searching effort upon which they are based.

Ideally one should record every single bird seen and heard, but in practice this is an impossible task. I recommend the following method which has been found very useful in Indonesia and other parts of SE Asia. Make a list by recording each new species seen until you reach 20 species; then start again with a new list. Any one species will only be recorded once in your first list of 20, but may be recorded again in subsequent lists. Analysis of ten or more lists for a given area will give a very good picture of its avifauna. If the cumulative total number of species recorded is plotted graphically against the number of lists made, this gives a species discovery curve whose steepness reflects species richness and indicates how many more species are still likely to be found in that locality. Species which occur on a high proportion of lists are clearly the most abundant or conspicuous species of the local avifauna. The graph below shows examples of field data of this sort for three reserves in Java and one in Borneo.

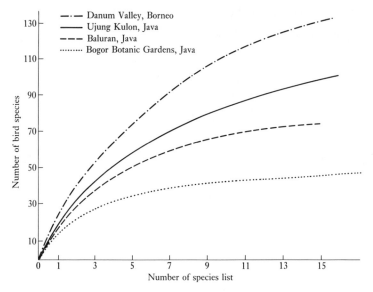

Species discovery curves for four Sundaic localities.

The great advantage of this method is that it is relatively independent of observer experience and expertise, and also independent of birdwatching intensity, weather conditions, or other factors. The only rules to keep in mind are that:

1. All species seen should be included in the lists even if they are not recognized, i.e. give your own name or code to unknown species so that you can at least recognize them as the same species when they are seen again.
2. The results reflect where the data were collected and will show any bias in the survey pattern. To reflect accurately the bird fauna of the whole area of a reserve or park, all habitat types should be surveyed in similar proportions to their abundance. Do not survey the same trail too many times.

Submitting records

Detailed notes of an unusual bird will give your record a much greater chance of being accepted by the relevant records screen body. A good description should include:

1. Name of observer and witness companions;
2. Name of species recorded;
3. Date of observation;
4. Time of day;
5. Location (include coordinates of remote areas);
6. Habitat (include estimated elevation);
7. Weather and light conditions;
8. Distance and observation conditions (include details of optics used);
9. Description of bird (size, shape, plumage, and bare parts);
10. Behavioural notes (flight, gait, posture, call, feeding, and associated species);
11. Additional supporting information (the prior familiarity of observer with this species, unusual weather conditions that may explain the unexpected observation, etc.)

Appendix 7 lists clubs, journals, etc. in the region. Submit your record to the most appropriate body.

Transcribing bird calls

The calls of most birds are as distinctive as their appearance. Indeed in some species, such as certain babblers, the call pattern may be the only really diagnostic field character. Birdwatchers wandering through a forest will hear far more species of birds than they will see. Keen ornithologists are therefore neglecting a great deal of information if they do not learn to recognize the calls of different birds.

To describe bird calls and record their form and structure is not easy. A standard method in many bird books, including this one, is to use human syllable mnemonics which approximate the form and sound of the bird call. This method of denoting bird calls is ancient in origin and almost worldwide. Many of the local names of birds are simply onomatopoeic representations of their calls, such as 'Kawau' for the Argus Pheasant, and 'Perkutut' for the Peaceful Dove. The Dutch familiar name 'Piet van Vliet' for the Plaintive Cuckoo is another example. This method has many drawbacks, however, as different people make quite different interpretations of the same birdsong and phrases, and letters that are meaningful or have one sound in one language may be unrecognizable in another.

Some authors have tried to improve on this method by writing bird tunes with musical notations, but this method is only possible for a few birds with very clear musical songs. It is also possible to give an idea of the pitch and duration of the notes in most songs with a series of horizontal strokes of different lengths and heights. By a further refinement the thickness of the stroke can indicate volume while the angle of the stroke can indicate inflection or whether the note is of rising, falling, or of constant pitch. Thus the familiar Plaintive Cuckoo call could be variously denoted as ⁄⁄⁄⁄⁄⁄⁄⁄ and ⁄﹀⁄. Examples of this latter method are given in Appendix 6.

More exact records of birdsong can of course be made with a tape recorder and, fortunately, the small, inexpensive, widely available ones are usually adequate. A directional microphone is necessary to amplify the sound of the bird and mask other irrelevant noises. Two types of directional microphones are available—amplifying unidirectional microphones and parabolic reflectors. The parabola gives better quality results but is larger and more clumsy to carry. Another improvement is to record in stereo, using two microphones set well apart or facing in divergent directions. Stereo recordings help to separate the subject sound spatially from any irrelevant sounds recorded at the same time. The secret of all methods, however, is to get as close as possible to the subject.

Taped sound has the disadvantage that it cannot be examined visually. However, this can be resolved with a sonogram or melogram machine, which can produce visual sonographs displaying the intensity and pitch of sound against a time scale. Such sonograms are very useful in the scientific analysis and comparison of bird sounds.

When and where to see birds

The avian year

ALTHOUGH the seasons of tropical regions are not as marked as in temperate countries, tropical birds are very sensitive to small changes in temperature and show quite seasonal patterns of breeding. The avifauna also reflects indirectly the more dramatic climatic changes of the temperate regions because about 25% of the entire avifauna is made up of temperate migrants. These are mostly from the North and arrive over the northern winter, but a few are southern migrants from Australasia. A few more are oceanic visitors.

The climate of the region is affected most strongly by the winter monsoon when there is a high pressure on the Asian mainland and cool wet winds blow south over the whole Greater Sundas area. November to April are the coldest and wettest months of the year.

Different groups of birds respond in different ways to this wet weather. Waterbirds nest at the end of the wet season in early spring when water level is highest and they are safest in their nesting trees which stand in the water. Many insectivorous birds also breed at the end of the wet season when insects are most numerous. Fruit-eaters breed a little later in the year when many of the forest trees and bushes are bearing fruit. The latest breeders are open country birds and seed eaters which breed in the drier times of the year. On Java, which has the greatest variation in climate, the breeding season in the east of the island, and on Bali, is generally one or two months after the breeding season in the west.

Autumn is the start of the arrival of northern migrants when tens of thousands of rails, shorebirds, raptors, swallows, flycatchers, pipits and wagtails move down the east coast of SE Asia to their wintering grounds on the Greater and Lesser Sundas. A major crossing point is from Cape Rachado on the Malay Peninsula to Sumatra. A few weeks later, the birds are crossing the Sunda Strait between Sumatra and Java, moving along the north coast of Java, and then across the narrow straits to Bali and the Lesser Sundas. Passage migrants move through quickly, but all along the route wintering birds settle and take up wintering grounds, spreading out the pressure on local resources. A separate migration route brings visitors to N Borneo and the Philippines. The process reverses in early spring, by which time many of the birds are already moulting into summer breeding plumage and even giving breeding song in anticipation of the temperate summer ahead.

In the Kelabit highlands of Borneo, the calendar is based on the migration of birds. The month during which Yellow Wagtails arrive is called *Sensulit mad'ting*; the next month *Sunsulit pererang* means the wagtail stays. October–November, *Neropa*, is the month of the Brown Shrike and the last month for timely planting, followed by *Kornio piting* for the Japanese Sparrowhawk and *Padawan*, the Dusky Thrush.

Where to see birds

There are over 300 official nature reserves in the region of this field guide. All are interesting to the birdwatcher. In addition, there is considerably more wild and semi-wild habitat in timber concessions, village forests, reforested areas, and scrublands where many more birds may be found. The swamps and coastal areas are frequently full of surprises and many of the new records for the region over the last few years have come from sea lanes far from land. However, to make this section of more practical use to the keen birdwatchers who have limited time in the region, I have selected 20 reserves that cover the complete range of bird habitats of the region. Brief descriptions of each reserve are given with directions on how to get there and what to look out for in each. The location of the reserves is shown on the map opposite. Appendix 3 gives a list of all regional endemics plus all species included in the ICBP book *Birds to watch* (Collar and Andrew, 1988) found in each of the 20 reserves so that the birdwatcher can plan an itinerary on what he or she hopes to see.

Important sites of Borneo

Mt. Kinabalu is Sabah's first and showcase national park. It is a must for all visitors who have the opportunity to visit the island. Kinabalu is the centre of distribution for all Borneo's montane birds except *Oriolus hosii* which, curiously, is absent. The 754 km^2 park contains the highest peak in SE Asia (4101 m); is easily accessible and has excellent trail and accommodation facilities. The park headquarters at 1500 m offer a great starting

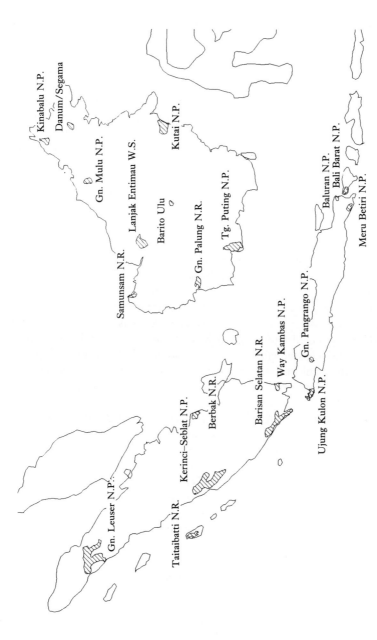

Location of major parks and reserves.

place to explore the montane fauna and flora, while a second lowland resort at Poring is in upper dipterocarp forest and offers an arboreal catwalk for those who want to watch canopy birds from a level vantage. A total of 289 birds has been recorded in the park, including many rare and endemic species found only on Borneo's highest peaks.

Visiting **Mt. Mulu** is an adventure that takes the visitor on a long boat trip from the coastal town of Miri up the Baram, Tutoh, and Malinau rivers to the headquarters at Long Pala. Various forest camps allow the visitor to spend several days exploring the magnificent scenery with huge caves, amazing limestone pinnacles, and superb views from the high ridges. The forests are varied in both altitude and form, depending on drainage and rock type, so that the diversity of birds is very rich. Over 260 species have been recorded, including 20 endemic species and all the Bornean hornbills, broadbills, and barbets. Plan enough time to work the area thoroughly—birdwatching in dense forests is not easy. A local speciality is the Bat Hawk which preys on the columns of bats emerging from the Deer Cave at dusk.

Danum Valley Conservation Area is an area of 438 km^2 of lowland rainforest within the concession areas of the Sabah Foundation. The reserve consists mostly of lowland and hill dipterocarp forests with some secondary and riverine habitat. A small area of montane forest is centred on Mt. Danum (1093 m). At a field station established for scientific research, the birds have been extensively studied. Access is by road from Lahad Datu, or if you have time, by the scenic boat trip up the Segama river, past the interesting caves of Tapadong and the orang-utan country of the Bole river. Most of the lowland forest birds of Borneo can be seen here. It is especially good for seeing hornbills, pittas, trogons, Bornean Bristlehead, and Black-throated Wren-Babbler. A total of 240 species of bird have been recorded here and in the Ulu Segama forests adjacent to the reserve. Orang-utans and the occasional elephant, rhinoceros, and banteng (tembadau) occur in the area.

Samunsam Wildlife Sanctuary is situated at the extreme western tip of Sarawak, Tanjung Datu, against the Kalimantan border. It is a wild and remote corner with an extensive and varied coastline where turtles nest and waders feed. Inland are various types of forest rising to about 1200 m in the enlarged area of 209 km^2. The reserve is in a restricted area and visitors must get a special permit to enter. The area offers an excellent introduction to the Borneo fauna with excellent viewing of trogons, barbets, hornbills, broadbills, woodpeckers, and nightjars, and a few rarities such as the Borneo Bristlehead, Storm's Stork, Rufous-bellied Eagle, and Sunda Ground-Cuckoo. Proboscis monkeys are a local speciality.

Lanjak Entimau Wildlife Sanctuary is a large, wild area of 1688 km^2 along Sarawak's southern border adjacent to the even larger Bentuang and Karimun reserve (6000 km^2) on the Indonesian side. The Indonesian reserve has never been surveyed. The Sarawak side was surveyed only with the aid of helicopters, and the bird list is far from complete. Nevertheless, it is clear that this is an area of extremely high conservation importance where there has been very little human disturbance, and orang-utans and maybe even rhinoceros still survive. The Sarawak reserve can be reached by road, then river, from Kanowit. The area is normally closed to tourists and visitors need a special permit. Interesting birds already recorded from the area include Bulwer's Pheasant, Crimson-headed Partridge, Wallace's Hawk-Eagle, Everett's Flowerpecker, and Large-billed Blue-Flycatcher.

Kutai National Park is the largest reserve on the east coast of Kalimantan (2000 km^2) and contains a good range of habitats with mangrove, riverine forest, lowland evergreen forest, ironwood forest, and some areas of heath forest. It was a magnificent area rich in wildlife, including orang-utan, rhino, and banteng. Sadly, the reserve has been partly logged, partly burnt, and partly invaded by farmers, greatly devaluing it as a nature reserve. The bird fauna is largely intact, however, and most of the island's lowland birds can still be found in the relative ease of an opened-up forest. The park can be reached by road from Samarinda and Bontan. A total of 236 bird species have been recorded.

Gunung Palung Nature Reserve is a unique feature of the west coast of Borneo. This horseshoe-shaped peak of 1100 m rises from the surrounding flat and swampy forests to provide an island of tall, dry dipterocarp forest capped with a small but interesting montane forest. The reserve of 900 km^2 is home to many orang-utans and other primates as well as a rich bird fauna of some 240 species. There is a scientific field station on the southern aspect of the mountain. Visitors can also rent a boat and travel up the Matan river to the village of Kampong Baru to explore the mountain from the north. As well as good representation of most of Borneo's lowland bird families, the reserve is home to certain rarities such as White-crowned Hornbill, Bornean Peacock-Pheasant, Bornean Bristlehead, and Black Laughing-Thrush. In addition there is a major roost of Long-tailed Parakeets close to the coast and every morning and evening the visitor can watch dozens of noisy flocks winging their way over the town of Sukadana.

Tanjung Puting National Park can be reached by boat from the town and airport of Palangkaraya. Most famous for its orang-utans, the 300 km^2 reserve is also a fascinating area for birds. The area is low-lying with a mosaic of heath forests, peat swamps, and dry, sandy, open padangs. Travel is mostly by boat, through a network of waterlogged trails which extend around the orang-utan research station. The reserve contains a group of lakes which are important nesting habitat for waterbirds but accessible only by well-planned expeditions. The forests are more stunted and open than the dipterocarp forests which cover most of the rest of Borneo, and they lack the giant fig-trees. Hornbills are rare, but other birds are more easily seen; these include Javan Frogmouth, Bornean Bristlehead, Bulwer's Pheasant, Storm's Stork, Bornean Peacock-Pheasant, and Black Partridge. A total of 218 species has been recorded in the reserve, but this list is still not complete.

Barito Ulu Research Area is not yet gazetted as a nature reserve but this has been proposed, and it has also been the site of extensive fieldwork. The area is approached by river taxi up the Barito as far as Muaratewe, then hired sampan to Muarajulai and the foothills of the Mueller Hills. This is as near the geographical centre of Borneo as you will get and the area was for a long time a big question mark as far as bird distributions were concerned. Recent studies have revealed it to be surprisingly rich, with over 230 species recorded including 15 island endemics. Several species have been found that were previously thought to live only in North Borneo or only in mountains; these include Blyth's Hawk-Eagle, Red-breasted Partridge, and Black-throated Barbet. The area also has several birds thought to be extreme lowland birds including Bornean Bristlehead and Black Partridge. Other interesting species include Blue-headed Pitta, Bulwer's Pheasant, Bat Hawk, Mountain Barbet, Rail-Babbler, Hook-billed Bulbul, Pygmy White-eye, and the rare White-shouldered Ibis. Take malaria pills—several visitors have become quite ill here.

Important sites of Sumatra

Mt. Leuser (Gunung Leuser) National Park is a huge area of magnificent forests and mountains with a small coastal and lowland extension. The park covers a total of almost 9000 km^2 and can be approached from several directions. Most visitors use one of three entrances—Bohorok-Bukit Lawang, to the north of Medan, where there is an orang-utan rehabilitation station; Berastagi, which is a hill resort some 50 km west of Medan, from which one can walk north to the southern borders of the park; or Ketambe, way up the Alas valley north of Kotacane where there is an orang-utan research field station. All are excellent places to explore the forest birds of northern Sumatra, but many of the Sumatran endemics recorded from the park are at higher altitudes on the slopes of Mt. Leuser (3419 m). This peak is not easily accessible and would require a camping expedition to be explored. Most of the montane endemics are more easily found on Mt. Kerinci, but Mt. Leuser is a must to find Hoogerwerf's Pheasant, Ground-Cuckoo, and Mountain Serin. This is also the locality to look for some of Sumatra's rarest birds that have not been seen for many years such as Rueck's Blue-Flycatcher, White-fronted Scops-Owl, Blue-Wattled Bulbul, Sumatran Cochoa, and Vanderbilt's Babbler.

Kerinci–Seblat National Park is another very large park of about 15000 km^2. It covers large parts of four provinces and includes the important peak of Mt. Kerinci (3085 m) and, in the Kerinci valley, the highest marshland and wetlands in Sumatra. The park includes a full range of continuous forest from lowlands to montane, including some natural stands of tropical pine. It is a wild area where rhinoceros and tapir still roam and it has a wonderful wealth of birds. The best place from which to explore is the town of Sungai Penuh in the Kerinci Valley, an enclave in the centre of the park near an interesting lake. This is the best reserve to see the montane Sumatran avifauna, including most of the island's endemic species such as the beautiful little Blue-masked Leafbird, Sumatran Peacock-Pheasant, Rufous Woodcock, Sumatran Green-Pigeon, Rajah Scops-Owl, Blue-tailed Trogon, Schneider's Pitta, Rusty-breasted Wren-Babbler, Shiny Whistling-Thrush, and Sumatran Cochoa.

Taitaibatti Nature Reserve is a large reserve on the island of Siberut in the Mentawai group. The island can be reached by regular boats from Padang and the reserve can be reached, with some difficulty, overland from the port of Muara Siberut. In fact, however, there are more easily accessible forests just as good for birdwatching. Mentawai supports several endemic mammals including four primates and one endemic bird, the Mentawai Scops-Owl. Many other birds on the island are represented by endemic races. The Siberut people are fascinating to visit, being great hunters with bow and arrow and having a deep understanding of and respect for the forests and wildlife. Some 105 bird species are recorded for the island.

Berbak Game Reserve includes extensive areas of peat swamp, coastal forest, and mangroves on the east coast of Sumatra in Jambi Province. The reserve totals 1900 km^2 and can be reached from Jambi and Tanjung Jabung by road or boat. A few tigers still persist. The reserve is important for coastal and wetland birds as well as being representative of the eastern swamp forests. Birds of interest in the reserve include Milky Stork, Storm's Stork,

Lesser Adjutant, Wallace's Hawk-Eagle, Asian Dowitcher, Silvery Wood-Pigeon, and Buettikofer's Babbler. A total of 245 species has been recorded.

Barisan Selatan Nature Reserve consists of the southern end of the Barisan mountains. It is a long narrow reserve of 3650 km^2 covering all forest types rising from the sea at Belimbing to the peak of Gn. Pulung (1964 m). The reserve can be reached by road from Tanjung Agung and several roads cross the reserve to the west coast. The southern peninsula can be visited by boat from Tanjung Agung to Belimbing where feral water-buffalo have become established. The reserve is being nibbled on all sides by logging and coffee farmers but much of the area is rugged and wild, with tigers, elephants, wild dogs, and a great variety of monkeys. It is rather underrated and would certainly reveal many of the montane birds if properly surveyed. The current list of 121 species is probably less than half complete and birdwatchers should be encouraged to fill out the rest of the list, which could prove very interesting.

Way Kambas National Park is a large coastal reserve of 1235 km^2 in the south-east of Sumatra in Lampung Province. It can be reached by road from Tanjung Karang. Old logging roads provide access deep into the reserve, or you can go round the coast from the Penet River. The reserve consists of non-peaty swamp forest, coastal forest, and patches of original lowland rainforest. The area has had a history of logging and fires and the vegetation is now a mosaic of open *Imperata* grassland, secondary scrub, and original forest. This mosaic proves very rich for birds and over 230 species have been recorded in the reserve which also harbours important populations of tapir and elephant.

Important sites of Java and Bali

Ujung Kulon National Park is the famous last home of the Javan Rhinoceros. It is a beautiful and unique National Park situated at the extreme west point of Java. The reserve comprises expanses of flat swampy forest with two small mountains of excellent evergreen rainforest, some opened grazing areas, coastal scrub, beaches, cliffs, and several offshore islands. The park can be reached from the port of Labuan by boat or road to the headquarters at Taman Jaya, or by boat only to resort posts on the islands of Handeulum and Peucang. The park offers excellent birdwatching in all habitats and is a great introduction to Java's special bird fauna with Green Junglefowl, Green Peafowl, Javan Lorikeet, Sunda Coucal, Javan Kingfisher, plus endemic babblers, sunbirds, and many other interesting species.

Gunung-Pangrango National Park is easily reached from the main Bogor-to-Bandung road at the Puncak Pass. The headquarters is at the famous Cibodas Botanic Gardens at a chilly 1200 m, and the park rises to 3026 m at Pangrango Peak in a mossy elfin forest with alpine meadows. Most of the 140 km^2 park is composed of evergreen sub-montane forests. Gales have damaged many large trees near the entrance to the park but otherwise the forests are lush and grand. The park is a must for the twitcher in Java as it is here that most of the Javan endemics and other rarities can be seen—Javan Hawk-Eagle, Javan Partridge, Rufous Woodcock, Javan Scops-Owl, Volcano Swiftlet, Blue-tailed Trogon, three endemic barbets, Pygmy Tit, all the endemic babblers, Javan Cochoa, Javan Tesia, Rufous-tailed Fantail, Javan White-eye, and Mountain Serin. However, birdwatching is difficult. The forest is tall, dense, dark, and often in rain. It takes several days to do the reserve justice.

Meru Betiri National Park is a remarkably wild lowland area on the south-east coast of Java where the last Javan tigers survived until a few years ago. The park covers an area of 495 km^2 and has some rugged coastline, beautiful beaches, and extensive primary and secondary forests, mangroves, and rubber plantations. The highest peak is 1223 m. The reserve is reached by road from Banyuwangi. Over 180 bird species have been recorded but the area is still poorly explored, and more species can undoubtedly be found here. The forest is surprisingly moist for such an eastern locality and thus supports many birds not found in drier Baluran.

Baluran National Park lies in the extreme north-east corner of Java and is easily accessible from Besuki. It is a pretty area of 250 km^2 with savannah monsoon forests centred on the dormant volcano of Mt. Baluran (1250 m). The semi-open forests provide excellent conditions for birdwatching and the reserve boasts an impressive list of about 200 species, including Green Peafowl, Green Junglefowl, and Black-winged Starling. There is a small moist forest inside the volcano crater that may still contain some surprise species, and the extensive coastline and mangroves are a good place to see waders and waterbirds.

West Bali (Bali Barat) National Park is a beautiful large wild area at the extreme western end of Bali with a total area of 777 km^2 and with a highest elevation of 1414 m. Forests range from montane to dry monsoonal with moist forests in sheltered gulleys and coastal savanna, scrub, and some mangroves. The park is famous as the only wild home of the endemic white Bali Myna, but also contains a wide range of other interesting birds such as the rare Mangrove White-eye of Menjangan Island, Banded Pitta, Green

Junglefowl, and many more. In autumn this is a good point to watch migrating raptors coming across the Javan Straits to their winter quarters in the Lesser Sundas.

Entry permits for Indonesian National Parks can be obtained at the respective Park headquarters offices but permits for other reserves should be applied for in advance to the Directorate Jenderal P.H.P.A. (Perlindungan Hutan dan Pelestarian Alam), Jl. Ir. H. Juanda, Bogor, Indonesia.

THE PLATES

Abbreviations used on the plates
Juv.   juvenile
Br.   breeding plumage
Non-br.   non-breeding plumage

Abbreviations used in the captions
Occurrence:
B   Borneo
J   Java/Bali
S   Sumatra

AG   See p. 3 for explanation
>   See p. 3 for explanation

PLATE 1. **Petrels, Shearwaters, and Storm-Petrels**

3. **Barau's Petrel** *Pterodroma baraui*   S
   Head pattern; dark diagonal bar on white underwing; pale W pattern on brown-grey upperwing.
4. **Slender-billed Prion** *Pachyptila belcheri*   J
   Slender bill. In flight like 5. Often in large flocks.
   (a) Undertail pattern.
5. **Antarctic Prion** *Pachyptila desolata*   J
   Thicker bill than 4; black W pattern on upperwings; larger than 4; short eye-line.
   (a) Undertail pattern.
6. **Bulwer's Petrel** *Bulweria bulwerii*   SBJ
   Smaller than 7; pale underwing bar; flies low over waves. Wedge-shaped tail looks long and pointed in flight.
7. **Jouanin's Petrel** *Bulweria fallax*   S
   Larger and heavier than 6; short, thick, black bill; flies high over waves.
8. **Streaked Shearwater** *Calonectris leucomelas*   SBJ
   Large size; slow, unhurried flight; can occur in large flocks.
9. **Flesh-footed Shearwater** *Puffinus carneipes*   S
   Straw coloured bill with brown tip; short, round tail; pale feet; straight wings; underside of primaries have paler bases; gregarious habits.
10. **Wedge-tailed Shearwater** *Puffinus pacificus*   SJ
    Wedged-shaped tail; wings curved forward.
    (a) Pale morph: pale brown above.
    (b) Dark morph: all dark.
11. **Wilson's Storm-Petrel** *Oceanites oceanicus*   SJ
    White rump, flanks and thighs; square tail.
12. **Swinhoe's Storm-Petrel** *Oceanodroma monorhis*   SBJ
    Dark rump, forked tail.
13. **Matsudaira's Storm-Petrel** *Oceanodroma matsudairae*   J
    Larger than 12; broader-based wings; forked tail; white shafts of outer primaries show as pale patch.
14. **White-faced Storm-Petrel** *Pelagodroma marina*   S
    Forehead, eyebrow, chin, and underparts white; pale upperwing bar.

Plate 1

PLATE 2. **Tropicbirds and Boobies**

15. **Red-tailed Tropicbird** *Phaethon rubricauda* SJ
    Adult: red bill, red streamers. Fresh plumage with pink flush, becomes white with wear.
    Juvenile: black bill.
16. **White-tailed Tropicbird** *Phaethon lepturus* SJ
    Adult: yellow bill; pale streamers.
    (a) *P. l. fulvus* (breeds Christmas Island).
    (b) *P. l. dorothea* (breeds Pacific islands).
    Juvenile: yellow bill.
20. **Red-footed Booby** *Sula sula* SBJ
    Adult: blue bill; red feet; white tail.
    (a) White-tailed brown morph.
    (b) White morph (Christmas Island): black patch on underwing.
    (c) Juvenile: all morphs fledge in this plumage.
21. **Masked Booby** *Sula dactylatra* SBJ
    Adult: white head and body; yellow bill; black tail; green-grey feet.
    Juvenile: white collar; underwing pattern.
22. **Abbott's Booby** *Papasula abbotti* SJ
    Wing shape differs from other boobies.
    Male: blue bill; black thigh patch.
    Female: pink bill.
    Juvenile: like adult but with grey bill.
23. **Brown Booby** *Sula leucogaster* SBJ
    Adult: brown neck; yellow bill and feet; white belly and underwing.
    Juvenile: like adult but bill, facial skin and feet grey; underparts mottled brown.

PLATE 3. **Pelicans and Cormorants**

17. **Great White Pelican** *Pelecanus onocrotalus*   J
    Adult: all white except black primaries and secondaries and yellow throat-patch.
    Immature: brownish.
18. **Spot-billed Pelican** *Pelecanus philippensis*   SJ
    Silvery grey with darker shaft streaks; short, brownish crest and ridge along neck; spotted bill.
    Immature: brownish.
19. **Australian Pelican** *Pelecanus conspicillatus*   SJ
    Black and white; sometimes with a rosy tinge; black tail and upperwing coverts.
    Immature: pale brown and grey.
24. **Little Black Cormorant** *Phalacrocorax sulcirostris*   SBJ
    All dark; grey bill.
    Breeding: glossy black with white tufts on side of head.
    Non-breeding: dull; dark throat.
    Immature: brown.
25. **Great Cormorant** *Phalacrocorax carbo*   SB
    Large size; long hooked bill; white on face and whitish flank patch.
    Immature: white more extensive.
26. **Little Pied Cormorant** *Phalacrocorax melanoleucus*   J
    Black above, white below; yellow bill. Sometimes with yellow staining.
    Immature: eye-stripe and crown black; black flank patch.
27. **Little Cormorant** *Phalacrocorax niger*   SBJ
    Small size; compressed brown-yellow bill.
    Breeding: white flecks on side of head; black throat.
    Non-breeding: white throat patch.
    Immature: as non-breeding but browner.
28. **Oriental Darter** *Anhinga melanogaster*   SBJ
    Long snake-like neck; long plumes on wing coverts; narrow pointed bill.

Plate 3

Plate 4. **Frigatebirds**

29. **Christmas Frigatebird** *Fregata andrewsi*   SBJ
    Male: white belly.
    Female: black chin; white underparts with convex rear edge.
    Immature: similar to adult but lower breast is heavily scaled black and white.

30. **Great Frigatebird** *Fregata minor*   SBJ
    Male: no white on belly.
    Female: pale chin; breast patch with concave rear edge.
    Immature: similar to adult. Belly white, mottled brownish-black and white.

31. **Lesser Frigatebird** *Fregata ariel*   SBJ
    Male: white spots on flanks.
    Female: dark chin and throat; breast patch with concave rear edge.
    Immature: very variable.

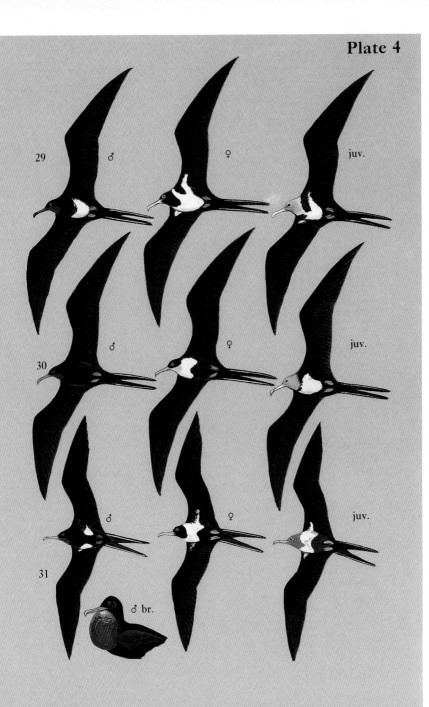

PLATE 5. Herons and Egrets

32. Great-billed Heron *Ardea sumatrana* — SBJ
Very large; dusky plumage.

33. Grey Heron *Ardea cinerea* — SBJ
White head and neck; black shoulders; black eye-stripe and plumes.
Juvenile: black cap; greyer neck; less contrast.

34. Purple Heron *Ardea purpurea* — SBJ
Adult: rufous and purplish plumage.
Juvenile: chestnut plumage; lacks plumes.

35. White-faced Heron *Ardea novaehollandiae* — J
White-face and black flight feathers distinguish from 40. (Nusa Penida Island, off Bali only.)

39. Cattle Egret *Bubulcus ibis* — SBJ
Bulbous head and heavier bill than other egrets.
Breeding: orange wash; legs sometimes pink.
Non-breeding; yellow bill; dark legs.

40. Pacific Reef-Egret *Egretta sacra* — SBJ
Greenish legs; grey facial skin; coastal habitat.
(a) Dark phase (commoner).
(b) White phase.

41. Chinese Egret *Egretta eulophotes* — SB
Non-breeding: medium size; greenish legs; dull bill.

42. Great Egret *Egretta alba* — SBJ
Large size; kink in neck.
Breeding: black bill; red legs; grey facial skin.
Non-breeding: yellow bill and facial skin; black legs.

43. Intermediate Egret *Egretta intermedia* — SBJ
Medium-sized; no kink in neck.

44. Little Egret *Egretta garzetta nigripes* — SBJ
E. g. nigripes shown: slim shape; bill greyish-black; legs black (migrant race *E. g. garzetta* to N Borneo has yellow toes).
Breeding: slender head plumes.

Plate 5

PLATE 6. **Smaller Herons and Bitterns**

36. **Striated Heron** *Butorides striatus*   SBJ
    Male: black cap; greyish plumage; yellow or orange legs. Several subspecies vary.
    Female: smaller than male.
    Immature: streaky greenish brown with white spots.
37. **Chinese Pond-Heron** *Ardeola bacchus*   SB
    Breeding: similar to breeding 38 but head and back dark chestnut and breast maroon.
    Non-breeding: streaky brown; no white spots.
38. **Javan Pond-Heron** *Ardeola speciosa*   SBJ
    Breeding: white wings and tail; blackish maroon back; buff head and breast.
    Non-breeding and juvenile: as non-breeding 37.
45. **Black-crowned Night-Heron** *Nycticorax nycticorax*   SBJ
    Adult: black, grey, and white plumage; white plumes.
    Breeding: red legs and lores; red iris.
    Immature: looks like immature 46, more spotted.
46. **Rufous Night-Heron** *Nycticorax caledonicus*   BJ
    Adult: black cap; rufous upperparts; white plumes.
    Immature: mottled brown, streaked, and spotted white.
47. **Malayan Night-Heron** *Gorsachius melanolophus*   SBJ
    Adult: long black crest; blue-green lores; barred back.
    Immature: similar to immature 45 but has stubbier bill.
48. **Japanese Night-Heron** *Gorsachius goisagi*   B
    Short, chestnut crest; yellow facial skin; stubbier bill than 47; back mottled with black, not barred.
49. **Schrenck's Bittern** *Ixobrychus eurhythmus*   SBJ
    Male: black cap; maroon back; black flight feathers; grey underwing.
    Female and immature: duller, streaked and spotted black and white.
50. **Yellow Bittern** *Ixobrychus sinensis*   SBJ
    Adult: yellowish with black cap and flight feathers.
    Immature: no white spotting on upperparts.
51. **Cinnamon Bittern** *Ixobrychus cinnamomeus*   SBJ
    Male: bright chestnut upperparts.
    Female and immature: black cap, duller, blacker throat.
52. **Black Bittern** *Dupetor flavicollis*   J
    Male: black upperparts; long dagger-like bill.
    Female: uniform brown upperparts.
    Immature: black crown; buffy feather tips to upperparts.

Plate 6

PLATE 7. **Storks, Ibises, and Spoonbills**

54. **Milky Stork** *Mycteria cinerea* SJ
    White plumage with black flight feathers; red bare facial skin; pink bill.
55. **Asian Openbill** *Anastomus oscitans* S
    Grey plumage with black flight feathers; yellowish 'open' bill.
56. **Woolly-necked Stork** *Ciconia episcopus* SJ
    Full neck woolly white; grey facial skin; blackish bill with red tip.
57. **Storm's Stork** *Ciconia stormi* SB
    Black pattern on side of white neck; reddish, slightly upturned bill; yellow eye-ring.
59. **Greater Adjutant** *Leptoptilos dubius* S
    Huge size; massive bill; gular pouch; contrasting grey bar on black wing.
60. **Lesser Adjutant** *Leptoptilos javanicus* SBJ
    Huge size; massive bill; no gular pouch; all dark wing.
61. **Black-headed Ibis** *Threskiornis melanocephalus* SBJ
    Black head and neck; decurved bill.
62. **White-shouldered Ibis** *Pseudibis davisoni* B
    Brown head; decurved bill; red nape; white shoulder flash.
63. **Glossy Ibis** *Plegadis falcinellus* SBJ
    Purple glossy plumage, decurved bill.
64. **Black-faced Spoonbill** *Platelea minor* B
    White plumage; grey spatulate bill; less extensive bare black face than 65.
65. **Royal Spoonbill** *Platalea regia* BJ
    White plumage; yellowish-grey spatulate bill; extensive bare black face.

Plate 7

Plate 8. **Ducks**

66. **Lesser Whistling-Duck** *Dendrocygna javanica* SBJ
    Small size; buff underparts; lack of flank plumes.
67. **Wandering Whistling-Duck** *Dendrocygna arcuata* SBJ
    Chestnut belly; black and white flank plumes.
68. **Northern Pintail** *Anas acuta* SBJ
    Male: neck pattern; long pointed tail.
    Female: brown speculum; pointed tail.
69. **Green-winged Teal** *Anas crecca* B
    Male: small size, head pattern. Female: mottled.
70. **Sunda Teal** *Anas gibberifrons* SBJ
    Green speculum; white wing bar.
    Male has lump on forehead.
71. **Mallard** *Anas platyrhynchos* B
    Male: green head; yellow bill.
    Female: blue speculum; orange feet; yellow bill.
72. **Eurasian Wigeon** *Anas penelope* B
    Male: chestnut head; golden crown.
    Female: rufous brown; grey bill, round head.
73. **Pacific Black Duck** *Anas superciliosa* SJ
    Striped head; green speculum; dark; white underwing.
74. **Garganey** *Anas querquedula* SBJ
    Male: broad white eyebrow; white belly.
    Female: pale eyebrow; white patch at base of bill; olive speculum.
75. **Northern Shoveler** *Anas clypeata* B
    Male: chestnut belly; white breast; long, broad, spatulate bill.
    Female: white tail; long, broad, spatulate bill.
76. **Tufted Duck** *Aythya fuligula* SB
    Male: black plumage; white flanks; long crest; yellow eye.
    Female: brown plumage; short crest; yellow eye.
77. **Australian White-eyed Pochard** *Aythya australis* J
    Brown plumage; white undertail; black bill with white tip. Eye is white in drake, dark in female.
78. **Cotton Pygmy-Goose** *Nettapus coromandelianus* SBJ
    Small size; whitish head and belly.
79. **White-winged Duck** *Cairina scutulata* SJ
    Dark plumage with whitish head, neck, and underwing coverts. Sumatran males whiter headed than mainland form.

Plate 8

Plate 9. Osprey, Hawks, Kites, and Eagles

80. **Osprey** *Pandion haliaetus* — SBJ
    Black eye-stripe; crest; black shoulder patch on underwing.
    P. h. melvillensis of N Australia goes west to Sumatra, has whiter head and less streaking on underparts.

81. **Jerdon's Baza** *Aviceda jerdoni* — SB
    Rufous, banded plumage and long crest; white throat with mesial stripe.
    A. j. borneensis shown: head and sides of neck rufous; belly heavily barred rufous.

82. **Black Baza** *Aviceda leuphotes* — S
    Long crest; dark and white barred underparts.
    A. l. syama shown. Other races more chestnut on underparts.

83. **Oriental Honey-Buzzard** *Pernis ptilorhynchus* — SBJ
    Variable crest; pale throat patch with black mesial stripe.
    (a) *P. p. torquatus* — Sumatra, Borneo.
    (b) *P. p. ptilorhynchus* — Java only.
    (c) *P. p. orientalis* — winter migrant.

84. **Bat Hawk** *Machaeramphus alcinus* — SB
    Black plumage; white throat; black mesial stripe; long, broad, pointed wings; short, square tail.
    M. a. westermanni shown.

85. **Black-winged Kite** *Elanus caeruleus* — SBJ
    Adult: white below; grey above; black shoulder and wing tips; red eye.
    Juvenile: tinged brownish.

86. **Black Kite** *Milvus migrans* — SB
    Dull brown plumage; grey feet and cere; forked tail.
    M. m. lineatus shown: migrant.
    Juvenile: mottled and less forked tail.

87. **Brahminy Kite** *Haliastur indus* — SBJ
    Adult: white head; chestnut upperparts; black wing tips; rounded tail.
    Juvenile: browner; pale base of primaries.

88. **White-bellied Fish-Eagle** *Haliaeetus leucogaster* — SBJ
    White head and underparts; grey feet; short wedge-shaped tail.

89. **Lesser Fish-Eagle** *Ichthyophaga humilis* — SB
    Grey head; brownish upperparts; grey feet; dark tail with slight white mottling at base.

90. **Grey-headed Fish-Eagle** *Ichthyophaga ichthyaetus* — SBJ
    Grey head; brownish upperparts; yellow feet; short rounded tail; dark terminal band on pale tail; short broad wings.

Plate 9

PLATE 10. **Eagles, Harriers, and Hawks**

92. **Short-toed Eagle** *Circaetus gallicus*   J
 Large rounded head; head and throat can be dark brown. Four tail bands, subterminal ones broadest; very long wings; pale carpals of underwing.
95. **Eastern Marsh Harrier** *Circus spilonotus*   SB
 Male: greyish; black throat and breast streaked white; white underwing; pale grey unbarred tail. Female: brown rump; buff crown streaked; rufous underparts; sometimes buffish breast and nape patches absent; usually barred tail. Immature: as female but darker with pale crown and nape; pale rump.
96. **Western Marsh Harrier** *Circus aeruginosus*   S
 Male: buffy head, brown back and wing coverts, grey secondaries and tail; brownish rump; underparts dark rufous. Female: buff, unstreaked cap; dark brown upper back; unbarred tail; brown rump. Immature: dark brown plumage with pale buff crown, nape, and throat; pale brown rump; unbarred tail.
97. **Northern Harrier** *Circus cyaneus*   B
 Male: grey and white; black wing tips; white rump. Female: fulvous brown; tail with 4 bars and white tip; heavily streaked underparts; white rump. Immature: as female but darker.
98. **Pied Harrier** *Circus melanoleucos*   B
 Male: striking black, grey, and white plumage. Female: buffy brown head and breast lightly streaked; brown back; greyish barred wings and tail with 5 bars; white rump. Immature: similar to female but more rufous, with pale rufous rump.
99. **Japanese Sparrowhawk** *Accipiter gularis*   SBJ
 Four tail bands; mesial stripe faint or absent.
100. **Besra** *Accipiter virgatus*   SBJ
 Male: dark uniform grey upperparts; 3 tail bars; black mesial stripe on white throat; some streaking on breast; rufous barred underparts. Female: brownish upperparts; whitish underparts broadly barred rufous.
 A. v. rufotibialis (N. Borneo): shown.
 A. v. vanbemmeli (Sumatra): even more rufous.
101. **Eurasian Sparrowhawk** *Accipiter nisus*   B
 No mesial throat stripe; barred tail. Male has rufous cheeks; female has prominent white eyebrow and paler head.
102. **Crested Goshawk** *Accipiter trivirgatus*   SBJ
 Larger size; crest; dark mesial and malar stripes; tail with 4 bars; barred and streaked underparts.
 A. t. javanicus (Java) shown: greyer above, tawny breast.
 A. t. trivirgatus (Sumatra): pale bars, less rufous.
 A. t. microstictus (Borneo): paler, rufous streaks and barring.
103. **Chinese Goshawk** *Accipiter soloensis*   SBJ
 Unbarred underparts; no mesial stripe. Sometimes orange eye. Female larger.
104. **Shikra** *Accipiter badius*   S
 Plain pale grey upperparts; faint mesial stripe; light rufous barring below.

Plate 10

PLATE 11. Eagles, Buzzards, and Hawk-Eagles

93. Crested Serpent-Eagle *Spilornis cheela*   SBJ
Spotted plumage; bold barring on underwing and tail. Races vary.
S. c. bido (Java and Bali) shown.
S. c. malayensis (N. Sumatra) lacks white tips to secondaries. White markings clearer.
S. c. pallidus of Bornean lowlands is smaller, paler. Uniform rufous brown breast.

94. Mountain Serpent-Eagle *Spilornis kinabaluensis*   B
Black throat; broader white tail bands than 93.

105. Rufous-winged Buzzard *Butastur liventer*   J
No mesial streak; grey throat; paler underparts than 106.

106. Grey-faced Buzzard *Butastur indicus*   SBJ
White throat; mesial stripe on throat; bold tail bars.

107. Common Buzzard *Buteo buteo*   SJ
Highly variable between and within races. Black carpal patches on underwing.
B. b. japonicus shown—more buff below than *B. b. vulpinus*.

108. Black Eagle *Ictinaetus malayensis*   SBJ
Long tail; long wings; very dark.

109. Booted Eagle *Hieraaetus pennatus*   J
(a) Dark morph: rufous underparts.
(b) Pale morph: pale underparts.

110. Rufous-bellied Eagle *Hieraaetus kienerii*   SBJ
Adult: 'Hobby' colour pattern; small crest.
Juvenile: pale underparts.

111. Changeable Hawk-Eagle *Spizaetus cirrhatus*   SBJ
Very variable; short crest.
(a) Light morph.
(b) Dark morph—predominates in Borneo.

112. Javan Hawk-Eagle *Spizaetus bartelsi*   J
Large size; prominent crest; barred underparts.

113. Blyth's Hawk-Eagle *Spizaetus alboniger*   SB
Adult: black and white plumage; long crest; mesial throat stripe; darker than 114; broad, white tail bar.
Juvenile: can be mistaken for 114 but tail has several dark bands.

114. Wallace's Hawk-Eagle *Spizaetus nanus*   SB
Brown plumage; long crest; three black bands in tail.

Plate 11

PLATE 12. **Falcons**

---

115. **Black-thighed Falconet** *Microhierax fringillarius*   SBJ
 Narrow pale forehead; pale stripe behind eye.
 Juvenile: face washed rufous.

116. **White-fronted Falconet** *Microhierax latifrons*   B
 Broad pale forehead; no pale stripe behind eye.
 Female: chestnut forehead.
 Juvenile: same as adult. No spots on tail.

117. **Common Kestrel** *Falco tinnunculus*   SB
 Male: grey crown; grey unbarred tail.
 Female and immature: brown barred tail.
 F. t. interstinctus shown: darker chestnut and more heavily spotted.

118. **Spotted Kestrel** *Falco moluccensis*   BJ
 Darker and more heavily spotted than 117.
 Male: brown crown; grey unbarred tail.
 Female: grey barred tail.
 F. m. javensis shown: heavily spotted and dark rufous below.

119. **Australian Kestrel** *Falco cenchroides*   J
 Paler; more cinnamon colour than 117.
 Male: forehead grey; nape brown; grey unbarred tail.
 F. c. baru shown: male greyer on head and tail; pale underparts.

120. **Eurasian Hobby** *Falco subbuteo*   J
 Breast white, streaked with black; rufous thighs and vent.
 Female similar to male but browner and more streaked on thighs and undertail coverts.
 F. s. streichi shown: smaller than *F. subbuteo*.

121. **Oriental Hobby** *Falco severus*   SBJ
 Rufous underparts; streaked on immature.

122. **Peregrine Falcon** *Falco peregrinus*   SBJ
 Large size; black cheek pattern. Adult belly barred, immature belly streaked.
 (a) *F. p. ernesti*: resident; very dark underparts.
 (b) *F .p. calidus*: migrant in winter; larger and paler.

Plate 12

PLATE 13. **Hawks and Falcons in flight (not to scale)**

95. **Eastern Marsh Harrier** *Circus spilonotus* — SB
Larger than other harriers with broader wings.
96. **Western Marsh Harrier** *Circus aeruginosus* — S
Male: rufous wing lining; dark vent.
Female: all dark; pale head.
97. **Northern Harrier** *Circus cyaneus* — B
White rump in all plumages.
Four dark bands on tail of female.
98. **Pied Harrier** *Circus melanoleucos* — B
Narrow wings.
Female has 5 narrow bands on tail.
Juvenile: very rufous underparts.
99. **Japanese Sparrowhawk** *Accipiter gularis* — SBJ
Small size; longer wings than 100; less barring on body.
100. **Besra** *Accipiter virgatus* — SBJ
Prominent mesial streak; heavy rufous wash underneath.
101. **Eurasian Sparrowhawk** *Accipiter nisus* — B
Long, slender wings; long, square-cut tail.
102. **Crested Goshawk** *Accipiter trivirgatus* — SBJ
Wings rounded, relatively short; long tail showing 4 bands; prominent mesial streak.
103. **Chinese Goshawk** *Accipiter soloensis* — SBJ
Whitish underwing; long, black wing tips.
Juvenile: more rufous with heavier barring.
104. **Shikra** *Accipiter badius* — S
Generally very pale below.
Juvenile: heavier streaking than female.
117. **Common Kestrel** *Falco tinnunculus* — SB
Longer-tailed than other falcons.
F. moluccensis is much darker rufous below and more heavily spotted.
119. **Australian Kestrel** *Falco cenchroides* — J
Palest of the three kestrels; thin streaks on breast.
120. **Eurasian Hobby** *Falco subbuteo* — J
Long, pointed wings; relatively short tail; bright rufous thighs and undertail coverts.
121. **Oriental Hobby** *Falco severus* — SBJ
Smaller, shorter-winged than 120; dark rufous underparts.
122. **Peregrine Falcon** *Falco peregrinus* — SBJ
Large size; pointed wings, broad at base.
F. p. ernsti shown.

Plate 13

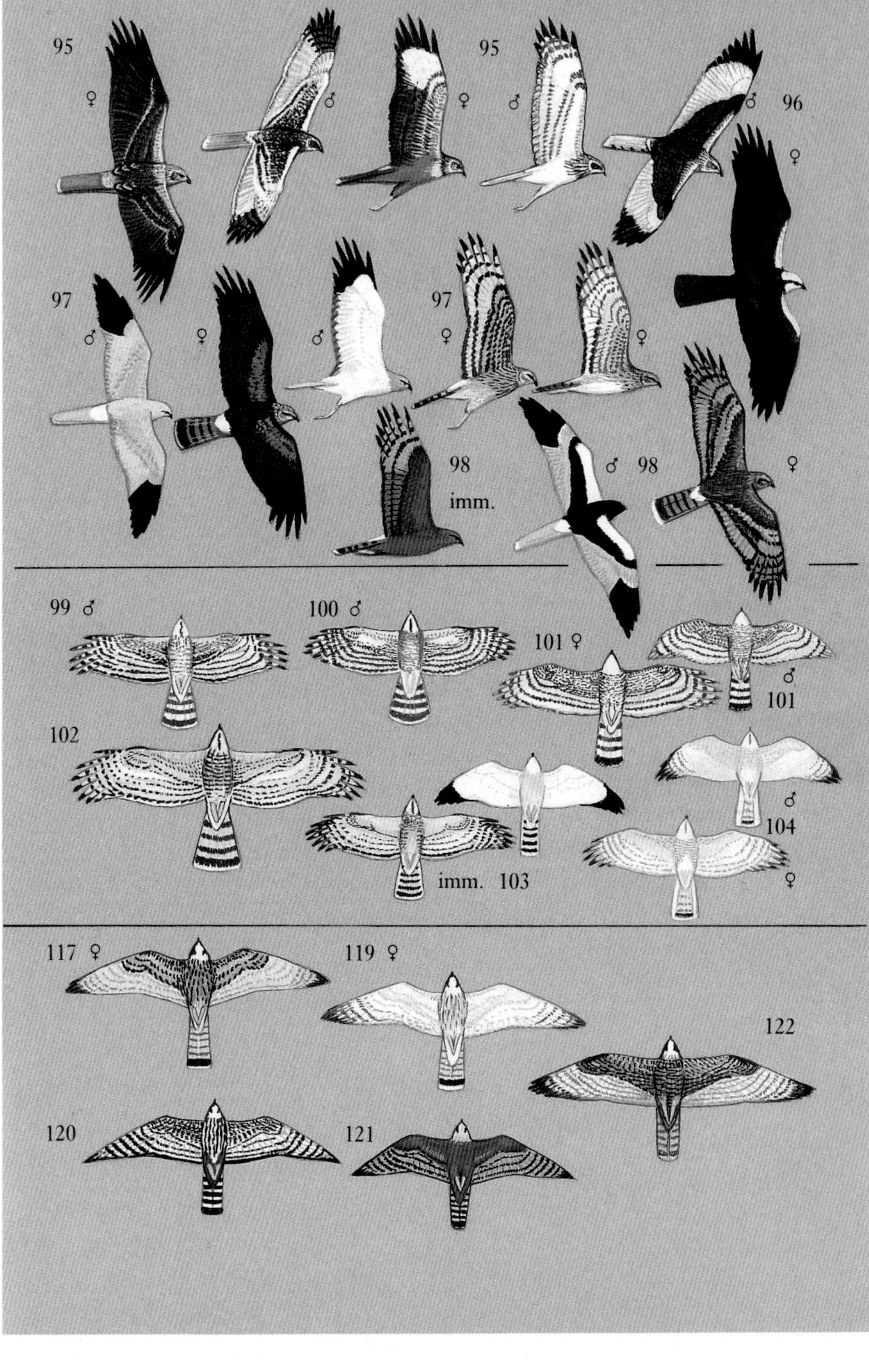
95
♀
♂
♀
95
♂
♂
96
♀
97
♂
♀
♂
97
♀
♀
98
imm.
♂
98
♀
99 ♂
100 ♂
101 ♀
♂
101
102
♂
104
♀
imm.
103
117 ♀
119 ♀
122
120
121

PLATE 14. **Megapodes and Partridges**

123. **Orange-footed Scrubfowl** *Megapodius reinwardt*   J
 Orange feet; small crest; reddish face.
124. **Tabon Scrubfowl** *Megapodius cumingii*   B
 Grey brown plumage; blackish red legs; scarlet face; no crest.
125. **Long-billed Partridge** *Rhizothera longirostris*   SB
 Large grey bill; grey breast.
 Female is more fulvous and lacks grey breast.
126. **Black Partridge** *Melanoperdix nigra*   SB
 Male: all blackish plumage
 Female: brown with black scalloping on wing coverts.
128. **Grey-breasted Partridge** *Arborophila orientalis*   SJ
 Grey breast; whitish belly; black blotched flanks; white pattern on head variable.
 (a) *A. o. orientalis* (E. Java).
 (b) *A. o. rolli* (N. Sumatra).
129. **Chestnut-bellied Partridge** *Arborophila javanica*   J
 Grey breast, chestnut belly and flanks; buff patterned head.
130. **Red-billed Partridge** *Arborophila rubrirostris*   S
 Red bill; black and white head; brownish breast; black and white flanks. Birds have varying extent of white on head.
131. **Red-breasted Partridge** *Arborophila hyperythra*   B
 Grey bill; brown breast; black and white flanks. Some birds in Sarawak have grey eye stripe and cheeks, darker breast.
132. **Chestnut-necklaced Partridge** *Arborophila charltonii*   SB
 Brown lower-breast band; brownish flanks; greenish-yellow legs.
134. **Crimson-headed Partridge** *Haematortyx sanguiniceps*   B
 Crimson head and vent.

# Plate 14

PLATE 15. **Partridges, Quail, Buttonquails, and Pheasants**

127. **Blue-breasted Quail** *Coturnix chinensis* SBJ
 Male: dark plumage; black and white throat pattern.
 Female: larger than buttonquails, yellow feet.
133. **Ferruginous Partridge** *Caloperdix oculea* SB
 Black spots on brown wings; scaly black and white mantle.
135. **Crested Partridge** *Rollulus rouloul* SB
 Male: crimson crest.
 Female: green plumage; chestnut wings.
145. **Great Argus** *Argusianus argus* SB
 Male: very long wings and tail; secondaries decorated with eye-spots; bare blue head.
146. **Green Peafowl** *Pavo muticus* J
 Crest; iridescent green plumage.
 Male: 'fan' tail.
147. **Small Buttonquail** *Turnix sylvatica* J
 Unbarred underparts; chestnut spots on sides of breast; rufous upper breast.
148. **Barred Buttonquail** *Turnix suscitator* SJ
 Barred underparts.
 Female: black throat and black and white mottled head.

Plate 15

PLATE 16. **Fireback Pheasants**

136. **Crestless Fireback** *Lophura erythrophthalma*   SB
 (a) *L. e. erythrophthalma* (Sumatra).
 (b) *L. e. pyronota* (Borneo).
137. **Crested Fireback** *Lophura ignita*   SB
 Male: black crest; blue facial skin; chestnut rump; two coloured tail.
 Female: brown plumage; blue facial skin; underparts scaled white.
 Races vary: (a) *L. i. rufa* (Sumatra).
             (b) *L. i. nobilis* (N. Borneo).
138. **Bulwer's Pheasant** *Lophura bulweri*   B
 Male: blue wattles; white tail.
 Female: brown plumage; red legs; blue facial skin.
 (a) Male in display.
139. **Salvadori's Pheasant** *Lophura inornata*   S
 Male: iridescent blue-black plumage; red facial skin.
 Female: brown mottled plumage; red facial skin; grey legs.

Plate 16

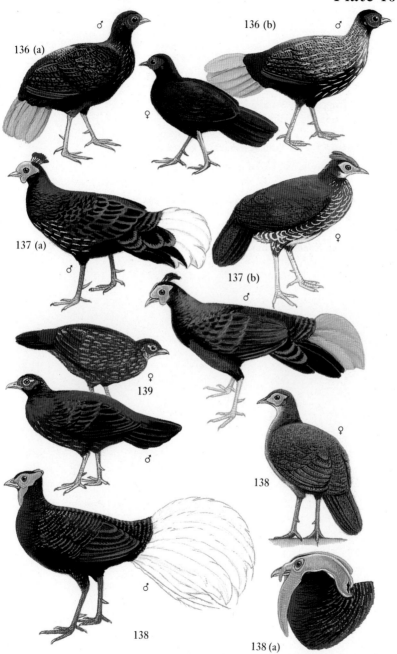

PLATE 17. **Pheasants**

140. **Hoogerwerf's Pheasant** *Lophura hoogerwerfi*   S
 Male supposedly like 139. None collected.
 Female browner than 139, more uniform, without pale shaft streaks.
141. **Red Junglefowl** *Gallus gallus*   SJ
 Male: serrated comb; hackles.
 G. g. bankiva (S. Sumatra, Java, Bali): shown.
 G. g. spadiceus (N. Sumatra): hackles longer and pointed.
142. **Green Junglefowl** *Gallus varius*   J
 Male: purple rounded comb, green iridescent hackles.
143. **Sumatran Peacock-Pheasant** *Polyplectron chalcurum*   S
 Brown plumage, finely barred black; purple iridescent patches on long tail.
144. **Bornean Peacock-Pheasant** *Polyplectron schleiermacheri*   B
 Male: plumage with green 'eye'-spots; crest.
 Female: blue 'eye'-spots.

Plate 17

PLATE 18. **Rails**

---

149. Water Rail *Rallus aquaticus*   B
Large size; red bill; pale eyebrow.

150. Slaty-breasted Rail *Gallirallus striatus*   SBJ
Chestnut crown; back and wings finely barred white.
G. s. striatus (Borneo) shown.

151. Red-legged Crake *Rallina fasciata*   SBJ
Rufous head; white barring on chestnut wings; red legs; heavily barred belly.

152. Slaty-Legged Crake *Rallina eurizonoides*   SJ
Grey legs; upperparts brown; almost no white on wing; underparts less heavily barred than 151 or 155.

153. Baillon's Crake *Porzana pusilla*   SBJ
Small; brown back, streaked black and white.
Immature: pale ochre where adult is grey.
P. p. mira (Borneo, Sumatra) is more rufous above.

154. Ruddy-breasted Crake *Porzana fusca*   SBJ
Bright chestnut-red head and breast; finely barred belly; red legs; lacks white on wings.

155. Band-bellied Crake *Porzana paykullii*   SBJ
Red legs; brown above with a few white bars in wings; boldly barred belly.

156. White-browed Crake *Porzana cinerea*   SBJ
Small; black eye-stripe; short, white eyebrow and white line below eye.

157. White-breasted Waterhen *Amaurornis phoenicurus*   SBJ
Large size; dark crown and upperparts; white face and underparts; orange undertail.

Plate 18

PLATE 19. **Rails, Finfoot, and Jacanas**

158. **Watercock** *Gallicrex cinerea* — SBJ
 Breeding male: black plumage and protruding red shield.
 Female: as non-breeding male but smaller and lacks frontal shield.
159. **Common Moorhen** *Gallinula chloropus* — SBJ
 White stripes along flanks, yellow lower legs.
160. **Dusky Moorhen** *Gallinula tenebrosa* — B
 Lacks white flank stripes. Red frontal shield larger than 159; red lower legs.
161. **Purple Swamphen** *Porphyrio porphyrio* — SBJ
 Large size; purple plumage; long red legs; powerful red bill.
162. **Common Coot** *Fulica atra* — BJ
 All blackish; white frontal shield and bill.
163. **Masked Finfoot** *Heliopais personata* — SJ
 Long neck; yellow bill; black and white striped neck.
164. **Comb-crested Jacana** *Irediparra gallinacea* — B
 Red comb like wattle on crown; black breast bar.
165. **Pheasant-tailed Jacana** *Hydrophasianus chirurgus* — SBJ
 Breeding: very long tail; white wing patch; golden nape patch.
166. **Bronze-winged Jacana** *Metopidius indicus* — SJ
 Adult: bronzy plumage; white eyebrow.
 Immature: mainly brown.

Plate 19

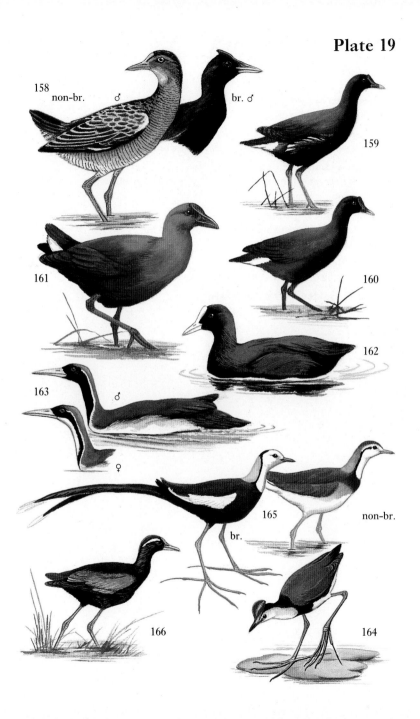

PLATE 20. **Large Plovers**

168. **Northern Lapwing** *Vanellus vanellus*   B
 Long up-turned crest; rufous undertail coverts.
169. **Grey-headed Lapwing** *Vanellus cinereus*   B
 Long yellow legs; grey breast. Very broad white flash on upper wing.
170. **Javanese Lapwing** *Vanellus macropterus*   J
 Black head and throat; creamy pink wattles; no white in upper wing; black spur at bend of wing.
171. **Red-wattled Lapwing** *Vanellus indicus*   S
 White bar on upperwing; white collar; white ear patch; red facial skin and base of bill.
172. **Grey Plover** *Pluvialis squatarola*   SBJ
 White wing bar; white rump; black axillaries visible in flight.
173. **Pacific Golden-Plover** *Pluvialis fulva*   SBJ
 Mottled golden-brown plumage.

Plate 20

168 non-br.

169 non-br.

170

171

172 br.

non-br.

173 br.

non-br.

PLATE 21. **Small Plovers**

174. **Common Ringed Plover** *Charadrius hiaticula*
Complete white and black collars; yellow legs; black breast bar; no white fringe above black on crown; white wing bar in flight.

175. **Little Ringed Plover** *Charadrius dubius* SBJ
(a) August female: blackish breast bar
Breeding: complete black and white collars; yellow legs; yellow eye-ring; white fringe above black on crown; no wing bar in flight.

176. **Kentish Plover** *Charadrius alexandrinus* SBJ
Non-breeding: blackish legs; incomplete dark breast band; white hind collar.

177. **Javan Plover** *Charadrius javanicus* J
Breeding: breast bar sometimes complete.
Non-breeding: as 176. White hind collar sometimes incomplete; legs grey.
(a) February male, (b) May male, (c) September female (Kangean).

178. **Red-capped Plover** *Charadrius ruficapillus* J
Like 176 but no white hind collar and cap more rufous.
(a) July female (E. Java), (b) January male.

179. **Malaysian Plover** *Charadrius peronii* SBJ
Breeding: black breast bar and forehead bar.
Non-breeding: legs pinkish; white hind collar; rufous eyebrow extension.
(a) December male, (b) August female.

180. **Long-billed Plover** *Charadrius placidus* BJ
Non-breeding: long bill; yellow legs; narrow black and white hind collar; black breast band; narrow white wing bar in flight.

181. **Mongolian Plover** *Charadrius mongolus* SBJ
Breeding: broad chestnut bar on breast; blackish legs; forehead totally black in some races.
Non-breeding: small bill; blackish legs.
(a) January female *C. m. stegmanni*

182. **Greater Sand-Plover** *Charadrius leschenaultii* SBJ
Breeding: narrow chestnut breast band; heavy, long bill; prominent white forehead patch. Greenish legs extend beyond tail in flight.
Non-breeding: greenish legs.
(a) October female, (b) April adult.

183. **Oriental Plover** *Charadrius veredus* SBJ
Breeding: broad orange breast band; underlined black, dark cap; long eyebrow; no white wing bar.
Non-breeding: broad brown breast band; dark cap; no white wing bar in flight; yellow legs.

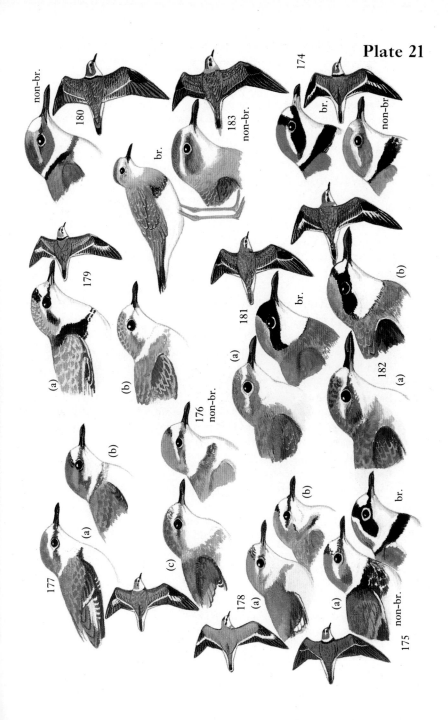

Plate 21

PLATE 22. **Curlews, Godwits, Tattler, Turnstone, and Dowitcher**

184. **Eurasian Curlew** *Numenius arquata*   SBJ
 Very long decurved bill; white lower back; whitish underwing. Male has shorter bill than female.
 N. a. orientalis shown: mainly white underwing and rump.
185. **Whimbrel** *Numenius phaeopus*   SBJ
 Smaller; shorter-billed than 184 or 187; striped head; grey or white back.
 (a) *N. p. phaeopus*: white rump and lower back; mainly white underwing.
 (b) *N. p. variegatus*: brownish rump and barred underwing.
186. **Little Curlew** *Numenius minutus*   BJ
 Small; bill shorter than 185; fulvous plumage.
187. **Far-Eastern Curlew** *Numenius madagascariensis*   SBJ
 Very long decurved bill; dark rump; barred underwing; underparts buffish. Male has shorter bill than female.
188. **Black-tailed Godwit** *Limosa limosa*   SBJ
 Slightly upturned long bill; white rump; black tail.
 (a) *L. l. melanuroides*: smaller than *L. l. limosa*, darker upperparts, only thin white wing bar.
 (b) *L. l. limosa*: prominent white wing bar.
189. **Bar-tailed Godwit** *Limosa lapponica*   SBJ
 Markedly upturned long bill; some barring on tail.
 (a) *L. l. baueri*: brownish rump and barred underwing.
 (b) *L. l. lapponica*: whitish rump and underwing.
200. **Grey-tailed Tattler** *Tringa brevipes*   SBJ
 Breeding: barred underparts.
 Non-breeding: black eye-stripe; straight bill; short yellow legs.
201. **Ruddy Turnstone** *Arenaria interpres*   SBJ
 Breeding: black and white head, rufous upperparts.
 Non-breeding: short upturned bill; head pattern; brown breast; short orange legs.
202. **Asian Dowitcher** *Limnodromus semipalmatus*   SBJ
 Straight black bill; long dark legs; whitish rump; sometimes has white bar on primaries.
203. **Long-billed Dowitcher** *Limnodromus scolopaceus*   BJ
 White lower back; white trailing wing edge; grey-green legs; no white bar on primaries; yellow base of bill.

Plate 22

PLATE 23. Sandpipers

190. Spotted Redshank *Tringa erythropus*   B
Breeding: unmistakable, all black.
Non-breeding: long red legs; long slim bill; white eyebrow; upperwings barred; white back.

191. Common Redshank *Tringa totanus*   SBJ
Red legs, medium bill, faint eyebrow; broad white trailing edge to wing. *T. t. ussuriensis* shown.

192. Marsh Sandpiper *Tringa stagnatilis*   SBJ
Long-legged; slim; thin black bill; white back and rump.

193. Common Greenshank *Tringa nebularia*   SBJ
Large size; shorter build than 192; upturned bill; white back and rump contrast with dark wings; toes project beyond tail. Underwing faintly barred brown.

194. Nordmann's Greenshank *Tringa guttifer*   SB
Breeding: intense black spotting on underparts.
Non-breeding: straight two-toned bill; toes do not project beyond tail in flight; underwing pure white.

195. Lesser Yellowlegs *Tringa flavipes*   S
Long yellow legs; white rump, cut off square above tail coverts; no white wing flash.

196. Green Sandpiper *Tringa ochropus*   SBJ
Tips of toes extend well beyond tail in flight. Broad, pointed wings; underwing very dark; legs dark.

197. Wood Sandpiper *Tringa glareola*   SBJ
Slim; feet project beyond tail; pale greyish underwing; prominent white eyebrow.

198. Terek Sandpiper *Tringa cinereus*   SBJ
Small; short orange-yellow legs; long upcurved bill; white trailing edge of secondaries.

199. Common Sandpiper *Tringa hypoleucos*   SBJ
Small; white 'patch' in front of wing; prominent white wing bar.

Plate 23

PLATE 24. **Painted Snipe, Snipes, Woodcocks, and Ruff**

167. **Greater Painted Snipe** *Rostratula benghalensis* SBJ
 Female: larger and more brightly coloured than male; striking pale mark around eye; white half 'collar'.
204. **Pintail Snipe** *Gallinago stenura* SBJ
 Tail only slightly longer than folded wings; slightly smaller and paler than other snipes; toes project beyond tail tip in flight; 'pin-like' outer tail feathers.
205. **Swinhoe's Snipe** *Gallinago megala* SBJ
 Larger, heavier snipe; tips of toes project only slightly beyond tail tip.
206. **Common Snipe** *Gallinago gallinago* SBJ
 White trailing edge of secondaries; whitish underwing; tail projects further beyond folded wings than 205.
207. **Eurasian Woodcock** *Scolopax rusticola* B
 Very long straight bill; transverse dark head bands. Paler back pattern, larger, and with longer bill than 208.
208. **Rufous Woodcock** *Scolopax saturata* SJ
 Dark rufous plumage; long straight bill; transverse bars on head.
220. **Ruff** *Philomachus pugnax* SBJ
 Non-breeding: long-legged, long-necked. Only a narrow white wing bar.
 (a) Male in breeding plumage. Some birds passing through may show beginnings of moult change.

Plate 24

PLATE 25. **Knots, Stints, and Sandpipers**

209. **Red Knot** *Calidris canutus* SBJ
 Short straight bill; short legs.
 Breeding female shows more white on belly than male (shown).
210. **Great Knot** *Calidris tenuirostris* SBJ
 Larger than 209 with longer bill and more distinct white rump.
211. **Rufous-necked Stint** *Calidris ruficollis* SBJ
 Tiny plump wader. Shows more white in wing than other stints; dark legs.
212. **Temminck's Stint** *Calidris temminckii* B
 Legs greenish or yellowish; white sides of tail; tail projects behind folded wings.
213. **Long-toed Stint** *Calidris subminuta* SBJ
 Pale legs; relatively long neck and legs.
214. **Sharp-tailed Sandpiper** *Calidris acuminata* BJ
 Pale legs; shortish bill; brownish plumage; prominent white eyebrow and chestnut crown even in winter; narrow white wing bar.
215. **Dunlin** *Calidris alpina* J
 Decurved bill tip (male has shorter bill than female); clear white wing bar; dark rump.
216. **Curlew Sandpiper** *Calidris ferruginea* SBJ
 Long, decurved, black bill (male shorter bill than female); clear white wing bar; white rump.
217. **Sanderling** *Calidris alba* SBJ
 Chunky small size; broad white wing bar; blackish trailing edge to wing. Non-breeding: black shoulder conspicuous.
218. **Broad-billed Sandpiper** *Limicola falcinellus* SBJ
 'Double' eyebrow; terminally decurved bill; short legs.
219. **Spoon-billed Sandpiper** *Eurynorhynchus pygmaeus* B
 Spatulate bill. In flight looks like a stint.

Plate 25

PLATE 26. **Grebes, Stilts, Phalaropes, Thick-knee, and Pratincoles**

1. **Little Grebe** *Tachybaptus ruficollis* — SBJ
 Breeding: chestnut throat and sides of neck; less white in wing than 2.
 Non-breeding and immature: look the same as 2.
2. **Australasian Grebe** *Tachybaptus novaehollandiae* — J
 Breeding: black throat; white of wing extends further than 1.
221. **White-headed Stilt** *Himantopus leucocephalus* — SBJ
 Adult: black nape and hindneck, sometimes extends to collar.
 Immature: similar to 222.
222. **Black-winged Stilt** *Himantopus himantopus* — SBJ
 Adult: crown and hindneck varies from white to grey.
 Immature: grey brown above, greyish crown and hindneck.
223. **Red Phalarope** *Phalaropus fulicaria* — B
 Non-breeding: larger, heavier, and broader-billed than 224; pale grey upperparts.
224. **Red-necked Phalarope** *Phalaropus lobatus* — BJ
 Non-breeding: needle-like bill smaller than 223. Dark centres to feathers on upperparts.
225. **Beach Thick-knee** *Burhinus giganteus* — SBJ
 Very large; banded head and wings.
 Immature: paler.
226. **Oriental Pratincole** *Glareola maldivarum* — SBJ
 Forked tail; chestnut underwing coverts, black primaries.
 Breeding: cream throat outlined with black.
227. **Australian Pratincole** *Stiltia isabella* — SBJ
 Short tail; very long wings; black underwing coverts and primaries.

Plate 26

PLATE 27. **Jaegers and Gulls**

228. **Pomarine Jaeger** *Stercorarius pomarinus*   SBJ
 Blunt tail extensions.
 (a) Adult breeding, dark morph.
 (b) Adult breeding, light morph.
 (c) Adult non-breeding.
 (d) Juvenile, light morph.

229. **Parasitic Jaeger** *Stercorarius parasiticus*   BJ
 Sharp tail extensions.
 (a) Adult breeding, light morph.
 (b) Non-breeding, light morph.
 (c) Juvenile intermediate morph.
 (d) Adult breeding, dark morph.

230. **Long-tailed Jaeger** *Stercorarius longicaudus*
 Longer central tail feathers, no dark breast band.
 (a) Adult moulting.
 (b) Juvenile, light morph.
 (c) Adult, dark morph.
 (d) Juvenile, intermediate morph.

231. **South Polar Skua** *Catharacta maccormicki*   S
 Heavy; prominent white wing flash; pale cap; yellowish nape.
 (a) Adult, light morph.
 (b) Adult, light morph.
 (c) Adult, intermediate morph.
 (d) Juvenile, light morph.

232. **Common Black-headed Gull** *Larus ridibundus*   SB
 (a) Adult non-breeding.
 (b) Adult non-breeding.
 (c) 1st winter.
 (d) 1st summer.
 (e) Adult breeding: white eye-ring, red bill.

233. **Sabine's Gull** *Xema sabini*   S
 Tricolour wing pattern in flight.
 Breeding: black head; red eye-ring; yellow tip of bill.

234. **Brown-headed Gull** *Larus brunnicephalus*   S
 Heavy red bill; white iris; white spots in black wing tip.
 (a) 1st winter.

PLATE 28. **Terns**

235. **Whiskered Tern** *Chlidonias hybridus*   SBJ
 Breeding: black cap; red bill; dark breast.
 Non-breeding: reddish bill, grey rump.
 C. h. javanicus shown.
236. **White-winged Tern** *Chlidonias leucopterus*   SBJ
 Breeding: black body; white upperwing coverts; red bill.
 Non-breeding: black bill; white rump.
237. **Gull-billed Tern** *Sterna nilotica*   SBJ
 Thick black bill; notched tail.
238. **Caspian Tern** *Sterna caspia*   S
 Large size; heavy red bill; slight crest.
250. **Common White Tern** *Gygis alba*   SJ
 Small size; pure white plumage; black eye-spot; slender bill; graceful flight.

PLATE 29. **Terns**

239. **Common Tern** *Sterna hirundo* — SBJ
 Breeding: black cap; greyish belly; dark wing tips.
 Non-breeding: short white forehead.
240. **Roseate Tern** *Sterna dougallii* — SBJ
 Breeding: red feet; long, pale, deeply forked tail; pinkish wash.
 Non-breeding: extensive white forehead.
241. **Black-naped Tern** *Sterna sumatrana* — SBJ
 Pale colour; narrow black bill; narrow black nape.
242. **Bridled Tern** *Sterna anaethetus* — SBJ
 Adult: smaller; browner than 243; forehead narrowly white; white collar.
 Juvenile: scalloped upperparts; head and throat white with blackish nape.
243. **Sooty Tern** *Sterna fuscata* — SBJ
 Adult: forehead broadly white; darker and larger than 242.
 Juvenile: dark head and breast.
244. **Little Tern** *Sterna albifrons* — SBJ
 Breeding: yellow bill with black tip; yellow legs; pale rump.

Plate 29

Plate 30. Crested-Terns and Noddies

245. **Great Crested-Tern** *Sterna bergii* — SBJ
 Large size; yellow bill; crest.
 Non-breeding: crown streaked with black.
246. **Lesser Crested-Tern** *Sterna bengalensis* — SBJ
 Slender; orange bill.
 Non-breeding: crown white.
 (a) *S. b. bengalensis* (breeds E to Singapore).
 (b) *S. b. torresii* (breeds in Sundas, Wallacea) has darker mantle.
247. **Chinese Crested-Tern** *Sterna bernsteini* — B
 Long crest; yellow bill with black tip.
248. **Brown Noddy** *Anous stolidus* — SBJ
 Forecrown whitish; tail wedge-shaped.
249. **Black Noddy** *Anous minutus* — SBJ
 Pure white crown and nape; more slender and tail more forked than 248.

Plate 30

Plate 31. Green-Pigeons

251. **Sumatran Green-Pigeon** *Treron oxyura* — SJ
 Male: blue patch on nape; 'pin' tail.
 Female: duller.
252. **Wedge-tailed Green-Pigeon** *Treron sphenura* — SJ
 Male: golden cap and breast; chestnut vent.
 Female: pale green head and underparts.
 Javan form *T. s. korthalsi* shown.
 Sumatran form *T. s. etorques* lacks orange and chestnut.
253. **Thick-billed Green-Pigeon** *Treron curvirostra* — SBJ
 Heavy bill with red base; grey crown; central green tail feathers; outer tail feathers grey with black subterminal band.
254. **Grey-cheeked Green-Pigeon** *Treron griseicauda* — SJ
 Male: grey cheeks; orange shoulder patch; green base to bill.
 Female: like female 253 but green base to bill; grey on head extends to face.
255. **Cinnamon-headed Green-Pigeon** *Treron fulvicollis* — SB
 Male: cinnamon head.
 Female: duller green than female 253.
 T. f. baramensis (N Borneo) shown.
 In nominate subspecies breast is ochraceous orange and belly greyish green.
256. **Little Green-Pigeon** *Treron olax* — SBJ
 Small size; dark tail with paler grey terminal band.
 Male: maroon back; grey head; orange breast.
 Female: grey crown; white chin; buff vent.

Plate 31

PLATE 32. **Green-Pigeons and Fruit-Doves**

257. **Pink-necked Green-Pigeon** *Treron vernans* SBJ
 Black subterminal bar on grey tail.
 Male: grey head; pink neck; orange breast.
258. **Orange-breasted Green-Pigeon** *Treron bicincta* J
 Green head; grey nape; pale tip to blackish tail.
 Male: pink and orange breast.
259. **Large Green-Pigeon** *Treron capellei* SBJ
 Male: large size; orange breast; chestnut vent.
 Female: duller.
260. **Jambu Fruit-Dove** *Ptilinopus jambu* SBJ
 Crimson face (male); dull purple face (female); white belly.
261. **Pink-headed Fruit-Dove** *Ptilinopus porphyreus* SJ
 Male: pink head and breast; black and white breast bands.
 Female: duller.
262. **Black-naped Fruit-Dove** *Ptilinopus melanospila* SBJ
 Male: whitish head with black nape; yellow belly and chin; red undertail.
 Female: green plumage; red undertail.
263. **Black-backed Fruit-Dove** *Ptilinopus cinctus* J
 White head and throat; black breast bar; grey underparts.

Plate 32

PLATE 33. **Pigeons**

264. **Green Imperial-Pigeon** *Ducula aenea*   SBJ
 Large; bronze-green back; grey underparts; uniform tail; chestnut vent.
265. **Pied Imperial-Pigeon** *Ducula bicolor*   SBJ
 White plumage; limited black on wing; no red eye-ring.
266. **Mountain Imperial-Pigeon** *Ducula badia*   SBJ
 Dark maroon upperparts; whitish throat; buff vent; pale tip to grey tail.
267. **Dark-backed Imperial-Pigeon** *Ducula lacernulata*   J
 Dark grey back; chestnut vent; grey terminal bar to dark tail.
268. **Grey Imperial-Pigeon** *Ducula pickeringi*   B
 No chestnut on undertail coverts. Some birds are paler underneath.
269. **Pink-headed Imperial-Pigeon** *Ducula rosacea*   J
 Pinkish head and underparts; chestnut vent
 D. r. rosacea from Romah Island shown. Underparts suffused with pinkish purple.
270. **Metallic Pigeon** *Columba vitiensis*   B
 White cheeks and throat, iridescent purple underparts; red eye-ring.
271. **Silvery Wood-Pigeon** *Columba argentina*   SB
 Grey-white plumage; red eye-ring; extensive black on wing.
272. **Rock Pigeon** *Columba livia*   B
 Black wing bars; grey and black tail; iridescent neck.

Plate 33

PLATE 34. **Pigeons**

---

273. **Barred Cuckoo-Dove** *Macropygia unchall* SJ
 Tail and back heavily barred with black. Female lacks green gloss of nape.
274. **Ruddy Cuckoo-Dove** *Macropygia emiliana* J
 Male: unbarred with pink iridescence.
 Female: lacks iridescence; has slight barring.
 (a) Female of *M. e. elassa* (Mentawai): smaller with dark throat.
275. **Little Cuckoo-Dove** *Macropygia ruficeps* SBJ
 Small size; buff underparts; blackish blotching on breast.
276. **Island Collared-Dove** *Streptopelia bitorquata* SBJ
 Black nape patch with white outline.
277. **Spotted Dove** *Streptopelia chinensis* SBJ
 Black nape patch, spotted with white.
278. **Zebra Dove** *Geopelia striata* SBJ
 Small size; fine barring on nape and flanks.
279. **Emerald Dove** *Chalcophaps indica* SBJ
 Green wings; pink underparts; barred back.
280. **Nicobar Pigeon** *Caloenas nicobarica* SBJ
 Long hackles; short white tail.

Plate 34

PLATE 35. **Parrots**

281. **Red-breasted Parakeet** *Psittacula alexandri* SBJ
Grey head; pink breast.
Female slightly duller.
282. **Long-tailed Parakeet** *Psittacula longicauda* SB
Very long tail; green cap and breast.
Female has blue-grey nape, shorter tail.
283. **Rainbow Lorikeet** *Trichoglossus haematodus* J
Races vary; *T. h. mitchelli* shown. Female is duller.
284. **Yellow-crested Cockatoo** *Cacatua sulphurea* J
White plumage; long yellow crest.
285. **Blue-rumped Parrot** *Psittinus cyanurus* SB
Small; short tail; blue rump; crimson shoulder patch.
286. **Great-billed Parrot** *Tanygnathus megalorhynchos*
Large size; massive red bill; blue rump.
287. **Blue-naped Parrot** *Tanygnathus lucionensis* B
Extent of blue on crown and nape varies; red bill.
Female: slightly duller.
288. **Blue-crowned Hanging-Parrot** *Loriculus galgulus* SB
Very small; red rump.
Male: blue crown and red throat patch.
Female: green throat.
289. **Yellow-throated Hanging-Parrot** *Loriculus pusillus* J
Very small; red rump; yellow throat patch; pale iris.

Plate 35

PLATE 36. **Large Cuckoos**

290. **Chestnut-winged Cuckoo** *Clamator coromandus*   SBJ
 Black head and crest; rufous wings; orange throat and breast.
291. **Large Hawk-Cuckoo** *Cuculus sparverioides*   SBJ
 Large; white throat and rufous breast streaked black; barred belly; large yellow eye-ring.
292. **Moustached Hawk-Cuckoo** *Cuculus vagans*   SBJ
 Blackish moustachial stripe; brown back; black streaks on whitish underparts.
293. **Hodgson's Hawk-Cuckoo** *Cuculus fugax*   SBJ
 Grey head and back; whitish underparts with rufous and dark streaks.
294. **Indian Cuckoo** *Cuculus micropterus*   SBJ
 Greyish-brown; dull grey eye-ring; subterminal black tail bar.
295. **Common Cuckoo** *Cuculus canorus*   SJ
 Bright yellow eye-ring; underparts thinly barred black.
 Hepatic female has unbarred rump.
296. **Oriental Cuckoo** *Cuculus saturatus*   SBJ
 Bright yellow eye-ring; grey throat; underparts broadly barred black.
 Hepatic female has barred rump.
306. **Asian Koel** *Eudynamys scolopacea*   SBJ
 Male: glossy blue-black, green bill, red eyes.
 Female: barred and spotted brown and buff.

Plate 36

PLATE 37. **Small Cuckoos**

297. **Banded Bay Cuckoo** *Cacomantis sonneratii* SBJ
 Brown above, white below, finely barred; pale eyebrow stripe and cheek patch.
 Immature: with black streaks and blotches.
298. **Plaintive Cuckoo** *Cacomantis merulinus* SBJ
 Red eye; grey head and upper breast.
 Immature: like 297 but more buffy underparts and lacks white eyebrow.
299. **Rusty-breasted Cuckoo** *Cacomantis sepulcralis* SBJ
 Adult: grey head; dirty orange underparts extend up to greyish throat; prominent yellow eye-ring.
 Immature: heavily barred black.
300. **Asian Emerald Cuckoo** *Chrysococcyx maculatus* S
 Male: emerald green upperparts.
 Female: yellow base to bill; rufous crown.
301. **Violet Cuckoo** *Chrysococcyx xanthorhynchus* SBJ
 Male: glossy purple upperparts.
 Female: red base to bill; brown crown.
302. **Horsfield's Bronze-Cuckoo** *Chrysococcyx basalis* SBJ
 Head browner than back; white eyebrow and dark ear patch; grey eye-ring.
303. **Little Bronze-Cuckoo** *Chrysococcyx minutillus* J
 Crown deeper green than back. Outer tail feathers entirely barred black and white.
 C. m. cleis (N. and E Borneo) shown. Ventral barring heavier than *C. m. albifrons* (elsewhere).
304. **Gould's Bronze-Cuckoo** *Chrysococcyx russatus* B
 Purplish iridescence. White forehead duller than 303; outer tail feathers have rufous outer web.
 P. r. ahencus (Borneo) shown.
305. **Drongo Cuckoo** *Surniculus lugubris* SBJ
 All greenish black with white barring on underside of tail and vent; white spot on nape usually invisible.
 Immature: spotted white.

Plate 37

Plate 38. Malkohas, Coucals, and Ground-Cuckoo

307. **Black-bellied Malkoha** *Phaenicophaeus diardi* — SB
 Dark belly; green bill; dark red skin around eye.
308. **Chestnut-bellied Malkoha** *Phaenicophaeus sumatranus* — SB
 Rufous belly; green bill; orange-red skin around eye.
309. **Green-billed Malkoha** *Phaenicophaeus tristis* — SJ
 Large size; long tail; often has whitish ring around bare eye-patch.
310. **Raffles's Malkoha** *Phaenicophaeus chlorophaeus* — SB
 Small; chestnut mantle.
 Male: chestnut breast; grey vent.
 Female: grey breast; chestnut vent.
311. **Red-billed Malkoha** *Phaenicophaeus javanicus* — SBJ
 Red bill; chestnut underparts with grey breast bar; grey tail.
312. **Chestnut-breasted Malkoha** *Phaenicophaeus curvirostris* — SBJ
 Rufous underparts; rufous-tipped tail.
 (a) *P. c. microrhinus* (Borneo): chin and cheeks rufous.
 (b) *P. c. erythrognathus* (Sumatra): chin and cheeks grey, belly black.
313. **Sunda Ground-Cuckoo** *Carpococcyx radiceus* — SB
 C. r. radiceus (Borneo) shown.
 C. r. viridis (Sumatra): smaller and greener.
314. **Short-toed Coucal** *Centropus rectunguis* — SB
 Like 315 but tail shorter.
315. **Greater Coucal** *Centropus sinensis* — SBJ
 Large; glossy black with chestnut wing. Juvenile has black barring on wing.
 (a) Pale form from Bawean.
316. **Lesser Coucal** *Centropus bengalensis* — SBJ
 Smaller than 315; no blue gloss.
317. **Sunda Coucal** *Centropus nigrorufus* — J
 Like 316 but darker, with black back; purple gloss.

Plate 38

PLATE 39. **Barn Owls and Owls**

318. **Barn Owl** *Tyto alba* SJ
Large pale owl; white, heart-shaped facial disc.
319. **Oriental Bay Owl** *Phodilus badius* SBJ
Smaller than Barn Owl; reddish brown, very short tail.
332. **Barred Eagle-Owl** *Bubo sumatranus* SBJ
Large size; grey-brown barred underparts; prominent ear tufts; dark eyes.
333. **Buffy Fish-Owl** *Ketupa ketupu* SBJ
Large size; brown colour; streaked underparts; ear tufts; eyes yellow.
334. **Collared Owlet** *Glaucidium brodiei* SB
Very small; spotted head; white collar. Black spots on nape form a 'face' seen from behind.
335. **Javan Barred Owlet** *Glaucidium castanopterum* J
Small; barred head; white front collar; prominent white streaking on body.
336. **Brown Hawk-Owl** *Ninox scutulata* SBJ
Dark hawk-like shape; long barred tail.
337. **Spotted Wood-Owl** *Strix seloputo* SJ
Large size; spotted crown; pale face disc; pale underparts, scalloped dark.
338. **Brown Wood-Owl** *Strix leptogrammica* SBJ
Large; brown crown; dark face disc; white eyebrows. Geographically variable. Southern birds are smaller, redder, with plain coloured breast. *S. l. leptogrammica* (Borneo) shown.
339. **Short-eared Owl** *Asio flammeus* B
Round pale facial disc; black feathers around bright yellow eyes.

Plate 39

Plate 40. Scops-Owls

320. **White-fronted Scops-Owl** *Otus sagittatus* — S
 Large size; forecrown and base of eyebrows white, as are throat patches; ear tufts.
321. **Reddish Scops-Owl** *Otus rufescens* — SBJ
 Rufous brown; underparts with buff and black spots; ear tufts.
322. **Mountain Scops-Owl** *Otus spilocephalus* — SB
 (a) *O. s. luciae* (Borneo): buff.
 (b) *O. s. vandewateri* (Sumatra): collar of white feathers with black tips.
323. **Stresemann's Scops-Owl** *Otus stresemanni* — S
 Paler than 322. Spotted with black and white underneath; ear tufts.
324. **Javan Scops-Owl** *Otus angelinae* — J
 Dark; rufous underparts, streaked black; prominent ear tufts.
325. **Oriental Scops-Owl** *Otus sunia* — S
 Buffy; underparts streaked black.
 O. s. malayanus shown.
326. **Mantanani Scops-Owl** *Otus mantananensis* — B
 Mottled dark brown, white, and black, with black streaking on breast. Some birds are much paler.
327. **Simeulue Scops-Owl** *Otus umbra* — S
 Dark reddish brown; thinly peppered black; barred white underparts; ear tufts.
328. **Enggano Scops-Owl** *Otus enganensis* — S
 Like 327 but larger.
329. **Rajah Scops-Owl** *Otus brookii* — SBJ
 Medium-sized; brown; broad whitish collar and yellow eyes; conspicuous ear tufts.
330. **Collared Scops-Owl** *Otus lempiji* — SBJ
 Smaller than 329 and greyer; eyes dark brown.
331. **Mentawai Scops-Owl** *Otus mentawi* — S
 Dark rufous brown; yellow eyes; black streaks on belly. Some birds are blackish.

Plate 40

PLATE 41. **Frogmouths and Nightjars**

340. **Large Frogmouth** *Batrachostomus auritus* SB
 Orange protruding eyelids.
 Sumatran female shown. Male brighter chestnut.
 Females from Borneo have almost pure cream lower breast and underparts with no markings.
341. **Dulit Frogmouth** *Batrachostomus harterti* B
 Large size; yellow eyelids; pale nape collar.
 Sarawak female shown. Male is more chestnut with whiter collar.
342. **Gould's Frogmouth** *Batrachostomus stellatus* SB
 Bold rufous scaling on underparts; no collar.
 Male from Borneo shown. Female is less reddish.
343. **Short-tailed Frogmouth** *Batrachostomus poliolophus* SB
 Borneo female shown. Male is more greyish brown.
344. **Javan Frogmouth** *Batrachostomus javensis* SBJ
 Long tail, long ear tufts. Much variation in colouring between sexes.
345. **Sunda Frogmouth** *Batrachostomus cornutus* SBJ
 Dimorphic within sexes.
 (a) Female from Sarawak.
 (b) Female from Sumatra.
346. **Malaysian Eared-Nightjar** *Eurostopodus temminckii* SB
 Medium size; ear tufts; dark plumage.
347. **Great Eared-Nightjar** *Eurostopodus macrotis* S
 Large size; long ear tufts.
348. **Grey Nightjar** *Caprimulgus indicus* SBJ
 White patches at base of primaries in both sexes.
 Female: tail marked buff not white.
349. **Large-tailed Nightjar** *Caprimulgus macrurus* SBJ
 Greyish; base of primaries and tail spots white in males, buff in females.
350. **Savanna Nightjar** *Caprimulgus affinis* SBJ
 Small size; short tail. Female lacks white in tail.
351. **Bonaparte's Nightjar** *Caprimulgus concretus* SB
 Small size; no white in wings.
352. **Salvadori's Nightjar** *Caprimulgus pulchellus* SJ
 Similar to 351 but has white spots at base of outer primaries.

Plate 41

Plate 42. Swifts and Treeswifts

353. **Giant Swiftlet** *Hydrochous gigas* — SBJ
 Large, all dark plumage; notched tail.
354. **Edible-nest Swiftlet** *Collocalia fuciphaga* — SBJ
 Rump variable from same colour as back to paler grey or brown; tail notched. Some races indistinguishable from Moss-nest Swiftlet.
355. **Black-nest Swiftlet** *Collocalia maxima* — SBJ
 Rump brown, square tail.
357. **Volcano Swiftlet** *Collocalia vulcanorum* — SJ
 Smaller than 353 with more slender shape.
358. **Glossy Swiftlet** *Collocalia esculenta* — SB
 Whitish lower belly grading to dark breast and vent.
 C. linchii is indistinguishable but has greener gloss in hand.
360. **White-throated Needletail** *Hirundapus caudacutus* — SBJ
 White throat, white flecks on side of neck, and white tips to tertials.
361. **Silver-backed Needletail** *Hirundapus cochinchinensis* — J
 Black lores; white chin; grey throat.
362. **Brown-backed Needletail** *Hirundapus giganteus* — SBJ
 White lores and chin.
363. **Silver-rumped Swift** *Raphidura leucopygialis* — SBJ
 White rump and uppertail; broad wings.
364. **Fork-tailed Swift** *Apus pacificus* — SBJ
 Narrow crescent white rump; forked tail.
365. **Little Swift** *Apus affinis* — SBJ
 Square white rump, squarish tail.
366. **Asian Palm-Swift** *Cypsiurus balasiensis* — SBJ
 Small size; long, deeply forked tail; slim build.
367. **Grey-rumped Treeswift** *Hemiprocne longipennis* — SBJ
 Crest; very long wings and tail; whitish tertials.
 Male: chestnut cheeks.
368. **Whiskered Treeswift** *Hemiprocne comata* — SB
 Adult: long white 'whiskers'.
 Immature speckled like 367.

Plate 42

PLATE 43. Trogons

369. Blue-tailed Trogon *Harpactes reinwardtii* — SJ
Greenish above; blue tail, red bill; yellowish underparts.
(a) *H. r. reinwardtii* (Java).
(b) *H. r. mackloti* (Sumatra): maroon rump; shorter tail.

370. Red-naped Trogon *Harpactes kasumba* — SB
Male: black head; red nape collar.
Female: grey throat; tawny belly.

371. Diard's Trogon *Harpactes diardii* — SB
Male: pinkish crescent on breast and nape; mauve skin around the eye.
Female: brown breast; brown rump.

372. Whitehead's Trogon *Harpactes whiteheadi* — B
Male: red head; grey throat and breast.
Female: cinnamon crown and belly; grey breast.

373. Cinnamon-rumped Trogon *Harpactes orrhophaeus* — SB
Male: cinnamon rump; no breast bar or neck collar.
Female: rusty mask; buff belly.

374. Scarlet-rumped Trogon *Harpactes duvaucelii* — SB
Male: scarlet rump; small size.
Female: pink rump; brownish breast.

375. Orange-breasted Trogon *Harpactes oreskios* — SBJ
Male: olive head; orange belly.
Female: grey head; cinnamon belly.

376. Red-headed Trogon *Harpactes erythrocephalus* — S
Male: red head; white breast crescent.
Female: cinnamon head; pink belly; white breast crescent.

Plate 43

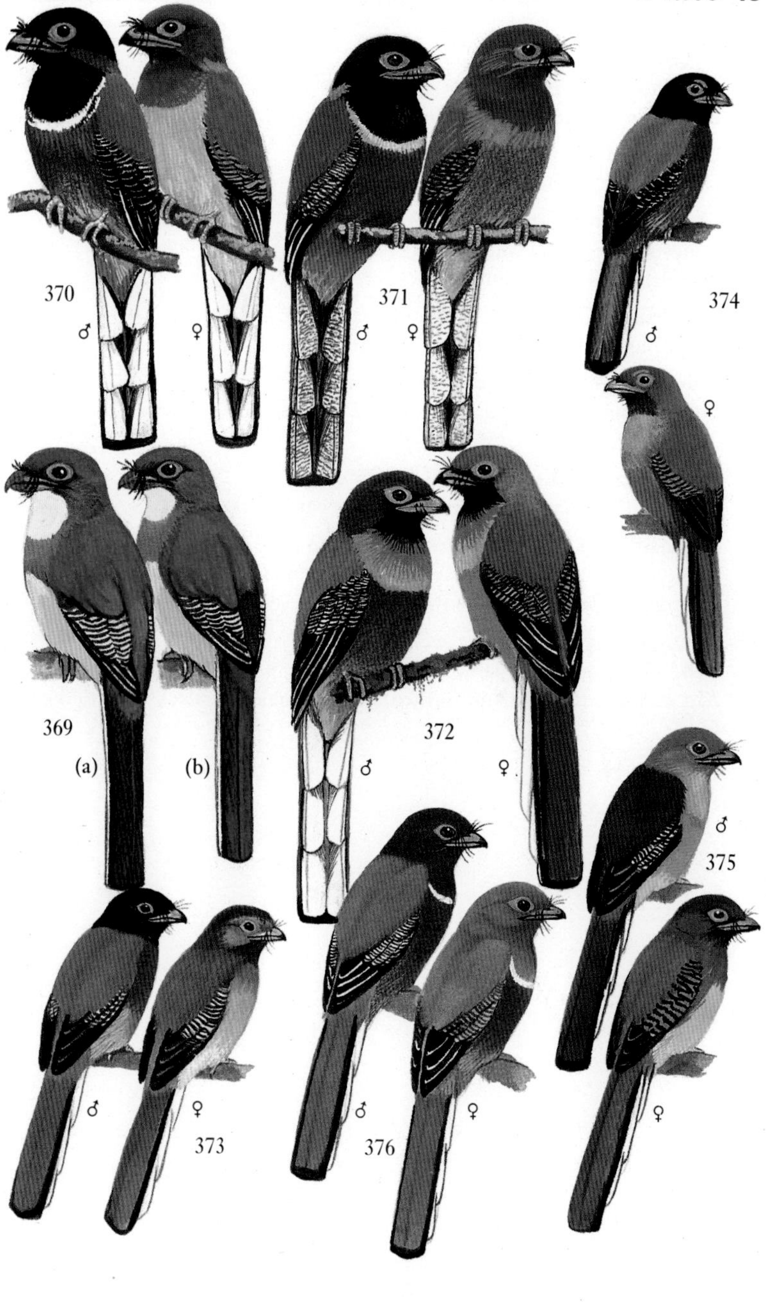

370 ♂ ♀

371 ♂ ♀

374 ♂
♀

369 (a) (b)

372 ♂ ♀

375 ♂
♀

373 ♂ ♀

376 ♂ ♀

PLATE 44. **Kingfishers**

377. **Common Kingfisher** *Alcedo atthis* SBJ
     Blue-green above; orange ear coverts.
378. **Blue-eared Kingfisher** *Alcedo meninting* SBJ
     Dark royal blue; blue ear coverts.
379. **Blue-banded Kingfisher** *Alcedo euryzona* SBJ
     Male: dull head and wings; light blue back; dark breast band.
     Female: rufous breast and belly.
380. **Small Blue Kingfisher** *Alcedo coerulescens* SJ
     Shiny blue and white plumage; blue breast band.
381. **Black-backed Kingfisher** *Ceyx erithacus* SBJ
     Small; generally darker above than 382. Intermediates between 381 and 382 can occur.
382. **Rufous-backed Kingfisher** *Ceyx rufidorsa* B
     Small; rufous wings and back; lacks blue ear coverts.
383. **Stork-billed Kingfisher** *Pelargopsis capensis* SBJ
     Large size; blue back; massive red bill. Several subspecies occur. *P. c. capensis* from Borneo shown.
384. **Banded Kingfisher** *Lacedo pulchella* SBJ
     Red bill, barred upperparts. Sumatran race *L. p. pulchellus* shown.
     L. p. melanops (Bangka and Borneo): darker; male has black sides to the head.

Plate 44

PLATE 45. Kingfishers

385. **Ruddy Kingfisher** *Halcyon coromanda* — SBJ
 Violet upperparts; rufous below. Some races with pale blue rump.
 Immature: paler underneath.
386. **White-throated Kingfisher** *Halcyon smyrnensis* — SJ
 Brown head; white bib.
387. **Javan Kingfisher** *Halcyon cyanoventris* — J
 Brown head and throat, purple on nape and back.
388. **Black-capped Kingfisher** *Halcyon pileata* — SBJ
 Black cap; white collar.
 (a) Some birds have dark fringes to breast feathers as do 389 and 390. Their status is uncertain.
389. **Collared Kingfisher** *Todirhamphus chloris* — SBJ
 Clean white collar and underparts. Female duller and greener.
390. **Sacred Kingfisher** *Todirhamphus sanctus* — SBJ
 Smaller than 389, duller greenish blue with buffy wash to underparts and lores.
391. **Rufous-collared Kingfisher** *Actenoides concretus* — SB
 Green cap; blue malar stripe.
 H. c. borneana (Borneo) shown.

Plate 45

PLATE 46. **Bee-Eaters, Roller, and Hoopoe**

392. **Chestnut-headed Bee-Eater** *Merops leschenaulti* — SJ
 Chestnut cap and mantle; green tail without elongate central feathers; black bar on upper breast.
393. **Blue-tailed Bee-Eater** *Merops philippinus* — SBJ
 Blue tail; green back and crown; adult has elongate central tail feathers; no black bar on breast.
394. **Blue-throated Bee-Eater** *Merops viridis* — SBJ
 Blue throat; brown cap and mantle; elongate central tail feathers.
395. **Rainbow Bee-Eater** *Merops ornatus* — J
 Chestnut crown; green back; black tail.
 Adult: black throat bar, elongate central tail feathers.
396. **Red-bearded Bee-Eater** *Nyctyornis amictus* — SB
 Scarlet breast. Crown lilac in male, red in female, green in immature.
397. **Dollarbird** *Eurystomus orientalis* — SBJ
 Red bill; dusky plumage; pale blue patch under wing.
398. **Eurasian Hoopoe** *Upupa epops* — SB
 Long crest, barred pied wings.

Plate 46

PLATE 47. **Hornbills**

399. **Bushy-crested Hornbill** *Anorrhinus galeritus*   SB
Black plumage with two-toned tail.
Female occasionally has dull yellowish bill.

400. **White-crowned Hornbill** *Aceros comatus*   SB
Long white tail; white bushy crest; white wing tip.
Female: black underparts.

401. **Wrinkled Hornbill** *Aceros corrugatus*   SB
Male: red casque; black crown; black base of tail. Tail often stained yellowish.

402. **Wreathed Hornbill** *Aceros undulatus*   SBJ
Pink skin around eye; corrugations on lower mandible; dark blue-black line on gular pouch.

403. **Plain-pouched Hornbill** *Aceros subruficollis*   S
Smaller than 402. No corrugations on lower bill; no dark line on gular pouch.

404. **Asian Black Hornbill** *Anthracoceros malayanus*   SB
Black plumage; white outer tips to tail feathers.
Casque: ivory in male, black in female.
Sometimes has broad white stripe over eye to nape.

405. **Oriental Pied Hornbill** *Anthracoceros albirostris*   SBJ
White belly and spot under eye.
Female casque smaller and more blackish.

406. **Rhinoceros Hornbill** *Buceros rhinoceros*   SBJ
Male: upturned casque; black tail bar.
Female: similar, but eye whitish to pale blue.

407. **Great Hornbill** *Buceros bicornis*   S
Male: flat yellow casque; pale wing bar; black tail bar.
Female: similar, eye white; more black on casque.

408. **Helmeted Hornbill** *Buceros vigil*   SB
Male: very long central tail feathers; bare red skin on neck.
Female: pale blue neck.

Plate 47

PLATE 48. **Barbets**

409. **Fire-tufted Barbet** *Psilopogon pyrolophus* S
 Creamy bill; prominent tuft of red feathers at base of lores; black breast bar.
410. **Lineated Barbet** *Megalaima lineata* J
 Brown streaked head; pale pink bill.
412. **Gold-whiskered Barbet** *Megalaima chrysopogon* SB
 Yellow cheek patch; yellow fore-crown.
 M. c. chrysopsis (Borneo) shown.
 M. c. chrysopogon (Sumatra): dirty brown eyebrow and eye-stripe.
413. **Red-crowned Barbet** *Megalaima rafflesii* SB
 Red crown; yellow cheek patch; blue throat.
414. **Red-throated Barbet** *Megalaima mystacophanos* SB
 Male: yellow forehead; red throat.
 Female: red lores and hind-crown.
416. **Black-browed Barbet** *Megalaima oorti* S
 Yellow chin and throat; black eyebrow; blue cheeks.
418. **Yellow-crowned Barbet** *Megalaima henricii* SB
 Yellow forehead and eyebrow; blue throat.
422. **Bornean Barbet** *Megalaima eximia* B
 Small; black forehead; blue eyebrow.
 (a) *M. e. exima* (Bornean mountains).
 (b) *M. e. cyanea* (Mt. Kinabalu only).

Plate 48

PLATE 49. **Barbets and Honeyguide**

411. **Brown-throated Barbet** *Megalaima corvina* — J
 Dark brown head and throat, streaked crown.
415. **Black-banded Barbet** *Megalaima javensis* — J
 Yellow crown; black breast band.
417. **Mountain Barbet** *Megalaima monticola* — B
 Dull coloration.
419. **Orange-fronted Barbet** *Megalaima armillaris* — J
 Orange forehead and upper breast; blue crown.
420. **Golden-naped Barbet** *Megalaima pulcherrima* — B
 Bright blue cap and throat.
421. **Blue-eared Barbet** *Megalaima australis* — SBJ
 Blue crown and throat with black breast bar.
 (a) *M. a. duvaucelii* (Borneo and Sumatra).
 (b) *M. a. australis* (Java).
423. **Coppersmith Barbet** *Megalaima haemacephala* — SJ
 Small size; red forehead; streaked underparts.
 (a) *M. h. rosea* (Java and Bali).
 (b) *M. h. delica* (Sumatra).
424. **Brown Barbet** *Calorhamphus fuliginosus* — SB
 Brown upperparts, red feet.
 Male: dark bill; female: pinkish bill.
 (a) *C. f. tertius* (N Borneo): orange throat.
 (b) *C. f. hayi* (Sumatra).
425. **Malaysian Honeyguide** *Indicator archipelagicus* — SB
 Drab colour; fat bill; yellow shoulder patch (lacking in female).

Plate 49

PLATE 50. **Woodpeckers**

426. **Speckled Piculet** *Picumnus innominatus* — SB
 Male: tiny; two white stripes on head; spotted below. Female lacks orange forehead.
427. **Rufous Piculet** *Sasia abnormis* — SBJ
 Male: tiny; forehead golden; orange underparts; 3 toes. Female forehead same colour as underparts.
428. **Rufous Woodpecker** *Celeus brachyurus* — SBJ
 Male: black bill; brown, barred black upperparts. Crimson cheeks lacking in female.
436. **Common Goldenback** *Dinopium javanense* — SBJ
 Male: red crown and crest; golden mantle; heavy black malar stripe; 3 toes.
 Female: black crown with white streaks.
437. **Olive-backed Woodpecker** *Dinopium rafflesii* — SB
 Green back; striped head; white spots on flanks.
438. **Buff-rumped Woodpecker** *Meiglyptes tristis* — SBJ
 Male: small; barred black; buff rump.
 Female: lacks red malar stripe.
439. **Buff-necked Woodpecker** *Meiglyptes tukki* — SB
 Male: barred plumage; brown head; buff neck patch; black throat bar. Female lacks red.
441. **White-bellied Woodpecker** *Dryocopus javensis* — SBJ
 Huge; black above, white below. Female lacks red.
442. **Fulvous-breasted Woodpecker** *Dendrocopus macei* — SJ
 Male: white face; buffy breast, streaked black; red rump. Female: lacks red.
443. **Grey-capped Woodpecker** *Dendrocopus canicapillus* — SBJ
 Male: grey cap; striped face; fulvous breast; streaked underparts. Female lacks red.
444. **Sunda Woodpecker** *Picoides moluccensis* — SBJ
 Male: brown cap; white underparts streaked.
 Female: lacks red.
445. **Grey-and-buff Woodpecker** *Hemicircus concretus* — SBJ
 Male: tiny; crest; red forehead; scalloped back; white rump. Female: lacks red.
446. **Maroon Woodpecker** *Blythipicus rubiginosus* — SB
 Male: ivory bill; red nape; dark maroon back.
 Female: lacks red; pale greyish barring on wings.

Plate 50

PLATE 51. **Woodpeckers**

429. **Laced Woodpecker** *Picus vittatus* — SJ
 Male: red cap; bluish cheeks; yellow rump; scalloped olive on underparts. Female: black crown.
430. **Grey-headed Woodpecker** *Picus canus* — S
 Grey face; black nape; maroon upperparts; orange rump. Crown red in male, black in female.
431. **Greater Yellownape** *Picus flavinucha* — S
 No red in plumage; yellow throat and crest; brown wings barred black. Female has duller malar area and more streaking on throat.
432. **Crimson-winged Woodpecker** *Picus puniceus* — SBJ
 Crimson wings; yellow throat; faint barring on flanks. Female lacks red malar stripe, crimson crown, and yellow crest.
433. **Lesser Yellownape** *Picus chlorolophus* — S
 Red forehead; red and white malar stripe; yellow crest; barred white flanks. Female lacks red forehead and malar stripe.
434. **Checker-throated Woodpecker** *Picus mentalis* — SBJ
 Male: chestnut collar; yellow crest; red wings; checkered throat. Female has brown malar area.
435. **Banded Woodpecker** *Picus miniaceus* — SBJ
 Rufous face; yellow crest; red wings; barred underparts. Female has reddish sides to face.
440. **Great Slaty Woodpecker** *Mulleripicus pulverulentus* — SBJ
 Large size; grey plumage. Female lacks red cheek patch.
447. **Orange-backed Woodpecker** *Reinwardtipicus validus* — SBJ
 Male: red crest and underparts; orange back.
 Female: brown head and underparts; cream back.
448. **Greater Goldenback** *Chrysocolaptes lucidus* — SBJ
 Striped head; two narrow black malar stripes; golden back. Male: red crown and crest. Female: black crown with white spots. (a) *C. l. strictus* (E Java and Bali): female has yellowish cap.

Plate 51

PLATE 52. **Broadbills**

449. **Dusky Broadbill** *Corydon sumatranus* — SB
 Dark; large, pink, hooked bill; orange patch on back.
450. **Black-and-red Broadbill** *Cymbirhynchus macrorhynchos* — SB
 Dark red below; white wing stripe; black breast band.
451. **Banded Broadbill** *Eurylaimus javanicus* — SBJ
 Purplish underparts; yellow plumes and rump; narrow black breast bar.
452. **Black-and-yellow Broadbill** *Eurylaimus ochromalus* — SB
 Pink breast; white collar; black breast band; yellow plumes and rump.
453. **Silver-breasted Broadbill** *Serilophus lunatus* — S
 Black eyebrow; blue wing patch; rufous rump. Male lacks white breast band.
454. **Long-tailed Broadbill** *Psarisomus dalhousiae* — SB
 Yellow throat; long, blue tail; black cap with blue.
455. **Green Broadbill** *Calyptomena viridis* — SB
 Small, rounded shape; bright iridescent green plumage.
456. **Hose's Broadbill** *Calyptomena hosii* — B
 Resembles 455 but bright blue breast and abdomen.
457. **Whitehead's Broadbill** *Calyptomena whiteheadi* — B
 Large size; chunky shape; green plumage; black throat.

Plate 52

PLATE 53. **Pittas**

458. **Schneider's Pitta** *Pitta schneideri*   S
 Large size; rufous crown.
 Male: blue back. Female: brown back.
459. **Giant Pitta** *Pitta caerulea*   SB
 Very large size; blackish, mottled crown, bright blue rump and tail.
 Male: blue back. Female: brown back.
460. **Blue-banded Pitta** *Pitta arquata*   B
 Scarlet breast and belly, chestnut head.
461. **Blue-headed Pitta** *Pitta baudii*   B
 Crimson back; shiny blue crown on male.
462. **Blue-winged Pitta** *Pitta moluccensis*   SBJ
 Longer white bar on wing than 463.
 Note: Head pattern.
463. **Fairy Pitta** *Pitta nympha*   B
 Generally paler coloration, shorter white wing bar than 462.
 Note: Head pattern.
464. **Mangrove Pitta** *Pitta megarhyncha*   SB
 Large heavy bill; shorter white wing bar than 462. Note: Head pattern.
465. **Elegant Pitta** *Pitta elegans*   J
 Black throat; blue rear eyebrow, black on belly.
 P. e. concinna (Nusa Penida) shown.
466. **Garnet Pitta** *Pitta granatina*   SB
 Purple back, red belly.
 (a) *P. g. ussheri* (Sabah).
 (b) *P. g. granatina* (Sarawak).
 (c) *P. g. coccinea* (E Sumatra).
467. **Black-crowned Pitta** *Pitta venusta*   SB
 Black back, red belly.
468. **Hooded Pitta** *Pitta sordida*   SBJ
 Green breast, black head.
 (a) *P. s. mulleri* (Sumatra, Borneo, Java): resident.
 (b) *P. s. cucullata* (Sumatra, Java): migrant.
469. **Banded Pitta** *Pitta guajana*   SBJ
 Pale throat, heavy eye-stripe, barred underparts.
 (a) *P. g. guajana* (E Java and Bali).
 (b) *P. g. ripleyi* (Sumatra).

Plate 53

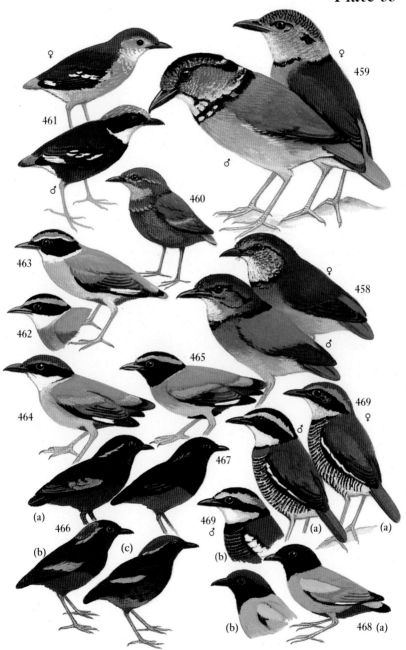

PLATE 54. **Larks and Swallows**

470. **Australian Lark** *Mirafra javanica*   BJ
 Rufous wing patch, reddish brown above, no crest.
471. **Oriental Skylark** *Alauda gulgula*   B
 Small crest; more buffy than 470; no rufous wing patch.
472. **Sand-Martin** *Riparia riparia*   B
 Brown above; squarish tail; brown breast band.
473. **Barn Swallow** *Hirundo rustica*   SBJ
 H. r. gutturalis shown: long tail deeply forked, dark blue breast band, clean white belly.
474. **Pacific Swallow** *Hirundo tahitica*   SBJ
 Tail moderately forked; shorter than 473. No breast band, underparts buffy.
475. **Red-rumped Swallow** *Hirundo daurica*   SB
 Tail deeply forked; chestnut rump; underparts streaked.
476. **Striated Swallow** *Hirundo striolata*   SBJ
 As 475 but paler, no rufous on head. Some forms of 475 can be as pale, making field observations doubtful.
477. **Asian House-Martin** *Delichon dasypus*   SBJ
 Small size; white rump; greyish breast.

Plate 54

PLATE 55. **Cuckoo-Shrikes and Fruit-Hunter**

478. **Bar-winged Flycatcher-Shrike** *Hemipus picatus* — SB
 White wing bar and white on tail tip.
479. **Black-winged Flycatcher-Shrike** *Hemipus hirundinaceus* — SBJ
 No white in wings or tail tip.
480. **Large Woodshrike** *Tephrodornis gularis* — SBJ
 Heavy, hooked bill; black mask; no white in tail; white rump.
481. **Malaysian Cuckoo-Shrike** *Coracina javensis* — J
 Large size; grey colour; white vent. Black lores and eye-ring only in male.
482. **Sunda Cuckoo-Shrike** *Coracina larvata* — SBJ
 Dark face; no barring below; grey vent.
483. **Bar-bellied Cuckoo-Shrike** *Coracina striata* — SBJ
 Yellow eye; pale barred rump; no mask.
484. **Lesser Cuckoo-Shrike** *Coracina fimbriata* — SBJ
 Small size; white tips to outer tail feathers.
 Male: black wings.
 Female: underparts entirely barred.
485. **Pied Triller** *Lalage nigra* — SBJ
 Male: broad white eyebrow.
 Female: brownish above, lightly barred black below.
 Immature: mottled brownish above, streaked below.
486. **White-shouldered Triller** *Lalage sueurii* — J
 Less white in wing and tail than 485.
 Male: narrow white eyebrow.
629. **Black-breasted Fruit-Hunter** *Chlamydochaera jefferyi* — B
 Rounded tail. Rufous face with heavy black eye-stripe and black breast patch.

Plate 55

PLATE 56. **Minivets**

487. **Ashy Minivet** *Pericrocotus divaricatus*   SB
 White forehead; grey rump; no wing bar.
488. **Small Minivet** *Pericrocotus cinnamomeus*   J
 Male: grey head and mantle; orange belly.
 Female: paler; whitish breast.
489. **Fiery Minivet** *Pericrocotus igneus*   SB
 Smaller than 492; lacks second colour patch in secondaries; female more orange.
490. **Grey-chinned Minivet** *Pericrocotus solaris*   SB
 Male: grey sides of head and throat.
 Female: grey sides of head; no yellow on forehead.
491. **Sunda Minivet** *Pericrocotus miniatus*   SJ
 Male: black throat; single red patch in wing.
 Female: reddish mantle and face.
492. **Scarlet Minivet** *Pericrocotus flammeus*   SBJ
 Larger than 489; second colour flash patch in secondaries.
 Female: yellower than 489.

Plate 56

PLATE 57. **Leafbirds**

493. **Green Iora** *Aegithina viridissima*   SB
 Green underparts, two pale wing bars, grey eye.
494. **Common Iora** *Aegithina tiphia*   SBJ
 Yellow underparts, two pale wing bars, pale eye. Variation in colour according to season, age and subspecies.
 A. t. aequanimis (N Borneo) shown.
495. **Lesser Green Leafbird** *Chloropsis cyanopogon*   SB
 No silky blue patch on shoulder. Female has green throat.
496. **Greater Green Leafbird** *Chloropsis sonnerati*   SBJ
 Silky blue patch on shoulder. Female has yellow throat.
 C. s. sonnerati (Java) shown.
497. **Golden-fronted Leafbird** *Chloropsis aurifrons*   S
 Golden forehead, blue on wing only at shoulder.
 C. a. media (Sumatra) shown.
498. **Blue-winged Leafbird** *Chloropsis cochinchinensis*   SBJ
 Blue wing and tail.
 (a) *C. c. flavocincta* (mountains of N Borneo): female has black throat.
 (b) *C. c. icterocephala* (Sumatra).
 (c) *C. c. nigricollis* (Java).
499. **Blue-masked Leafbird** *Chloropsis venusta*   S
 Small size, pale blue on face; male: purple malar stripe.

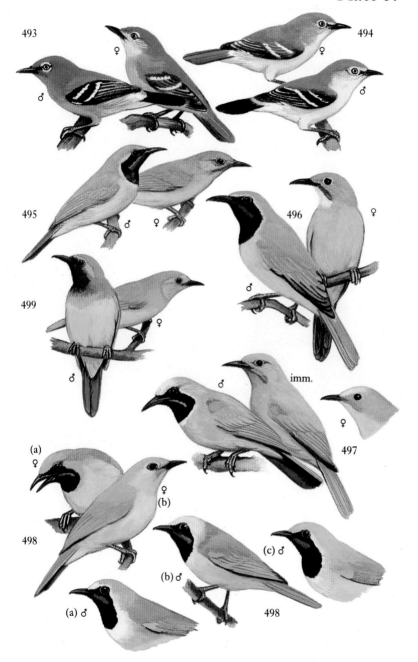

Plate 57

PLATE 58. **Bulbuls**

500. **Straw-headed Bulbul** *Pycnonotus zeylanicus* — SBJ
 Large size; straw-coloured head; black moustache.
501. **Cream-striped Bulbul** *Pycnonotus leucogrammicus* — S
 Small crest; white streaking on upper body; pale eye.
503. **Black-and-white Bulbul** *Pycnonotus melanoleucos* — SB
 Black plumage; white wing bar.
504. **Black-headed Bulbul** *Pycnonotus atriceps* — SBJ
 Black head; no crest; blue eye; black subterminal tail bar.
 (a) Normal adult.
 (b) Rare colour form.
505. **Black-crested Bulbul** *Pycnonotus melanicterus* — SBJ
 Black crest; plain tail.
 (a) *P. m. dispar* (Sumatra, Java, Bali).
 (b) *P. m. montis* (Borneo).
506. **Scaly-breasted Bulbul** *Pycnonotus squamatus* — SBJ
 Black head; scaly breast; white throat.
507. **Grey-bellied Bulbul** *Pycnonotus cyaniventris* — SB
 Grey head and breast; yellow vent.
508. **Red-whiskered Bulbul** *Pycnonotus jocosus* — S
 Long crest; red vent and ear patch.
509. **Sooty-headed Bulbul** *Pycnonotus aurigaster* — SJ
 Black head; whitish rump and belly; vent yellow or orange.

Plate 58

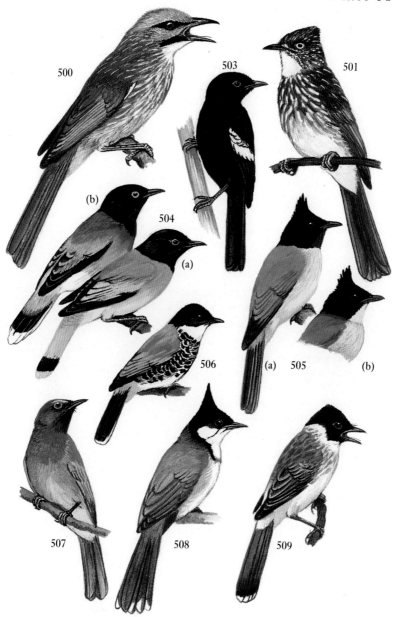

PLATE 59. **Bulbuls**

502. **Spot-necked Bulbul** *Pycnonotus tympanistrigus* — S
 Black skin around eye, yellow ear coverts; heavily streaked breast and belly; barred ochre vent.
510. **Puff-backed Bulbul** *Pycnonotus eutilotus* — SB
 Brown above, whitish below; tail with white tips.
 (a) Borneo: short crest; browner.
 (b) Sumatra: long crest.
511. **Blue-wattled Bulbul** *Pycnonotus nieuwenhuisi* — SB
 Dark head; blue wattles around eyelids.
512. **Orange-spotted Bulbul** *Pycnonotus bimaculatus* — SJ
 Orange spots on face; yellow ear coverts.
 P. b. snouckaerti (N Sumatra) has dusky belly.
513. **Flavescent Bulbul** *Pycnonotus flavescens* — B
 Yellow vent, olive-yellow on wing, lacks black eye-stripe.
 (a) *P. f. leucops* (Borneo): whiter face.
 (b) Elsewhere: yellow streaking on underparts.
514. **Yellow-vented Bulbul** *Pycnonotus goiavier* — SBJ
 Brown upperparts; yellow vent; white face; black eye-stripe and cap.
515. **Olive-winged Bulbul** *Pycnonotus plumosus* — SBJ
 Red eye; olive wings; buffy underparts; olive vent.
516. **Cream-vented Bulbul** *Pycnonotus simplex* — SBJ
 Browner than 515; cream vent.
 (a) *P. simplex*: white eye.
 (b) *P. s. perplexus* (Borneo): red eyes.
517. **Red-eyed Bulbul** *Pycnonotus brunneus* — SBJ
 Red eye; plain brown with no olive tinge; brownish-buff vent.
518. **Spectacled Bulbul** *Pycnonotus erythrophthalmos* — SB
 Orange-yellow eye-ring; whitish throat.

Plate 59

PLATE 60. **Bulbuls**

519. **Finsch's Bulbul** *Criniger finschii* — SB
 Small size; stocky; yellow throat; no crest.
520. **Ochraceous Bulbul** *Alophoixus ochraceus* — SB
 Crest; white throat; dull underparts; cinnamon vent.
 A. o. fowleri (Borneo) shown.
 Sumatran birds are greenish yellow below.
521. **Grey-cheeked Bulbul** *Alophoixus bres* — SBJ
 Crest; white throat; yellowish underparts.
 (a) *A. b. gutturalis* (Borneo).
 (b) *A. b. balicus* (E Java and Bali).
 Several other races differ in colour intensity of underparts.
522. **Yellow-bellied Bulbul** *Alophoixus phaeocephalus* — SB
 Grey crown; lack of crest; bright yellow underparts.
 (a) *A. p. connectens* (N Borneo) lacks yellow tip to tail.
523. **Hook-billed Bulbul** *Setornis criniger* — SB
 Long hooked bill; whitish eyebrow.
524. **Hairy-backed Bulbul** *Tricholestes criniger* — SB
 Small size; yellowish sides of head and underparts.
525. **Buff-vented Bulbul** *Iole olivacea* — SB
 Short crest; pale eye.
526. **Sunda Bulbul** *Iole virescens* — SJ
 Whitish streaking on grey breast; short crest; creamy belly.
 (a) *I. v. virescens* (Java): olive green.
 (b) *I. v. sumatranus* (Sumatra): browner.
527. **Streaked Bulbul** *Ixos malaccensis* — SB
 Plain olive above; pale below; white streaking on grey breast.
528. **Ashy Bulbul** *Hypsipetes flavala* — SB
 Short crest; white throat and belly.
 (a) *H. f. connectens* (Borneo): olive wings; yellow vent.
 (b) *H. f. cinereus* (elsewhere): no green in plumage.

Plate 60

Plate 61. Drongos

529. Black Drongo *Dicrurus macrocercus* — SJ
   Small bill. Long, deeply forked tail.

530. Ashy Drongo *Dicrurus leucophaeus* — SBJ
   Greyish with long, deeply forked tail. Many facial variations.
   (a) *D. l. stigmatops* (Bornean mountains).
   (b) *D. l. leucophaeus* (Java and Bali).

531. Crow-billed Drongo *Dicrurus annectans* — SBJ
   Heavy bill, glossy tail slightly forked.

532. Bronzed Drongo *Dicrurus aeneus* — SB
   Small size, glossy plumage, forked tail.

533. Lesser Racket-tailed Drongo *Dicrurus remifer* — SJ
   Small size, glossy plumage, square tail with long outer streamers and long, flat rackets.

534. Hair-crested Drongo *Dicrurus hottentottus* — SBJ
   Crest of long hairs; 'lyre'-shaped tail. Bornean form *borneenis* shown.
   (a) *D. h. jentincki* (E Java, Bali, Kangean Island): red- and white-eyed forms occur.

535. Sumatran Drongo *Dicrurus sumatranus* — S
   Lacks hair crest; fewer spangles on breast, and smaller in size than 534. Tail only slightly forked.

536. Greater Racket-tailed Drongo *Dicrurus paradiseus* — SBJ
   Short frontal crest, glossy plumage, forked tail with long outer shafts and twisted rackets.

Plate 61

PLATE 62. **Orioles and Fairy-Bluebird**

537. **Dark-throated Oriole** *Oriolus xanthonotus*   SBJ
     Male: black head and throat, white belly streaked with black.
     Female: green above; streaked underparts.
538. **Black-naped Oriole** *Oriolus chinensis*   SBJ
     Male: yellow plumage with black eye-stripe.
     Female: similar but duller.
539. **Black-hooded Oriole** *Oriolus xanthornus*   SB
     Adult: sexes alike, black head and yellow underparts.
     Immature: yellow forehead, throat and breast streaked with white.
540. **Black Oriole** *Oriolus hosii*   B
     Black plumage with chestnut vent; female and immature as male.
541. **Black-and-crimson Oriole** *Oriolus cruentus*   SBJ
     Male: black plumage and crimson belly.
     Female: as male (Borneo, Java) or breast and belly grey, mottled black (Sumatra).
542. **Asian Fairy-Bluebird** *Irena puella*   SBJ
     Male: blue and black plumage, red eye.
     Female: mostly greenish, bright rump, red eye.

Plate 62

PLATE 63. **Crows and Bristlehead**

543. **Crested Jay** *Platylophus galericulatus* — SBJ
 Long crest, white neck patch.
 (a) *P. g. coronatus* (Sumatra, Borneo).
 (b) *P. g. galericulatus* (Java).
544. **Short-tailed Magpie** *Cissa thalassina* — BJ
 Short tail, no yellow on head, tertials not pied.
 C. thalassina (Java) shown.
545. **Green Magpie** *Cissa chinensis* — SB
 Long graduated tail with pale tips. Yellowish on forehead, tertials pied.
546. **Sumatran Treepie** *Dendrocitta occipitalis* — SB
 Brownish mantle, white nape.
547. **Bornean Treepie** *Dendrocitta cinerascens* — B
 Grey mantle, silvery crown.
548. **Racket-tailed Treepie** *Crypsirina temia* — SBJ
 Greenish-black colour; central tail feathers spatulate, blue eye.
549. **Black Magpie** *Platysmurus leucopterus* — S
 Black plumage; crest. Races vary.
 (a) *P. l. aterrimus* (Borneo): crest, all black.
 (b) *P. l. leucopterus* (Sumatra): smaller crest, white wing patch.
550. **Slender-billed Crow** *Corvus enca* — SJ
 Longer, thinner bill than 552, greyer underparts.
551. **House Crow** *Corvus splendens* — J
 Nape and breast greyish.
552. **Large-billed Crow** *Corvus macrorhynchos* — SJ
 Large, heavy bill; purplish gloss.
553. **Bornean Bristlehead** *Pityriasis gymnocephala* — B
 Very heavy, hooked bill. Brightly coloured head.
 Males have no red on flanks.

Plate 63

PLATE 64. **Tits, Nuthatches, Shortwings, and Robins**

554. **Pygmy Tit** *Psaltria exilis*   J
 Tiny size, long tail, dull plumage.
555. **Great Tit** *Parus major*   SBJ
 Bold black and white pattern, white nape spot.
556. **Sultan Tit** *Melanochlora sultanea*   S
 Male: yellow crest and belly.
 Female: olive instead of black.
557. **Velvet-fronted Nuthatch** *Sitta frontalis*   SBJ
 Red bill, black forehead, pinkish underparts.
 Female lacks black eyebrows.
 S. f. corallipes (Borneo) shown.
558. **Blue Nuthatch** *Sitta azurea*   SJ
 Black head, white breast. Subspecies differ in amount of blue and black.
 S. a. nigriventer (W Java) shown.
616. **Lesser Shortwing** *Brachypteryx leucophrys*   SJ
 Skulking, babbler-like. White eyebrow is often not visible.
 Javan female shown.
617. **White-browed Shortwing** *Brachypteryx montana*   SBJ
 (a) *B. m. montana* (Java).
 (b) *B. m. saturata* (Sumatra).
 (c) *B. m. erythroggma* (Borneo).
618. **Siberian Rubythroat** *Luscinia calliope*   B
 Male: red throat and striped head.
 Female: whitish throat with only small, faint red patch.
619. **Siberian Blue Robin** *Erithacus cyane*   SB
 Male: blue upperparts, white underparts.
 Immature male also shows blue on mantle.
 Immature female can be all dull brown on back.
620. **Orange-flanked Bush-Robin** *Tarsiger cyanurus*   B
 Male: orange flanks, blue upperparts.
 Female and immature have dull brown upperparts; brownish breast; blue rump and tail.

Plate 64

PLATE 65. **Jungle-Babblers**

559. **Black-capped Babbler** *Pellorneum capistratum*   SBJ
 Dark head, white eyebrow, white throat; rufous underparts. *P. c. morrelli* (N Borneo) shown.
560. **Temminck's Babbler** *Pellorneum pyrrogenys*   BJ
 Dark brown head and rufous upperparts, streaked with white; rufous flanks and breast band. *P. p. longstaffi* (NW Borneo) shown.
561. **Buettikofer's Babbler** *Pellorneum buettikoferi*   S
 Like 560 but no rufous in plumage.
562. **White-chested Babbler** *Trichastoma rostratum*   SB
 Brown above with rufous rump; white underparts with grey sides. *T. r. macroptera* (Borneo) shown.
563. **Ferruginous Babbler** *Trichastoma bicolor*   SB
 Rufous upperparts, longish tail; white underparts, washed with rufous.
564. **Short-tailed Babbler** *Malacocincla malaccense*   SB
 Short tail, grey cheeks, fulvous belly. *M. m. feriata* (N Borneo) shown.
565. **Horsfield's Babbler** *Malacocincla sepiarium*   SBJ
 Heavy bill, dark above, streaked greyish breast. Rufous flanks and undertail coverts, short tail. *M. s. harterti* (N Borneo) shown.
566. **Abbott's Babbler** *Malacocincla abbotti*   SBJ
 Longer tail than 565. Short pale eyebrow and shaft lines on cheeks. *M. a. finschi* (Borneo) shown.
567. **Black-browed Babbler** *Malacocincla perspicillata*   B
 Blackish eyebrow, yellow eye, grey breast and belly, streaked with white.
568. **Vanderbilt's Babbler** *Malacocincla vanderbilti*   S
 Like 567 but eye dark, bill heavier. Grey breast and belly unstreaked. Like 565 but found at higher altitude.
569. **Moustached Babbler** *Malacopteron magnirostre*   SB
 Chestnut tail, grey cheeks, dark moustachial stripe, brownish-grey breast. (a) *M. m. cinereocapillum* (Borneo): black crown.
570. **Sooty-capped Babbler** *Malacopteron affine*   SB
 Black crown, chestnut tail, greyish breast and flanks. (a) *M. a. phoeniceum* (Borneo): brown crown.
571. **Scaly-crowned Babbler** *Malacopteron cinereum*   SBJ
 Rufous crown with black scaling, black nape. (a) *M. c. rufifrons* (Java): no black nape.
572. **Rufous-crowned Babbler** *Malacopteron magnum*   SB
 Like 571 but larger. No scaling on rufous crown. Throat and breast streaked with brown.
573. **Grey-breasted Babbler** *Malacopteron albogulare*   SB
 Dark grey head, white eyebrow and throat, grey breast, rufous flanks and vent. *M. a. moultoni* (Borneo) shown.

Plate 65

PLATE 66. **Scimitar-Babbler and Wren-Babblers**

574. **Chestnut-backed Scimitar-Babbler** *Pomatorhinus montanus* SBJ
 Black head; white eyebrow; long, decurved, yellow bill.
575. **Long-billed Wren-Babbler** *Rimator malacoptilus* S
 Short tail; fluffy plumage; white streaks; long, decurved bill.
576. **Bornean Wren-Babbler** *Ptilocichla leucogrammica* B
 Short tail; dark chestnut plumage; whitish face; heavy white streaking on underparts.
577. **Striped Wren-Babbler** *Kenopia striata* SB
 White streaks on dark upperparts; buffy lores and flanks.
578. **Rusty-breasted Wren-Babbler** *Napothera rufipectus* S
 Large size; streaked dark brown with black and buff; white throat.
579. **Black-throated Wren-Babbler** *Napothera atrigularis* B
 Large size; dark brown with black face and throat. Head, back, and underparts scaled with black.
580. **Large Wren-Babbler** *Napothera macrodactyla* SJ
 Large size; upperparts scaled with buff, black, and brown; greyish underparts streaked with white.
581. **Marbled Wren-Babbler** *Napothera marmorata* S
 Large size; upperparts dark brown scaled with black; throat white; black underparts scaled with white; rufous ear coverts.
582. **Mountain Wren-Babbler** *Napothera crassa* B
 Short tail; white throat and eyebrow. Buffy lores and streaking.
583. **Eye-browed Wren-Babbler** *Napothera epilepidota* SBJ
 Like 582 but lores dark; smaller; paler underneath.
584. **Pygmy Wren-Babbler** *Pnoepyga pusilla* SJ
 Tiny; almost tailless; pale scalloped underparts.

Plate 66

PLATE 67. **Tree-Babblers**

585. **Rufous-fronted Babbler** *Stachyris rufifrons* — SB
Chestnut cap; white throat streaked with black; orange breast.
586. **Golden Babbler** *Stachyris chrysaea* — S
Crown yellow, streaked black; black eye-stripe; yellowish below.
S. c. frigida (Sumatra) shown.
587. **White-breasted Babbler** *Stachyris grammiceps* — J
Black crown and moustache stripe streaked with white; rufous back and tail; grey flanks on white underparts.
588. **Grey-throated Babbler** *Stachyris nigriceps* — SB
Nape grey streaked with white; white moustache stripe; tawny brown above and below.
589. **Grey-headed Babbler** *Stachyris poliocephala* — SB
Head dark grey streaked with white on forehead and throat; chestnut above and below.
590. **Spot-necked Babbler** *Stachyris striolata* — S
Black eye-stripe bordered with white. Black moustache; spotted on neck; white throat; rufous underparts.
591. **Chestnut-rumped Babbler** *Stachyris maculata* — SB
Lower back and rump rufous; throat black; streaked crown and breast; blue eye-ring.
592. **White-necked Babbler** *Stachyris leucotis* — SB
Resembles 593 but duller; no malar patch; white spots on side of neck and buffy lores.
S. l. goodsoni (Borneo) shown.
593. **Black-throated Babbler** *Stachyris nigricollis* — SB
Crown dark streaked with white; white line behind eye and malar patch; throat and breast black.
594. **White-bibbed Babbler** *Stachyris thoracica* — SJ
Rufous with dark face and white breast band.
(a) *S. t. orientalis* (E Java): grey crown.
595. **Chestnut-winged Babbler** *Stachyris erythroptera* — SB
Blue skin around eye; upperparts chestnut with dark grey face and underparts.
S. e. bicolor (N Borneo) with grey crown shown.
596. **Crescent-chested Babbler** *Stachyris melanothorax* — J
Crown and wings bright chestnut; cheeks pale grey; throat white with black patch at side and small crescent on breast.

Plate 67

PLATE 68. **Tit-Babblers, Fulvettas, Sibias, Yuhinas, and Rail-Babbler**

597. **Grey-cheeked Tit-Babbler** *Macronous flavicollis* — J
 Grey head; rufous wings and tail; streaked breast and throat.
598. **Striped Tit-Babbler** *Macronous gularis* — SBJ
 Brown cap; streaked throat. Races vary.
 (a) *M. g. borneensis* (Borneo except NE).
 (b) *M. g. javanica* (Java).
599. **Fluffy-backed Tit-Babbler** *Macronous ptilosus* — SB
 Chestnut cap; bluish lores and eye-ring; black throat.
600. **Chestnut-capped Babbler** *Timalia pileata* — J
 Long tail; chestnut crown separated from black lores and eye-stripe by white eyebrow.
609. **Brown Fulvetta** *Alcippe brunneicauda* — SB
 Greyish head and underparts.
610. **Javan Fulvetta** *Alcippe pyrrhoptera* — J
 Chestnut brown above; buffy breast and flanks; rufous rump.
611. **Spotted Crocias** *Crocias albonotatus* — J
 Black head; graduated tail with white tips. Mantle and back chestnut, streaked with white.
612. **Long-tailed Sibia** *Heterophasia picaoides* — S
 Grey plumage; long, graduated tail with whitish tips; white wing patch.
613. **Chestnut-crested Yuhina** *Yuhina everetti* — B
 Chestnut crest and head, white lores and eye-ring, outer tail feathers white.
614. **White-bellied Yuhina** *Yuhina zantholeuca* — SB
 Full crest; yellow vent.
615. **Malaysian Rail-Babbler** *Eupetes macrocerus* — SB
 Long thin neck and tail; long black eye-stripe and white eyebrow.

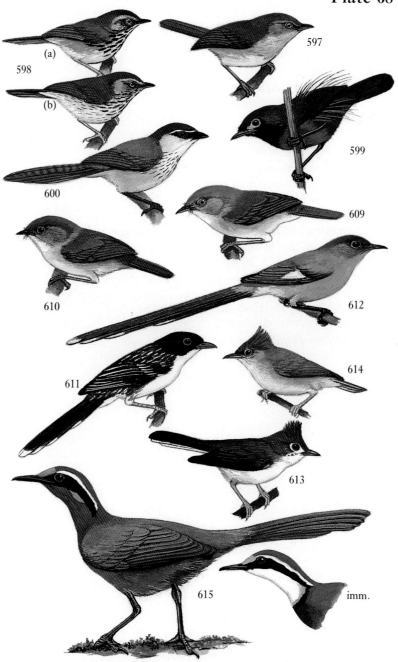

Plate 68

PLATE 69. Laughingthrushes, Mesia, and Shrike-Babblers

601. **Sunda Laughingthrush** *Garrulax palliatus* — SB
 Pale blue orbital skin, grey foreparts.
602. **Rufous-fronted Laughingthrush** *Garrulax rufifrons* — J
 Long tail, brownish with rufous forehead.
603. **White-crested Laughingthrush** *Garrulax leucolophus* — S
 White head with black mask.
 G. p. bicolor (Sumatra) shown.
604. **Black Laughingthrush** *Garrulax lugubris* — SB
 (a) *G. l. calvus* (Borneo): head and throat bald.
 Immatures have feathers on head; browner.
 (b) *G. l. lugubris* (Sumatra): bare skin only around eye; dark.
605. **Chestnut-capped Laughingthrush** *Garrulax mitratus* — SB
 Chestnut head, grey plumage.
 (a) *G. m. damnatus* (NW Borneo).
 (b) *G. m. mitratus* (Sumatra).
606. **Silver-eared Mesia** *Leiothrix argentauris* — S
 Black head, silver cheek, red and yellow pattern.
 L. a. laurinae (Sumatra) shown.
607. **White-browed Shrike-Babbler** *Pteruthius flaviscapis* — SBJ
 Shrike-like head, pale eyebrow, tertials chestnut and gold.
608. **Chestnut-fronted Shrike-Babbler** *Pteruthius aenobarbus* — J
 Small size; chestnut forehead; yellow underparts.

Plate 69

PLATE 70. **Shamas and Forktails**

621. **Magpie Robin** *Copsychus saularis* — SBJ
 Very variable in amount of black and white in plumage.
 (a) *C. s. musicus* (Sumatra, W Java, W Borneo).
 (b) *C. s. pluto* (E Java, N and E Borneo).
622. **White-rumped Shama** *Copsychus malabaricus* — SBJ
 Male: white rump, long black tail.
 Female has shorter tail.
623. **White-browed Shama** *Copsychus stricklandi* — B
 Differs from 622 in white crown.
 Intermediates can occur.
624. **Rufous-tailed Shama** *Trichixos pyrrhopygus* — SB
 Rufous rump and tail.
625. **Sunda Blue Robin** *Cinclidium diana* — SJ
 Male sometimes has pale forehead. White flash at base of tail.
626. **Lesser Forktail** *Enicurus velatus* — SJ
 Male: grey crown and nape; white frontal stripe extends past eye.
 Female: chestnut crown and nape.
627. **Chestnut-naped Forktail** *Enicurus ruficapillus* — SB
 Crown and nape bright chestnut; mottling on breast.
 Chestnut extends to mantle in females.
628. **White-crowned Forktail** *Enicurus leschenaulti* — SBJ
 White forehead, black nape.
 Length of tail differs in subspecies.

Plate 70

PLATE 71. **Cochoas, Chats, Rock-Thrush, and Whistling-Thrushes**

630. **Javan Cochoa** *Cochoa azurea* — SJ
 Male: dark iridescent blue upperparts.
 Female is duller.
 Immature: scaled brown and black underneath.

631. **Sumatran Cochoa** *Cochoa beccarii* — S
 Male: larger and brighter than 630 with light blue patches in wings and tail.
 Female: unknown.

632. **Stonechat** *Saxicola torquata* — SB
 Black head, orange breast, white wing bar.

633. **Pied Bushchat** *Saxicola caprata* — BJ
 Female much darker than 632 with no white in wing, rump rufous.

634. **Wheatear** *Oenanthe oenanthe* — B
 Male shown in winter plumage. White rump and bases to outer tail feathers.

635. **Blue Rock-Thrush** *Monticola solitarius* — SB
 Males can be all blue grey or have rufous underparts.

636. **Shiny Whistling-Thrush** *Myiophoneus melanurus* — S
 Small size; black bill; bright blue eyebrow and shoulders; spangled plumage.

637. **Sunda Whistling-Thrush** *Myiophoneus glaucinus* — SBJ
 Black bill, no spangles. Variable in colour.
 (a) *M. g. castaneus* (Sumatra): Female is dark chestnut with blue shoulders and black crown.
 (b) *M. g. glaucinus* (Java): dark blue all over. Female: duller.

638. **Blue Whistling-Thrush** *Myiophoneus caeruleus* — SJ
 Large, violet, upper feathers spangled; thick, yellow bill.

Plate 71

PLATE 72. **Thrushes**

639. Chestnut-capped Thrush *Zoothera interpres*   SBJ
Small, chestnut cap, spotted flanks, wing bars.

640. Orange-headed Thrush *Zoothera citrina*   SBJ
Orange head and underparts.

641. Everett's Thrush *Zoothera everetti*   B
Long bill; chestnut breast.

642. Sunda Thrush *Zoothera andromedae*   SJ
Large size, short tail, long bill.
Immatures are brownish spotted with buff above.

643. Siberian Thrush *Zoothera sibirica*   SBJ
White eyebrow.
Z. s. sibirica (shown) is paler than *Z. s. davisoni*. Both are winter visitors.

644. Scaly Thrush *Zoothera dauma*   SBJ
Large size; bold scaly markings with whitish underparts.

645. Eyebrowed Thrush *Turdus obscurus*   SBJ
Prominent white eyebrow; orange breast and flanks.

646. Island Thrush *Turdus poliocephalus*   SBJ
Dark colour, yellow eye-ring and bill. Very variable according to subspecies.
(a) *T. p. javanicus* (C Java).
(b) *T. p. seebohmi* (Borneo).

Plate 72

639

640

642

643

644

641

645

646
(a)
(b)

PLATE 73. **Flyeater, Leaf-Warblers, and Reed-Warblers**

647. **Golden-bellied Gerygone** *Gerygone sulphurea* — SBJ
Small size; white lores; yellow belly.

648. **Chestnut-crowned Warbler** *Seicercus castaniceps* — S
Striped chestnut crown; grey cheeks and breast band; yellow rump.

649. **Sunda Warbler** *Seicercus grammiceps* — SJ
Rufous head; black eyebrow; white underparts; white rump.

650. **Yellow-breasted Warbler** *Seicercus montis* — SB
Rufous head; black eyebrow; yellow underparts; yellow rump.

651. **Yellow-bellied Warbler** *Abroscopus superciliaris* — SBJ
White eyebrow; white throat; yellow underparts.

652. **Inornate Warbler** *Phylloscopus inornatus* — S
Two pale wing bars and pale eyebrow. Occasionally has a pale crown stripe.

653. **Arctic Warbler** *Phylloscopus borealis* — SBJ
Usually one, sometimes two thin wing bars; heavier bill than 654.

654. **Eastern Crowned-Warbler** *Phylloscopus coronatus* — SJ
Brighter green than 653, with yellowish crown stripe; one, sometimes two yellowish wing bars.

655. **Mountain Leaf-Warbler** *Phylloscopus trivirgatus* — SBJ
Very prominent crown and eyebrow stripes; no wing bars.

656. **Clamorous Reed-Warbler** *Acrocephalus stentoreus* — BJ
Slightly longer, thinner bill than 657; buffy underparts.

657. **Eastern Reed-Warbler** *Acrocephalus orientalis* — SBJ
Longer, broader eyebrow and heavier bill than 656; faint streaking on upper breast; pale tips of tail feathers.

658. **Black-browed Reed-Warbler** *Acrocephalus bistrigiceps* — S
Pale eyebrow bordered above and below by black stripes; rufous rump and flanks.

Plate 73

PLATE 74. **Warblers, Cisticolas, Tesia, and Bush-Warblers**

659. **Pallas's Warbler** *Locustella certhiola* — SBJ
 Streaked back; rufous rump; black subterminal tail bar.
660. **Middendorff's Warbler** *Locustella ochotensis* — B
 Plain olive-brown above; conspicuous eyebrow; tail with white tip.
 Smaller than reed-warblers, with white throat and brown breast band.
661. **Lanceolated Warbler** *Locustella lanceolata* — SBJ
 Smaller than 659; plain brown tail; streaked underparts.
662. **Striated Grassbird** *Megalurus palustris* — BJ
 Large; long, pointed tail; long pale eyebrow; heavily streaked black.
674. **Zitting Cisticola** *Cisticola juncidis* — SJ
 Browner than 675; paler underparts; white tips of tail.
675. **Golden-headed Cisticola** *Cisticola exilis* — SJ
 Streaked orange and black upperparts.
 Breeding male: all golden head.
676. **Javan Tesia** *Tesia superciliaris* — J
 Tiny; looks tailless. Black head with grey eyebrow.
677. **Bornean Stubtail** *Urosphena whiteheadi* — B
 Small, with short tail. Buffy eyebrow and sides of face.
678. **Sunda Bush-Warbler** *Cettia vulcania* — SBJ
 Brown; paler underneath; whitish eyebrow; brownish sides to breast.
 (a) *C. v. oreophila* (Mt. Kinabalu): darker, black blotching on throat and breast.
679. **Russet Bush-Warbler** *Bradypterus seebohmi* — J
 Rufous; greyish face and eyebrow, white throat streaked with black.
680. **Friendly Bush-Warbler** *Bradypterus accentor* — B
 Dark reddish brown; throat spotted with white, black, and grey; rufous eyebrow.

Plate 74

PLATE 75. **Tailorbirds and Prinias**

663. **Common Tailorbird** *Orthotomus sutorius* — J
 Central tail feathers of male very long in breeding plumage; buffy eyebrow and underparts.
664. **Dark-necked Tailorbird** *Orthotomus atrogularis* — SBJ
 Rufous crown; grey cheeks; yellowish vent; upper breast streaked with black.
665. **Ashy Tailorbird** *Orthotomus ruficeps* — SBJ
 Reddish face; grey back; greyish underparts.
666. **Olive-backed Tailorbird** *Orthotomus sepium* — J
 Red head; upperparts washed olive; underparts yellowish.
667. **Rufous-tailed Tailorbird** *Orthotomus sericeus* — SB
 Rufous tail; buffy underparts.
668. **Mountain Tailorbird** *Orthotomus cuculatus* — SBJ
 Orange forehead; bright olive back and wings; yellow belly and vent.
669. **Hill Prinia** *Prinia atrogularis* — S
 Very long tail; white eyebrow; grey cheeks; streaked breast; buffish flanks.
670. **Plain Prinia** *Prinia inornata* — J
 Buffy eyebrow, cheeks, and underparts; pale brown above.
671. **Yellow-bellied Prinia** *Prinia flaviventris* — SBJ
 Grey head; short pale eyebrow; white throat; yellow belly; no wing bars.
672. **Bar-winged Prinia** *Prinia familiaris* — SJ
 Olive above; yellow belly; two white wing bars.
673. **Brown Prinia** *Prinia polychroa* — J
 Upperparts mottled brown; pale greyish breast; buffish belly.

Plate 75

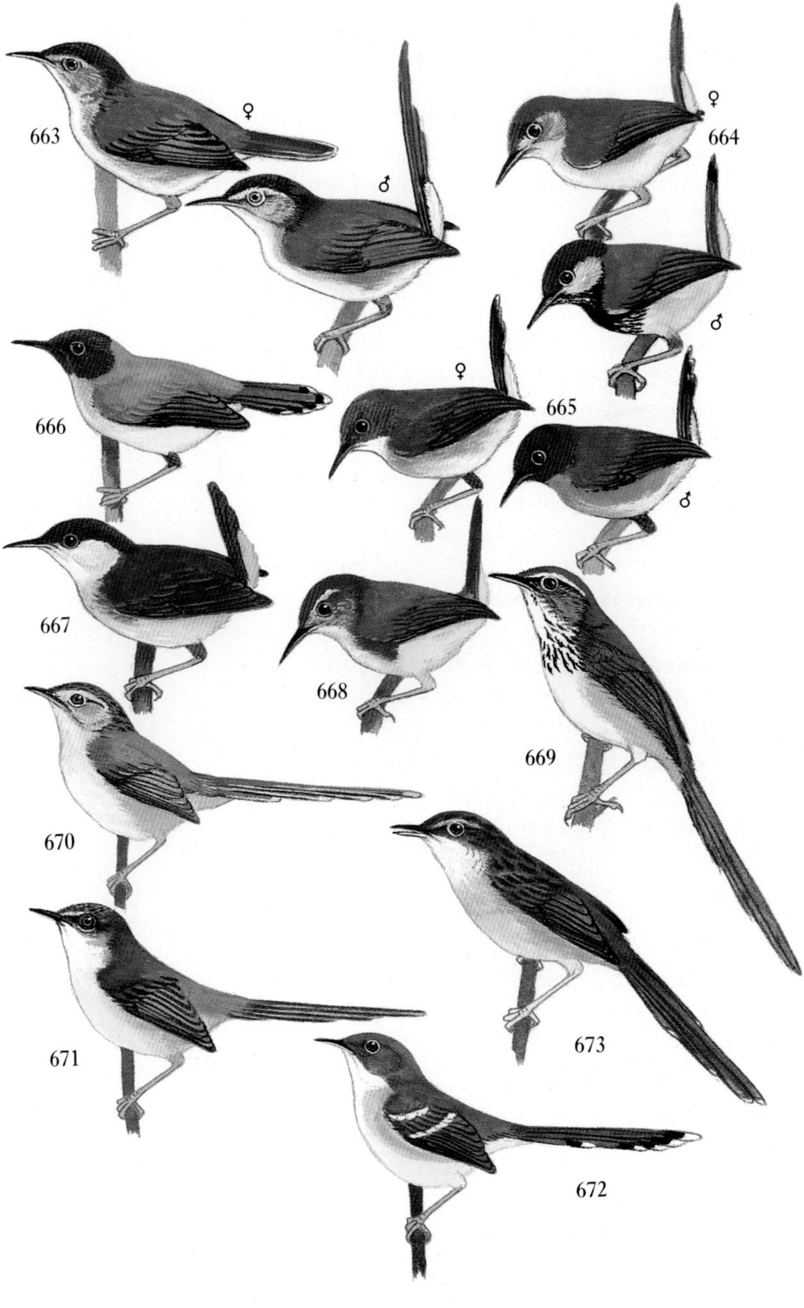

PLATE 76. **Jungle-Flycatchers and Flycatchers**

681. **Fulvous-chested Jungle-Flycatcher** *Rhinomyias olivacea*   SBJ
 Long bill and tail; buffy lores; brownish-buff breast.
682. **Brown-chested Jungle-Flycatcher** *Rhinomyias brunneata*   B
 Lower mandible white; light scalloping on throat; chestnut wings and tail.
683. **Grey-chested Jungle-Flycatcher** *Rhinomyias umbratilis*   SB
 Dark olive-brown above; underparts very white with grey-brown breast band.
684. **Rufous-tailed Jungle-Flycatcher** *Rhinomyias ruficauda*   B
 Greyish sides of breast; tail bright rufous.
685. **Eyebrowed Jungle-Flycatcher** *Rhinomyias gularis*   B
 Short tail; rufous brown above with whitish lores and eyebrow; breast grey-brown; belly white.
686. **Dark-sided Flycatcher** *Muscicapa sibirica*   SBJ
 White eye-ring; white half collar.
 Immature: spotted white on face and back.
687. **Grey-streaked Flycatcher** *Muscicapa griseisticta*   B
 White eye-ring; white lores and wing bar; streaky underparts.
688. **Asian Brown Flycatcher** *Muscicapa dauurica*   SBJ
 (a) *M. d. williamsoni* (migrant): streaky flanks and sides of breast.
 (b) *M. d. latirostris* (migrant): uniform grey flanks and sides of breast.
689. **Ferruginous Flycatcher** *Muscicapa ferruginea*   SBJ
 Grey head; underparts white with chestnut breast and flanks.
690. **Verditer Flycatcher** *Eumyias thalassina*   SB
 Adult: greenish-blue colour; black lores.
 Immature: greyish brown, mottled buff and blackish.
691. **Indigo Flycatcher** *Eumyias indigo*   SBJ
 Adult: bright indigo; paler eyebrow; buffish vent.
 Immature: breast and throat mottled buffy-pink.

Plate 76

PLATE 77. **Flycatchers**

692. Yellow-rumped Flycatcher *Ficedula zanthopygia*   SJ
Male: white eyebrow; yellow rump.
Female: yellow rump.

693. Narcissus Flycatcher *Ficedula narcissina*   B
Male: yellow rump; yellow eyebrow.
Female: reddish tail; no yellow rump.

694. Mugimaki Flycatcher *Ficedula mugimaki*   SBJ
Small bill; orange chin, throat, and breast.
Male: small white eyebrow.

695. Red-breasted Flycatcher *Ficedula parva*   B
White flashes at base of dark tail.

696. Rufous-browed Flycatcher *Ficedula solitaris*   S
Rufous head; buffy eye-ring and lores; white throat sometimes outlined black.

697. Snowy-browed Flycatcher *Ficedula hyperythra*   SBJ
Male: slaty blue upperparts; short white eyebrow.
Female: buffy eyebrow.

698. Rufous-chested Flycatcher *Ficedula dumetoria*   SBJ
Longer bill than 694; darker back; paler chin.
Female has buffy lores.

699. Little Pied Flycatcher *Ficedula westermanni*   SBJ
Male: black and white.
Female: brownish; rufous tail.

712. Pygmy Blue-Flycatcher *Muscicapella hodgsoni*   SB
Small size; tiny bill; short tail like a flowerpecker.

713. Grey-headed Flycatcher *Culicicapa ceylonensis*   SBJ
Grey head and breast; olive back; yellow underparts.

Plate 77

PLATE 78. **Flycatchers**

700. **Blue-and-white Flycatcher** *Cyanoptila cyanomelana* — SBJ
 Male: black face and breast; clean white belly and tail flash patches.
 Female: whitish throat and belly.
701. **Large Niltava** *Niltava grandis* — S
 Male: large size; dark underparts.
 Female: blue shoulder flash.
702. **Rufous-vented Niltava** *Niltava sumatrana* — S
 Male: black face; entire orange belly.
 Female: brownish throat; rufous vent; narrow white throat band.
703. **White-tailed Blue-Flycatcher** *Cyornis concretus* — SB
 Male: dark breast; white belly and white in tail.
 Female: broad white throat band and white in tail.
704. **Rueck's Blue-Flycatcher** *Cyornis ruckii* — S
 Male: bright rump; no white in tail.
 Female: lacks throat band; rufous rump and tail.
705. **Pale Blue-Flycatcher** *Cyornis unicolor* — SBJ
 Male: turquoise blue with black lores.
 Female: fulvous eye-ring and lores.
706. **Hill Blue-Flycatcher** *Cyornis banyumas* — BJ
 Male: black lores and chin spot; orange breast grades to white belly.
 Female: brown with rufous rump and tail.
707. **Large-billed Blue-Flycatcher** *Cyornis caerulatus* — SB
 Male: shining blue rump; no shoulder flash; rufous vent.
 Female: buff eye-ring; blue tail and back; buff throat.
708. **Bornean Blue-Flycatcher** *Cyornis superbus* — B
 Male: shining blue forehead, eyebrow, nape, and lower back.
 Female: rufous forehead and tail.
709. **Malaysian Blue-Flycatcher** *Cyornis turcosus* — SB
 Male: blue throat.
 Female: as male but duller and with white throat.
710. **Tickell's Blue-Flycatcher** *Cyornis tickelliae* — S
 Male: clean separation between orange breast and white belly.
 Female: dull blue upperparts.
711. **Mangrove Blue-Flycatcher** *Cyornis rufigastra* — SBJ
 Male: dull forehead.
 Female: white lores and chin, blue back.

Plate 78

Plate 79. Paradise Flycatchers, Fantails, and Monarchs

714. **Rufous-tailed Fantail** *Rhipidura phoenicura*   J
 Rufous wings; tail and belly.
715. **White-bellied Fantail** *Rhipidura euryura*   J
 Grey throat and breast.
716. **White-throated Fantail** *Rhipidura albicollis*   SB
 White throat; grey breast.
717. **Spotted Fantail** *Rhipidura perlata*   SBJ
 Throat and upper breast spotted.
718. **Pied Fantail** *Rhipidura javanica*   SBJ
 White underparts with grey breast band.
719. **Black-naped Monarch** *Hypothymis azurea*   SBJ
 Blue head. Male has small black nape crest and black throat band.
720. **Maroon-breasted Philentoma** *Philentoma velatum*   SBJ
 Black face and maroon breast distinctive; red eyes.
721. **Rufous-winged Philentoma** *Philentoma pyrhopterum*   SB
 Red eyes; rufous wings and tail.
 (a) Blue phase male: smaller than female 720 with white spotting on belly.
722. **Japanese Paradise-Flycatcher** *Terpsiphone atrocaudata*   S
 Dark maroon mantle and back; very long tail streamers. Female looks like female 723 but much darker.
723. **Asian Paradise-Flycatcher** *Terpsiphone paradisi*   SBJ
 Male has two colour phases, can also be chestnut like female.
 Immature: like female with short tail.

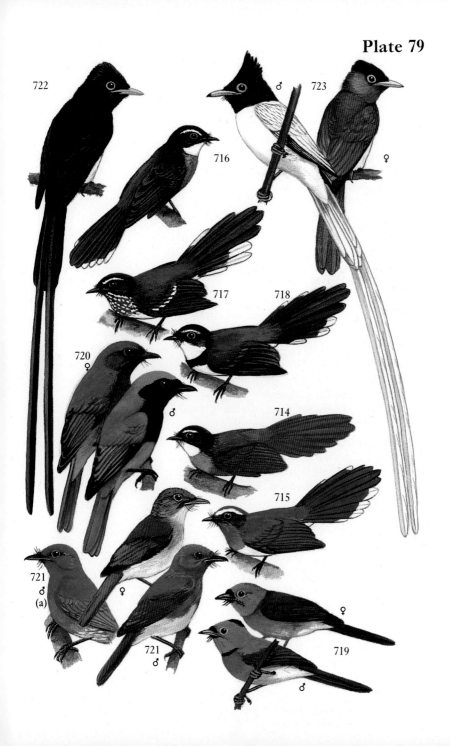

Plate 79

PLATE 80. **Whistlers, Wood-Swallow, and Shrikes**

724. **Bornean Whistler** *Pachycephala hypoxantha*   B
 Olive green above; yellow below; black lores.
 P. h. hypoxantha (Kinabalu) shown.

725. **Mangrove Whistler** *Pachycephala grisola*   SBJ
 Greyish head; lack of rufous on rump.
 P. c. vandepolli shown.
 P. c. secedens (N Borneo) is smaller.

726. **White-vented Whistler** *Pachycephala homeyeri*   B
 Rufous above; white below with faint streaking.

727. **Golden Whistler** *Pachycephala pectoralis*   J
 Male: black head, white throat; yellow below.
 P. p. javana (E Java, Bali) shown.

736. **White-breasted Wood-Swallow** *Artamus leucorhynchus*   SBJ
 White rump; thick bluish bill. Broad triangular wings and unforked tail in flight.

737. **Brown Shrike** *Lanius cristatus*   SBJ
 Adult: white eyebrow; unbarred back. Some subspecies are more rufous.
 Immature: lightly barred above and below.

738. **Tiger Shrike** *Lanius tigrinus*   SBJ
 Male: grey crown and nape; no white eyebrow; barred back.
 Immature: black eye-stripe indistinct.

739. **Long-tailed Shrike** *Lanius schach*   SBJ
 Long tail, white wing spot.
 (a) *L. s. nasutus* (N Borneo).
 (b) *L. s. bentet* (elsewhere in Sundas).

Plate 80

PLATE 81. **Wagtails and Pipits**

728. **Pied Wagtail** *Motacilla alba*   B
 Varies according to race.
 Immatures are grey where adults are black.
729. **Grey Wagtail** *Motacilla cinerea*   SBJ
 Adult: grey head and mantle, long tail, yellow rump.
 Immature has whiter underparts.
730. **Yellow Wagtail** *Motacilla flava*   SBJ
 Shorter tail than 729. Upperparts olive-green, rump same as back.
 (a) *M. f. simillima*: grey crown.
 (b) *M. f. taivana*: olive crown.
 Immature: browner upperparts, whiter underparts.
731. **Forest Wagtail** *Dendronanthus indicus*   SBJ
 Brownish; two black partial breast bands.
732. **Olive-backed Pipit** *Anthus hodgsoni*   B
 Olive upperparts, heavy streaking on breast.
 A. h. yunnanensis shown.
733. **Common Pipit** *Anthus novaeseelandiae*   SBJ
 Pale brownish; sides of body lightly streaked; longer legs than other pipits.
734. **Petchora Pipit** *Anthus gustavi*   B
 Darkish; heavily marked with black on upperparts; heavy streaking on breast; white throat and belly.
735. **Red-throated Pipit** *Anthus cervinus*   B
 Breeding: rufous throat.
 Non-breeding: like 734 but more buffy.
 Rump and uppertail coverts heavily streaked.

Plate 81

PLATE 82. **Starlings**

741. **Short-tailed Starling** *Aplonis minor*   J
 Purplish gloss on head.
742. **Asian Glossy Starling** *Aplonis panayensis*   SBJ
 Greenish gloss on head; longer tail than 741.
743. **White-shouldered Starling** *Sturnus sinensis*   B
 Wholly white wing coverts; grey head and back.
744. **Chestnut-cheeked Starling** *Sturnus philippensis*   B
 Male: chestnut cheek patch. Both sexes lack nape spot.
745. **Purple-backed Starling** *Sturnus sturninus*   SBJ
 Male: no cheek patch; both sexes have nape spot (brown in female).
746. **Asian Pied Starling** *Sturnus contra*   SJ
 Pied plumage with black throat.
747. **Black-winged Starling** *Sturnus melanopterus*   J
 Pied plumage with white throat.
 (a) *S. m. tertius* (Bali).
 (b) E Java.
 (c) W Java.
748. **Bali Myna** *Leucopsar rothschildi*
 White plumage, crest, blue skin around eye, black tips only on wings and tail. On Bali only.
749. **Common Myna** *Acridotheres tristis*   SB
 Brownish colour; no crest; yellow skin around eye.
750. **Javan Myna** *Acridotheres javanicus*   SJ
 Slight crest; yellow bill; white vent.
751. **Crested Myna** *Acridotheres cristatellus*   B
 Prominent crest; red base to bill; white scaling on vent.
752. **Hill Myna** *Gracula religiosa*   SBJ
 Yellow wattles on head. Island races are larger.

Plate 82

PLATE 83. **Sunbirds**

754. Plain Sunbird *Anthreptes simplex* SB
Forehead metallic purple in male; green in female.

755. Plain-throated Sunbird *Anthreptes malacensis* SBJ
Male: brown throat; yellow belly.
Female: yellowish eye-ring; no white on tail.
A. m. malacensis shown.

756. Red-throated Sunbird *Anthreptes rhodolaema* SB
Male: reddish wing coverts; maroon cheeks; reddish throat.
Female: as female 755 but duller and greener.

757. Ruby-cheeked Sunbird *Anthreptes singalensis* SBJ
Both sexes buffy rufous throat.
Several races in region. *A. s. phoenicotis* (Java) shown.

758. Purple-naped Sunbird *Hypogramma hypogrammicum* SB
Large size; purple rump; streaked underparts.
Female: lacks purple.

759. Purple-throated Sunbird *Nectarinia sperata* SBJ
Male: small size; dark, brilliant metallic plumage.

760. Copper-throated Sunbird *Nectarinia calcostetha* SBJ
Male: large size; graduated blue tail; yellow flank feathers.

761. Olive-backed Sunbird *Nectarinia jugularis* SBJ
Breeding male: dark throat; yellow belly; olive back. (In eclipse, black only down centre of throat.)
Female: white on edge of tail.

762. White-flanked Sunbird *Aethopyga eximia* J
Male: olive back; yellow rump; long, fluffy white flank feathers.

763. Crimson Sunbird *Aethopyga siparaja* SBJ
Male: purple forehead; yellow rump; blue tail; grey belly.

764. Scarlet Sunbird *Aethopyga mystacalis* SBJ
Male: pointed purple tail; purple crown; whitish belly.
Female: much smaller than male; rufous undertail.

765. Temminck's Sunbird *Aethopyga temminckii* J
Male: long red tail; whitish belly; yellow and purple rump.
Female: as 764.

Plate 83

Plate 84. Spiderhunters and Honeyeater

766. **Little Spiderhunter** *Arachnothera longirostra* — SBJ
 Throat and breast pale greyish. Yellow belly.

767. **Thick-billed Spiderhunter** *Arachnothera crassirostris* — SB
 Thick, very curved bill; orange pectoral tufts in males.

768. **Long-billed Spiderhunter** *Arachnothera robusta* — SBJ
 Very long bill; throat streaked with green; orange tufts in males.
 A. r. robusta (Sumatra and Borneo) shown.
 A. r. armata (Java) is more whitish grey underneath.

769. **Spectacled Spiderhunter** *Arachnothera flavigaster* — SB
 Large size; heavy bill; broad yellow eye-ring.

770. **Yellow-eared Spiderhunter** *Arachnothera chrysogenys* — SBJ
 Slender bill; yellow ear coverts; thin yellow eye-ring.

771. **Grey-breasted Spiderhunter** *Arachnothera affinis* — SBJ
 Grey underparts with dark streaks.
 (a) *A. a. affinis* (Java and Bali).
 (b) *A. a. concolor* (Sumatra).

772. **Bornean Spiderhunter** *Arachnothera everetti* — B
 Similar to 771 but larger; heavily streaked breast. In N Borneo only.

773. **Whitehead's Spiderhunter** *Arachnothera juliae* — B
 Dark brown streaked with white; yellow rump and undertail coverts.

774. **Indonesian Honeyeater** *Lichmera limbata*
 Adult: green with yellow ear patch.
 Immature: browner and lacks ear patch.
 On Bali only.

Plate 84

PLATE 85. **Flowerpeckers**

775. **Scarlet-breasted Flowerpecker** *Prionochilus thoracicus* — SB
Yellow rump; green back.
Male: red crown and breast patch.
776. **Yellow-breasted Flowerpecker** *Prionochilus maculatus* — SB
Sexes alike. Yellow breast with olive streaks; orange crown patch.
777. **Yellow-rumped Flowerpecker** *Prionochilus xanthopygius* — B
Male: yellow rump; blue back; red crown patch.
Female: as 778 with yellowish rump; no white malar stripe.
778. **Crimson-breasted Flowerpecker** *Prionochilus percussus* — SBJ
Male: blue back; blue rump; red crown patch.
779. **Thick-billed Flowerpecker** *Dicaeum agile* — SJ
Dull colour; streaky breast; red eye.
780. **Brown-backed Flowerpecker** *Dicaeum everetti* — SB
Dull colour; uniform breast; pale eye.
781. **Yellow-vented Flowerpecker** *Dicaeum chrysorrheum* — SBJ
Yellow vent; streaked breast; sexes alike.
782. **Orange-bellied Flowerpecker** *Dicaeum trigonostigma* — SBJ
Orange rump.
783. **Plain Flowerpecker** *Dicaeum concolor* — SBJ
Sexes alike. Dull colour; small size; slender bill.
784. **Scarlet-backed Flowerpecker** *Dicaeum cruentatum* — SB
(a) *D. c. sumatranum* (Sumatra): white throat.
(b) *D. c. nigrimentum* (Borneo): black throat.
785. **Red-chested Flowerpecker** *Dicaeum maugei* — J
Red throat and rump; small size.
786. **Blood-breasted Flowerpecker** *Dicaeum sanguinolentum* — SJ
Male: scarlet breast; white throat; small size.
787. **Black-sided Flowerpecker** *Dicaeum monticolum* — B
Male: scarlet throat; black flanks and breast bar.
Female: as 783 but larger.
788. **Fire-breasted Flowerpecker** *Dicaeum ignipectus* — S
Male: cinnamon breast band.
D. i. beccarii (Sumatra) shown.
789. **Scarlet-headed Flowerpecker** *Dicaeum trochileum* — SBJ
Male: orange-red head and back.
Female: red rump; reddish wash on head and mantle.

Plate 85

PLATE 86. **White-eyes**

790. **Oriental White-eye** *Zosterops palpebrosus* SBJ
 (a) Montane race (Java, Sumatra): entire belly yellow.
 (b) Lowland form (W Java, Borneo, Sumatra): narrow yellow band down centre of belly.
791. **Enggano White-eye** *Zosterops salvadorii* S
 Like 790(b) but belly creamy white.
792. **Black-capped White-eye** *Zosterops atricapilla* SB
 Forehead and crown blackish.
793. **Everett's White-eye** *Zosterops everetti* B
 Like 790(b) but broader yellow band down centre of breast; yellowish forecrown.
794. **Mountain White-eye** *Zosterops montanus* SJ
 Pale iris; no yellow on belly; flanks brownish.
795. **Javan White-eye** *Zosterops flavus* BJ
 Smaller and yellower than 796; lacks black loral spot.
796. **Lemon-bellied White-eye** *Zosterops chloris* SBJ
 Larger and paler than 795; black loral spot.
797. **Javan Grey-throated White-eye** *Lophozosterops javanicus* J
 Grey throat, incomplete eye-ring, buffy lores.
 (a) *L. j. javanica* (C Java).
 (b) *L. j. frontalis* (W Java).
798. **Pygmy White-eye** *Oculocincta squamifrons* B
 Very small; narrow white eye-ring; forecrown spotted.
799. **Mountain Blackeye** *Chlorocharis emiliae* B
 Large; long pinkish bill; eye-ring and lores black.

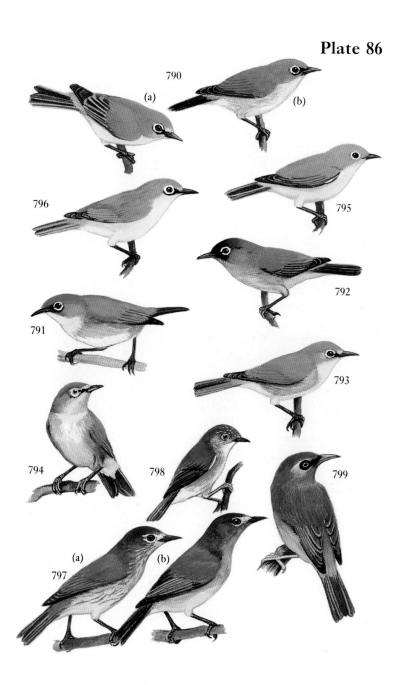

Plate 86

PLATE 87. **Finches and Munias**

804. **Red Avadavat** *Amandava amandava* — SBJ
 Red bill and rump.
 Male in eclipse plumage, as female.
805. **Pin-tailed Parrotfinch** *Erythrura prasina* — SBJ
 Red 'pin' tail and green upperparts.
 E. p. prasina (Sumatra and Java) shown.
806. **Tawny-breasted Parrotfinch** *Erythrura hyperythra* — BJ
 Blunt green tail and green rump.
 E. h. hyperythra (W Java) shown.
807. **Java Sparrow** *Padda oryzivora* — SBJ
 Red bill; white cheek patch.
808. **White-rumped Munia** *Lonchura striata* — S
 White rump; sharp tail; whitish belly.
809. **White-bellied Munia** *Lonchura leucogastra* — SBJ
 Adult: dark rump; paler tail.
810. **Javan Munia** *Lonchura leucogastroides* — SJ
 Adult: clean white belly; dark rump and vent.
811. **Dusky Munia** *Lonchura fuscans* — B
 All dark plumage; bluish bill.
812. **Black-faced Munia** *Lonchura molucca* — J
 White rump; whitish belly; blunt tail.
813. **Scaly-breasted Munia** *Lonchura punctulata* — SJ
 Scaly breast; pale brown rump.
814. **Black-headed Munia** *Lonchura malacca* — SBJ
 Black head; chestnut body.
815. **Chestnut Munia** *Lonchura ferruginosa* — J
 White head; black chin and throat; chestnut body.
816. **White-headed Munia** *Lonchura maja* — SJ
 White head and throat; chestnut body.

Plate 87

PLATE 88. **Sparrow, Weavers, and Buntings**

800. **Eurasian Tree Sparrow** *Passer montanus* — SBJ
 Chestnut crown; black throat and cheek patch.
801. **Baya Weaver** *Ploceus philippinus* — SJ
 Breeding male: crown and nape yellow; plain buff underparts.
802. **Streaked Weaver** *Ploceus manyar* — J
 Breeding male: gold crown; black head; streaked underparts.
803. **Asian Golden Weaver** *Ploceus hypoxanthus* — SJ
 Breeding male: gold and black head and underparts.
817. **Mountain Serin** *Serinus estherae* — SJ
 Yellow rump and 3 wing bars; forehead and breast yellow, streaked with black.
818. **Little Bunting** *Emberiza pusilla* — B
 Head pattern distinctive; one white wing bar; whitish underparts.
819. **Yellow-breasted Bunting** *Emberiza aureola* — B
 (a) Winter male: two white wing bars; chestnut rump and partial breast band; blackish face.
820. **Black-headed Bunting** *Emberiza melanocephala* — B
 (a) 1st winter male: fulvous brown streaked upperparts; rump tinged with yellow. Note lack of eyebrow.

Plate 88

FAMILY AND SPECIES DESCRIPTIONS

Grebes—Family Podicipedidae

A WORLDWIDE family of small to medium-sized duck-like waterbirds. Grebes have pointed bills, short wings, very short tails, erect necks, lobed rather than webbed feet, and long, silky feathers.
Grebes are excellent divers, able to stay under water for several minutes at a time. They feed on fish and water insects and make nests on rafts of floating vegetation.
Two species occur in the Greater Sundas.

1. LITTLE GREBE *Tachybaptus ruficollis* PLATE 26
(Dabchick/Red-throated Grebe)

Description: Small (25 cm), dark, squat, duck-like bird, swimming high in the water and diving repeatedly, remaining submerged for long periods. In breeding plumage, throat and sides of neck are reddish; crown and nape dark brown; upperparts brown; underparts greyish; conspicuous yellow rictal patch. In non-breeding plumage upperparts are greyish brown.
Iris—yellow; bill—black; feet—bluish grey.
Voice: Repeated high-pitched chittering *ke-ke-ke-ke* particularly during courtship chases.
Range: Africa, Eurasia, India, China, SE Asia, Philippines, Sundas, Sulawesi, and through eastern Indonesia to northern New Guinea.
Distribution and status: Recorded in Sumatra (north only), vagrant in Borneo (recorded once), and rare resident on both lowland and montane lakes in Java and Bali.
Habits: Frequents lakes, swamps, and flooded ricefields, where there are clear water and plenty of water plants. Dives under water to feed and in alarm at the slightest provocation. Generally single or in small groups of dispersed individuals. In breeding season birds chase each other, running over the water and calling.

2. AUSTRALASIAN GREBE *Tachybaptus novaehollandiae* PLATE 26
(Australian Little/Black-throated Grebe)

Description: Small (24 cm), squat, diving waterbird, very similar to Little Grebe but differs in having white underparts and, in breeding season, black rather than

46 GREBES

rufous throat. There is a whitish bar on the wing in flight. A chestnut brown stripe behind the eye separates the black crown and nape from greyish-black chin and throat.
Iris—white to orange; bill—black with white tip; feet—olive to black.
Voice: Shrill chittering *chee-ee-ee-ee*, similar to Little Grebe.
Range: From Java and Bali through Lesser Sundas to New Guinea, Australia, and New Zealand.
Distribution and status: One doubtful record from Banjarmasin region of Borneo in the last century; and a rare local resident on lowland and montane lakes in Java and Bali.
Habits: As Little Grebe.

SHEARWATERS—FAMILY PROCELLARIIDAE

A MODERATELY large family of oceanic gull-like birds with curiously structured tube-nose bills, hooked at the tip and with the nostrils opening in a double tube on top of the bill.

Shearwaters get their English name from their habit of flying just above the water surface or even touching the water. They dive for fish, squid, plankton, and crustaceans and nest on cliff ledges or in burrows on rocky islets. They are silent at sea.

Shearwaters are predominantly temperate in distribution, with 8 species seen in Greater Sundas waters.

3. BARAU'S PETREL *Pterodroma baraui* PLATE 1

Description: Medium-sized (38 cm), slender, white-bellied petrel with long wings and short rounded tail. Upperparts generally sooty brown with white forehead and black hind-crown, tail, flight feathers and leading edge to wing. Underparts white with blackish edges to wing.
Iris—brown; bill—black; feet—yellow with black toes.
Range: Breeds in Reunion Islands and ranges eastwards through Indian Ocean.
Distribution and status: Recorded at sea off Enggano, SW Sumatra.
Habits: Flies fast over water surface, banking and swooping with ease.

[4. SLENDER-BILLED PRION *Pachyptila belcheri* PLATE 1
(Thin-billed/Narrow-billed Prion)

Description: Small (26 cm), grey, black, and white petrel. Similar to Antarctic Prion but with thinner bill.

Distribution and status: Sub-antarctic and southern oceans including S Australia and New Zealand. A possible candidate for the Javan prion specimen mentioned under Antarctic Prion.
Habits: As Dove Prion.
Note: Suggested by some authors as the likely prion of Javan record, but the specimen is lost and there are no sound grounds for revising the earlier identification.]

5. ANTARCTIC PRION *Pachyptila desolata* PLATE 1
(Dove Prion)

Description: Small (27 cm), grey, black, and white petrel. Tail wedge-shaped with black tip. Upperparts bluish grey with a black W shape formed by outer primaries, scapulars, and a dark band across the lower back. A blackish line extends across the eye. Underparts white.
Iris—dark; bill—grey; feet—greyish blue.
Range: Breeds on Antarctic and sub-Antarctic islands, dispersing north in non-breeding season to subtropical regions of Atlantic and Indian Oceans.
Distribution and status: A single, exhausted individual beached on the S Java coast in July 1938.
Habits: Flies with fast, zig-zagging, erratic flight low over the sea, often in flocks, banking from side to side.

6. BULWER'S PETREL *Bulweria bulwerii* PLATE 1

Description: Small (27 cm), sooty brown petrel with paler brown underparts. Pale bar on upperwing coverts usually visible. Distinguished from Swinhoe's Storm-petrel by larger size and long, wedge-shaped tail (looks long and pointed in flight).
Iris—brown; bill—black; feet—pinkish.
Range: Breeds on islands in Atlantic and Pacific Oceans.
Distribution and status: Some of the W Pacific populations regularly disperse south into Indonesian waters and the Indian Ocean.
Habits: Flies more strongly than storm-petrels with fluttery erratic swooping and high-wheeling loops.

7. JOUANIN'S PETREL *Bulweria fallax* PLATE 1

Description: Medium-sized (31 cm), dark brown, long-winged petrel with long, wedge-shaped, and slightly notched tail, and pale upper wing coverts. Bill is short. Similar, but much larger than Bulwer's Petrel and generally found in different oceans. Distinguished from dark morph of Wedge-tailed Shearwater by smaller

size, shorter and thicker bill, held downward at an angle of 45° and fast, dainty flight.
Iris—brown; bill—black; feet—pink.
Range: Breeds on islands in Arabian Sea and Gulf of Aden.
Distribution and status: Ranges into Indian Ocean and once recorded off Enggano, SW Sumatra.
Habits: Flight is fast and agile, dipping and rising with the swell, often with bill down-turned to view sea.

8. STREAKED SHEARWATER *Calonectris leucomelas* PLATE 1
(White-faced/White-fronted/Streak-headed Shearwater)

Description: Large (48 cm) shearwater, dark brown above, white face and underparts, with dark streaks on head and sides. Distinguished from the pale form of the Wedge-tailed Shearwater by white face and bill colour.
Iris—brown; bill—horn; feet—pinkish.
Range: Breeds on islands in NW Pacific and winters south to the Equator.
Distribution and status: Not uncommon in Indonesian waters; more common in the east but at least sporadically common west to Sumatra and NW Indian Ocean.
Habits: Similar to Wedge-tailed Shearwater.

9. FLESH-FOOTED SHEARWATER *Puffinus carneipes* PLATE 1
(Pale-footed Shearwater)

Description: Largish (43 cm), heavily built, chocolate brown shearwater with long wings and short, rounded tail. Often confused with Wedge-tailed Shearwater but distinguished by thicker, dark-tipped, pale bill and whitish bases to underside of primaries.
Iris—brown; bill—straw with brown tip; feet—yellow to pink.
Range: Breeds on islands off Australia and New Zealand.
Distribution and status: Indian Ocean populations disperse north in winter and there are several records off the west and north coasts of Sumatra.
Habits: Flies low over sea with deliberate beat and long glides; more graceful than Wedge-tailed Shearwater.

10. WEDGE-TAILED SHEARWATER *Puffinus pacificus* PLATE 1

Description: Medium-sized (43 cm), long-winged, wedge-tailed shearwater. Occurs in light and dark colour phases. In dark phase it is dark chocolate all over. In pale phase it is brown above and whitish below, with underwings showing dark borders and undertail coverts.
Bill—dark grey; feet—flesh-coloured.

Range: Breeds on islands in tropical Indian and Pacific Oceans.
Distribution and status: Stray birds can occur anywhere in SE Asian waters. Regularly recorded off Sumatra and Java, once off Borneo.
Habits: Flies low over the sea, banking from side to side with occasional swoops and frequently gliding so low over water that the wing tips touch the surface at the bottom of the beat.

Storm-petrels—FAMILY HYDROBATIDAE

A FAMILY of oceanic birds similar to shearwaters but smaller with a more fluttery flight, and with the nostril tubes joined into a single aperture.

Storm-petrels are the smallest oceanic birds and their weak butterfly-like flight, and habit of hovering and treading the water with their webbed feet makes them easy to distinguish from other seabirds.

Storm-petrels feed on small crustaceans or floating organic debris. They nest in rock crevices and burrows on rocky shores and islands.

Only 3 species are recorded for the Greater Sundas waters but individual species are often difficult to distinguish. A fourth species is expected but has never been confirmed.

11. WILSON'S STORM-PETREL *Oceanites oceanicus*   PLATE 1
(Wilson's Petrel)

Description: Very small (17 cm), black and white petrel. Upperparts sooty brown with narrow, greyish wing bar and a conspicuous white rump bar formed by white uppertail coverts; underparts sooty brown. The feet protrude just beyond the short square tail in flight.
Iris—dark; bill—black; feet—black with yellow webs (difficult to see at sea).
Voice: Twittering and piping calls at breeding sites but only a faint peeping by feeding birds at sea.
Range: Breeds on islands in southern oceans but disperses north over all seas.
Distribution and status: Regularly seen over waters off Sumatra, Java, Bali.
Habits: Flies singly or in small parties, low over sea with short glides interspersed with loose wing-beats, tilting and rolling from side to side. When feeding, it hovers and paddles on the water with its long feet. Often follows ships.

12. SWINHOE'S STORM-PETREL *Oceanodroma monorhis*   PLATE 1

Description: Small (20 cm), dark petrel. Plumage all dark brown with inconspicuous paler grey wing bar, and slightly forked tail.
Iris—dark; bill—black; feet—black.

Range: Breeds on islands off Japan, Korea, and China and migrates westward (presumably) through Indonesian waters to the northern Indian Ocean.
Distribution and status: There are surprisingly few records apart from a substantial passage through the Sunda Straits in September.
Habits: The flight is distinctive and tern-like with much bounding and swooping over the water, never pattering along the surface. Sometimes follows ships.
Note: May be conspecific with Leach's Storm-petrel *O. leucorhoa*.

[13. **MATSUDAIRA'S STORM-PETREL** *Oceanodroma matsudairae*
PLATE 1

Description: Smallish (25 cm), dark, fork-tailed petrel, similar to Swinhoe's Storm-Petrel but larger, and with whitish patch near wing tips caused by white shafts of primaries. Flight is slower.
Iris—dark; bill—black; feet—black.
Distribution and status: Breeds in S Japan but winters to South China Sea, Philippines, and probably N Indian Ocean. Must pass through Greater Sundas but not yet confirmed.
Habits: The flight is heavier than Swinhoe's Storm-Petrel.
Note: May be conspecific with Black Storm-Petrel *O. melania*.]

14. **WHITE-FACED STORM-PETREL** *Pelagodroma marina* PLATE 1
(Frigate Petrel)

Description: Small (20 cm), dainty, black, grey and white petrel. Upperparts greyish with dark brownish-grey crown, white forehead and eyebrow, black ear coverts and flight feathers. Throat, breast, belly, and wing lining white. The tail is short and squarish.
Iris—brown; bill—black; feet—black.
Range: Atlantic, Indian and Pacific Oceans, breeding in southern latitudes but migrating north of equator in northern summer.
Distribution and status: Rarely visits Greater Sundas but one was collected off the west coast of Aceh, Sumatra, in 1930.
Habits: Flight is fast and erratic, hopping over water with dangling legs, darting suddenly from side to side.

TROPICBIRDS—FAMILY PHAETHONTIDAE

A SMALL family comprising three species of elegant white seabirds characterized by a wedge-shaped tail with two elongated central streamers. These birds range far

out to sea, are excellent divers, feeding largely on squids so are often active at night. They swim with cocked tail.
Two species are recorded in the Greater Sundas waters.

15. RED-TAILED TROPICBIRD *Phaethon rubricauda* PLATE 2

Description: Largish (46 cm, excluding tail streamers), white or pinkish seabird. Birds are pink in fresh plumage but bleach quickly to white. Adults distinguished from White-tailed Tropicbird by red bill, less black in plumage, and red tail streamers (very difficult to see from a moving boat). Immature usually has blackish bill and black-barred upperparts.
Iris—dark; bill—red; feet—blue with black webs.
Voice: Ratchet-like *pirr-igh* call in flight, and loud screams at nest.
Range: Tropical and subtropical Indian and Pacific oceans.
Distribution and status: Not known to breed in the Greater Sundas but there are colonies on nearby Cocos Keeling and Christmas Islands in the Indian Ocean, and on Manuk and Gunung Api in the Banda Sea. Recorded off the west coast of Sumatra, in the seas south of Java, and rarely in the Java Sea.
Habits: Keeps well to sea. Flight similar to White-tailed Tropicbird.

16. WHITE-TAILED TROPICBIRD *Phaethon lepturus* PLATE 2
(Yellow-billed Tropicbird)

Description: Medium-sized (39 cm, excluding tail streamers), white or yellowish seabird with trailing, long, white tail streamers. Adult: mainly white with black eyebrow, black wing tips, and black bar on upperwing. Christmas Island race *fulvus*, seen off Sumatra, is apricot-golden colour. Immature: lacks streamers and has black barring on upperparts, but more black on primaries than immature Red-tailed Tropicbird.
Iris—dark; bill—orange or yellow; feet—greyish with black webs.
Voice: Rattling *tetetete* and *tik* call in flight and loud screams at nest.
Range: Tropical and subtropical seas of Atlantic, Indian and Pacific oceans.
Distribution and status: Recorded off west Sumatra (*fulvus*) from Christmas Island, and there are (probable) breeding colonies on the cliffs of the south coast of Java, Bali (Ulu Watu), and Pulau Nusa Penida.
Habits: Flies high over sea with fast wing-beat circling, or twisting and turning sharply to plunge onto food in the sea.

Pelicans—FAMILY PELECANIDAE

A SMALL family comprising eight species of unmistakable, huge waterbirds with very large bills and large distensible gular pouches stretching the full length of the bill. Pelicans are generally found in gregarious parties and cooperate in scoop-net fishing by side sweeping their beaks. They also fish by plunge-diving in flight. The flight, with recurved neck, appears laboured but is powerful, and some species undertake long migrations.

Three species have been recorded as visitors to the Greater Sundas. No breeding has been confirmed but one species may be partly resident in Sumatra.

17. GREAT WHITE PELICAN *Pelecanus onocrotalus*   PLATE 3
(Eastern White/European White/Rosy Pelican)

Description: Very large (157 cm) white pelican. Plumage white except for black primaries and secondaries. There is a small crest on back of head and a tuft of yellow on the breast. Immature birds are brown.
Iris—red; bill—purplish; gular pouch—yellow; naked facial skin—pink; feet—pink.
Voice: Generally silent but can make throaty grunts.
Range: Breeds in Africa, SC Eurasia, S Asia. Migrates south.
Distribution and status: One doubtful record from Sumatra. Vagrant to marshland of NW Java.
Habits: Typical of family with cooperative fishing, flying in formation and plunge-diving. Frequents lakes and large rivers.

18. SPOT-BILLED PELICAN *Pelecanus philippensis*   PLATE 3
(Grey/Philippine Pelican)

Description: Very large (140 cm) grey pelican. Grey plumage and blue-spotted pink bill distinctive. Wings darkish grey. Gular pouch is purple.
Iris—pale brown; naked facial skin—pink; bill—pink; feet—brown.
Voice: Husky, hissing calls only during breeding period.
Range: Breeds in SW India, Sri Lanka, Burma, and SE China; doubtfully SE Asia, Philippines. Migrant south.
Distribution and status: Winters to N Sumatra. Recorded S Sumatra where possibly breeding. Pelicans reported by Spenser St John on Balambangan Island, off Sabah, in the last century are probably this species. One record from Java without precise locality.
Habits: Lives in large flocks. Inhabits sheltered coastal bays, estuaries, lakes, and large rivers.

BOOBIES 53

19. AUSTRALIAN PELICAN *Pelecanus conspicillatus* PLATE 3
(Spectacled Pelican)

Description: Very large (150 cm) black and white pelican. Distinguished from Great White Pelican by more extensive black on secondaries and inner wing coverts, black uppertail coverts, and mostly black tail, also by pinkish gular pouch and lack of bare facial skin.
Iris—brown; bill—pink or yellow; feet—slate blue.
Voice: In breeding season, a high-pitched *peep-pee-pee* or low-pitched *pep-pep-pur*. Otherwise generally silent.
Range: Breeds in Australia. Erupts N to New Guinea, and occasionally to W Indonesia and SW Pacific.
Distribution and status: Probably irregular. In 1978 birds reached Sumatra (probably) and Java. A few subsequent records are unconfirmed.
Habits: Generally in groups but single birds also occur.

Boobies—FAMILY SULIDAE

THIS is a small, worldwide family of oceanic diving birds characterized by large size; long, thin pointed wings; cigar-shaped bodies, and sharp powerful bills. They wander far out to sea in flocks and make the most spectacular vertical plunge-dives on fish shoals.
Four species have been recorded in Greater Sundas waters.

20. RED-FOOTED BOOBY *Sula sula* PLATE 2

Description: Large (72 cm), black and white, or ashy-brown booby with diagnostic red feet and white tail. Light, dark, and intermediate phases occur. Light phase: plumage mostly white with black primaries and secondaries. Dark phase: head, back, and chest ashy brown but tail white. Immature: ashy brown all over.
Iris—brown; bill—greyish with pink base; naked skin at base of bill—blue; naked skin under bill—black; feet—bright red (diagnostic), but juveniles have yellowish grey feet in all phases.
Voice: Silent at sea. Calls only when nesting.
Range: Tropical oceans. Nearest breeding sites are Christmas Island in Indian Ocean, Bankoran and Tubbataha Reef in Sulu Sea and Manuk and Gunung Api in Banda Sea.
Distribution and status: Recorded off coasts of W Sumatra, N Borneo and Java.
Habits: As Masked Booby.

21. MASKED BOOBY *Sula dactylatra* PLATE 2
(Blue-faced/White Booby)

Description: Very large (85 cm) black and white booby. General colour white with black tail, primaries, and secondaries, and compact black face mask. Facial skin blue-black. Immature: brown head and upperparts with white collar, underparts white with pattern of black bands on underwing.
Iris—yellow; bill—yellow or greenish; feet—yellow or greenish.
Voice: Silent at sea. Calls only when breeding.
Range: Tropical and subtropical oceans. Nearest breeding areas on Cocos Keeling in Indian Ocean, Tubbataha Reef in Sulu Sea and Gunung Api in Banda Sea.
Distribution and status: Probably the least common booby in Sundaic waters. Recorded in Sumatra (Malacca and Sunda Straits), N Borneo, and off S Java.
Habits: Flies with direct powerful beats, low over sea, in ragged formation. Forms wheeling clouds above prey when located. Uses ships as resting places.

22. ABBOTT'S BOOBY *Papasula abbotti* PLATE 2

Description: Large (71 cm) black and white booby. Wings and tail black above, but underwing white except for black tips. Naked skin around eye blue-black; rest of plumage white with some black spots on rump and flanks. Female has pink bill. Immature: like adult but with grey bill.
Iris—brown; bill—male: bluish grey, female: pink, juv.: grey (all with black tip); feet—grey.
Voice: Silent at sea.
Range: Breeds on Christmas Island (and formerly on other islands in Indian Ocean). Forages widely in Indian Ocean.
Distribution and status: Ranges regularly N to upwelling off SW Sumatra and SW Java; recorded once in Sunda Straits.
Habits: As other boobies.

23. BROWN BOOBY *Sula leucogaster* PLATE 2
(White-bellied Booby)

Description: Large (72 cm) dark brown and white booby with dark head and tail. Adult is dark sooty brown with a white belly. Immature birds are light ashy brown instead of white. Naked skin of face is red-yellow in female or bluish in male.
Iris—grey; bill—yellow in adult, grey in juv.; feet—yellowish green.
Voice: Silent at sea. Calls only when nesting; crows, quacks, and hisses.
Range: Tropical and subtropical oceans. Nearest known breeding colonies are on

Christmas Island in Indian Ocean, Bankoran in Sulu Sea, and Gunung Api in Banda Sea.

Distribution and status: The commonest booby in the Greater Sundas and the only species regularly recorded inshore. Probably breeds on islands in Malacca Straits. Regularly reported off all coasts of Java, particularly in the Sunda Straits where breeding is also possible. Occasional records off Sumatran coasts.

Habits: As Masked Booby but a more coastal bird.

Cormorants—FAMILY PHALACROCORACIDAE

A MEDIUM-SIZED, worldwide family of large, fish-eating birds with long sharp bills hooked at the tip. Cormorants generally chase their prey by swimming for long periods under water. This is made easier for them by the absence of waterproofing oils on the feathers so that they become quickly waterlogged and have a low buoyancy. As a consequence, they have to spend long periods after fishing with their wings spread out, drying in the sun.

Two breeding species and two visitors occur in the Greater Sundas.

24. LITTLE BLACK CORMORANT *Phalacrocorax sulcirostris*
(Black Cormorant)　　　　　　　　　　　　　　　　　　　　　　　PLATE 3

Description: Medium-sized (61 cm) black cormorant. Entire plumage black with green or purple gloss. In breeding season, there is a white tuft of plumes on the side of the head behind the eye. Wing coverts greyer, edged with black, giving scaly appearance. Immature: duller and mottled brownish. Bare skin of face and gular pouch blue-grey.

Iris—green; bill—greyish; feet—black.

Voice: Harsh, guttural, creaking calls at nest.

Range: Australia, New Guinea, and Indonesia west to Greater Sundas.

Distribution and status: Recorded in S Sumatra, where probably a visitor from Java. Known from S Borneo by four specimens collected last century. Breeds on Java (first published record 1947, first published breeding record 1954) and now the commonest cormorant at Pulau Rambut in W Java.

Habits: Frequents ponds, lakes, and estuaries, and occasionally seacoasts and brackish fish ponds; generally in small groups but solitary individuals also seen.

56 CORMORANTS

25. GREAT CORMORANT *Phalacrocorax carbo* PLATE 3
(Common/Big Black/Large Black Cormorant)

Description: Large (90 cm) blackish cormorant with heavy bill and whitish cheeks and throat. In breeding season, displays white silky plumes on neck and head, and white patches on flanks. Immature is dark brown with dirty whitish underparts.
Iris—blue; bill—black; feet—black.
Voice: Guttural groans when breeding; otherwise generally silent.
Range: Cosmopolitan. In E Asia, winters south to Malay Peninsula.
Distribution and status: Formerly a breeding resident in Sumatra (Tawar and Singarak and probably other large lakes) but is now locally extinct. Vagrants might reach N Sumatra but there are no recent records. Recorded on N Bornean coasts (and possibly resident). Not recorded in Java.
Habits: Chases fish under water. Swims half-submerged like other cormorants and frequently stands on rock or branch perch drying outstretched wings. Flies in V formation or lines.

26. LITTLE PIED CORMORANT *Phalacrocorax melanoleucus* PLATE 3
(White-throated Cormorant)

Description: Medium-sized (60 cm) black and white cormorant. Distinguished from other cormorants by entirely white underparts. Yellow bill and facial skin diagnostic. Immature birds have a black eye-stripe and crown and a black flank patch.
Iris—blue-green; bill—yellow with black line on top; feet-black.
Voice: Cooing at nest but silent when not breeding.
Range: Resident in New Zealand, Australia, New Guinea, and Eastern Indonesia.
Distribution and status: Vagrant E Java and Bali, usually in coastal areas.
Habits: Frequents ponds, ditches, estuaries, lagoons, and seacoasts. Habits as other cormorants.

27. LITTLE CORMORANT *Phalacrocorax niger* PLATE 3
(Javan Cormorant)

Description: Smallish (56 cm) black cormorant. Breeding plumage blackish green with a few tiny white plumes on sides of head, over eye and on sides of neck. In non-breeding plumage, loses plumes but has whitish chin and sometimes throat. Distinguished from Little Black Cormorant by smaller size, uniform wing coverts, and compressed bill. Immatures are whiter on breast and browner on upperparts.
Iris—blue-green; bill—brown with black tip and purplish base; feet—black.

Voice: Long drawn calls *keh-eh-eh-eh-eh-e* at breeding site.
Range: India, SW China, SE Asia, and Greater Sundas.
Distribution and status: Sight records from Sumatra probably visitors from Java. Known from S Borneo by specimens collected last century. In Java it is (or was formerly) a fairly common bird of coastal and lowland areas. The relative abundances of this species and the Little Black Cormorant are uncertain.
Habits: Inhabits mangroves, lakes, flooded marshes, and river estuaries. Generally in small flocks, swimming about with only head exposed and diving repeatedly for fish.
Note: Some authors place this species together with Pygmy Cormorant *P. pygmaeus* of Eurasia.

Anhingas—FAMILY ANHINGIDAE

A small family of only four species of cormorant-like birds, one in the Neotropics, one in Africa, one in Asia, and one in Australasia.

Anhingas chase fish underwater and can remain submerged for long periods. The neck is very long and snake-like. Unlike cormorants, anhingas have straight, dagger-shaped bills, but like cormorants, their feathers become waterlogged and they spend a lot of time standing in the sun drying their outstretched wings .

One species is resident in the Greater Sundas.

28. ORIENTAL DARTER *Anhinga melanogaster* PLATE 3
(Snakebird)

Description: Unmistakable, large (84 cm), cormorant-like waterbird with very long, slender neck and small narrow head. Head and neck brown with white chin stripe extending down side of neck. Rest of plumage blackish with white plume-like covert feathers, with black edges.
Iris—brown; bill—yellowish brown, black culmen ridge; feet—grey.
Voice: Rattling and clicking calls. Screams during courtship.
Range: India, SE Asia, Philippines, Sulawesi, and Sundas.
Distribution and status: Recorded in S Sumatra as a visitor from Java or probably resident. In Borneo it remains a common resident on rivers but has apparently disappeared from most coastal areas. Once a common bird on Java (where recorded to 1400 m), it is now local, and there are few recent records from Bali.
Habits: Lives in large stretches of clean fresh water in lakes and big rivers where it is an amazing diver, spending long periods under water. It can reduce buoyancy so that only the head comes out of the water but, thus waterlogged, has difficulty

58 ANHINGAS

running and flapping over the water to get airborne. Spends many hours sitting on an exposed perch with wings held out to dry; roosts communally in open trees.

Frigatebirds—FAMILY FREGATIDAE

A SMALL family comprising five species of large, tropical, oceanic birds characterized by gliding flight and unique silhouette with bow-shaped, long, pointed wings and long, forked tail (often closed and pointed). These birds are superb gliders, soaring and spiralling effortlessly on thermals or circling and diving over fish shoals. They frequently harry other seabirds to piratize disgorged food.

Three species occur in Greater Sundas waters.

29. CHRISTMAS FRIGATEBIRD *Fregata andrewsi* PLATE 4
(Christmas Island Frigatebird)

Description: Large (95 cm) dark frigatebird. Male: glossy green-black with red gular pouch and diagnostic white belly. Female: breast and belly white with white 'spur' extending onto underwing, and white collar; pink eye-ring. Juvenile: browner, head pale rusty brown, broad darkish band across breast.
Iris—dark brown; bill—black (male) or pinkish (female and juvenile); feet—purplish grey with flesh-coloured soles.
Voice: Silent at sea, yodelling and clappering calls only at nest.
Range: Breeds on Christmas Island in the Indian Ocean. Recorded north to Malay Peninsula.
Distribution and status: Regularly recorded on Sumatran and Bornean coasts. Not uncommon in Java Sea and rather common on S coast of Java, particularly SW Java.
Habits: Pelagic, soaring high over sea on thermals or spiralling over fish. Snatches food from surface without landing and harries other seabirds for food. Birds roost or rest on fish platforms and trees on small islands.

30. GREAT FRIGATEBIRD *Fregata minor* PLATE 4
(Greater Frigatebird)

Description: Large (95 cm) dark frigatebird. Male: distinct all black plumage except for pale bar across upperwing coverts and crimson gular pouch. Female: distinct with greyish white chin and throat; white upper breast; little or no white on base of underwing; pinkish red eye-ring. Immature: dark brown above with whitish head, neck, and underparts stained rusty, and distinguished from Lesser

Frigatebird only by larger size, posteriorly convex white belly patch, and less white at base of underwing.
Iris—brown; bill—slate blue in male, pinkish in female; feet—reddish in adult or blue in juvenile.
Voice: Braying, clappering, and rattling calls recorded from nest areas but silent at sea.
Range: Tropical oceans. Nearest confirmed breeding colonies are on Christmas Island in the Indian Ocean and Manuk and Gunung Api in Banda Sea.
Distribution and status: Throughout the Greater Sundas, but less common than Lesser Frigatebird.
Habits: Pelagic habits similar to Christmas Island Frigatebird, but regular visitor along the coastline.

31. LESSER FRIGATEBIRD *Fregata ariel* PLATE 4
(Least Frigatebird)

Description: Large (76 cm) dark frigatebird. Male: all blackish with white patches on flanks and under base of each wing, and red gular pouch. Female: black with brownish head, white breast and concave belly patch, with some white on base of underwing; pink or bluish grey eye-ring; black chin. Immature: brownish black upperparts but head, neck, breast, and flanks whitish and rufous-stained. Distinguished from immature Great Frigatebird by smaller size, concave shape to white underparts, and more white on base of underwing.
Iris—brown: bill—grey; feet—red-black.
Voice: Clappering calls when breeding.
Range: Tropical oceans. Nearest confirmed breeding colonies are on Cocos Keeling and Christmas Island in Indian Ocean and islands of N Australia. Breeding is suspected in Indonesian region but not proved.
Distribution and status: Throughout the Greater Sundas; sometimes occurs in large numbers suggesting migratory movement, possibly from the W Pacific.
Habits: Pelagic; birds soar high in the air on air currents or wheel in circles over surface-feeding fish shoals. Sometimes fly fast and low over the water with heavy deliberate strokes. Snatch food from surface without landing or harry other seabirds' nest colonies. Birds sometimes roost or rest on bamboo fish platforms or trees on small islands.

HERONS—FAMILY ARDEIDAE

A LARGE, worldwide family of long-legged wading birds. Herons have long necks and long straight, spear-like bills used for striking at fish, small vertebrates, and invertebrate prey. Several species exhibit long, fine, erectile plumes during the breeding season. Nests are generally large twig platforms built in trees. Most of the 22 species occurring in the Greater Sundas are fairly distinctive, but care must be taken to separate the white egrets.

32. GREAT-BILLED HERON *Ardea sumatrana* PLATE 5
(Dusky Grey/Giant Heron)

Description: Very large (115 cm) dark grey heron. Plumage grizzled brownish grey all over; has a slight crest.
Iris—yellow; bill—blackish; feet—grey.
Voice: A harsh croak or angry repeated roar.
Range: Coastal SE Asia, Philippines, and Indonesia to Australia.
Distribution and status: Uncommon but widely distributed around the coasts of the Greater Sundas. Inhabits mangrove and associated mud-flats, locally coastal reefs particularly on small islands, and rarely large rivers (Borneo).
Habits: Inhabits beaches, coral reefs, and mangroves. Generally seen stalking singly along the seashore, hunting coral reef fish, or at the edge of a mangrove creek.

33. GREY HERON *Ardea cinerea* PLATE 5
(The Heron)

Description: Large (92 cm), white, grey, and black heron. Adult: black eye-stripe and crest; black flight feathers, shoulder, and two breast bars; head, neck, breast, and back white with some black streaks down the throat; otherwise grey. Young birds are greyer on head and neck, and lack black on head.
Iris—yellow; bill—greenish yellow; feet—blackish.
Voice: Deep guttural croaks *kroak* and goose-like honk.
Range: Africa, Eurasia to Philippines, and Sundas.
Distribution and status: Throughout the Greater Sundas in wetland habitats, generally near the sea but also found on inland lakes, up to 900 m. Possibly only a visitor to Borneo.
Habits: Solitary hunter of shallow water, stalking fish with poised head and beak, or standing on one leg waiting for fish to pass. Ponderous wing-beat in flight. Roosts in trees.

34. PURPLE HERON *Ardea purpurea* PLATE 5

Description: Large (80 cm), grey, chestnut, and black heron. Black cap with droopy crest, and black stripe down side of rufous neck diagnostic. Back and wing coverts grey; flight feathers black; rest of plumage reddish brown.

Iris—yellow; bill—brown; feet—reddish brown.
Voice: Harsh croaks.
Range: Africa, Eurasia to Philippines, Sulawesi, and Sundas.
Distribution and status: Throughout the Greater Sundas in wetlands, particularly freshwater habitats, in lowlands, and occasionally hills to 1500 m.
Habits: Frequents mangroves, paddy fields, lakes, and streams, and is less restricted to coastal areas than the Grey Heron. Solitary birds creep through shallow, weedy water with head cocked low and sideways to strike at fish and other food. Flies with a slow heavy beat. Nests in large colonies.

35. WHITE-FACED HERON *Ardea novaehollandiae* PLATE 5

Description: Large (68 cm), greyish heron. Distinguished from Pacific Reef-Egret and Purple Heron by white face and black flight feathers. Upperparts grey, underparts pinkish grey.
Iris—yellow; bill—dark grey; feet—greenish yellow.
Voice: Harsh, gravelly croaks.
Range: Australasian region and eastern Indonesia; resident population in Lesser Sundas probably augmented by non-breeding birds.
Distribution and status: Irregular visitor; recorded on the coast of Bali and Nusa Penida.
Habits: Hunts fish in shallow waters, singly or in small parties.
Note: Sometimes placed in genus *Egretta*.

36. STRIATED HERON *Butorides striatus* PLATE 6
(Little/Green-backed/Mangrove Heron)

Description: Small (45 cm), dark grey heron. Adult: glossy, greenish black crown and long floppy crest with a black line running from base of bill, under eye, and across cheek. Wings and tail slaty blue, with green gloss and buff edge. Belly pinkish grey, chin white. Female slightly smaller than male. Young birds are streaky brown with white spots.
Iris—yellow; bill—black; feet—greenish.
Voice: Alarm call is a loud explosive *kweuk*, also rattling *kee-kee-kee-kee*.
Range: Cosmopolitan.
Distribution and status: Resident throughout the Greater Sundas in coastal mangroves, estuaries, reefs, and dense vegetation along rivers and lakes; the population in N Sumatra and N Borneo augmented in winter by northern migrants.
Habits: A solitary, shy bird staying in or close to thick cover of reeds, bushes, or mangroves. Sometimes comes out onto the rocks and exposed coral reef at low tide. Nests in small colonies.

37. CHINESE POND-HERON *Ardeola bacchus* PLATE 6

Description: Smallish (45 cm), white-winged, streaky brown heron. Similar to Javan Pond-Heron, but in breeding season the head and neck are dark chestnut and the breast is maroon. Winter birds are indistinguishable from Javan Pond-Heron in field, looking streaky brown when standing, or white with a dark brown back in flight.
Iris—brown; bill—yellow (in winter); legs and feet—greenish grey.
Voice: Normally silent; low deep croaks in disputes.
Range: Bangladesh to China and SE Asia. Winters in Malay Peninsula, Indochina, and Greater Sundas.
Distribution and status: Uncommon winter visitor to N Sumatra (including W Sumatran Islands and Lingga Archipelago) and N Borneo.
Habits: As Javan Pond-Heron. It is an occasional visitor to paddy fields and fishponds.

38. JAVAN POND-HERON *Ardeola speciosa* PLATE 6
(Javanese Pond-Heron)

Description: Small (45 cm), white-winged, streaky brown heron. In breeding season, head and breast buffish, back nearly black, rest of upperparts streaky brown. Underparts white. In flight the white wings contrast strongly with the dark back. Non-breeding adults and juveniles look like non-breeding Chinese Pond-Heron.
Iris—yellow; bill—yellow with black tip; feet—dull green.
Voice: Creaky *krak* when disturbed.
Range: Malay Peninsula, Indochina, Sulawesi, and Sundas.
Distribution and status: Recorded in S Sumatra as a non-breeding visitor from Java. Breeds in SE Borneo, but a rare visitor in the north. On Java and Bali still a common bird of freshwater marshes.
Habits: Lives in paddy fields or other flooded areas, singly or in small dispersed flock. Stands motionless with body poised low and head retracted, waiting for prey. Flies in twos and threes to communal roosts each evening with a slow, short-winged beat. Nests in colonies with other waterbirds.

39. CATTLE EGRET *Bubulcus ibis* PLATE 5
(Puff-backed Heron)

Description: Smallish (50 cm) white heron. Breeding: white with head, neck, and breast washed orange; iris, bill, legs, and lores briefly bright red. Non-breeding: white except for an orange wash on forehead of some birds. Distinguished from

other egrets by stockier shape with shorter neck, rounder head, and thicker, shorter bill.
Iris—yellow; bill—yellow; feet—black.
Voice: Silent apart from croaks at nesting colonies.
Range: Cosmopolitan.
Distribution and status: Throughout the Greater Sundas. Resident in Sumatra and Java (probably also winter visitor); common winter visitor to Borneo but not known to breed. A common bird of freshwater swamps and grassland.
Habits: Associates with grazing cattle, water buffalos, and wild banteng, catching the flies attracted or disturbed by these animals as they walk through the grass. Small flocks fly in formation low over water courses each evening to communal roosting sites. Nests in colonies over water.

40. PACIFIC REEF-EGRET *Egretta sacra* PLATE 5
(Eastern Reef-Egret/Reef heron)

Description: Largish (58 cm) white or charcoal grey heron. Dimorphic: the commoner grey form is distinguished by uniform grey plumage with a short crest and whitish chin, often invisible in field. White form is distinguished from Cattle Egret by larger size and narrow head and neck; from other egrets by relatively shorter greenish legs and pale bill.
Iris—yellow; bill—pale yellow; feet—green.
Voice: A hoarse grunted croak when feeding, and harsher *arrk* when alarmed.
Range: Coasts of E Asia, W Pacific, and Indonesia to New Guinea, Australia, and New Zealand.
Distribution and status: Throughout the Greater Sundas. Common on reefs; particularly reefs and sand beaches on offshore islands.
Habits: Almost always encountered on the shoreline, resting on rocks or cliff-sides, or hunting at water's edge, either poised or actively chasing small fish in shallows. Rarely encountered up rivers on sand bars. Nests on rock stacks.

41. CHINESE EGRET *Egretta eulophotes* PLATE 5
(Swinhoe's Egret)

Description: Medium-sized (68 cm) white egret with greenish legs and black bill. Base of lower mandible yellow. In winter, distinguished from Little Egret by larger size and leg colour, and from pale form of Pacific Reef-Egret by longer legs and duskier bill. Birds coming into breeding plumage have yellow bill and black legs.
Iris—yellow-brown; bill—black with yellow base below; feet—yellow-green to blue-green.
Voice: Generally silent. Low croaks when disturbed.

Range: Breeds on islands off W coast of North Korea and islands off Shanghai, China. Apparently winters mainly in Philippines.
Distribution and status: One sight record from Sumatra (and one specimen collected in Mentawai but now lost). Specimens collected earlier this century in N Borneo; still regularly reported from Borneo including S Borneo (Mahakam delta). One sight record from Java (Pangandaran). Sight records should be treated with some caution, except when birds in breeding plumage.
Habits: Like Little Egret, actively chasing prey through shallow water.

42. GREAT EGRET *Egretta alba* PLATE 5
(Great White/Large Egret/Great White Heron)

Description: Large (95 cm) white heron. Much larger than other white egrets, with heavier bill and characteristic kink in neck. In breeding season the bare facial skin is blue-green; bill—black; bare thighs red and feet black. In non-breeding plumage the bare facial skin is yellowish; bill yellow, usually with dark tip; feet and legs black.
Iris—yellow.
Voice: A low croaked *kraa-a* given in alarm.
Range: Cosmopolitan.
Distribution and status: Throughout the Greater Sundas. In Sumatra and Borneo, possibly a non-breeding visitor; but may be resident on both islands. Known to breed on Java. Uncommon in coastal swamps, mangroves, and mudflats.
Habits: Generally singly or in small parties in mangroves, along mud and sand flats, or in paddy fields and lagoons. The birds stand rather upright, stabbing down on prey from above. In courtship, pair dances and chases each other about in an elegant manner. Flies with a graceful and powerful slow stroke.

43. INTERMEDIATE EGRET *Egretta intermedia* PLATE 5
(Plumed/Lesser/Yellow-billed/Smaller Egret)

Description: Large (69 cm) white heron, intermediate in size between Little and Great Egrets, and distinguished by rather short bill and S-shaped, unkinked neck. In breeding plumage it has long fluffy plumes on back and breast; bill and thighs briefly pink and facial skin grey.
Iris—yellow; bill—yellow often tipped with brown; legs and feet—black.
Voice: Fairly silent, but a deep rasping *kroa-kr* on take-off when disturbed.
Range: Africa, India, E Asia to Australasia.
Distribution and status: Throughout the Greater Sundas: common resident on Sumatra (including Nias and Belitung) and Java, in lowlands up to 1000 m, non-breeding visitor (some year round) to Borneo.

Habits: Occasionally solitary but more usually found in small flocks in paddy fields, lake sides, swampy areas, mangroves, and mud-flats. Flocks disperse to feed but congregate again when disturbed or when flying to and from feeding areas. Nests in colonies with other waterbirds. Sometimes feeds in association with cattle like a Cattle Egret.

44. LITTLE EGRET *Egretta garzetta* PLATE 5

Description: Medium-sized (60 cm) white heron. Distinguished from Cattle Egret by larger size, slimmer build, black bill, black legs with or without yellow toes, and in breeding season pure white, long tapering feathers on nape, and floppy plumes on back and breast.

Iris—yellow; facial skin—greenish yellow but pinkish in breeding season; bill—always black; legs and feet—black (with yellow toes in migrant race).

Voice: Silent apart from croaking calls at breeding colonies.
Range: Africa, Europe, Asia, and Australasia.
Distribution and status: A non-breeding visitor to Sumatra and Borneo; most commonly the yellow-toed subspecies from Asia, but resident black-footed *nigripes* of Java reaches both islands. It is not uncommon in coastal areas (and up to 900 m around Lake Toba in Sumatra).
Habits: Frequents paddy fields, riverbanks, sand and mud-bars, and small coastal streams. Feeds in scattered flocks, often mixed with other species. Sometimes dashes after prey across coastal shallows. Birds fly in V-shaped formation when returning to night roosts. Nests in colonies with other waterbirds.

45. BLACK-CROWNED NIGHT-HERON *Nycticorax nycticorax*
(Common Night-Heron) PLATE 6

Description: Medium-sized (61 cm), large-headed, stocky, black and white heron. Adult: crown black; neck and breast white; two long plumes from nape white; back black; wings and tail grey. Female smaller than male. During breeding season, legs and lores become red. Immature: streaked and spotted brown, and can only be distinguished from young of Rufous Night-Heron in the hand.

Iris—yellow in immature, bright red in adult; bill—black; feet—dirty yellow.

Voice: Deep throaty croak *wok* or *kowak-kowak* uttered in flight, and hoarse croaks when disturbed.
Range: Cosmopolitan.
Distribution and status: A non-breeding visitor to Sumatra and N Borneo; resident in Borneo and Java. In Javan colonies breeds alongside Rufous Night-Heron.
Habits: Rests by day in tree roosts or colonies. At dusk the birds circle the roost

and fly out to their feeding areas giving deep croak calls. Feeds at night in paddy fields, pastures, and along waterways. Nests in noisy colonies in trees over water. Flight resembles *Kalong* fruit-bat.

46. RUFOUS NIGHT-HERON *Nycticorax caledonicus* PLATE 6
(Nankeen Night-Heron)

Description: Medium-sized (59 cm), large-headed, rufous-brown heron. Adult: crown black with two long white streamers from nape; upperparts dark chestnut; underparts rufous buff. Immature: brown, heavily streaked and spotted, with pinkish wash on tail and wings.
Iris—yellow; bill—black above, yellowish below; feet—greenish.
Voice: Loud deep *kyok* when leaving roost in evening.
Range: Philippines, E Indonesia, and Australasia.
Distribution and status: In N Borneo a straggler from the Philippines (occasional breeding Sabah and Brunei). In Java bred formerly Pulau Dua, but now known only from E Java, in Brantas delta. It is not recorded for Bali but could occur there.
Habits: By day birds hide in roost or colony in leafy trees. In the evening they fly out to feed along streams, paddy fields, grazing areas, and pastures. The flight is slow with short, rounded wings. Sometimes roosts and nests with Black-crowned Night-Heron and interbreeding is recorded.

47. MALAYAN NIGHT-HERON *Gorsachius melanolophus* PLATE 6
(Tiger Bittern/Malay Night-Heron)

Description: Medium-sized (49 cm), stoutly built, dark reddish brown and black heron with diagnostic stubby bill with downcurved culmen. Adult: crown and short crest black: upperparts chestnut brown, finely speckled black; underparts rufous buff, streaked black and white; chin white with central row of black streaks. In flight black flight feathers and white tips to wings distinguish it from Cinnamon Bittern. Immature: dark brown upperparts, spotted with white and barred buff; underparts whitish, spotted, and barred brown. Similar to immature of Black-crowned Night-Heron but with a stubbier bill.
Iris—yellow; naked eye-ring—olive; bill—olive; feet—olive.
Voice: The call is a series of deep *oo* notes, at about 1.5 second intervals, at dawn and dusk from high canopy; also hoarse croaks and a rasping *arh, arh, arh.*
Range: India, S China, SE Asia, and Philippines. Winters S to Greater Sundas.
Distribution and status: An uncommon winter visitor to inland swamps and swamp forests in Sumatra (including islands), N Borneo, and Java (rarely).
Habits: This is a solitary, shy, retiring, nocturnal bird which hides up in reeds, bamboo, or other dense vegetation by day and feeds at night in open areas rather

than waterways. In defensive display it raises crest, opens wings, and stabs repeatedly with sharp beak. Feeds along forest trails by day, flies up into nearby trees when disturbed.

48. JAPANESE NIGHT-HERON *Goraschius goisagi*   PLATE 6
(Japanese Bittern)

Description: Smallish (49 cm), squat, brown heron. Similar to Malayan Night-Heron but differs in having smaller bill and crown; nape slaty brown to chestnut, not black; wing tips not white. Has a characteristic black and white shoulder patch. Upperparts dark brown with paler, fine steaking; underparts buff with central line of dark brown streaks. Grey flight feathers contrast with brown coverts in flight.
Iris—yellow; bare skin of face—yellow; bill—horn; feet—dark green.
Voice: Silent in winter.
Range: Breeds in Japan. Winters in the Philippines.
Distribution and status Recorded in Brunei in 1988.
Habits: Favours woody areas but feeds in open grassy areas in early morning and evening. Rather tame.

49. SCHRENCK'S BITTERN *Ixobrychus eurhythmus*   PLATE 6
(Von Schrenck's Bittern)

Description: Small (34 cm) dark brown heron. Male: crown black; upperparts purplish chestnut, underparts streaky buff with a line of dark streaks down the throat and breast. Female and immature browner with white, black, and brown flecks above, and streaks below. In flight the grey underwing is characteristic.
Iris—yellow; bill—greenish yellow; feet—green.
Voice: Low squawks in flight.
Range: Breeds in SE Siberia, E China, Korea, and Japan. Winters south to SE Asia, Philippines, and Indonesia.
Distribution and status: Recorded throughout Greater Sundas but scarce, particularly south to Java (not recorded Bali).
Habits: A solitary, secretive bird inhabiting reed beds, rice paddy, and grassy marshes. Freezes when disturbed, with beak held vertically upward.

50. YELLOW BITTERN *Ixobrychus sinensis*   PLATE 6
(Little Yellow/Chinese Little Bittern)

Description: Small (38 cm) buff and black heron. Adult: black cap, light fulvous-brown upperparts, and buff underparts. Black flight feathers contrast strongly

with buff wing coverts. Immature is as adult, but browner and heavily streaked all over with black wings and tail.
Iris—yellow; bare eye-ring—greenish yellow; bill—greenish brown; feet—greenish yellow.
Voice: Generally silent. Slight screeching and soft, staccato *kakak kakak* in flight.
Range: India, E Asia to Philippines, Micronesia, and Sumatra. In winter to Indonesia and New Guinea.
Distribution and status: Resident (and probably winter visitor) in Sumatra where it is uncommon; up to 1200 m. In Borneo, Java, and Bali it is a locally common winter visitor.
Habits: Favours *Pandanus* and reed thickets along rivers and canals in swampy areas, also rice paddy. An active hunter, clambering about among reeds and over obstacles, flicking its tail, erecting its crest, and flapping one wing at a time. When alarmed it freezes with bill vertical and eyes peering forwards strangely, watching the source of its alarm.

51. CINNAMON BITTERN *Ixobrychus cinnamomeus*   PLATE 6
(Chestnut Bittern)

Description: Small (41 cm) cinnamon-orange heron. Adult male: upperparts chestnut; underparts buffy orange with a central stripe of black streaks, and black streaks on flanks; whitish streaks on side of neck. Female: duller and browner, with a black cap. Immature: streaky below, and barred and spotted above, with a black cap.
Iris—yellow; cere—orange; bill—yellow; feet—green.
Voice: A croak when disturbed into flight, and a low courtship call *kokokokoko* or *geg-geg*.
Range: India, China, SE Asia, Sulawesi, and Sundas.
Distribution and status: Common resident of freshwater swamp and rice paddy throughout the Greater Sundas.
Habits: A shy, solitary bird which stalks prey in the cover of rice or grass by day, more active by night. When disturbed it jumps up and flies low with a slow, powerful rhythm. Nests in reeds or long grass.

52. BLACK BITTERN *Dupetor flavicollis*   PLATE 6
(Mangrove Bittern)

Description: Medium-sized (58 cm) blackish heron. Adult male: general colour slaty grey (looks black in the field) with yellow side of neck, and black- and yellow-streaked throat. Female: browner and more whitish below. Immature: black crown and buffy rufous feather tips of back and wings give brown scaly appearance. Long dagger-like bill separates this species from others of similar colour.
Iris—red or brown; bill—yellow-brown; feet—black-brown, variable.

Voice: Loud, hoarse croak in flight, and a deep booming sound during the breeding season.
Range: India, S China, SE Asia, Philippines, and Indonesia to Australasia.
Distribution and status: Status uncertain. A scarce resident in Sumatra (including Mentawai and Belitung) and uncommon winter visitor. In Borneo, Java, and Bali probably a non-breeding visitor from SE Asia, common in N Borneo but scarce elsewhere.
Habits: A shy bird favouring forest and dense tangled swamps by day, but climbing into higher vegetation and flying to other feeding areas at night. Nests in bushes above water in dense swamps.

53. GREAT BITTERN *Botaurus stellaris*
(Common/The Bittern)

Description: Large (76 cm), golden brown and black heron with black cap, and white chin and throat bordered by conspicuous black malar stripe. Sides of head golden but rest of plumage barred and mottled with black. In flight brown-barred flight feathers contrast with gold of coverts and back.
Iris—yellow; bill—yellow; feet—greenish yellow.
Voice: The famous booming call is heard only in breeding season. Wintering birds are silent.
Range: Africa, Eurasia. Migrant in SE Asia, rarely south to Malay Peninsula and Philippines.
Distribution and status: Recorded in Sabah in 1966 and Brunei in 1987. Reports from Way Kambas, S Sumatra, are unreliable.
Habits: Secretive bird of tall reeds. Freezes when alarmed and holds bill vertically. Sometimes flushed and seen flying low over reeds.

Great Bittern, *Botaurus stellaris* (53)

70 STORKS

Storks—FAMILY CICONIIDAE

A SMALL, worldwide family of very large birds with long powerful beaks. They have long legs, broad wings, and short tails. They feed mostly on fish or small animals, which they catch while stalking quietly in open wet areas.

Storks are strong fliers and several species migrate over large distances. They are experts at soaring on thermals and often circle high in the sky, gaining height for easy travel, or searching for likely feeding places.

There are four resident species in the Greater Sundas and three doubtful vagrants.

54. MILKY STORK *Mycteria cinerea* PLATE 7
(Milky Wood-Stork)

Description: Very large (92 cm) white stork with black wings. Plumage white except for black flight feathers; naked skin of face pink to red. Immature birds are greyish brown with a white rump and black flight feathers.
Iris—brown; bill—long, decurved, yellowish; feet—grey.
Voice: Generally silent except for croaking of young birds and clatter of mandibles.
Range: Indochina, Malay Peninsula, Sulawesi, Sumatra, Java, and Sumbawa.
Distribution and status: Breeding colonies are known from Riau and South Sumatra provinces on the east coast of Sumatra, and Pulau Rambut in W Java. Recorded in some numbers in apparently suitable areas in SC and E Java, but breeding is not confirmed. Forages widely and recently added to the Bali list.
Habits: Frequents muddy, flooded areas including swamps, coastal mud-flats, and paddy fields. Generally single or in small parties, usually near the coast but up to 900 m in Sumatra. Associates with herons and Adjutant Storks; sometimes soars in circles high in the air. When feeding the clapping of its mandibles is audible at a distance. Nests in mixed colonies with other waterbirds.

[55. ASIAN OPENBILL *Anastomus oscitans* PLATE 7
(Open-billed Stork/Asiatic/Oriental Openbill)

Description: Largish (81 cm) grey stork with diagnostic gap between mandibles of closed bill. General plumage in winter smoky grey with black wings and tail.
Iris—whitish brown; bill—greenish, creamy grey; legs and feet—pinkish.
Voice: Generally silent; occasional deep moans and bill clattering.
Range: India and SE Asia.

Distribution and status: Sight records of flocks along the coast of northern Sumatra in 1977 and 1979 reported, but not accepted as conclusive.
Habits: Stands in marshy coastal grounds or mud-flats searching for molluscs.]

56. WOOLLY-NECKED STORK *Ciconia episcopus* PLATE 7
(White-necked Stork)

Description: Very large (86 cm) black and white stork. Crown glossy black with white forehead and narrow eyebrow; entire neck fluffy white; wings and tail glossy black; breast band and black thighs; lower belly and undertail white; facial skin grey.
Iris—red-brown; bill—black with red tip (redder in Java); feet—dull red.
Voice: Generally silent apart from audible clapping of bill.
Range: Africa, India, SE Asia, Philippines, Sulawesi, Sumatra, Java, and Lesser Sundas.
Distribution and status: A non-breeding visitor from SE Asia in N Sumatra; birds in S Sumatra are visitors from Java, or probably resident. In Java and Bali this is an uncommon bird in the lowlands and hills, probably more common in the East.
Habits: Frequents paddy fields and grassy pastures in small flocks. Roosts in tall trees, often with other storks or, where present, peafowl. Often soars high in the sky on thermals. This species is a non-colonial breeder.

57. STORM'S STORK *Ciconia stormi* PLATE 7

Description: Large (80 cm) black and white stork with slightly upturned, red bill. Wings, back, crown, and breast black; throat, nape, belly, and tail white. Bare facial skin pinkish red, especially in breeding season. Bare eye-ring is yellow. Distinguished from Woolly-necked Stork by black side of neck, yellow eye-ring, lack of white forehead, and redder bill, but note Javan form of Woolly-necked Stork also has reddish bill and reaches S Sumatra. Immature has black plumage replaced by brown.
Iris—red; bill—red; legs and feet—pink.
Voice: Bill clapping.
Range: Malay Peninsula, Sumatra, and Borneo.
Distribution and status: Sparsely distributed in freshwater swamp and swamp forests in lowlands of Sumatra (including Mentawai) and Borneo. One record from W Java in 1920 (specimen in Bogor museum).
Habits: Frequents dense swamp forests and nests in colonies.

72 STORKS

[**58. BLACK-NECKED STORK** *Ephippiorhynchus asiaticus*
(Green-necked Stork)

Description: Very tall (130 cm), black and white stork. Iridescent black head and neck of adult distinctive. Immature is brown.
Iris—yellow; bill—black; legs and feet—orange.
Voice: Bill clapping.
Range: India, SE Asia, New Guinea, and Australia.
Distribution and status: Not recorded in recent years in the Sunda region but one skull was reported for Java in 1908. Must once have been present but now certainly locally extinct.
Habits: Inhabits open and lightly wooded country, usually close to water but sometimes in dry areas.]

Black-necked stork, *Ephippiorhynchus asiaticus* (58)

[**59. GREATER ADJUTANT** *Leptoptilos dubius*   PLATE 7
(Greater Adjutant Stork)

Description: Huge (142 cm), black and white stork with massive bill. Distinguished from Lesser Adjutant by greater size, presence of pendulous yellow or pink gular pouch, and broad contrasting grey band on black wings.
Iris—brown; bill—grey; legs and feet—pinkish.
Voice: Bill clapping.
Range: India and SE Asia.
Distribution and status: A possible straggler to Sumatra. Reported but not confirmed from N Sumatra mud-flats in 1979.
Habits: Similar to Lesser Adjutant.]

60. LESSER ADJUTANT *Leptoptilos javanicus*   PLATE 7
(Lesser Adjutant Stork)

Description: Huge (110 cm) black and white stork with massive bill. Wings, back, and tail black; underparts and neck collar white; naked head, neck, and throat

pink with some fine white downy feathers on crown. Distinguished from Greater Adjutant by smaller size, uniform wing, and lack of gular pouch.
Iris—bluish grey; bill—grey; feet—dark brown.
Voice: Silent, apart from buzzing sound at nesting sites and audible wing-beats and bill clapping.
Range: India, S China, SE Asia, and Greater Sundas.
Distribution and status: Not uncommon in E Sumatra, particularly S Sumatra where groups of 40 or 50 are regularly reported. In Borneo this is a local and rather scarce bird though nesting is recently recorded in SC Kalimantan. In Java and Bali this once common bird is now rare in open lowland areas.
Habits: Frequents paddy fields, open burnt or flooded grassy areas, and mud banks and mangroves. Often seen soaring in thermals solitarily or in small parties with other storks, and even with eagles. Nests in colonies in well-wooded areas.

IBISES—FAMILY THRESKIORNITHIDAE

A SMALL worldwide, tropical family; similar and closely related to the storks, but generally slightly smaller and with bills modified for probing in water or mud, rather than stabbing and smashing prey. Ibises detect their food by touch rather than sight. The feet are partly webbed. The flight of most species consists of slow, flapping bouts alternated with short glides.

Five species are found in the Greater Sundas.

61. BLACK-HEADED IBIS *Threskiornis melanocephalus* PLATE 7
(Oriental/Indian Black-necked Ibis)

Description: Large (80 cm), unmistakable white ibis with black head, and long, decurved beak, and bushy 'tail' of elongated grey tertiary plumes.
Iris—red-brown; bill—black; feet—black.
Voice: Generally silent but makes curious grunts in the breeding season.
Range: India, S and E China, Japan, SE Asia, and Greater Sundas.
Distribution and status: Scarce, non-breeding visitor in Sumatra and N Borneo. Numerous breeding colonies were known from Java earlier this century; now local, the main extant colonies are in Pulau Dua, W Java, and Brantas delta, E Java.
Habits: Frequents reedy swamps, edges of lakes or reservoirs, and flooded grassy areas. Generally in small flocks, stalking actively in search of food, or flying in formation. Partly nocturnal; often roosts in trees by day. Nests in colonies with storks and other waterbirds.

62. WHITE-SHOULDERED IBIS *Pseudibis davisoni* PLATE 7

Description: Medium-sized (75 cm), black ibis with bare head, white patch on shoulder, red legs, and pale blue nape patch. General plumage is dark brown with glossy black wings and tail. Lacks chestnut on underparts.
Iris—dark; bill—black; feet—red.
Voice: Harsh *kyee-ahh*.
Range: Formerly in SW China and SE Asia; now restricted to Indochina and Borneo.
Distribution and status: Once recorded in Kuching. Local in SE Borneo, where recorded in Seruyan, Barito, and Mahakam drainages, in swamp and riverine forests.
Habits: As Glossy Ibis but favouring forested swamp forests and streams.
Note: Treated by some authors as a subspecies of Black Ibis *P. papillosa* of the Indian subcontinent.

63. GLOSSY IBIS *Plegadis falcinellus* PLATE 7

Description: Smallish (60 cm), glossy, blackish chestnut ibis, looking like a large dark curlew. Upperparts have a green and purple gloss.
Iris—brown; bill—blackish; feet—greenish brown.
Voice: Nasal grunts, and bleating and cooing sounds at nest.
Range: Cosmopolitan.
Distribution and status: Doubtfully recorded in Sumatra and once recorded in S Borneo (1851). In Java this is an uncommon and local lowland bird; the only extant colony known is on Pulau Dua, W Java.
Habits: Inhabits marshes, paddy fields, the edges of lakes, and flooded grasslands. Generally in small flocks, working slowly forward, probing deeply into mud with long bills. In the evening, flies in line or formation to communal roosts. Nests in colonies with egrets and herons.

64. BLACK-FACED SPOONBILL *Platalea minor* PLATE 7
(Lesser Spoonbill)

Description: Large (75 cm) white ibis with long, blackish grey spatulate bill. Similar to wintering Royal Spoonbill but bill entirely grey, and facial skin black and less extensive. Lacks red spot on crown and yellow eye-spot of Royal Spoonbill.
Iris—brown; bill—dark grey; legs and feet—black.
Voice: Silent outside breeding season.
Range: Breeds in E China and islands of N Korea. Recorded S and E to Philippines and Borneo.

Distribution and status: Recorded in Brunei; probably regular in small numbers.
Habits: As Royal Spoonbill.

65. ROYAL SPOONBILL *Platalea regia* PLATE 7

Description: Large (80 cm) white spoonbill with long, grey spatulate bill. Naked skin on head black with a reddish patch on forehead. Distinguished with difficulty from wintering Black-faced Spoonbill by more black on face and by red forehead spot, yellow eye-spot, and bill colour.
Iris—red or yellow; bill—grey with yellowish 'spoon'; feet—black.
Voice: Silent when not breeding.
Range: Australia and New Zealand; probably a non-breeding visitor to Indonesia, but may breed in Irian Jaya.
Distribution and status: A small breeding population on Pulau Dua, W Java, no longer extant. Now a summer vagrant recorded in E Java and Bali. Records from N Borneo probably refer to Black-faced Spoonbill.
Habits: Frequents muddy pools, lakes, or mud bars where it wades slowly, scything its bill from side to side in the water, sifting out food. Generally occurs singly or in small parties, and is partly nocturnal.

Ducks—FAMILY ANATIDAE

A FAMILIAR large, worldwide family of swimming waterbirds with webbed feet and rather specialized, characteristic, broad, flat bills. They have rather short legs, narrow pointed wings set well back on the body, and generally short tails. They fly fast with a continuous, audible, whistling flap.

Ducks are divided taxonomically into several tribes of which four are represented in the Greater Sundas: Tree ducks or Whistling ducks (*Dendrocygna*) which have clear whistling calls; Surface or Dabbling ducks (*Anas*) which swim high in the water; Diving ducks (*Aythya*) which dive for food and to escape when alarmed; and Perching ducks (*Nettapus and Cairina*).

There are a total of 14 species in the Greater Sundas, of which seven are winter visitors only.

66. LESSER WHISTLING-DUCK *Dendrocygna javanica* PLATE 8
(Lesser Tree-Duck)

Description: Medium-sized (41 cm) reddish brown duck. Very similar to Wandering Whistling-Duck with dark crown, buffy head and neck, brown back, and

76 DUCKS

reddish brown underparts, but smaller and lacking black and white side plumes.
Iris—brown; bill—black; feet—dark grey.
Voice: Shrill, musical whistle *seasick seasick* on wing.
Range: India, S China, SE Asia, and Greater Sundas.
Distribution and status: Resident and locally quite common in Sumatra, S Borneo, and W Java.
Habits: Found in small parties on lakes, swamps, mangroves, and paddy fields.

67. WANDERING WHISTLING-DUCK *Dendrocygna arcuata* PLATE 8
(Whistling Tree-Duck)

Description: Medium-sized (45 cm) rich red-brown duck. Top of head and back of neck dark brown, rest of head and neck paler. Back and tail brown, breast chestnut. White plumes with black edges protrude below wings; rump and undertail white.
Iris—brown; bill—black; legs—grey-brown.
Voice: High-pitched, twittering whistles given during flight.
Range: Philippines, Sulawesi, and Sundas to S New Guinea, Australia, and Fiji.
Distribution and status: Locally common in Sumatra, Java, and Bali. Probably resident S Borneo but there are few records.
Habits: Found on freshwater lakes and marshes, in flocks. Rests on bare or grassy margins but usually feeds in water, diving repeatedly.

68. NORTHERN PINTAIL *Anas acuta* PLATE 8
(Pintail)

Description: Medium-sized (55 cm) duck with long pointed tail. Male has brown head, white throat, scalloped grey flanks, black tail with extremely elongated central feathers, grey wings with a bronze-coloured speculum, and white underparts. Female is drabber—brown, speckled with black above; buff below with black spots on breast; wings grey with brown speculum.
Iris—brown; bill—blue-grey; feet—grey.
Voice: Rather silent. Female gives low guttural *kwuk-kwuk*.
Range: Breeds throughout Holarctic; winters south. Vagrant on Malay Peninsula.
Distribution and status: Doubtfully recorded in Sumatra; vagrant in N Borneo and Java.
Habits: Frequents marshes, lakes, large rivers, and sea coasts. On water it is a surface feeder, but sometimes up-ends for long periods in shallow water to search the muddy bottom for food.

69. GREEN-WINGED TEAL *Anas crecca* PLATE 8
(Common Teal)

Description: Small (37 cm), fast-flying duck with conspicuous green speculum in flight. Male has distinct metallic green, buff-bordered eye-stripe through chestnut head, a long white stripe on scapulars, and buff patch on side of rump; dark undertail; rest of plumage generally greyish. Female mottled brown with whitish belly. Distinguished from female Garganey by bright green speculum.
Iris—brown; bill—grey; feet—grey.
Voice: Male call is a metallic rattling *kirrik*; female a thinner, higher, short *quack*.
Range: Breeds throughout Palaearctic; winters south. Vagrant on Malay Peninsula.
Distribution and status: Regular winter visitor to coast of N Borneo.
Habits: Generally in pairs or flocks, on small lakes or ponds, often mixing with other waterfowl. Flies with very fast wing-beat.

70. SUNDA TEAL *Anas gibberifrons* PLATE 8
(Sunda/Indonesian Grey Teal)

Description: Smallish (42 cm) grey-brown duck. Crown dark reddish brown, face and neck buffish, sometimes almost white, sides and back reddish brown. Wing has a blackish speculum with blue-green metallic gloss, and in flight a conspicuous white patch in front of speculum. Males are slightly larger than females and have a bony lump on the forehead.
Iris—red-brown; bill—bluish grey with yellow patch near tip; legs and feet—grey.
Voice: Male clear *pip* note; female wild laughing chuckle; often at night.
Range: Andamans, Sulawesi, and Sundas.
Distribution and status: Common in S Sumatra but no breeding record; recently recorded in S and E Borneo. It is probably the commonest duck on Java and Bali.
Habits: Found in pairs or small flocks in mangroves, marshes, ponds, and rivers, often far inland.
Note: *A. gracilis* of Australia, formerly included in this species, has been recorded in the Moluccas and could reach the Greater Sundas. It lacks the lump on the forehead.

71. MALLARD *Anas platyrhynchos* PLATE 8

Description: Medium-sized (58 cm), wild form of domestic duck. Male has distinctive, dark, shiny green head and neck separated from chestnut breast by

white collar. Female is mottled brown with dark eye-stripe and blue speculum. Distinguished from female Pintail by shorter and blunter tail.
Iris—brown; bill—yellow; feet—orange.
Voice: Familiar *quack quack* of domestic duck.
Range: Holarctic; winters south.
Distribution and status: Rare winter visitor to N Borneo.
Habits: Frequents lakes, ponds, and river estuaries.

72. EURASIAN WIGEON *Anas penelope* PLATE 8
(Wigeon)

Description: Medium-sized (47 cm), rather compact, big-headed duck. Male has diagnostic chestnut head and buff forehead and cap. Rest of plumage grey with white patch on flanks, white belly, and black undertail coverts. In flight white wing coverts contrast against dark flight feathers and green speculum. Female is rich, uniform, rufous brown with white belly. In flight, pale grey wing coverts contrast with darker flight feathers.
Iris—brown; bill—blue-grey; feet—grey.
Voice: Male gives musical piping whistle *whee-oo*; female gives short growled *quack*.
Range: Palaearctic; winters south.
Distribution and status: Vagrant to N Borneo.
Habits: Mixes with other waterbirds on lakes, marshes, and among mangroves.

73. PACIFIC BLACK DUCK *Anas superciliosa* PLATE 8
(Australian Grey Duck)

Description: Large (55 cm), dark brown, surface-feeding duck. Head with diagnostic black stripes. Speculum dark green to purple. White under wings contrast conspicuously with dark plumage in flight.
Iris—brown; bill—dark grey; feet—dull yellow or brown.
Voice: Male like domestic duck gives quick *raab raaraab* or, when flushed, a hoarse whispered *fraank fraank*; female gives a deep, descending *kwark kwark*.
Range: Sumatra and Java to Australia and Polynesia.
Distribution and status: Status in Sumatra uncertain; formerly locally common on mountain lakes but few recent records. Local breeding bird on mountain lakes in E Java and Bali.
Habits: Found in reedy lakes and marshes; dabbles in shallow water and also feeds on land.

DUCKS 79

74. GARGANEY *Anas querquedula* PLATE 8
(Garganey Teal)

Description: Medium-sized (40 cm) surface-feeding duck. Male with distinctive chocolate head and broad white eyebrow. Brown back and breast contrasts with white belly. Scapulars are long, black, and white. Speculum glossy green with white margins. Female is brown with black striped head and olive speculum. Eclipse male is like female except wing coverts in flight. In flight the blue-grey wing coverts of the drake are a good characteristic.
Iris—hazel; bill—black; feet—bluish grey.
Voice: Generally silent. Drake gives a rattle-like croak. Female gives slight *kwak*.
Range: Breeds through Palaearctic; winters south.
Distribution and status: Recorded in N Sumatra, Borneo, and rarely to the N coast of Java.
Habits: Frequents coastal lagoons. Sleeps on water during day, flying inland at night to feed.

75. NORTHERN SHOVELER *Anas clypeata* PLATE 8
(Common Shoveler)

Description: Unmistakable, large (50 cm) duck with extremely long, broad spatulate bill. Male: chestnut belly, white breast, and dark glossy green head distinctive. Female is mottled brown with whitish tail and dark eye-stripe. Coloration similar to female Mallard but bill unmistakable. In flight pale upperwing coverts contrast with dark flight feathers and green speculum.
Iris—brown; bill—dark in male, orange-brown in female; feet—orange.
Voice: Similar to Mallard but softer and lower, also chuckling *quack*.
Range: Breeds throughout Holarctic; winters south.
Distribution and status: Occasional winter visitor to coast of N Borneo.
Habits: Favours coastal lagoons, ponds, lakes, and mangrove swamps.

76. TUFTED DUCK *Aythya fuligula* PLATE 8

Description: Medium-sized (42 cm), squat, chunky duck with distinctive long crest. Male is black with white belly and sides. Female is dark brown with brown flanks and short crest. In flight shows a white band on secondaries.
Iris—yellow; bill and feet—black.
Voice: Generally silent in winter. Harsh, low *kur-r-r, kur-r-r* given in flight.
Range: Breeds throughout N Palaearctic; winters south.
Distribution and status: Occasional visitors reach Borneo. An old report lists Sumatra.
Habits: Frequents lakes and deep ponds, diving for food. Flight is fast.

77. AUSTRALIAN WHITE-EYED POCHARD *Aythya australis*
(Hardhead) PLATE 8

Description: Medium-sized (50 cm), rich brown, diving duck. Male: mahogany with white eye; female paler brown with brown eye. Belly is white. Both sexes have conspicuous white bar on upperwing and white underwing in flight.
Iris—white in male, brown in female; bill—black with whitish band near tip; feet—grey.
Voice: Seldom heard; male: soft wheezy whistle; female: soft harsh croak.
Range: Australia and New Guinea; vagrant in W Indonesia.
Distribution and status: In Java known by a breeding record from the mountains of E Java. There are no recent records, however, so the status of this population is now uncertain.
Habits: Inhabits deep montane lakes with emergent vegetation. Dives under water to search for food for up to 30 seconds at a time. Fast flight with shallow flickering of pied wings.

78. COTTON PYGMY-GOOSE *Nettapus coromandelianus* PLATE 8
(White Pygmy-Goose/Cotton Teal)

Description: Small (30 cm) black and white duck. Male has glossy black crown and black neck band; back, wings, and tail black with green iridescent plumage, but otherwise white. In flight has conspicuous white patch on wing. Female is drabber with brown replacing the glossy black, and buff instead of white; has a brown eye-stripe and no white wing patch.
Iris—reddish; bill—grey above, yellow below; feet—grey.
Voice: In flight utters a soft, musical, laughing note *kar kar kar wark* several times in succession, also a soft *kwak*.
Range: India, S China, SE Asia, and locally to New Guinea and Australia.
Distribution and status: Status uncertain; an occasional bird of swamps and lakes in S Sumatra and Borneo. In Java it used to occur regularly in the Jakarta area; only recent records from Pulau Dua.
Habits: Usually on the water in ponds, canals, grassy backwaters, or paddy fields. Nests in tree holes and regularly perches in trees.

79. WHITE-WINGED DUCK *Cairina scutulata* PLATE 8
(White-winged Wood-Duck)

Description: Large (75 cm) black and white duck. Whitish head and neck; white lesser wing coverts, blue-grey median wing coverts and speculum; black back with a green gloss; underparts dark brown. In flight, when seen from below, white wing lining contrasts sharply with black flight feathers. Sumatran birds are often partly albino.

Iris—brown; bill—yellow with black tip; feet—yellow or orange.

Voice: Usually calls in flight. A goose or crane-like honk from the drake, accompanied by a curious whistle from the duck. Calls are diagnostic (D.A.H.). Bleating nest calls.

Range: Local in Assam, SE Asia, Sumatra, and formerly Java.

Distribution and status: Now extremely rare or extinct in many former haunts. Still widespread in lowland forest of E Sumatra and still occurs in NW Sumatra, but there are no recent records from W Java though a few nests have been recorded in the past.

Habits: A duck of swamp forest, but comes out at night to feed in paddy fields.

Osprey—FAMILY PANDIONIDAE

A MONO-SPECIES family. The osprey is a fishing hawk with characteristic long, narrow, angled wings, specialized for diving quite deep into water to catch fish. Otherwise much like other hawks. The recent review of Sibley and Monroe (1990) reduces the family to a sub-family of Accipitridae.

80. OSPREY *Pandion haliaetus* PLATE 9
(Fish Hawk)

Description: Medium-sized (55 cm), brown, black, and white hawk. White head and underparts with black eye-stripe diagnostic. Upperparts mostly dull brown. Has a short, dark, erectile crest. Races differ in whiteness of head and extent of streaking on underparts.

Iris—yellow; bill—black with grey cere; naked tarsus and feet—grey.

Voice: A loud plaintive whistle in breeding season. Young in nest scream loudly when parents sighted.

Range: Cosmopolitan.

Distribution and status: A regular and widespread winter visitor in Sumatra and Borneo. In Java and Bali an uncommon resident along coasts, and an irregular winter visitor further inland.

Habits: A dramatic fishing hawk, plunging deep into the water to catch prey, sometimes staying totally submerged for several seconds. It hunts from tree perches over the sea or lakes, or circles slowly over water, hovers or glides slowly, almost at stalling speed, into a headwind. Fish are carried to a perch to eat.

82 HAWKS, EAGLES, AND VULTURES

Hawks, Eagles, and Vultures—FAMILY ACCIPITRIDAE

Hawks and eagles are largish to very large, predatory, hook-billed birds with powerful talons or claws, specialized for killing and tearing up vertebrate prey. They differ from the falcons in having generally blunter, more rounded wings and paler eyes (yellow or red).

Eagles and vultures are specialized for soaring on air currents and hunt largely from the air, whereas the other hawks hunt from branches, but also sometimes soar; some species even hover over intended prey. Vultures feed mostly on carrion and have long barish necks for probing into carcasses. Members of the family make very large, stick pile nests in trees or on cliffs and the young give characteristic screaming calls.

There are 34 species in the Greater Sundas, including several migrants.

81. JERDON'S BAZA *Aviceda jerdoni* PLATE 9
(Asian Baza/Lizard Hawk)

Description: Medium-sized (45 cm) brown hawk with long crest often held vertically above head. Upperparts brown, underparts white with black mesial stripe, dark rufous streaked breast, and dark rufous-barred belly. Bornean race has rufous head and sides of neck. Distinguished from Crested Goshawk by much longer crest and wing tips almost reaching tip of tail. Distinguished from Blyth's Hawk-Eagle by smaller size and lack of white tail bar. In flight distinguished by very long broad wings, noticeably broad near tip, and square-cut tail.
Iris—yellow-red; bill—black with pale blue-grey cere; feet and legs—yellow.
Voice: Plaintive mewing on wing *pee-weeoh*, second note fading away, similar to Crested Serpent-Eagle.
Range: Himalayas, India, S China, SE Asia, and Sulawesi. Vagrant in Malay Peninsula.
Distribution and status: Few records from Sumatra where resident or winter visitor. Resident but local throughout Borneo.
Habits: Hunts from tree perches, favouring forest edge and coastal forests.

82. BLACK BAZA *Aviceda leuphotes* PLATE 9
(Black-crested Baza)

Description: Smallish (32 cm), unmistakable, black and white hawk with long black crest often held vertically above head. General plumage black with broad white breast bar, white wing patches, and dark banded belly. In flight, short

rounded wings are conspicuously patterned with black lining, grey secondaries, and pale contrasting primaries with black tips. Has crow-like flapping flight and glides with wings held flat.
Iris—dark brown; bill—horn with grey cere; feet—dark grey.
Voice: Weak, 1 to 3 noted airy screams, like a seagull mewing.
Range: Himalayas, S India, S China, SE Asia; winters to Greater Sundas.
Distribution and status: Regular winter visitor to Sumatra and rarely W Java.
Habits: Lives in pairs or small flocks making short flapping flights to catch large insects in air or on ground. Rather tame. Often found along waterways or in open forests and villages. Migrating birds soar high in the sky.

83. ORIENTAL HONEY-BUZZARD *Pernis ptilorhynchus* PLATE 9
(Crested/Asian/Eastern Honey-Buzzard)

Description: Medium-sized (50 cm) dark hawk with small crest. Colour very variable with light, normal, and dark forms of two distinct races each mimicking hawk-eagles and buzzards. Upperparts from white to rufous to dark brown, heavily spotted and barred. Tail has irregular barring. All forms have a contrasting pale throat patch outlined with heavy black streaks and often a black mesial streak. In flight, the relatively small head, longish neck and long narrow wings and tail are distinctive.
Iris—orange; bill—grey; feet—yellow; at close range scale-like feathers in front of eye are diagnostic.
Voice: Loud, high-pitched ringing four syllable note *wee-wey-uho* or *weehey-weehey*.
Range: E Palaearctic, India, and SE Asia to Greater Sundas.
Distribution and status: Resident (long-crested races *torquatus* and *ptilorhynchus*) are sparsely distributed through Sumatra, Borneo, and W Java. E Palaearctic short-crested race (*orientalis*) turns up as an occasional winter visitor throughout the Greater Sundas to 1200 m.
Habits: Frequents forested hills. Characteristic flight of a few wing-beats followed by a long glide. Soars high in the sky with wings held level. Has the curious habit of raiding bee and wasp nests.

84. BAT HAWK *Machaeramphus alcinus* PLATE 9
(Bat Kite)

Description: Medium-sized (45 cm) black hawk with short droopy crest, white throat, and black mesial stripe. In flight appears falcon-like with long, pointed, but rather broad-based wings, but short, square-cut tail. Flight distinctly lazier than similar-sized Peregrine.
Iris—yellow; bill—black; feet—grey.

84   HAWKS, EAGLES, AND VULTURES

Voice: Hoarse falcon-like cry of up to 7 piercing whistles, given by day from roost tree.
Range: Africa, Malay Peninsula, Sulawesi, Sumatra, Borneo, and New Guinea.
Distribution and status: On Sumatra (including Bangka) and Borneo this is a rare and local bird of lowlands up to 1000 m.
Habits: Frequents open, wooded, limestone country, close to bat caves. Mostly crepuscular attacking bat flocks as they exit from and return to cave roosts. Flies with rapid wing-beats.

85. BLACK-WINGED KITE   *Elanus caeruleus*   PLATE 9
(Black-shouldered/Indonesian Kite)

Description: Small (30 cm), white, grey, and black kite. Black shoulder patch and long black primaries diagnostic. Adult: grey crown, back, wing coverts and base of tail. Face, neck, and underparts white. Common white hawk which hovers when looking for prey. Immature similar to adults but tinged brown.
Iris—red; bill—black with yellow cere; feet—yellow.
Voice: Soft whistle *wheep, wheep*
Range: Africa, S Eurasia, India, Philippines, and Indonesia to New Guinea.
Distribution and status: On Sumatra, Java, and Bali, an occasional resident of open lowland and hilly habitats up to 2000 m. A straggler on Borneo recorded from most parts.
Habits: Sits on exposed perches such as dead trees, telegraph poles, etc. or hovers like a kestrel. Prefers rather dry, open, lightly wooded hunting areas.

86. BLACK KITE   *Milvus migrans*   PLATE 9
(Common/Pariah/Yellow-billed Kite)

Description: Largish (65 cm) dark brown hawk with diagnostic forked tail. In flight a pale patch at base of primaries contrasts with blackish wing tips. Head sometimes paler than back. Immature birds are streaked buff on head and underparts.
Iris—brown; bill—grey; cere—blue-grey; feet—blue-grey.
Voice: Shrill, whinnying *ewe-wir-r-r-r-r*
Range: Africa, Eurasia to Australia.
Distribution and status: Rare winter visitor from E Asia, has reached N Sumatra and N Borneo.
Habits: Frequents open country, coasts, harbours, and cities. Circles gracefully or flies with slow deliberate beat. Perches on poles, wires, trees, buildings, or ground.

HAWKS, EAGLES, AND VULTURES 85

87. BRAHMINY KITE *Haliastur indus* PLATE 9
(Red-backed/White-headed Kite)

Description: Medium-sized (45 cm) white and russet-brown kite. Adults: head, neck, and breast white; wings, back, tail, and belly rich reddish brown; contrasting black primaries. Immature brownish all over with streaked breast, becoming greyish white in second year, and reaching fully adult plumage in third year. Distinguished from Black Kite by rounded tail.
Iris—brown; bill and cere—greenish grey; legs and feet—dull yellow.
Voice: Shrill, querulous, mewing cries *shee-ee-ee* or *kweeaa*.
Range: India and S China to Australia.
Distribution and status: Common throughout Sumatra and Borneo. Rare in W Java but still common on Bali.
Habits: This is a common bird over much of the region, wheeling singly or several together over waterways or close to water. Frequents coasts, rivers, swamps, and lakes up to 3000 m.

88. WHITE-BELLIED FISH-EAGLE *Haliaeetus leucogaster* PLATE 9
(White-bellied Sea-Eagle)

Description: Large (70 cm), white, grey, and black eagle. Adult: head, neck, and underparts white; wings, back, and tail grey; primaries black. Immature: white areas of adult are pale brown and grey areas are dark brown. Wedge-shaped tail characteristic.
Iris—brown; bill and cere—grey; bare tarsus and feet—pale grey.
Voice: Loud honking cry *ah-ah-ah-ah*.
Range: India, SE Asia, Philippines, and Indonesia to Australia.
Distribution and status: A common resident around the coast and on large lakes and rivers near coasts throughout the Greater Sundas.
Habits: This magnificent eagle is frequently seen sitting very upright in a waterside tree, on a cliff, or on fishing platforms. Soars and glides beautifully and gracefully with wings held in pronounced dihedral. Flying with slow, powerful wing-beats, it catches surfacing fish in spectacular dive, barely getting wet as fish is seized with talons. Builds huge nest of branches and twigs in tall tree and often used year after year.

89. LESSER FISH-EAGLE *Ichthyophaga humilis* PLATE 9
(AG: Fishing-Eagle)

Description: Medium-sized (60 cm) brownish eagle with grey head, and neck and white belly. Distinguished from Grey-headed Fish-Eagle by smaller size and dark tail. Immature paler brown with buffy unstreaked underparts.

Iris—yellow or brown; bill—dark grey; feet—grey.
Voice: Occasional querulous cries *hak hak.* Juvenile gives weak plaintive *ngheea.*
Range: Himalayas, SE Asia, Malay Peninsula, Borneo, and Sumatra.
Distribution and status: In Sumatra and Borneo an uncommon bird of rivers in the lowlands and hills up to 1000 m.
Habits: Frequents forested rivers and wooded swamps, and catches fish from near surface.

90. GREY-HEADED FISH-EAGLE *Ichthyophaga ichthyaetus* PLATE 9
(Greater Fish-Eagle, AG: Fishing-Eagle)

Description: Large (70 cm), grey, brown, and white eagle. Adult: head and neck grey, grading to brown on breast. Wings and back dark brown; belly, thighs, and base of tail white; terminal black bar on tail. Immature: upperparts buffy brown; underparts streaked brown and white; tail light brown with dark brown terminal bar.
Iris—brown to yellow; bill and cere—grey; bare tarsus and feet—white to yellow.
Voice: A loud, harsh cry *awh-awhrr.*
Range: India, SE Asia, Philippines, Sulawesi, and Greater Sundas.
Distribution and status: Uncommon but widely distributed along rivers in Sumatra and Borneo; now rare in W Java. There are no recent records from E Java where it formerly occurred, and not recorded from Bali.
Habits: Frequents lowland forested waterways, lakes, rivers, and swamps. Dives on fish while flying or from a tree perch, but rarely soars.

91. WHITE-RUMPED VULTURE *Gyps bengalensis*
(White-backed/Indian White-backed Vulture)

Description: Huge (87 cm) raptor with short rounded tail and diagnostic long naked neck. The only vulture recorded in this region. When seen from below in flight, whitish wing lining and white neck collar contrast with blackish flight feathers and dark brown underparts; from above, generally blackish with dark grey secondaries and whitish rump. Immature is brownish grey with no contrasting white. Soars with wings in slight 'V'.
Iris—reddish brown; bill—grey; feet—grey.
Voice: Harsh grunts and shrieks on carcasses.
Range: India, SW China, and SE Asia.
Distribution and status: Recorded once Brunei.
Habits: Circles high in sky searching for carrion. Roosts in trees when fed or drinks at water's edge.

White-rumped Vulture, *Gyps bengalensis* (91)

92. SHORT-TOED EAGLE *Circaetus gallicus*   PLATE 10
(Short-toed Snake-Eagle)

Description: Large (65 cm), pale, heavy bodied eagle. Upperparts greyish brown; underparts white with dark streaked or uniform brown throat and breast, and indistinct bars on belly; tail with 4 broad indistinct bars. Immature birds are even paler than adults. In flight: long, broad wings with distinctive longitudinal barring on coverts and flight feathers.
Iris—yellow; bill—black; cere—grey; feet—greenish.
Voice: Generally silent in winter, occasional plaintive mewing.
Range: Africa, Eurasia, India, N China, and Lesser Sundas.
Distribution and status: Recorded in E Java (regular at Baluran) and Bali. The origin of these birds is uncertain; probably stragglers from the resident population in the Lesser Sundas, but migrants from Asia might reach the Sundas regularly.
Habits: Inhabits forest edge and secondary scrub. Circles and glides with wings held very straight and level. Often hovers like a giant kestrel.

93. CRESTED SERPENT-EAGLE *Spilornis cheela*   PLATE 11

Description: Medium-sized (50 cm), dark eagle with very broad, rounded wings and rather short tail. Adult: upperparts dark brown/grey, underparts brown with belly, flanks, and vent spotted with white. Tail has broad whitish-grey bar between black bars; short, puffy, broad crest of black and white feathers; yellow naked area between eye and bill characteristic. In flight, broad white tail bar and white

trailing edge of wings diagnostic. Bornean race is paler and browner. Immature: similar to adult but browner with more white in plumage.
Iris—yellow, bill—grey-brown; feet—yellow.
Voice: A very vocal eagle, frequently soaring over forest canopy, uttering a loud shrill cry *kiu-liu* or *kwee-kwee, kwee-kwee, kwee-kwee-kwee*.
Range: India, S China, SE Asia, Palawan, and Greater Sundas.
Distribution and status: Throughout the Greater Sundas and probably the commonest eagle over wooded and forested hills, up to 1900 m.
Habits: Frequently seen circling over forest and plantations, pairs often calling to each other. In courtship, pairs perform spectacular but usually sluggish aerobatics. Often perches on large branches in shady parts of the forest where it can watch the ground.

94. MOUNTAIN SERPENT-EAGLE *Spilornis kinabaluensis* PLATE 11
(Kinabalu Serpent-Eagle)

Description: Largish (55 cm), dark brown eagle with short full crest and pale band in dark tail. Distinguished from Crested Serpent-Eagle by longer wing, darker colour, smaller white spots, and rich umber brown nape band, black throat, and bolder white band on tail.
Iris—lemon yellow; cere—yellow; bill—creamy white; feet—yellow.
Voice: Similar to Crested Serpent-Eagle but distinctive.
Range: Endemic to Borneo.
Distribution and status: Confined to Mts Kinabalu, Murud, and Mulu in N Borneo. On Mt. Kinabalu it is an occasional bird above 1000 m.
Habits: Similar to Crested Serpent-Eagle but prefers montane forests.

95. EASTERN MARSH HARRIER *Circus spilonotus* PLATE 10, 13
(Spot-backed Harrier/Spotted Marsh Harrier)

Description: Medium-sized (50 cm), dark harrier. Male (rarely seen in Greater Sundas) is similar to male Pied Harrier, but black throat and breast heavily streaked white. Female is distinct among all female harriers except Western Marsh Harrier in having brown uppertail coverts; plumage dark brown with buff crown, nape, throat, and leading edge of wing; crown and nape streaked dark brown; tail is barred; whitish patch on base of primaries boldly mottled dark when viewed from below. Some individuals have entire head buffy and buff patches on breast. Immature like female but darker, with only crown and nape buff.
Iris—male: yellow, female light brown; bill—grey; feet—yellow.
Voice: Generally silent.
Range: Breeds in E Asia; in winter south to SE Asia and Philippines. Also resident New Guinea.

HAWKS, EAGLES, AND VULTURES

Distribution and status: Uncommon winter visitor to Sumatra and N Borneo.
Habits: Frequents open areas, especially grassy marshes or reed beds. When hunting, glides gracefully low over vegetation. Sometimes hovers. Flight heavier and less buoyant than Pied Harrier.
Note: Treated by some authors as a race of the Western Marsh Harrier *C. aeruginosus*.

96. WESTERN MARSH HARRIER *Circus aeruginosus* PLATE 10, 13

Description: Medium-sized (50 cm) dark harrier. Male: similar to subadult male Eastern Marsh Harrier but head buffier with less dark streaking. Female and immature: similar to Eastern Marsh Harrier but back darker brown, tail unbarred, and crown lacks bold dark streaking. White patch under primaries (if present) lacks dark mottling.
Iris—male: yellow, female: light brown; bill—grey; feet—yellow.
Voice: Generally silent.
Range: Breeds in W and C Palaearctic to W China; winters south to Africa, India, and Malay Peninsula.
Distribution and status: One specimen taken in N Sumatra. The relative abundance of the two Marsh Harriers in the Greater Sundas is not known but clearly both can occur.
Habits: As Eastern Marsh Harrier.

97. NORTHERN HARRIER *Circus cyaneus* PLATE 10, 13
(Hen/White-rumped Harrier)

Description: Medium-sized (45 cm), pale grey or brown harrier. Distinctive male is grey above with white rump and black primaries. Underparts mostly white. Female and immature: brown above with conspicuous white rump, grey tinge on barred tail; underparts buff to rufous, streaked dark brown. Tail with pale tip.
Iris—pale brown; bill—grey; feet—yellow.
Voice: Generally silent.
Range: Breeds in Holarctic; migrates south in winter to S China, SE Asia, Malay Peninsula, and Borneo.
Distribution and status: A scarce winter visitor to N Borneo.
Habits: As Eastern Marsh Harrier.

98. PIED HARRIER *Circus melanoleucos* PLATE 10, 13

Description: Smallish (42 cm) slender-winged harrier. Male: black, white, and grey plumage. Unstreaked black head, throat, and breast diagnostic. Female: streaky brown upperparts tinged grey, white rump, barred tail, buff underparts streaked with rufous; underside of flight feathers barred blackish. Immature: dark brown upperparts with whitish band on uppertail coverts. Underparts rufous chestnut with rufous buff streaks.

Iris—pale brown; bill—horn; feet—yellow.

Voice: Generally silent.

Range: Breeds in NE Asia; migrating south in winter to SE Asia, Philippines, and N Borneo.

Distribution and status: Locally common winter visitor to N Borneo, south to Kelabit highlands, and occasionally to Kalimantan.

Habits: Glides low over open country, marshes, reed beds, and rice fields.

99. JAPANESE SPARROWHAWK *Accipiter gularis* PLATE 10, 13
(Japanese Lesser/Asiatic Sparrowhawk)

Description: Small (27 cm) hawk, very similar in appearance to Crested Goshawk and Besra, but noticeably smaller and more dashing. Adult male: upperparts grey; tail grey with several dark bands; pale rufous breast and belly with very thin vertical mesial stripe; no pronounced moustachial stripe. Female: upperparts brown instead of grey; lacks rufous on underparts which are heavily barred brown. Immature: streaked rather than barred on breast and more rufous.

Iris—yellow to red; bill—blue-grey with black tip, cere—green-yellow; feet—green-yellow.

Voice: Occasional harsh cries.

Range: Breeds in Palaearctic E Asia; south in winter to the Greater Sundas.

Distribution and status: Regular passage migrant and winter visitor to Sumatra, Borneo, Java, and Bali. Relatively common in open countryside of lowlands.

Habits: Hunts along forest edge and over secondary forests, and open country generally from a tree perch, but sometimes wheels in small circles scanning the ground with typical sparrowhawk 'flap-flap-glide' flight.

100. BESRA *Accipiter virgatus* PLATE 10, 13

Description: Medium-sized (33 cm) dark hawk. Similar to Crested Goshawk but smaller and lacking crest. Adult male: upperparts dark grey with strongly barred tail. Underparts white with brown bars and rufous flanks; throat white with vertical black mesial stripe and black moustachial stripe. Bornean race—rufous breast and more purplish thighs; Sumatran race—more rufous and barring less distinct. Female and immature: less rufous on flanks and less strongly barred, with reddish brown undersides; back brown, tail brown with dark bars. Immature has streaked breast.

Iris—yellow; bill—black with grey cere; legs and feet—yellow.

Voice: Young birds give repeated cry *shew-shew-shew* when hungry.

HAWKS, EAGLES, AND VULTURES   91

Range: India, S China, SE Asia, Philippines, and Greater Sundas.
Distribution and status: Resident on Sumatra, N Borneo, Java, and Bali. This species is widespread in wooded and forested hills and mountains, mostly 300 to 1200 m (up to 3000 m on Mt. Kerinci), but it is common nowhere.
Habits: Sits quietly in the forest watching for reptile or avian prey.

101. EURASIAN SPARROWHAWK *Accipiter nisus*   PLATE 10, 13
(Northern/Common Sparrowhawk)

Description: Medium-sized (male 32 cm, female 38 cm) short-winged hawk. Male: brownish grey above with white underparts finely barred rufous and with banded tail. Rufous cheeks diagnostic. Female: larger; brown above with white underparts, narrowly barred grey-brown on breast, belly, and thighs; prominent white eyebrow. No mesial stripe on throat. Cheek patch less rufous. Immature: distinctive from other immature *Accipiter* in having brown barring on breast and no streaks.
Iris—bright yellow; bill—horn with black tip; feet—yellow.
Voice: Occasional shrill cries.
Range: Breeds in Palaearctic; migrants reach Africa, India, SE Asia, and rarely northern Borneo.
Distribution and status: Recorded once in Kuching (N Borneo) in 1895.
Habits: Hunts from tree perch or on the wing in 'ambush- flight' along forest edge or in open wooded areas.

102. CRESTED GOSHAWK *Accipiter trivirgatus*   PLATE 10, 13

Description: Large (40 cm) powerful *Accipiter* with distinctive crest. Adult male: upperparts grey brown with banding on wings and tail; underparts rufous, chest streaked black; bold black bars on white belly and thighs. Neck is white with black, mesial stripe down the throat and two black moustachial stripes. Immature and female: as adult male but streaks and bars of underside are brown and upperparts are paler brown.
Iris—brown changing to greenish yellow in mature bird; bill—grey with yellow cere; legs and feet—yellow.
Voice: A shrill scream *he-he-he-he-he-he*, and prolonged yelp.
Range: S Asia, SE Asia, Philippines, and Greater Sundas.
Distribution and status: Not uncommon in lowland forest in Sumatra (including Nias) and Borneo (including N Natunas), up to 1000 m. In Java and Bali it was previously well distributed in lowland and hill forest but now becoming scarce.
Habits: Hunts from a low perch in the forest. Keeps to fairly thick forest cover. During breeding season, often soars over forest canopy calling loudly.

92 HAWKS, EAGLES, AND VULTURES

103. CHINESE GOSHAWK *Accipiter soloensis*   PLATE 10, 13.
(Horsfield's/Grey Goshawk/Frog-Hawk)

Description: Medium-sized (33 cm) hawk with very pale underparts. Adult: upperparts blue grey with sparse white tips to back feathers and faint black barring on outer tail feathers; underparts white with faint rufous wash on breast and flanks, some light grey barring on flanks and slight barring on thighs. Underwings of adult distinctive—almost entirely white except for black tip of primaries. Immature: brown upperparts with dark bars on tail and white underparts, streaked on throat and barred brown on chest and thighs.
Iris—red or brown; bill—grey with black tip; cere—orange; feet—orange.
Range: Breeds in NE Asia and China; migrating south in winter to SE Asia, Philippines, Indonesia and New Guinea.
Distribution and status: Not uncommon in winter throughout Greater Sundas up to 900 m.
Habits: Hawks after small birds in open wooded areas. Usually hunts from a perch but sometimes circles overhead.

104. SHIKRA *Accipiter badius*   PLATE 10, 13
(Little Banded Goshawk)

Description: Medium-sized (32 cm), rather pale *Accipiter*. Male: pale grey upperparts with contrasting black primaries; white throat with faint grey mesial stripe; breast and belly narrowly barred rufous and white. Female: like male but back is brown and throat greyer. Immature: grey-brown scalloped rufous, brown-streaked underparts, and black mesial stripe; generally indistinguishable in field from most other immature sparrowhawks.
Iris—yellow to brown; bill—brown; feet—yellow.
Voice: Generally silent. Piping *kyeew* in breeding area.
Range: Africa, India, S China, SE Asia.
Distribution and status: Uncommon winter visitor to Sumatra lowlands; few records but probably rarely identified.
Habits: Hunts from tree perches at forest edge, open wooded areas, and farmland, chasing birds and sometimes circling in the sky.

105. RUFOUS-WINGED BUZZARD *Butastur liventer*   PLATE 11

Description: Medium-sized (40 cm) buzzard with chestnut wings and tail, and pale underparts; head and nape brownish grey; upperparts brown, mottled, and streaked with black. Chin, throat, and breast grey, belly and vent white. Wings are long and rather pointed; tail is long, slender, and square-cut.
Iris—yellow; bill—yellow with black tip; cere—yellow; feet—yellow.

Voice: Silent except during breeding season when it noisily gives repeated, long-drawn shrill mews *pit-piu* with first note higher.
Range: China, SE Asia, Sulawesi, and Java.
Distribution and status: In Java a rare resident of lowland forest, mostly below 800 m.
Habits: Inhabits dry open forest near rivers or swamps. Generally hunts from tree perches near open water or in open cultivated patches.

106. GREY-FACED BUZZARD *Butastur indicus*   PLATE 11

Description: Medium-sized (45 cm) brownish buzzard with prominent white chin, and throat with black mesial stripe and black moustache. Sides of head blackish; upperparts brown, streaked and barred blackish; breast brown streaked black; rest of underparts barred rufous; tail with bold bars.
Iris—yellow; bill—grey; cere—yellow; feet—yellow.
Voice: Tremulous *chit-kwee* with second note rising.
Range: Breeds in NE Asia; winters south to SE Asia, Philippines, and Indonesia.
Distribution and status: A winter visitor or straggler to the Greater Sundas; not uncommon N Borneo and Sumatra, rare south to Java. Not recorded Bali.
Habits: Inhabits open wooded areas up to 1500 m. Rather slow and laboured in flight, preferring to hunt from a tree perch.

107. COMMON BUZZARD *Buteo buteo*   PLATE 11
(The/Steppe Buzzard)

Description: Largish (55 cm) reddish brown buzzard. Upperparts dark reddish brown; sides of face buff, streaked reddish with prominent chestnut moustachial stripe; underparts whitish with rufous streaks and washed rufous on flanks and thighs. In flight the broad, rounded wings and white patch at the base of the primaries are diagnostic. There is usually a black subterminal tail bar. Soars with wings in slight 'V'.
Iris—yellow to brown; bill—grey with black tip; cere—yellow; feet—yellow.
Voice: Loud mewing *peeioo*.
Range: Breeds in Palaearctic; winters south to Africa, India, and SE Asia.
Distribution and status: Recorded Java and Bali. Not formally recorded Sumatra but presumably also occurs as migrant or winter visitor.
Habits: Prefers open country where it circles on thermals, high overhead, or rests on exposed tree branches. One of the few hawks that hover regularly.

94 HAWKS, EAGLES, AND VULTURES

108. BLACK EAGLE *Ictinaetus malayensis* PLATE 11
(Indian/Asian Black Eagle)

Description: Large (70 cm), long-winged, and long-tailed black eagle, appearing very big in flight. There is a pale patch at the base of the primaries and slight barring on the tail, but general appearance in flight or at rest is all black. Immature: paler with buff streaks and edges to feathers, and pale thighs.
Iris—brown; bill—black with grey tip; cere—yellow; feet—yellow.
Voice: Repeated plaintive *kleeee-kee* or *hee-lee-leeeuw*.
Range: India, SE China, SE Asia, Sulawesi, Moluccas, and Greater Sundas.
Distribution and status: Sparsely but widely distributed throughout the Greater Sundas, in lowland and hill forests up to 1400 m, on Java up to 3000 m.
Habits: Inhabits forests where it is usually seen circling low over the canopy. It is a magnificent glider, quartering effortlessly the forested hillsides, often in pairs. Regularly raids the nests of other birds.

109. BOOTED EAGLE *Hieraaetus pennatus* PLATE 11

Description: Smallish (50 cm), rufous-breasted (dark phase) or buffy-white breasted (light phase), crestless eagle with feathered legs. Two colour phases differ in ventral colour from buffy to chocolate chin, cheeks, and undertail coverts; upperparts brown mottled with black and buff, darker brown on wings and tail. In flight dark primaries contrast strongly with buff (light phase) or rufous (dark phase) underwing coverts; undertail pale.
Iris—brown; bill—blackish with yellow cere; feet—yellow.
Voice: Thin, high *keee*.
Range: Breeds in Africa, SW Eurasia, NW India, and N China; migrates south in winter to Africa, India, and Malay Peninsula.
Distribution and status: Recorded as a vagrant to Bali.
Habits: Similar to Rufous-bellied Eagle

110. RUFOUS-BELLIED EAGLE *Hieraaetus kienerii* PLATE 11
(Chestnut-bellied Eagle)

Description: Smallish (50 cm), rufous, black, and white eagle with short crest. Adult: blackish crown, cheeks, and upperparts; tail dark brown with black bars and a white tip; chin, throat, and breast white, streaked black; flanks, belly, thighs, and undertail rufous with black streaks on belly. In flight shows conspicuous, rounded pale patch at base of primaries. Immature: upperparts blackish brown with blackish eye patch and whitish eyebrow; underparts whitish.
Iris—red; bill—blackish, cere—yellow; feet—yellow.

HAWKS, EAGLES, AND VULTURES   95

Voice: High-pitched scream *chirrup* preceded by several ascending preliminary notes. High-pitched *kliu* (M.& W.).
Range: S India, Himalayas, SE Asia, Philippines, Sulawesi, and Greater Sundas.
Distribution and status: In the Greater Sundas an uncommon resident of forested areas, up to 1500 m.
Habits: Inhabits forest and forest edge where it is generally seen circling or gliding fairly low over the trees.

111. CHANGEABLE HAWK-EAGLE   *Spizaetus cirrhatus*   PLATE 11
(Sunda/Marsh/Crested Hawk-Eagle)

Description: Large (70 cm) slender eagle with very broad wings and long, rounded tail. Has very short crest. Occurs in dark, pale, and intermediate phases. Commoner dark phase: dark brown all over with a black terminal tail bar contrasting with lighter, brown tail. Intermediate phase: brown, streaked and mottled black and white on upperparts; reddish brown tail with several black bars; chin, throat, and breast white with bold black streaks; lower belly, thighs, and undertail banded with rufous. Light phase: upperparts greyish brown; underparts whitish with darkish eye-stripe and moustache.
Iris—yellow to brown; bill—blackish; cere—blackish yellow; feet—greenish yellow.
Voice: Prolonged, shrill scream *yeep-yip-yip-yip*, *yeep-yip-yip-yip*, rising *kwip-kwip-kwip-kwee-ah* or penetrating *klee-leeuw*.
Range: India, SE Asia, Philippines, and Sundas.
Distribution and status: Throughout the Greater Sundas, uncommon at all altitudes, up to 2000 m.
Habits: Frequents forest and open woodlands, and even raids village fowl. Hunts from perches in dead trees and also from the air. Generally hunts over clearings.

112. JAVAN HAWK-EAGLE   *Spizaetus bartelsi*   PLATE 11

Description: Large (60 cm) eagle with a prominent crest. Adult: crest, crown, and moustachial stripe black; sides of head and nape chestnut; back and wings dark brown; tail brown, barred black; throat white with mesial stripe of black streaks; rest of underparts whitish, barred rufous. Immature: unmarked, rufous underparts. Intermediate stages also occur.
Iris—yellow to brown; bill—blackish; cere—black; feet—yellow; legs—feathered.
Voice: Harsh, high-pitched cries, quite distinct from those of the Changeable Hawk-Eagle.
Range: Endemic to Java.
Distribution and status: An uncommon resident recorded mostly in the mountains in W Java up to 3000 m, but at sea-level in E Java (Meru Betiri).
Habits: Inhabits forest and open wooded areas, in hilly and mountainous areas.

113. BLYTH'S HAWK-EAGLE *Spizaetus alboniger*   PLATE 11

Description: Large (52 cm), black and white eagle with long crest and broad white tail bar; streaked breast and barred belly, almost black in some individuals; throat white with black mesial stripe. Underparts are boldly barred, and black tail has a single, broad, white subterminal bar and white tip. Immature: brown upperparts, scaled buff; pale head; buffy brown-barred underparts and banded tail. Sub-adult: like adult but brown instead of black.
Iris—yellow; bill—grey; feet—yellow.
Voice: Various shrill whistles usually with stressed higher-pitched notes.
Range: Malay Peninsula, Sumatra, and Borneo.
Distribution and status: Probably widespread in Sumatra and Borneo; specimens are known from mainland Sumatra, Simeulue, Nias, Mentawai, and Belitung; on Borneo specimens are all from Sabah and Sarawak but there are reliable sight records from Kalimantan. This is an uncommon bird of primary and logged forest and forest edge, in the hills and mountains between 300 and 1200 m.
Habits: Rests in tall forest trees, sometimes mobbed by drongos and bulbuls. Circles over forest when hunting.
Note: Placed by some authors in Mountain Hawk-Eagle *S. nipalensis*.

114. WALLACE'S HAWK-EAGLE *Spizaetus nanus*   PLATE 11

Description: Smallish (45 cm), brown and white eagle with crest and three broad black bands in tail. Head and underparts pinkish buff with black streaks on breast and narrow black barring on belly. Immature: similar to immature Blyth's Hawk-Eagle, but smaller.
Iris—yellow; bill—grey; feet—yellow.
Voice: Shrill, high-pitched *yik-yee* inflected upwards on second note (P.R.).
Range: Malay Peninsula, Sumatra, and Borneo.
Distribution and status: An uncommon but widespread bird of lowland forests through Sumatra (including Nias and Bangka) and Borneo, up to 1000 m.
Habits: Similar to other hawk-eagles.

FALCONS—FAMILY FALCONIDAE

A MEDIUM-SIZED, worldwide family of fast-flying predatory birds with long, pointed, sickle-shaped wings and long, narrow tails. Falcons are the 'jet fighters' among avian raptors, stooping on prey with superior speed. The powerful beak is hooked at the tip and has two additional small lateral 'hook teeth' on the upper bill.

Eight species of falcon occur in the Greater Sundas, of which five are resident.

FALCONS 97

115. BLACK-THIGHED FALCONET *Microhierax fringillarius*
(Black-legged/Black-tailed/Black-sided Falconet)   PLATE 12

Description: Tiny (15 cm) black and white falcon. Upperparts black with white spots on innermost secondaries and on tail. Underparts white on breast, rufous on belly, and black thighs; side of face and ear coverts black, ringed with a white line or patch. Juvenile face suffused reddish.
Iris—dark brown; bill—greyish; feet—grey.
Voice: A hard, high-pitched cry *shiew* and a fast repeated *kli-kli-kli-kli*.
Range: Malay Peninsula and Greater Sundas.
Distribution and status: Common in the wooded lowlands of Sumatra and Borneo (except north) up to 1000 m; now scarce in Java and Bali.
Habits: Sits on exposed perches at edge of forest or in open country, including paddy fields. Makes sudden dashes to catch dragonflies and other insects, and sometimes boldly attacks small birds and other prey. Nests in tree holes.
Note: Some authors place this species in Collared Falconet *M. caerulescens*.

116. WHITE-FRONTED FALCONET *Microhierax latifrons*   PLATE 12
(White-browed/Bornean Falconet)

Description: Tiny (15 cm) black and white falcon. Similar to Black-thighed Falconet but has whiter face and white of forehead extends back as far as the eyes. No white spots on wings and tail. Female has chestnut forehead. Immature has fulvous crown and cheeks.
Iris—brown; bill—greyish; feet—grey.
Voice: As other falconets.
Range: Endemic to Borneo.
Distribution and status: Restricted to N Borneo; south to the Lawas river in the west and Darvel Bay in the east. An occasional bird of forest edge and farmland, up to 1200 m.
Habits: Perches on exposed dead branch on high tree and makes dashing chases after insects or birds. Flies with fast wing-beats. Nests in tree holes.

117. COMMON KESTREL *Falco tinnunculus*   PLATE 12, 13
(Eurasian/European/Rock/Old World/The Kestrel)

Description: Small (33 cm) brown falcon. Male has grey crown and nape, bluish grey unbarred tail and rufous upperparts, lightly barred black; underparts buff, streaked black. Larger female: all brown upperparts; less rufous and more boldly barred than male. Immature: like female but with heavier streaking. Distinguished

from Spotted Kestrel by grey cap of male and brown tail of female, and from Australian Kestrel by darker colour and heavier black markings.
Iris—brown; bill—grey with black tip; feet—yellow.
Voice: Piercing cries *yak yak yak yak yak.*
Range: Africa, Eurasia, India, and China; wintering south to Philippines and SE Asia.
Distribution and status: Vagrants reach Sumatra (recorded once in Nias) and regularly N Borneo.
Habits: Superb aerial grace, circling lazily or hovering motionless when hunting. Dives on prey, often taken on the ground. Perches on poles and in dead trees. Prefers open country.

118. SPOTTED KESTREL *Falco moluccensis* PLATE 12
(Moluccan Kestrel)

Description: Small (30 cm), dark brown, erect falcon. Male: tawny crown and upperparts, strongly banded and spotted with black; fulvous underparts, broadly streaked black. Tail bluish grey with broad, black subterminal bar and white tip. Female: slightly larger than male and with bolder barring on tail. Immature: similar to adult but paler, and tail brown with dark bands.
Iris—brown; bill—blue-grey with black tip and yellow cere; legs and feet—yellow.
Voice: Young birds give repeated cry when they see parents *kiri kiri kiri*; or when excited strong *kekekeke*, also used by adults in territorial display.
Range: Java, Sulawesi, Moluccas, and Lesser Sundas.
Distribution and status: Once collected in S Borneo. In Java and Bali a frequent resident of open habitats at all altitudes.
Habits: As Common Kestrel.

119. AUSTRALIAN KESTREL *Falco cenchroides* PLATE 12, 13
(Nankeen Kestrel)

Description: Small (30 cm), erect, long-tailed, brown falcon. Similar to Common Kestrel but much paler with less bold black markings. Male has grey forehead and brown nape. Immature like female but duller, more greyish rufous below, and heavier black barring above.
Iris—brown; bill—grey with a black tip and yellow cere; legs and feet—yellow.
Voice: As Common Kestrel.
Range: Resident in Australia and S New Guinea; migrates north in summer into Lesser Sundas.
Distribution and status: Occasional stray birds reach Java and Bali.
Habits: As Common Kestrel.

FALCONS 99

120. EURASIAN HOBBY *Falco subbuteo*   PLATE 12, 13
(European/Northern/The Hobby)

Description: Small (30 cm), long-winged, black and white falcon with rufous thighs and vent. Upperparts dark grey; breast creamy white, streaked with black. Female larger than male, browner, and with more streaking on thighs and undertail coverts. Readily distinguished from Oriental Hobby by whitish breast. Iris—brown; bill—grey with yellow cere; feet—yellow.
Voice: Repeated shrill *kick*.
Range: Africa, Eurasia, India, China, and Burma; migrating south in winter.
Distribution and status: Rare stragglers recorded for Java.
Habits: Catches insects and birds on the wing in fast flight, over open and wooded areas.

121. ORIENTAL HOBBY *Falco severus*   PLATE 12

Description: Small (25 cm), long-winged, rufous and black falcon. Head and upperparts dark grey with a bluish hue; underparts rich chestnut; chin buffy. Immature has black streaks on the rufous breast.
Iris—brown; bill—grey with yellow cere; legs and feet—yellow.
Voice: *Kekekeke cry*, similar to Spotted Kestrel.
Range: Widely distributed through tropical Asia to Indonesia, New Guinea, and the Solomon Islands.
Distribution and status: Status on Sumatra uncertain; possibly resident but presence based on a few sight records only. On Borneo a vagrant. On Java and Bali, a rather rare resident of lowland forests.
Habits: Flies very fast over forest, chasing insects, looking rather like a large swift. Rests on trees rather than rocks.

122. PEREGRINE FALCON *Falco peregrinus*   PLATE 12
(Peregrine)

Description: Large (45 cm), heavily built, dark falcon. Adult: crown and cheeks blackish or with black stripe; upperparts dark grey, spotted and barred with black; underparts white with black streaks on breast, and fine black bars across the belly, thighs, and undertail. Female notably larger than the male 'tercel'. Immature browner with streaked belly. Races differ in darkness. Differs from Bat Hawk in flight by paler underparts and less pointed wings.
Iris—black; bill—grey with yellow cere; legs and feet—yellow.
Voice: In breeding season a shrill *kek-kek-kek-kek*.
Range: Cosmopolitan.

100 FALCONS

Distribution and status: Winter visitor from N Asia to the coasts and lowlands throughout the Greater Sundas. Darker resident *ernesti* is rare and local in the mountains of N and W Sumatra, N Borneo, Java, and Bali.
Habits: Lives in pairs. Shows great speed in flight making breathtaking, spiralling stoops onto prey from high in sky. A candidate for the fastest bird in the world. Sometimes performs acrobatics. Nests on cliff ledges.

Megapodes — FAMILY MEGAPODIIDAE

MEGAPODES are an Australasian family of ground-living birds which use their large, powerful feet for dragging and piling mounds, and digging pits in which to lay their eggs. The parent birds do not brood the eggs which are incubated by passive, natural warmth (from solar or geothermal heat, or decaying vegetation). On hatching the young birds are fledged and can fly immediately. Adult birds have mournful growling calls and roost in trees.

There are two species of megapodes in the Greater Sundas.

123. ORANGE-FOOTED SCRUBFOWL *Megapodius reinwardt*
(Reinwardt's Scrubfowl)   PLATE 14

Description: Medium-sized (35 cm), terrestrial, greyish brown fowl with reddish face and a short crest. Upperparts chestnut brown; underparts greyish with neck and breast dark. Young birds are mottled and barred dark brown.
Iris—brown; bill—yellow with dark base; feet—orange.
Voice: Curious haunting wails at night and occasionally low gurgling calls.
Range: Islands in Java, Flores, and Banda Seas, and Lesser Sundas to New Guinea and N Australia.
Distribution and status: In the Greater Sundas it is found only on Kangean Islands, E Java, where it is becoming fairly scarce.
Habits: Single birds or pairs scurry about the forest floor, scrub, and even mangroves, scratching for insects on the ground. When disturbed, run away or fly low over the ground. Roost in low trees at night. Eggs are incubated in huge mounds of dry, rotting vegetation. The nests are built up over many years, sometimes several pairs using the same mound and helping to drag new leaves, twigs, and dead wood with their feet. The female bird tunnels into the top of the mound and lays a large, pinkish red egg which becomes pale pink as it ages. Pigs, monitor lizards, and man raid the nests for food. The fully feathered young finally emerges about 70 days later, already able to fly.

124. TABON SCRUBFOWL *Megapodius cumingii* PLATE 14
(Philippine Scrubfowl)

Description: Medium-sized (35 cm), brown and grey fowl with scarlet face. Upperparts brown, underparts grey. There is almost no crest and the tail is short. The sexes are alike. Immature birds are brown, barred and blotched with black.
Iris—brown; bill—yellowish; feet—blackish, reddish at back.
Voice: Repeated mournful *miaows* like a distressed cat, given in the evening.
Range: Philippines, Sulawesi, and N Borneo.
Distribution and status: Formerly common on most of the small islands around Borneo (Labuan, Tiga, Mengalum, Mantanani, Balambangan, Banggi, Malawali, and Maratuas), and on the mainland coast in a few localities in Sabah. Now rather scarce due to excessive collecting of eggs.
Habits: As Orange-footed Scrubfowl.
Note: Sometimes placed in super-species Common Scrubfowl, *M. freycinet.*

Pheasants—FAMILY PHASIANIDAE

PHEASANTS are a worldwide family of ground-living birds with short round wings, but often long tails. Males are usually very decorative while females are drably camouflaged. They nest on the ground but roost in trees. Some species have loud, clear calls. Many species have wing-whirring or shaking displays. Most species have fighting spurs on the males' legs. Flight is flurried and usually only for short distances, but the birds can run well.

There are 22 species in the Greater Sundas. None is migratory.

125. LONG-BILLED PARTRIDGE *Rhizothera longirostris* PLATE 14

Description: Large (30 cm) brown partridge with long, curved, black bill. Adult male: distinctive chestnut-brown head, grey breast, brown back, double spur. Female: lighter, more fulvous, and lacks grey breast; single spur. The montane race of Mt. Dulit is greyer than lowland forms.
Iris—light brown; bill—grey; feet—yellow.
Voice: Shrill whistle *kanking* following single note of partner in bell-like duet. (See Appendix 6.)
Range: Malay Peninsula, Sumatra, and Borneo.
Distribution and status: Few records from Sumatra and little information on habitat; presumably lowland and hill forest to 1000 m. In Borneo it has a patchy distribution; a rare montane bird of Mts Dulit and Batu Song (at 1000 m); also in lowlands of SW Sarawak and the Barito catchment.

Habits: Lives on the floor of primary forest and forest on limestone hills. Alights in trees when flushed.

126. BLACK PARTRIDGE *Melanoperdix nigra* PLATE 14
(Black Wood-Partridge)

Description: Medium-sized (25 cm), dark, thick-billed partridge. Sexes differ. Male is almost entirely black; female rich brown with bold black scalloping on wing coverts, and pale throat and breast.
Iris—dark brown; bill—black in male, brown in female; feet—pale blue.
Voice: Double whistle similar to Crested Partridge.
Range: Malay Peninsula, Sumatra, and Borneo.
Distribution and status: Few records from the lowland forest of mainland Sumatra. Not uncommon in the lowlands of S and W Borneo, north to at least Mulu; not recorded E Borneo.
Habits: A quiet bird of forest floor in primary forests and peat swamp forests.

127. BLUE-BREASTED QUAIL *Coturnix chinensis* PLATE 15
(Painted/Asian Blue Quail)

Description: Very small (15 cm) plump quail. Male: dark with white marks on black bib; side of head, breast, and flanks deep blue; belly and undertail chestnut. Female: paler, streaked brown above, buff below with barring on breast and flanks and buff eyebrow.
Iris—brown; bill—black; feet—yellow.
Voice: Usually silent, but has plaintive sweet whistle *tee-yew* and soft *tir-tir-tir-tir*.
Range: India, China, SE Asia, Philippines, and Indonesia to New Guinea and Australia.
Distribution and status: In Greater Sundas not uncommon in suitable lowland areas, up to 1300 m.
Habits: Frequents dry open grassland, harvested paddy fields, *Imperata* grasslands, and fallow agricultural land. Partly nocturnal and usually seen by day only when flushed from cover.

128. GREY-BREASTED PARTRIDGE *Arborophila orientalis* PLATE 14
(Grey-bellied/Sumatran Partridge, AG: Hill-Partridge)

Description: Medium-sized (25 cm) boldly marked partridge. Crown, nape, and eye-stripe extending down side of neck dark brown, eyebrow, chin, and ear coverts white; back brown, barred with black; wings brown with black spots and orange barring; breast greyish brown; vent and belly whitish; flanks boldly

patterned black and white. The Javan form has more white on head, and Sumatran forms have broader black and white barring on flanks.
Iris—yellow; **bill**—reddish brown; **feet**—yellow.
Voice: Double chirruping whistle *wut-wut*, *wut-wut* rising in volume. Male's full call preceded by single spaced whistles.
Range: SW China, SE Asia, Sumatra, and Java.
Distribution and status: Not uncommon between 500 and 2000 m in forest of the Barisan Range on Sumatra. In Java restricted to E Java where an uncommon montane resident, notably found in Yang and Ijen Highlands.
Habits: Similar to Chestnut-bellied Partridge.
Note: Includes Cambell's Partridge *A. o. cambelli* of Malay Peninsula and can be combined with Chestnut-bellied Partridge *A. javanica*. All can be combined with the Bar-backed Partridge *A. brunneopectus* of SE Asia, but the name *javanica* has precedence.

129. CHESTNUT-BELLIED PARTRIDGE *Arborophila javanica*
(Javan Partridge, AG: Hill-Partridge) PLATE 14

Description: Medium-sized (25 cm) drably marked partridge. Three subspecies are distinguished by differences of head pattern but all have reddish orange and black markings on head and blackish collar; breast, back, and tail greyish, barred black; wings brownish with black bars and spots; underparts reddish brown.
Iris—grey; **bill**—black; **feet**—red.
Voice: A rail-like call of one or more birds giving a monotonous series of soft double cries in ever-increasing tempo and rising in volume.
Range: Endemic to Java.
Distribution and status: Confined to W and C Java (east to Lawu) where three local races are recognized. It is not uncommon in montane forest from 1000 to 3000 m.
Habits: Inhabits montane forest and open patches. Generally in pairs or small coveys, often encountered crossing a forest path as they search for food in the ground litter.
Note: May be conspecific with Chestnut-backed and Bar-backed Partridges *A. brunneopectus*.

130. RED-BILLED PARTRIDGE *Arborophila rubrirostris* PLATE 14
(AG: Hill-Partridge)

Description: Medium-sized (30 cm) brown partridge with pied head. Crown and sides of head black with small white spots; throat white, spotted black. The amount of white on head varies greatly, being whiter in northern specimens.

Upperparts brown, narrowly barred black; large black markings on wings. Breast bright brown with small white spots; belly grey, spotted with black.
Iris—brown; bill—red; bare skin of eye-ring—red; feet—red.
Voice: Loud whistled couplet *keow* rising in pitch.
Range: Endemic to Sumatra.
Distribution and status: Common in the Barisan Range of Sumatra, in mountain forests from 900 to 2500 m.
Habits: Shy and ground-living in small groups, preferring mossy gulleys and dense undergrowth of hill ridges.

131. RED-BREASTED PARTRIDGE *Arborophila hyperythra* PLATE 14
(Bornean Partridge, AG: Hill-Partridge)

Description: Medium-sized (32 cm), brown partridge with bold, black and white pattern on flanks. Crown and nape blackish, streaked brown; eyebrow and cheeks pale chestnut. Some birds from Sarawak have grey eyebrow and cheeks, and darker breast. Upperparts brown, narrowly barred black; bold black marks on tertiaries. Breast cinnamon; vent white. Female has browner, smaller white spots.
Iris—brown; bill—grey; bare eye-ring skin—crimson; feet—pinkish.
Voice: A repeated, ringing, rising note *kuwar*, uttered at a rate of about 3 per second, answered by a loud, double note *cuckoo* dropping in pitch, at a rate of about one per second.
Range: Endemic to Borneo.
Distribution and status: Confined to hills of N Borneo; from Mt. Kinabalu south to Usun Apau and upper Kayan, and outlier Mt. Mulu. Not uncommon in sub-montane forests from 600 to 1200 m, notably common in Kelabit highlands, also reported from Ulu Barito.
Habits: Shy, groups keep to thickets and bamboo in forest.

132. CHESTNUT-NECKLACED PARTRIDGE *Arborophila charltonii*
(Chestnut-breasted Partridge, AG: Hill-Partridge) PLATE 14

Description: Medium-sized (32 cm), brown partridge with whitish eyebrow and throat. Crown and upperparts brown, finely barred black; eyebrow and throat white, flecked with black and with black edge; lores and eye-stripe blackish. Underparts cinnamon, spotted black on upper breast and barred black on flanks, with a distinctive brown band across the lower breast. Undertail coverts are buffy white.
Iris—brown; bill—greenish yellow with red base; feet—greenish yellow.
Voice: Melodious, clear whistled couplets or triplets of steady pitch and volume, generally answered in antiphonal duet.

Range: N Vietnam, Malay Peninsula, Sumatra, and Borneo
Distribution and status: Recorded from the lowlands of E Sumatra but few records. On Borneo restricted to Sabah and locally common in lowland forest and scrub.
Habits: Shy, groups prefer forest edge and secondary scrub.
Note: Populations with chestnut breast band now separated from Scaly-breasted Partridge *A. chloropus* of S China and SE Asia; also formerly included Annam Partridge *A. merlini*.

133. FERRUGINOUS PARTRIDGE *Caloperdix oculea* PLATE 15
(AG: Wood-Partridge)

Description: Medium-sized (26 cm) rufous partridge with black back and sides of breast distinctively scaled white or buff. Blackish line behind eye; wings and flanks are boldly spotted with round black marks.
Iris—olive-brown; bill—black; feet—olive-green.
Voice: Call of male rises up scale, accelerates, and is repeated eight or nine times before breaking into 2–4 *ee-terang* calls. Female call is rising scale of about twenty fast whistled notes (T.H.).
Range: Malay Peninsula, Sumatra, and Borneo.
Distribution and status: In Sumatra a rare bird of lowland and hill scrub and forest, to 1000 m; apparently widespread but there are few records. On Borneo a montane resident from the Kelabit uplands to the Usun Apau plateau and Dulit; also found on Mt. Magdalena but absent from Mt. Kinabalu.
Habits: Poorly known. In Borneo a bird of tall forested hillsides and sandy highland valley bottoms, but the Sumatran race prefers secondary scrub.

134. CRIMSON-HEADED PARTRIDGE *Haematortyx sanguiniceps*
(AG: Wood-Partridge) PLATE 14

Description: Medium-sized (25 cm) plump partridge with unmistakable crimson head, breast, and vent. Male otherwise blackish brown; female olive-brown.
Iris—brown; bill—yellow in male, brown in female; feet and legs—grey with two spurs in male and one in female.
Voice: High, harsh clucking *kak kak kak. Pom-prang*, *pom-prang* repeated several times in small, metallic voice.
Range: Endemic to Borneo.
Distribution and status: Widespread in the hills and mountains of Borneo. Formerly regarded as common (Smythies, 1981) but now uncommon between 500 and 1700 m, occasionally found down to 200 m.

Habits: Generally confined to heath forests and dry sandy forests of valley bottoms in mountains.

135. CRESTED PARTRIDGE *Rollulus rouloul* PLATE 15
(Roulroul, AG: Wood-Partridge)

Description: Plump (25 cm), crested, ground-living partridge. Male has distinctive spreading crimson crest, white crown patch, and dark crimson wings contrasting with metallic purplish blue body. Female has grey head and crest, chestnut wings, and green body.
Iris—red; bare eye-ring skin—red; bill—black with red base in male, black in female; feet—red.
Voice: Shrill, plaintive whistles *si-il*, repeated in steady series, usually at dawn.
Range: Malay Peninsula, Sumatra, and Borneo.
Distribution and status: A common resident in lowland forests on Sumatra (formerly including Bangka and Belitung) to 800 m, and Borneo, up to 1200 m.
Habits: Associates in parties of 5—15, scrabbling in the forest floor litter for food, often under fruiting trees where primates or birds are feeding. Parties scatter noisily when alarmed.

136. CRESTLESS FIREBACK *Lophura erythrophthalma* PLATE 16
(Rufous-tailed Fireback, AG: Pheasant)

Description: Large (49 cm) crestless pheasant. Bornean male has greyish mantle, wings, throat, and breast, streaked white; underparts and tail coverts glossy blue-black; rump deep red and tail buff. Sumatran male lacks white streaking and has mantle, wings, and throat glossy bluish black. Female has glossy blue-black back with greyish brown head and buffy throat. Both sexes have bare scarlet facial skin.
Iris—brown; bill—greyish; feet—grey.
Voice: Low croaking *took-taroo*, repeated (Medway). Male gives wing-whirring display.
Range: Malay Peninsula, Sumatra, and Borneo.
Distribution and status: A scarce resident of lowland forest; there are few records from Sumatra, and range restricted on Borneo where recorded from S (north to Brunei) and W.
Habits: A shy bird of the forest floor in dense primary forest. Habits presumed to be similar to Crested Fireback.

137. CRESTED FIREBACK *Lophura ignita* PLATE 16
(Vieillot's/Malaysian Fireback, AG: Pheasant)

Description: Male (55 cm) has dark purplish blue plumage (appears black in forest) with diagnostic bushy black crest and white (Sumatra/Malaya) or yellow (Borneo) central tail feathers. The lower back is deep red. Bornean male has chestnut belly, Sumatran is blue-black with white streaks. Female (40 cm) has brownish plumage with a bushy brown crest, and white stripes and scaling on underparts.
Iris—red; bill—horn; fleshy facial skin—blue; legs and feet—pinkish or greenish (male has long spur).
Voice: Male gives croaking call followed by a shrill chirp and whir of wings *woonk-k(whir), woonk-k(whir)* (Beebe). Both sexes give sharp alarm call *chukun, chukun*.
Range: Malay Peninsula, Sumatra, and Borneo.
Distribution and status: Recorded throughout the lowland forest of Sumatra (formerly including Bangka) and Borneo. Formerly a common bird of the forest floor but now with patchy distribution and rarely abundant due to habitat destruction and hunting.
Habits: One-male parties of 5–6 birds keep to dark forest, searching around rotten logs or under fruiting trees for food. Scratches for food like a chicken. Fast runner; sometimes makes a short noisy flight.

138. BULWER'S PHEASANT *Lophura bulweri* PLATE 16
(Wattled/White-tailed Wattled Pheasant)

Description: A large handsome pheasant (male 77 cm, female 50 cm). Male unmistakable with splendid blue facial wattles and a long, spreading, curved white tail. The general body colour is bluish black with blue edges to feathers; throat and upper breast purplish. Female: dull mottled brown with blue facial skin.
Iris—red; bill—dark horn; legs and feet—red (male has a small spur).
Voice: In mating season gives a shrill, piercing cry (Heinroth). Penetrating metallic *kook kook* given by both sexes, and a sharp nervous *kak kak* in alarm (Beebe).
Range: Endemic to Borneo.
Distribution and status: Found in hill forests to 1600 m in all regions. The species is locally common but getting rarer.
Habits: Similar to junglefowl (which is absent from Borneo). Lives in primary and old secondary forests. In display male extends wattles, and raises and spreads tail.

139. SALVADORI'S PHEASANT *Lophura inornata*   PLATE 16
(Salvadori's Fireback)

Description: Male (50 cm), glossy bluish black, crestless pheasant with short tail and crimson facial skin. Female (42 cm), mottled reddish chestnut with dark brown tail and red facial skin.
Iris—brown; bill—greenish white; legs and feet—greenish grey (male with long spur).
Voice: Quiet alarm call.
Range: Endemic to Sumatra.
Distribution and status: Confined to the lower montane forests of the C and S Barisan range (south of Ophir district), from 1000 to 1800 m; to 2200 m on Mt. Kerinci. Uncommon to rare.
Habits: Lives on the floor of dense mountain forests, in pairs or small parties. Habits similar to other fireback pheasants.

140. HOOGERWERF'S PHEASANT *Lophura hoogerwerfi*   PLATE 17
(Sumatran Pheasant, AG: Fireback)

Description: Large dark pheasant (40–50 cm). Male has never been collected but photos taken in Mamas Valley of Gn. Leuser Park show a glossy bluish black, crestless male like Salvadori's Pheasant. The female is similar to Salvadori's but with browner back and less chestnut underparts, finely streaked with black. Appears more uniform without the scaly pattern of pale centred feathers of Salvadori's. Underparts yellowish brown; throat whitish; tail black.
Iris—amber; bill—blue-grey; bare facial skin—red; feet—dark blue.
Voice: No information.
Range: Endemic to Sumatra.
Distribution and status: Known from N Sumatra in montane forests at 1200 to 2000 m. Few records; all from Gayo Highlands (including Leuser National Park).
Habits: Lives on the floor of dense montane forests, in small groups of one male with several females.
Note: Taxonomic status uncertain. The form is known only from one female. It is treated (probably correctly) by some authors as a race of Salvadori's Pheasant.

141. RED JUNGLEFOWL *Gallus gallus*   PLATE 17
(Wild Junglefowl)

Description: Familiar, largish (male 70 cm, female 42 cm) wild ancestor of domestic chicken. Male: comb, wattles, and face red; hackles, tail coverts, and primaries bronze; mantle chestnut; elongated tail feathers and wing coverts glossy

greenish black; underparts dark green. Female drab brown with black streaks on neck and nape. N Sumatran race has longer hackles.
Iris—red; bill—horn; feet—bluish grey.
Voice: Male calls at dawn, similar to domestic cock but sharper and more abbreviated *bookikoh*; cheeping of young as in domestic chickens; also clucking alarm when disturbed. (See Appendix 6.)
Range: Himalayas, S China, SE Asia, Sumatra, and Java. Introduced in Philippines, Sulawesi, Lesser Sundas, and Australia.
Distribution and status: Common on Sumatra up to 900 m. In Java it was formerly widespread but is scarce now in many lowland areas; most recent records are from mountains in W Java.
Habits: Prefers semi-open scrub habitats. Male may be solitary, associating with female harems or occasionally with other males. Feeds on the ground but flies well and roosts in trees.

142. GREEN JUNGLEFOWL *Gallus varius* PLATE 17

Description: Large (male 60 cm, female 42 cm), greenish black junglefowl. Like Red Junglefowl but comb not serrated and with purple hue; nape, neck, and mantle glossy green; tail coverts golden and hackles iridescent green; wing flight feathers black; underparts black. Female is buffy brown, irregularly banded and mottled black.
Iris—red; bill—creamy grey; feet—pinkish.
Voice: A harsh, unmelodious, nasal *kookroh*, repeated at regular intervals.
Range: Java and Lesser Sundas.
Distribution and status: In Java and Bali locally common in suitable open habitat, up to 1500 m in W Java, and up to 3000 m in E Java.
Habits: Similar to Red Junglefowl but with a greater preference for open grassy areas and rarely, if ever, found in thick forest. Often feeds near grazing ungulates, catching insects attracted to them or their faeces or disturbed by their passing.

143. SUMATRAN PEACOCK-PHEASANT *Polyplectron chalcurum*
(Lesson's/Bronze-tailed Peacock-Pheasant) PLATE 17

Description: Medium-sized (male 50 cm, female 35 cm), russet brown pheasant with long pointed tail. Male is rich brown, mottled and barred reddish; long, pointed tail, barred and spotted black and tawny, with a dark, metallic, purple subterminal patch; almost no crest. Female is smaller and duller with shorter tail.
Iris—yellow; bill—brown; legs and feet—black (male with two spurs).
Voice: Loud, clear, far-carrying *karau karau karau* with second note stressed and

slightly higher-pitched, repeated in short series; also clucking and churring noises. (See Appendix 6.)
Range: Endemic to Sumatra.
Distribution and status: Not uncommon resident of primary and logged forest in Barisan range on Sumatra, from 800 to 1700 m.
Habits: A shy, secretive bird of dense primary and logged, lower montane forests.

144. BORNEAN PEACOCK-PHEASANT *Polyplectron schleiermacheri*
PLATE 17

Description: Medium-sized (male 42 cm, female 38 cm) pheasant marked with metallic (green in male, blue in female) eye-spots on the wings and tail. Male has metallic green crest and purplish green iridescent breast with white throat and breast patch. Female is duller and bluer. Both sexes have contrasting pale buff chin, and throat.
Iris—yellow; bill—dark greenish; bare facial skin—red; legs and feet—black. Male has two spurs.
Voice: A melancholic, double whistle *hor-hor*. (See Appendix 6.)
Range: Endemic to Borneo.
Distribution and status: Sight records of this species from Sumatra are presumed to be erroneous as no specimens have ever been collected there. In Borneo a rare bird known only from scattered localities in lowland forest, to 1100 m.
Habits: A shy bird of primary forest. Roosts in trees but walks quietly about the forest floor by day. Males call and display their wings and tail but do not have dancing grounds.
Note: May be a subspecies of Malay Peacock-Pheasant *P. malacense*.

145. GREAT ARGUS *Argusianus argus*
PLATE 15
(Great Argus/Argus Pheasant)

Description: Unmistakable large pheasant. Male (120 cm) has enormously elongated secondary feathers and central tail feathers. The wing feathers are boldly decorated with large eye-spots. Plumage generally rusty brown with intricate buff and black spots and patterns; underparts darker rufous. Female (60 cm) has shorter tail and wing feathers, is darker rufous, and lacks the male's eye-spots. Both sexes have blue bare skin of head and neck and a short dark crest.
Iris—red brown; bill—yellow; feet—red.
Voice: Male makes explosive, clear, double note *kow wow* often in response to a tree fall or the calls of other males. Another call, made by either sex, is a series of 20 or so clear *wow* notes on the same pitch, rising and speeding up slightly at the end. (See Appendix 6.)

Range: Malay Peninsula, Sumatra, and Borneo.
Distribution and status: On Borneo and Sumatra a common resident of tall, dry, lowland primary and logged forests, up to 1200 m. Now becoming rarer (locally extirpated) in many areas due to habitat destruction and excessive trapping. Sight records of the Crested Argus *Rheinartia ocellata* from Sumatra are presumed to be erroneous identifications of this species.
Habits: Males clear dancing rings on the forest floor, removing all leaves, seedlings, and stones. They call from these dancing grounds in the morning, and give a visual display to visiting females by raising and fanning the tail and wings, somewhat like the display of a peacock. Birds roost in trees at night, and sometimes rest in and even call from trees in the day.
Note: The Double-banded Argus *A. bipunctatus*, known from a single male primary, is thought to have been a Javan bird though there is some reason to suspect it came from Tioman Island, off E Malaya.

146. GREEN PEAFOWL *Pavo muticus* PLATE 15

Description: Unmistakable, huge (male 210 cm, female 120 cm) pheasant with elongated tail (male only) and vertical plume crest on head. Male has iridescent green mantle, neck, and breast, and enormously elongated 'fan' tail of ocellated iridescent feathers. Female lacks long tail and is less finely coloured with whitish underparts.
Iris—brown; bill—brown; feet—greyish black.
Voice: Loud, trumpet-like *kay-yaw* given at dusk and dawn.
Range: Assam, SW China, Indochina, and Java.
Distribution and status: On Java this bird is now largely restricted to forest margins in Ujung Kulon National Park, extreme W Java, and savanna woodland in Baluran National Park, E Java; elsewhere either extinct or very local.
Habits: Frequents open woodlands with grassy pastures, tea and coffee plantations, wandering about on the ground. Males display with fanned tail coverts raised or chase other males to and fro in the mating season. Roosts in tall bare trees at night.

Buttonquails—FAMILY TURNICIDAE

The buttonquails are tiny, short-tailed, compact birds similar in general appearance to the true quails of the pheasant family but lacking the hind toe. They also show reversed sexual roles with the female more brightly coloured and more aggressively territorial than the male. She often mates with several males and leaves them to incubate the eggs and raise the chicks. There are only two species in the Greater Sundas.

147. SMALL BUTTONQUAIL *Turnix sylvatica* PLATE 15
(Little/Common Buttonquail)

Description: Very small (14 cm), rufous brown, quail-like bird. Distinguished by rufous, unbarred breast, white streaks on upperparts, and reddish and black blotches on flanks. Larger female is darker and redder than male.
Iris—yellow; bill—grey; feet—whitish.
Voice: Low mooing or crooning notes.
Range: Africa, S Eurasia, India, SE China, SE Asia, Philippines, and Java.
Distribution and status: Recorded in Java and Bali where an uncommon lowland bird.
Habits: Similar to the commoner Barred Buttonquail.

148. BARRED BUTTONQUAIL *Turnix suscitator* PLATE 15
(Lesser Sunda/Sunda Buttonquail)

Description: Small (16 cm), russet brown, quail-like bird. Larger female has black chin and throat, blackish crown with grey and white mottled head. Male has brown mottled crown, brown and white streaked face and chin, and black barring on breast and flanks. Both sexes have brown mottled upperparts and rufous breast and flanks.
Iris—brown; bill—grey; feet—grey.
Voice: Female courting call is purring *krrrr*, maintained for several seconds at a time, often at night.
Range: India, Japan, S China, SE Asia, Philippines, Sulawesi, Sumatra, Java, Bali, and Lesser Sundas.
Distribution and status: On Sumatra, Java, and Bali this is the common quail found in most suitable habitats, from sea level to about 1500 m.
Habits: Lives singly or in pairs in open grassy habitats. When flushed these birds jump up, fly low over the ground for 20 metres or so, then drop down into the grass and hide.

RAILS—FAMILY RALLIDAE

RAILS are a worldwide family of rather secretive, swamp-living birds of medium size. They have strong, straight bills and long legs with very long toes. The wings are short and the flight weak and flappy. Rails prefer to walk and are good runners, but dash for cover and hide in thick reed clumps rather than try to outrun predators. Most species can swim and some do so regularly, coots having lobed feet for this purpose.

Most rails have loud, cacophonous calls, sometimes with more than one bird joining in. Rails frequent a range of habitats including swamps, lake edges, reed and cane beds, grasslands, rice fields, and secondary forest; few species live in forest. Rails nest on the ground and feed on a mixture of plant shoots, seeds, and invertebrates.

There are 14 species recorded from the Greater Sundas, four as visitors only.

149. WATER RAIL *Rallus aquaticus* PLATE 18

Description: Medium-sized (31 cm) dark rail with streaked upperparts. Crown brown, face grey with pale grey eyebrow and dark eye-stripe. Chin white; neck and breast grey. Flanks and underparts barred black and white. Immature birds have some indistinct white barring on upperwing coverts.

Iris—red; bill—red to black; feet—red.

Voice: Soft *chip chip chip* call, and strange grunts and groans.

Range: Palaearctic; migrants reach SE Asia and Borneo.

Distribution and status: A rare winter visitor to N Borneo; now three records from Brunei and Sarawak.

Habits: A shy bird of thick waterside vegetation, marshes, and mangroves.

150. SLATY-BREASTED RAIL *Gallirallus striatus* PLATE 18
(Blue-breasted Banded Rail)

Description: Medium-sized (24 cm) rufous-crowned rail with fine, transverse, white barring on back. Crown chestnut; chin white; breast and back grey; wings and tail finely barred white; flanks and undertail more coarsely barred black and white.

Iris—red; bill—black above, reddish below; feet—grey.

Voice: Hard, sharp double note *terrek* or buzzing *kech, kech, kech*, repeated 10–15 times, starting weakly, getting stronger, then fading again.

Range: India, S China, SE Asia, Philippines, Sulawesi, and Greater Sundas.

Distribution and status: Occurs throughout Sumatra and its islands, Borneo, Java, and Bali. A common bird in suitable habitat up to 1000 m, rarely up to 1500 m.

Habits: Found in mangroves, swamps, rice paddy, *Imperata* grassland, and even on dry coral islands. Not seen often as it is retiring and partly nocturnal. Generally solitary.

151. RED-LEGGED CRAKE *Rallina fasciata* PLATE 18
(Malay Banded/Malay Crake)

Description: Medium-sized (23 cm), reddish brown, short-billed rail with red legs. Head, back, and chest chestnut; wings and tail reddish brown; belly and undertail black with white bars; chin white. Conspicuous white spotting on wing coverts and white barring on flight feathers. Similar to Band-bellied Crake but white bars on flank and belly much bolder.
Iris—red; bill—brown; feet—red.

Voice: Loud series of nasal *pek* calls at half second intervals, given at dawn and dusk in breeding season; also slow descending trill (P.R.).

Range: India, SE Asia, Philippines, and Indonesia.

Distribution and status: Local resident throughout Greater Sundas; numbers augmented by winter visitors from mainland Asia.

Habits: A rarely seen inhabitant of open swamps in low-lying areas; shy and poorly known.

152. SLATY-LEGGED CRAKE *Rallina eurizonoides* PLATE 18
(Slaty-legged Banded/Ryukyu/Philippine/Banded Crake)

Description: Medium-sized (25 cm) brownish rail with chestnut head and breast; whitish chin; narrow white bars on blackish belly and undertail. White on the wing limited to sparse barring on inner secondaries and primaries.
Iris—red; bill—greenish yellow; feet—grey.

Voice: Double *kek-kek* call given at night.

Range: India, SE China, SE Asia, Philippines, Sulawesi; winters to Sumatra and Java.

Distribution and status: An uncommon visitor to Sumatra and vagrants may reach W Java but there are no recent records.

Habits: Shy rail in mangrove, scrub forest, and dense bushes.

153. BAILLON'S CRAKE *Porzana pusilla* PLATE 18
(Marsh/Tiny Crake)

Description: Small (18 cm), greyish brown, short-billed rail with white streaks on back and fine white barring on flanks and undertail. Crown and upperparts brown, streaked black and white; chin white; breast and face grey; dark eye-stripe.
Iris—red; bill—yellow; legs—greenish yellow.

Voice: A jarring trill descending the scale, similar to Little Grebe.

Range: Africa, Eurasia, Indonesia, and Philippines to New Guinea and Australia.

Distribution and status: Status uncertain; doubtful resident race known from

Sumatra (Belitung) and S Borneo. Mostly a scarce winter visitor in lowlands and hills, rare south to Java, and not recorded Bali.
Habits: Inhabits swampy lake sides and grassy marshes. Threads its way quickly but delicately through reed beds, rarely flies.

154. RUDDY-BREASTED CRAKE *Porzana fusca* PLATE 18
(Ruddy Crake)

Description: Small (21 cm), reddish brown, short-billed rail. Head and breast deep chestnut; chin white; hindcrown and nape brown; belly and undertail blackish with fine white barring. Similar to Red-legged and Band-bellied Crakes but smaller and lacks any white on wings.
Iris—red; bill—brown; feet—red.
Voice: Rather quiet; in breeding season, a soft *tewk* or *keek* uttered every 2 or 3 seconds, followed by bubbling sound.
Range: India, China, SE Asia, Philippines, Sulawesi, and Sundas.
Distribution and status: Locally common and sometimes abundant resident on Sumatra, Java and Bali, in suitable habitat up to 1000 m. Few records Borneo where possibly only a scarce winter visitor.
Habits: Inhabits reed beds, rice fields, and dry bush land beside lakes, but shy and not often seen. Occasionally ventures out at edge of reed beds. Partly nocturnal. Calls at dawn and dusk.
Note: Sometimes placed in genus *Amaurornis*.

155. BAND-BELLIED CRAKE *Porzana paykullii* PLATE 18
(Siberian Ruddy/Chinese banded Crake)

Description: Medium-sized (22 cm), reddish brown, short-billed rail with red legs. Crown and upperparts reddish brown; chin white; sides of head and breast chestnut; flanks and undertail blackish with fine white bars. Distinguished from Ruddy-breasted Crake by white wing bars, and from Red-legged Crake by finer white barring on underparts. Less white on wing coverts than Red-legged Crake; no white on flight feathers; nape and neck dark.
Iris—red; bill—yellowish; feet—red.
Voice: Strange call given at night, sounds like drumbeats or a wooden rattle.
Range: Breeds in E Asia; migrates in winter to Philippines and Greater Sundas.
Distribution and status: A scarce winter visitor to the Greater Sundas; recorded N Sumatra, N Borneo, and Java.
Habits: Inhabits wet grassland and paddy fields.
Note: Sometimes placed in genus *Rallina*.

156. WHITE-BROWED CRAKE *Porzana cinerea* PLATE 18
(Ashy/Grey-bellied Crake)

Description: Smallish (20 cm), greyish brown, short-billed rail with conspicuous head pattern of white stripes above and below black eye-stripe. Crown, back, and breast grey; wings and tail greyish brown; belly whitish; flanks and undertail buffy-brown.
Iris—red; bill—blackish; feet—greenish yellow.

Voice: Noisy high-pitched, thin, reedy piping *cutchi cutchi cutchi* with 2 or more birds joining in at once.

Range: Malay Peninsula, Philippines, and Greater Sundas to New Guinea and Australia.

Distribution and status: Resident and widespread in the lowlands, on mainland Sumatra (locally to 1200 m), Borneo, and Java. Locally common on N coast of Java and Bali.

Habits: A shy bird of flooded grassland, marshes, and paddy fields. Lives in pairs. In breeding season it is conspicuous only by its incessant calling both by day and night. Often walks out onto floating vegetation like small jacana, dashing back to cover among emergent vegetation at the slightest disturbance.

157. WHITE-BREASTED WATERHEN *Amaurornis phoenicurus*
PLATE 18

Description: Largish (30 cm), unmistakable, grey and white rail. Grey crown and upperparts; white face, forehead, breast, and upper belly; rufous lower belly and undertail.
Iris—red; bill—greenish with red base; feet—yellow.

Voice: Monotonous *uwok-uwok* call. Noisy, weird chorusing of several birds with grunts, croaks, and chuckles *turr-kroowak, per-per-a-wak-wak-wak*, and others for up to 15 minutes at a time, at dawn or at night.

Range: India, S China, SE Asia, Philippines, Sulawesi, and Sundas.

Distribution and status: A common resident and winter visitor in suitable habitat from lowlands up to 1600 m throughout the Greater Sundas.

Habits: Generally single, occasionally in twos and threes, skulking in damp scrub, lake sides, riverbanks, mangrove, and paddy fields but only when there is dense cover to hide in. Comes out into the open to feed so is seen more often than other rails. It also clambers about in bushes and small trees.

RAILS 117

158. WATERCOCK *Gallicrex cinerea*   PLATE 19

Description: Large (40 cm), black or buffy brown rail with short green bill. Female brown with finely barred underparts. Black breeding plumage of male with pointed, red frontal shield likely to be seen only in Sumatra.
Iris—brown; bill—yellow green; feet—green.
Voice: Deep booming on summer nesting grounds but generally silent in winter.
Range: India, China, SE Asia, and Philippines; migrants reach Sulawesi and Sundas.
Distribution and status: Breeding reported from Sumatra but otherwise an uncommon winter visitor in suitable lowland habitats throughout the Greater Sundas.
Habits: A shy, skulking, and largely nocturnal bird of reedy swamps. Sometimes comes into adjacent rice fields to feed on paddy.

159. COMMON MOORHEN *Gallinula chloropus*   PLATE 19
(Moorhen)

Description: Unmistakable, medium-sized (31 cm), black-and-white aquatic rail with bright red frontal shield and short bill. Plumage entirely slaty black except for line of white streaks along flanks and two white patches under tail, conspicuously displayed when tail is cocked.
Iris—red; bill—dull greenish with red base; feet—green.
Voice: Loud, harsh, croaking calls *pruruk-pruuk-pruuk*.
Range: Almost worldwide except Australia.
Distribution and status: Common resident on Sumatra, Java, and Bali in wetland areas up to 1200 m. Breeding is recorded S Borneo but records from N Borneo are probably winter migrants.
Habits: Frequents lakes, pools, and canals. Largely aquatic, swimming slowly about, dabbling at surface vegetation and insects. Runs over water or swims to cover when alarmed but can dive for safety, staying under water for long periods if pursued by a hawk. Comes onto open land to feed in mornings and evenings, and also clambers about in small bushes. Constantly flicks tail, on land or in water. Flies rather weakly after a long paddling take-off run over water.

160. DUSKY MOORHEN *Gallinula tenebrosa*   PLATE 19
(Black Moorhen)

Description: Medium-sized (30 cm), black aquatic rail. Similar to Common Moorhen but distinguished by lack of white streaks on flanks and by red feet and lower legs (both species sometimes have red thighs).
Iris—brown; bare frontal shield—red; bill—greenish yellow; legs and feet—red.

118 RAILS

Voice: Similar to Common Moorhen—a harsh double croak.
Range: Australia, New Guinea, Wallacea to S Borneo.
Distribution and status: In Borneo it is recorded only from the Bankau lakes in South Kalimantan, but it has not been recorded for many years.
Habits: As Common Moorhen.

161. PURPLE SWAMPHEN *Porphyrio porphyrio* PLATE 19
(Purple Gallinule/Purple Waterhen/Purple Coot/Purple Moorhen)

Description: Unmistakable, large (42 cm), chunky, purplish blue rail with a massive red bill. Entire plumage blue black with a purple and green sheen except for white undertail coverts. Has a red frontal shield.
Iris—red; bill—red; feet—red
Voice: Cackling grunts and hoots; trumpeted nasal *wak*.
Range: Australia, New Guinea, and E Indonesia.
Distribution and status: Not uncommon in suitable wetland habitat throughout Sumatra, S Borneo, Java, and Bali.
Habits: Inhabits reed-lined swamps and lakes, walking over floating vegetation and through reed beds. Sometimes comes out into open flooded grasslands, paddy fields, or even burnt grass areas in small parties. Constantly flicks tail.

162. COMMON COOT *Fulica atra* PLATE 19
(Black/Eurasian/European/The Coot)

Description: Unmistakable, large (40 cm), black aquatic rail with conspicuous white bill and frontal shield. Plumage entirely slaty black except for narrow whitish trailing edge to wing, visible in flight.
Iris—red; bill—white; feet—green.
Voice: Various loud calls and sharp *kik kik*.
Range: Palaearctic; south in winter, rarely reaching Indonesia. Also New Guinea and Australia.
Distribution and status: Not recorded Sumatra and a rare visitor to Borneo, Java, and Bali. Formerly a breeding population was known from the mountain lakes of the Yang Plateau, E Java, but there are no recent records.
Habits: Highly aquatic and gregarious; regularly dives to collect water weeds from the lake bottom. Fights and chases occur in breeding season. Birds make long running take-off over water.

Finfoots — FAMILY HELIORNITHIDAE

A TROPICAL family of only three species, one each in America, Africa, and Asia. Finfoots are heavy-billed, grebe-like waterbirds living in swampy places where there is abundant tree cover but, unlike grebes, they roost in trees. The feet are lobed rather than webbed. Like grebes and cormorants they swim partly submerged in the water but are not such expert divers. All species make strange deep calls.

163. MASKED FINFOOT *Heliopais personata* PLATE 19
(Asian Finfoot)

Description: Large (52 cm), olive brown, aquatic bird, like a heavy-billed grebe, swimming low in the water and with, in male, a distinctive black face and throat, bordered by a white band. Female has white centre of throat. In flight the tail appears rather broad and rounded.
Iris—brown; bill—yellow; feet—green and lobed.
Voice: Curious bubbling or gargling sounds.
Range: Assam, SE Asia; wintering to Sumatra and Java.
Distribution and status: Status on Sumatra uncertain; few records mostly consistent with winter visitors. A single vagrant recorded from coastal mangroves on Pulau Rambut, W Java, in 1984.
Habits: This shy and secretive bird is generally found swimming along streams, creeks, or lake edges under overhanging vegetation of mangroves, freshwater swamps, and overgrown ponds. Bobs head back and forth as it swims partly submerged. Runs like a grebe treading water and vanishing into undergrowth. Flies low over the water with shallow wing-beats.

Jacanas — FAMILY JACANIDAE

A SMALL, pan-tropical family of medium-sized waterbirds, resembling rails in general appearance, but having immensely elongated toes which facilitate walking over water lily leaves and other floating plants on freshwater lakes and ponds. Several of the species are polyandrous (i.e. more than one cock mating with one hen)—a rare occurrence among birds generally.
Three species occur in the Greater Sundas, one as visitor only.

164. COMB-CRESTED JACANA *Irediparra gallinacea* PLATE 19
(Lotusbird)

Description: Smallish (23 cm) black and white jacana with distinctive red comb. Mantle brown with a bronze gloss; crown, upper back, rump, tail, breast, and flanks black; throat white and sides of face and neck golden-yellow; belly white. Iris—brown; bill—red with black tip; feet—grey.
Voice: Shrill piping calls.
Range: Australia, New Guinea, Wallacea, and SE Borneo.
Distribution and status: Recorded in the Greater Sundas only from the Barito drainage in SE Kalimantan. Still common in Rawa Negara and Binuang regions in 1978.
Habits: As other jacanas.

165. PHEASANT-TAILED JACANA *Hydrophasianus chirurgus* PLATE 19

Description: Largish (33 cm), long-tailed, black and white jacana. White wings distinctive in flight. In non-breeding plumage crown, back, and breast bar greyish brown; eyebrow, chin, throat, and belly white; wings whitish. Has a black stripe through the eye and down side of neck and a golden patch on lower nape.
Iris—yellow; bill—yellow-grey; feet—grey-brown.
Voice: Loud nasal mewing in alarm
Range: India to China, SE Asia, Philippines, and Greater Sundas.
Distribution and status: A regular though rare winter visitor to Sumatra and the north coast of W and C Java. Recorded once in Bali. Probably resident in the Barito drainage of SE Borneo but there are no breeding records.
Habits: Walks on top of floating vegetation often on water lily and lotus leaves, in small ponds and lakes. Picks about for food and makes short fluttering flights to new feeding spots.

166. BRONZE-WINGED JACANA *Metopidius indicus* PLATE 19

Description: Medium-sized (29 cm) brown and black jacana with a bold white eyebrow. Head, neck, and underparts black with green gloss. Upperparts olive-bronze; tail chestnut; forehead chestnut; eyebrow white; young birds have brown crown and some white on breast.
Iris—brown; bill—green with red base and yellow tip; feet—dull green.
Voice: Loud piping alarm calls and low guttural notes.
Range: India, S China, SE Asia, Sumatra, and Java.
Distribution and status: Resident in S Sumatra and locally common in lowland

wetlands. In Java this formerly common bird is now very rare and local in lowland swamps.
Habits: Like other jacanas. Shy and rarely seen.

Painted snipes—FAMILY ROSTRATULIDAE

A FAMILY of two species of rather specialized snipe-like waders. Painted snipes are boldly marked with white flash stripes on the head and shoulders, and ornate wings with many bars, stripes, and eye-spots. The bill is elongated and slightly decurved. Females are larger and more colourful than males and are most active in defence of territories and mate with several cock birds—a habit shared only with buttonquails, jacanas, and a few other birds that show sexually reversed roles. Birds nest on the ground in reed beds. The male incubates.
Only one species occurs in the Greater Sundas.

167. GREATER PAINTED SNIPE *Rostratula benghalensis*   PLATE 24
(Painted Snipe)

Description: Smallish (25 cm), colourful, plump, snipe-like wader with short tail. Female: head and breast dark chestnut with white eye-patch and yellow median crown-stripe; back and wings greenish with a white V-shaped marking over back and white band around shoulder to white underparts. Male: smaller and duller than female, more mottled, less buff, and wing coverts spotted golden; eye-patch yellow.
Iris—red; bill—yellow; feet—grey.
Voice: Generally silent but female courting call is a deep hollow note, also soft purring notes.
Range: Africa, India to China and Japan, SE Asia, Philippines, Sundas, and Australia.
Distribution and status: Locally common up to 900 m on Sumatra. Few records from Borneo but probably a widespread but local resident. On Java this formerly very common bird is now a rare lowland inhabitant.
Habits: Inhabits swampy grassland and paddy fields. Bobs its tail up and down as it walks. In flight it dangles its legs like a rail.

PLOVERS—FAMILY CHARADRIIDAE

A LARGE worldwide family of waders characterized by shortish, straight beaks with a hard swelling at the tip. The legs are long and powerful, and most lack the hind toe. The wings are longish but the tails are short. Most plovers are patterned brown, black, and white. They are birds of the water's edge or open space.
There are 16 species in the Greater Sundas. Most of these are winter visitors, only 3 are resident, of which 1 is probably extinct.

168. NORTHERN LAPWING *Vanellus vanellus* PLATE 20
(Common/The Lapwing/Green Plover/Peewit)

Description: Largish (30 cm) black and white plover with long, narrow, black, upward-curving crest. Upperparts glossy greenish black; tail white with broad black subterminal bar; head with dark crown and black ear coverts with dirty white sides of head and throat; breast blackish; belly white.
Iris—brown; bill—blackish; leg and feet—orange brown.
Voice: Nasal, drawn-out *pee-wit*.
Range: Palaearctic; south in winter to India and SE Asia.
Distribution and status: Recorded twice in Brunei.
Habits: Favours cultivated field, rice stubble, or short grass.

169. GREY-HEADED LAPWING *Vanellus cinereus* PLATE 20

Description: Large (35 cm) noisy, black, white, and grey plover. Head and breast grey; mantle and back brown; wing tips, a band across the breast and the centre of the tail are black, and the rest of hindwing, rump, tail, and belly are white. Immature like adult but browner and lacks black breast band.
Iris—brown; bill—yellow with black tip; feet—yellow.
Voice: Loud plaintive *chee-it, chee-it* alarm call and sharp *ping* given in flight (Smythies).
Range: Breeds in E Palaearctic; migrating in winter to India, SE Asia, and Philippines.
Distribution and status: Vagrant recorded in Brunei and Sarawak.
Habits: Inhabits open areas near water, river flats, rice fields, and marshes.

170. JAVANESE LAPWING *Vanellus macropterus* PLATE 20
(Javan/Javan Wattled Lapwing)

Description: Medium-sized (28 cm) grey-brown lapwing with black head. Back and breast greyish brown; belly black; vent white; flight feathers black; tail white

with broad subterminal black bar. Has a black spur at the bend of the wing. Iris—brown; bill—black with fleshy pinkish white wattles; legs—yellowish green or orange.
Voice: Not recorded.
Range: Endemic to Java; doubtfully recorded from S Sumatra.
Distribution and status: Formerly reported from the marshes of NW (until 1930) and river deltas of SE (until 1940) Java but there are no recent records, and the species is feared to be extinct.
Habits: Lived in pairs on open grassy patches along the north coast of West Java and south coast of East Java.

171. RED-WATTLED LAPWING *Vanellus indicus*   PLATE 20

Description: Large (33 cm), black, white, and brown plover with black head, throat, and centre of breast. White patch over ear coverts. Mantle, wing coverts, and back pale brown; wing tips, rear edge of tail, and a subterminal tail bar black; wing patches, base and tip of tail, and rest of underparts white. A small red wattle above the base of the bill gives name to the species.
Iris—brown; bill—red with black tip; feet—yellow.
Voice: Loud, shrill alarm calls *did-he-do-it*, *pity-to-do-it*, and a *ping* note (Smythies 1981).
Range: India to SW China and SE Asia.
Distribution and status: Occasional winter visitor to N Sumatra.
Habits: Inhabits open spaces, farmland, rice fields, marshes, and river flats. Conspicuous in flapping alarm display flight.

172. GREY PLOVER *Pluvialis squatarola*   PLATE 20
(Black-bellied Plover)

Description: Medium-sized (28 cm) stout wader with a short heavy bill. Distinguishable from Pacific Golden-Plover by larger size and bigger bill, by colour (brownish grey upperparts, whitish underparts), and in flight by whitish wing bar, rump, and tail on upperside, and black axillaries appearing as black patch at base of white underwing.
Iris—brown; bill—black; legs—grey.
Voice: Mournful three-noted whistle *kwee-u-ee*, slurred and both dipping and rising in pitch.
Range: Worldwide
Distribution and status: A common winter visitor to coastal areas of the Greater Sundas.
Habits: Feeds in small flocks on tidal mud-flats and sand bars.

173. PACIFIC GOLDEN-PLOVER *Pluvialis fulva* PLATE 20
(Eastern/Asiatic/Asian Golden Plover)

Description: Medium-sized (25 cm) stout wader with a large head and a short heavy bill. Colour buffy golden-brown with paler eye stripe, sides of face, and underparts. Wing lining shows no contrast in flight.
Iris—brown; bill—black; legs—grey.
Voice: A clear, shrill, single or double whistle *tu-ee*.
Range: Worldwide.
Distribution and status: A common winter visitor to coastal areas of the Sundas, up to 1000 m in Sumatra.
Habits: Feeds singly or in flocks on mud-flats, sand bars, open grassy areas, lawns, golf courses, airports near the coast.
Note: Sometimes included in Golden-Plover *P. dominica*.

174. COMMON RINGED PLOVER *Charadrius hiaticula* PLATE 21
(Ringed Plover)

Description: Medium-sized (19 cm), plumpish, black, brown, and white plover. Distinguished from Little Ringed Plover by larger size, lack of white on crown, orange legs, and conspicuous white wing bar in flight. Immature has black markings of adult replaced with brown, all black bill, and yellow legs.
Iris—brown; bill—black with yellow base; feet—orange.
Voice: Mellow whistle *tu-weet*, second note of higher pitch.
Range: Africa, Eurasia, Australia.
Distribution and status: A rare migrant to Malay Peninsula and N Borneo. Not recorded for Sumatra but can be expected.
Habits: As other plovers.

175. LITTLE RINGED PLOVER *Charadrius dubius* PLATE 21
(Little Dotterel)

Description: Small (16 cm), black, grey, and white, short-billed plover. Distinguished from Kentish and Malaysian Plovers by the full black or brown band across the breast and yellow legs. Distinguished from Common Ringed Plover by more conspicuous yellow eye-ring and lack of wing bar. In immature the black parts of the adult are brown. No white wing bar in flight.
Iris—brown; bill—grey; legs—yellow.
Voice: Clear, soft, drawn out, descending whistle *pee-oo*.
Range: Widespread from Africa, Eurasia, SE Asia to New Guinea.
Distribution and status: A regular winter visitor to the Greater Sundas. Recorded up to 1000 m on Lake Toba in Sumatra.

Habits: Usually encountered along the sandy banks of coastal streams and rivers, also marshes and mud-flats; sometimes found far inland.

176. KENTISH PLOVER *Charadrius alexandrinus* PLATE 21

Description: Small (15 cm), short-billed, brown and white plover. Distinguished from Little Ringed Plover by black legs, white wing bar in flight, and whiter outer tail. Males have a black patch on side of breast; in females this patch is brown.
Iris—brown; bill—black; legs—black.
Voice: A soft, single, unmusical rising note *pik*, repeated.
Range: Africa, Eurasia.
Distribution and status: Uncommon winter visitor to Greater Sundas; recorded Sumatra, N Borneo, Java, and Bali.
Habits: Feeds singly or in small flocks, often mixed with other waders, on beaches or sandy grassland areas near coast, also coastal rivers and marshes.

177. JAVAN PLOVER *Charadrius javanicus* PLATE 21
(Javan Sand-Plover)

Description: Small (15 cm), short-billed, brown and white wader. Similar to Kentish Plover with which this was formerly classified but breast whiter, feet grey, and breast band sometimes complete. White back collar more complete than Kentish Plover.
Iris—brown; bill—black; legs—olive-grey.
Voice: A soft, single rising note, *kweek*, repeated.
Range: Endemic to Java.
Distribution and status: Resident in coastal lowlands of Java (including Kangean) and possibly Bali.
Habits: As Kentish Plover
Note: Formerly treated by some authors as a race of Kentish or Malaysian Plover but now regarded as a separate species.

178. RED-CAPPED PLOVER *Charadrius ruficapillus* PLATE 21
(AG: Sand-Plover)

Description: Small (15 cm), short-billed, brown and white wader. Like Kentish and Javan Plovers with incomplete breast band and incomplete white collar, but with more uniform plumage and more creamy breast. Uniform rufous or brown cap is lined by blackish brow. Bill very narrow.
Iris—brown; bill—black; legs—olive-grey.
Voice: Sharp *wit* or trilled *prrrt* in flight.
Range: Australia; irregular migrant to New Guinea and E Indonesia.

Distribution and status: Recorded in E Java.
Habits: As Kentish Plover.
Note: Formerly treated as a race of Kentish Plover but now regarded as a separate species.

179. MALAYSIAN PLOVER *Charadrius peronii* PLATE 21
(Malay Plover, AG: Sand-Plover)

Description: Small (15 cm), black, brown, and white, short-billed plover. Differs from Mongolian Plover and Greater Sand-Plover by smaller size and the narrow black (male) or rufous (female) collar. Differs from Kentish Plover by generally complete white back collar and ear patches being separate rather than a continuous line through the eye.
Iris—brown; bill—black; legs—grey.
Voice: A quiet soft *kwik*, similar to Kentish Plover.
Range: Malay Peninsula, Indochina, Philippines, Sulawesi, and Sundas.
Distribution and status: Local, mostly on sand beaches, around the coasts of Sumatra (including many islands) and Borneo. Status on Java uncertain; breeds Kangean and Bali but the few records from mainland Java (e.g. Ujung Kulon) are doubtful.
Habits: Lives in pairs on sandy beaches. Prefers small bays of pure coralline sand. Does not form flocks nor mixes with other waders, and is generally rather tame.

180. LONG-BILLED PLOVER *Charadrius placidus* PLATE 21
(AG: Ringed Plover)

Description: Largish (22 cm), robust, black, brown, and white plover with longish, entirely black bill, and longer tail than Common Ringed Plover and Little Ringed Plover. White wing bar less bold than in Ringed Plover. Breeding plumage distinct with black forehead bar and complete breast band, but lacking black cheeks. Immature stages as in Common Ringed and Little Ringed Plovers. Never shows clear black eye-stripe.
Iris—brown; bill—black; legs and feet—dull yellow.
Voice: Loud, clear dissyllabic or trisyllabic piping.
Range: Breeds in NE Asia, E and S China; south in winter to SE Asia.
Distribution and status: Vagrants recorded N Borneo and Bali.
Habits: Similar to other plovers but preferring gravel areas along river edge and mud-flats.

181. MONGOLIAN PLOVER *Charadrius mongolus* PLATE 21
(Lesser Plover, AG: Sand-Plover/Dotterel)

Description: Medium-sized (20 cm), grey, brown, and white, short-billed wader. Very similar to Greater Sand-Plover with which it often mixes but distinguishable by smaller size and shorter, gracile bill; also fainter white wing bar in flight. Early arrivals may show distinctive breeding plumage of broad rufous breast bar and black mask with fully black forehead in race *atrifrons* which is commonest form.
Iris—brown; bill—black; legs—dark grey.
Voice: Short, quiet trill or a sharp *kip-ip*
Range: Breeds in NE Asia, migrating to Africa, SE Asia, Australia, and New Zealand.
Distribution and status: A regular winter visitor to the coasts of the Greater Sundas where it is often very common.
Habits: Found mixed with other waders on coastal mud-flats and sands, sometimes in large flocks numbering several hundreds.

182. GREATER SAND-PLOVER *Charadrius leschenaultii* PLATE 21
(Geoffrey's Plover/Large/Great/Large-billed Sand-Plover, AG: Sand-Dotterel)

Description: Medium-sized (23 cm), grey, brown, and white plover. Distinguishable from Mongolian Plover by larger size and longer, thicker bill, and from all other wintering plovers (except Mongolian) by lack of breast bar or collar bar. Early arrivals may show distinctive breeding plumage of rufous breast bar and black mask with white forehead.
Iris—brown; bill—black; legs—greenish grey.
Voice: Generally silent in winter. Low *trrrt* on take-off.
Range: Widespread from Africa and Asia to Australia and New Zealand.
Distribution and status: A common winter migrant to the coasts of the Greater Sundas.
Habits: Frequents coastal mud-flats and sandy patches, mixing with other waders especially Mongolian Plover.

183. ORIENTAL PLOVER *Charadrius veredus* PLATE 21

Description: Medium-sized (23 cm), brown, and white, short-billed wader. Winter plumage: broad brown breast band, narrow bill, and whitish face; upperparts uniformly brown with no wing bar. In summer plumage, breast band is orange underlined black. Distinguished from Pacific Golden, Mongolian, and Greater Sand Plovers by yellow legs. In flight underwing, including axillaries, white.
Iris—hazel; bill—olive-brown; legs—yellow to orange.

Voice: A sharp piping whistle *kwink*, and in flight a loud repeated *chip-chip-chip*.
Range: Breeds in Mongolia and North China, occasional birds visiting the Greater Sundas in winter.
Distribution and status: A rare bird in the Greater Sundas.
Habits: Feeds along coasts in grassy areas and beside rivers and marshes.
Note: May be conspecific with Caspian Plover *C. asiaticus*.

Sandpipers—FAMILY SCOLOPACIDAE

THE sandpiper family is a large, worldwide family of waders, generally found on the seashore or in wet, open places often close to the sea. A few species range inland into higher altitudes but of the Sundaic species, only the Rufous Woodcock is a regular forest dweller.

All members of the family have long legs, long pointed wings, and generally elongated, slender bills. In some species the bills are extraordinarily long and used for probing deep into mud to search for hidden worms or crustaceans. Most species are migratory; only the Rufous Woodcock is a breeding resident.

These migratory waders may pass through the Sunda region on their way to and from winter feeding grounds to the south and east, or may be regular winter visitors that settle along the coast for part of the year. Sometimes large mixed flocks are seen. Descriptions concentrate on winter plumage normally seen in the Greater Sundas but some birds may be found assuming breeding plumage. Both forms are shown. Waders are often difficult to identify as there are so many similar species and they are often seen at rather long range. Careful note should be taken of the general appearance and whether there are conspicuous white wing bars visible in flight. Calls are valuable in identification as many species have very distinctive flight calls.

Thirty-five confirmed and 2 unconfirmed species have been recorded in the Greater Sundas.

184. EURASIAN CURLEW *Numenius arquata*   PLATE 22
(Western/Common/European/The Curlew)

Description: Very large (55 cm) brown-streaked wader with long legs and very long decurved bill. Rump white grading into white and brown barred tail. Distinguished from Far-Eastern Curlew by whiter rump and tail and white underwing, and from Whimbrel by larger size, lack of head pattern, and proportionally longer bill.
Iris—brown; bill—brown; feet—slate blue.
Voice: Loud, plaintive, rising cry *cur-lew*.

Range: Breeds in N Eurasia but migrates south in winter as far as Indonesia and Australia.
Distribution and status: A regular visitor to the Greater Sundas but never numerous. However, it is more common than the Far-Eastern Curlew.
Habits: Frequents tidal estuaries and mud-flats, rarely far from the sea. Often seen singly, sometimes in small flocks, or mixing with other curlew species.

185. WHIMBREL *Numenius phaeopus* PLATE 22
(Hudsonian Curlew)

Description: Large (43 cm) brown-streaked wader with pale eyebrow, black crown stripes, long legs, and decurved bill. Similar to Eurasian Curlew but much smaller and bill proportionally shorter. Rump brownish in commoner race *variegatus*, but some individuals with white rump and underwing approach form of nominate *phaeopus*.
Iris—brown; bill—black; feet—blackish brown.
Voice: Loud whinnying whistle *ti-ti-ti-ti-ti-ti*.
Range: Breeds in N Europe and Asia but migrates south in winter to SE Asia, Australia, and New Zealand.
Distribution and status: In the Greater Sundas a widespread and common visitor. Some non-breeding birds can be found in summer.
Habits: Favours mud-flats, tidal estuaries, pastures near coast, marshes, and rocky beaches, generally in small to large flocks and often mixing with other waders.

186. LITTLE CURLEW *Numenius minutus* PLATE 22
(Little Whimbrel)

Description: Medium-sized (30 cm) brown-streaked wader with long legs and a medium length decurved bill; buff eyebrow. Differs from Whimbrel by smaller size and shorter, straighter bill. Rump never white.
Iris—brown; bill—brown with pink base; feet—blue grey.
Voice: Chattering *te-te-te* when flocks feed and sharp harsh *chay-chay-chay* in alarm.
Range: Breeds NE Asia; migrates south in winter to Australia and New Zealand.
Distribution and status: Passage is mainly east of the Greater Sundas but recorded in N Borneo and Java.
Habits: Prefers dry open grassy areas near coast such as airports, also coastal mud-flats.

187. FAR-EASTERN CURLEW *Numenius madagascariensis* PLATE 22
(Eastern/Australian Curlew)

Description: Very large (57 cm) brown-streaked wader with long legs, and very long and heavy decurved bill. Darker and browner than Eurasian Curlew, rump and tail brown; underparts buff. In flight barred underwing distinguishes it from Eurasian Curlew with white underwing.
Iris—brown; bill—black with pink base; feet—grey.
Voice: Similar to Eurasian Curlew
Range: Breeds in NE Asia but migrates south in winter as far as Australasia.
Distribution and status: In the Greater Sundas a regular but uncommon visitor.
Habits: As Eurasian Curlew; very shy.

188. BLACK-TAILED GODWIT *Limosa limosa* PLATE 22

Description: Large (40 cm), long-legged, long-billed wader. Similar to Bar-tailed Godwit but larger, bill less upturned, eye-stripe more pronounced, upperparts less mottled, terminal half of tail blackish with rump and tail base white. White wing bar distinctive; narrow in race *melanuroides*, broader in rare nominate form *limosa*.
Iris—brown; bill—pink at base with black tip; feet—greenish grey.
Voice: Generally quiet but gives occasionally a loud *wikka wikka wikka* or *kip-kip-kip* in flight.
Range: Breeds in N Europe and Asia, but migrates south in winter as far as Australia and New Zealand.
Distribution and status: In Sumatra and Borneo a locally common visitor, with flocks totalling thousands reported on the east coast of Sumatra. In Java and Bali a rare passage visitor.
Habits: Frequents coastal mud-flats and the shores of rivers and lakes. Feeds as Bar-tailed Godwit but in muddier places.

189. BAR-TAILED GODWIT *Limosa lapponica* PLATE 22

Description: Large (37 cm) long-legged wader with long, slightly upturned bill. Upperparts mottled grey and brown; conspicuous white eyebrow; underparts with some grey on breast; characterized by narrow pale wing bar, and narrow brown bars on white tail and rump. Normal form *baueri* has brownish lower back, rarer nominate form *lapponica* has white lower back and rump.
Iris—brown; bill—base pink with black tip; feet—dark green or grey.
Voice: Rather quiet but occasional low nasal *kurrunk* or clear double note *kew-kew*, and in flight a soft *kit-kit-kit-kit*.

Range: Breeds in N Europe and Asia, but migrates south in winter as far as Australia and New Zealand.
Distribution and status: In Sumatra and Borneo can be locally rather common. In Java and Bali a regular visitor but never common.
Habits: Frequents tidal waters, estuaries, sand-flats, and shallows.

190. SPOTTED REDSHANK *Tringa erythropus*   PLATE 23

Description: Medium-sized (30 cm), red-legged, grey wader with long straight bill. Black breeding plumage unmistakable. In winter, similar to Common Redshank, but larger and greyer with longer bill. Distinguished by white spotting on dark wings and more prominent eye-stripe. Differs in flight by lack of white trailing bar and legs which trail more beyond the tail.
Iris—brown; bill—black with red base; feet—orange.
Voice: Very distinctive sharp, explosive whistle *too-it* in flight and at rest.
Range: Breeds in Europe, migrating to Africa, India, and SE Asia in winter.
Distribution and status: A few individuals reach N Borneo.
Habits: Similar to Common Redshank. Prefers mud-flats and marshes.

191. COMMON REDSHANK *Tringa totanus*   PLATE 23
(Redshank)

Description: Medium-sized (28 cm) wader with reddish orange legs and red basal half of bill. Upperparts brownish grey; underparts white, streaked brown on breast. In flight white rump conspicuous and white secondaries give conspicuous trailing edge to wing. Tail finely barred black and white.
Iris—brown; bill—red at base, black at tip; feet—orange red.
Voice: Musical whistle *teu hu-hu* dropping in pitch, or single *teyuu*.
Range: Breeds in Africa and Eurasia; migrates south in winter as far as Sulawesi, Timor, and Australia.
Distribution and status: In the Greater Sundas a common visitor with flocks of over 10 000 reported from the east coast of Sumatra.
Habits: Frequent mud-banks, beaches, dried-up swamps and fish ponds, paddy fields near the sea or occasionally far inland. Generally in small parties, associating with other waders.

192. MARSH SANDPIPER *Tringa stagnatilis*   PLATE 23

Description: Medium-sized (23 cm) delicate sandpiper with white forehead and very thin, straight bill. Upperparts greyish brown with white rump and lower back; underparts white. Distinguished from Common Greenshank by smaller size, paler forehead, proportionally longer legs, and finer, straighter bill.

132 SANDPIPERS

Iris—brown; bill—black; feet—greenish.
Voice: Quiet *chew* or distinctive, sharp, thin *chewp* or *chip*.
Range: Breeds in Eurasia but migrates south in winter to Africa, S and SE Asia, and as far as Australia and New Zealand.
Distribution and status: In Java and Bali an uncommon visitor. In Borneo and Sumatra a regular and locally common visitor.
Habits: Frequents coastal mud-flats, marshes, and pools. Generally alone or in twos and threes.

193. COMMON GREENSHANK *Tringa nebularia* PLATE 23
(Greater/Eurasian/The Greenshank)

Description: Largish (32 cm) greyish wader with white rump, long, slightly upturned bill, and green legs. Upperparts greyish; underparts white. In flight: blackish wings, white rump and lower back, barred tail, and long legs.
Iris—brown; bill—black; feet—green.
Voice: Loud, ringing *tew tew tew*.
Range: Breeds in N Eurasia but migrates south in winter as far as Australia and New Zealand.
Distribution and status: In the Greater Sundas a regular visitor in small numbers.
Habits: Frequents coastal and inland marshes and mud-flats. Generally alone or in twos or threes. Forages with sideways sweeps of bill in water.

194. NORDMANN'S GREENSHANK *Tringa guttifer* PLATE 23
(Spotted Greenshank)

Description: Medium-sized (31 cm) grey wader with yellowish legs and bi-coloured bill. Very similar to Common Greenshank and distinguished with difficulty by yellow base of stouter bill, paler upperparts with more scaling and less streaking (in winter), paler barring on tail, shorter and yellower legs, feet extending less beyond the tail in flight, and by call. In the hand, shows webbing between all three toes compared with webbing between only two toes in Common Greenshank.
Iris—brown; bill—black with yellow base; legs and feet—yellow-green.
Voice: Loud, repeated, piercing screams *keyew*.
Range: Breeds in NE Asia and migrates in winter to Japan, China, SE Asia, and Philippines.
Distribution and status: Very rare individuals have reached the coasts of N Sumatra and N Borneo.
Habits: As Common Greenshank; prefers mud-flats.

195. LESSER YELLOWLEGS *Tringa flavipes*   PLATE 23

Description: Medium-sized (23 cm), greyish, brown-backed wader with straight bill and conspicuous yellow legs; smaller and more slender than Common Redshank. The white rump patch in flight is cut off square above the tail coverts, not wedge-shaped as in redshanks or greenshanks.
Iris—brown; bill—black; legs and feet—yellow.
Voice: Single or double notes like a Common Redshank but quieter. Also has *tuk-tuk-tuk* alarm call, and continuous *pill-e-wee, pill-e-wee, pill-e-wee* yodelling flight call.
Range: Breeds in Alaska and Canada, migrating south in winter to South America.
Distribution and status: Recorded as vagrant once from N Sumatra in 1983.
Habits: A lively and graceful wader; walks with strongly flexed legs.

196. GREEN SANDPIPER *Tringa ochropus*   PLATE 23

Description: Medium-sized (23 cm) greenish brown wader with conspicuous white rump. In flight black underwings, white rump, and barred tail characteristic. Upperparts greenish brown with white flecks; wings and lower back almost black; tail white with terminal black bars; underparts whitish. Feet extend well beyond tail in flight. Appears very black and white in field.
Iris—brown; bill—dusky olive; feet—olive-green
Voice: Loud, liquid *tlooit-ooit-ooit* with the second note more drawn out.
Range: Breeds in N Eurasia but migrates south in winter as far as N Borneo, Sumatra, and the Philippines.
Distribution and status: Rare vagrants have reached Java but there are no recent records.
Habits: Usually solitary, frequenting small pools and ponds, marshes and ditches. When surprised it flies off with a zig-zag, snipe-like flight.

197. WOOD SANDPIPER *Tringa glareola*   PLATE 23

Description: Medium-sized (20 cm) brownish grey wader with white rump. Upperparts greyish brown, rather spotted; white eyebrow; white tail, barred brown; underparts white. In flight, barred tail, white rump and underwing, and lack of wing bar characteristic; feet extend well beyond tail.
Iris—brown; bill—black; feet—yellowish to olive green.
Voice: High-pitched whistle *chee-chee-chee* or in alarm *chif-chif-chif*, less ringing than Common Greenshank.
Range: Breeds in N Eurasia but migrates south in winter as far as Australia.
Distribution and status: In the Greater Sundas a common and widespread visitor.

134 SANDPIPERS

Habits: Prefers muddy coastal habitats, and also occurs well inland, in paddy fields and freshwater swamps, up to 750 m. Generally in small loose flocks of up to 20, and regularly mixes with other waders.

198. TEREK SANDPIPER *Tringa cinereus*   PLATE 23

Description: Medium-sized (23 cm) grey wader with long, slightly upturned bill. Upperparts grey with white eyebrow; black primaries conspicuous; underparts white; legs relatively short. Narrow white trailing edge of wing conspicuous in flight.
Iris—brown; bill—black with yellow base; feet—orange.
Voice: Loud, high-pitched, melodious trill *tee-tee-tee* or *tit-ter-tee*.
Range: Breeds in N Eurasia but migrates south in winter as far as Australia and New Zealand.
Distribution and status: In the Greater Sundas a common coastal visitor, with numbers of over 6000 recorded along the east coast of Sumatra.
Habits: Frequents coastal mud-flats, creeks, and estuaries, mixing with other waders to feed but flying separately. Generally solitary or in ones and twos, rarely in flocks.

199. COMMON SANDPIPER *Tringa hypoleucos*   PLATE 23

Description: Smallish (20 cm), brown and white, short-billed, restless wader. Upperparts brown, flight feathers blackish; underparts white with a brown-grey patch on side of breast. In flight white wing bar, lack of white rump, and white, barred outer tail feathers are distinctive.
Iris—brown; bill—dark grey; feet—pale olive-green.
Voice: Thin, high-pitched piping *twee-wee-wee-wee*.
Range: Breeds in Africa and Eurasia, but migrates south in autumn as far as Australia.
Distribution and status: A very common visitor which can be seen almost all year around.
Habits: Frequents a wide range of habitats from coastal mud-flats and sand bars to upland paddy fields, up to 1500 m, and along streams and river sides. Walks with incessant bobbing gait and also flies in a characteristic manner, with stiff-winged glides.

200. GREY-TAILED TATTLER *Tringa brevipes*   PLATE 22
(Siberian/Polynesian/Grey-rumped Tattler)

Description: Medium-sized (25 cm) grey wader with conspicuous black eye-stripe and white eyebrow, and short yellow legs. Chin whitish; plumage

unmarked, grey on upperparts; pale grey on breast with white belly. Rump is finely barred. Underwing dark in flight.
Iris—brown; bill—black; feet—yellowish.
Voice: Sharp double whistle *too-weet* or soft trill.
Range: Breeds in Siberia but migrates south in winter as far as Australia and New Zealand.
Distribution and status: An uncommon to rare coastal visitor to the Sundas. In Java it prefers the south coast.
Habits: Frequents rocky beaches, coral banks, and sandy or shingle beaches rather than mud-flats. Generally solitary or in small groups. Does not mix with other waders. Runs in a characteristic crouched manner with tail high.
Note: May be conspecific with Wandering Tattler *T. incana*

201. RUDDY TURNSTONE *Arenaria interpres* PLATE 22
(Turnstone)

Description: Medium-sized (23 cm), unmistakable, short-billed wader with distinctive bright orange legs and feet. Complex pattern of black, brown, and white on head and breast distinctive. Bill shape is diagnostic. In flight from above, boldly patterned black and white.
Iris—brown; bill—black; feet—orange.
Voice: Metallic staccato rattle *ktititit*, or ringing *kee-oo*.
Range: Worldwide, breeding in northern latitudes and migrating south in winter through Indonesia to Australia.
Distribution and status: A fairly common visitor to Sundaic coasts. Some non-breeding birds may be seen in summer.
Habits: Frequents mud-flats, sandy shores, and rocky reeflets in small flocks. Sometimes feeds inland or close to the sea in open areas. Generally does not mix with other species. Gets its English name from curious habit of pushing or turning over stones on beach to search for crustaceans underneath.

202. ASIAN DOWITCHER *Limnodromus semipalmatus* PLATE 22
(Asiatic/Snipe-billed Dowitcher)

Description: Large (35 cm), grey wader with long straight bill. Back grey; rump, lower back, and tail white with fine black bars; underparts pale with buffy brown on breast; sometimes has white wing bar on primaries. Distinguished from godwits by smaller size and straight, all-black bill, swollen at tip.
Iris—brown; bill—black; legs—blackish.
Voice: Generally silent but occasionally gives a quiet plaintive *miau*.
Range: Breeds in N Asia; migrant to E India, SE Asia, Philippines, and Indonesia to N Australia.

Distribution and status: The coastal mud-flats of the Greater Sundas are the main wintering areas; up to 4000 winter on SE Sumatra (Jambi and South Sumatra) and up to 1000 winter NE Java. Probably largely overlooked on Borneo where there are relatively few records.

Habits: Inhabits mud-flats where it is recognized by its characteristic feeding method, walking forward stiffly, rocking to plunge its bill deep in the mud at each step like a mechanical toy.

203. LONG-BILLED DOWITCHER *Limnodromus scolopaceus*   PLATE 22

Description: Largish (30 cm) grey wader with long straight bill. Similar to Asian Dowitcher but distinguished by smaller size, paler legs and bill, and in flight by unbarred white wedge on back, lack of white bar on primaries, and more prominent white trailing edge of secondaries; whitish wing lining barred with black.

Iris—brown; bill—yellowish with dark tip; legs—greenish grey.

Voice: High, thin *keek* sometimes repeated; similar to Wood Sandpiper.

Range: Breeds in E Siberia and W Nearctic.

Distribution and status: Recorded in Brunei and Bali.

Habits: A bird of marshes and coastal mud-flats.

204. PINTAIL SNIPE *Gallinago stenura*   PLATE 24
(Pin-tailed Snipe)

Description: Medium-sized (24 cm), plump, short-legged wader with very long, straight bill. Upperparts brown, finely streaked white, yellow, and black; underparts white, washed rufous on breast, and finely barred black; head pale with three dark brown stripes above, through, and below eye. Distinguished from Common and Swinhoe's Snipes with difficulty by smaller size, shorter tail, yellow feet trailing further behind tail in flight, and by call. Differs from Common Snipe by lack of white trailing edge on wing.

Iris—brown; bill—brown with dark tip; feet—yellowish.

Voice: Rasped *squak-squak* with nasal twang in alarm.

Range: Breeds in NE Asia but migrates south in winter as far as Moluccas and Lesser Sundas.

Distribution and status: In Sumatra, W Borneo, and Kelabit highlands common. Elsewhere in Borneo uncommon. In Java and Bali an erratic visitor.

Habits: Frequents paddy fields and marshes and damp hollows in forest, and mangroves but usually in drier situations than Common Snipe. Habits similar to other snipes, including fast leap and 'zig-zag' flight, giving alarm call when disturbed.

SANDPIPERS 137

205. SWINHOE'S SNIPE *Gallinago megala* PLATE 24
(Chinese/Marsh Snipe)

Description: Medium-sized (28 cm), plump, short-legged wader with very long straight bill. Distinguished with difficulty in the field from Pintail Snipe by longer tail and feet trailing less far beyond tail in flight. Distinguished from Common Snipe by more white on sides of tail tip, and lack of white trailing edge of wing in flight.
Iris—brown; bill—brown; feet—olive-grey.
Voice: Harsh, rasping cry, similar to Common Snipe but higher-pitched.
Range: Breeds in NE Asia; migrates south in winter as far as Australia.
Distribution and status: A regular visitor to the Greater Sundas but far less numerous than the Pintail Snipe; doubtfully recorded in Sumatra but presumably overlooked.
Habits: Inhabits swamps and wet grasslands, including paddy fields. Habits as other snipes, but rises and flies slower and less erratically than the other two species.

206. COMMON SNIPE *Gallinago gallinago* PLATE 24
(Fantail/The Snipe)

Description: Medium-sized (27 cm), plump, short-legged wader with very long straight bill. Head buff with dark stripes above, below, and through eye; upperparts dark brown, finely streaked white and black, underparts buffish, streaked brown. Distinguished with difficulty from other snipes by white trailing edge of secondaries, and faster, more erratic flight.
Iris—brown; bill—brown; feet—olive.
Voice: Loud cry *snipe-snipe* on a rising note given in alarm when flushed.
Range: Breeds in Palaearctic; south in winter to Philippines and SE Asia.
Distribution and status: Scarce winter visitor to the Greater Sundas. Doubtfully recorded in Sumatra but probably overlooked.
Habits: Found in marshes and paddy fields, generally keeping under cover of long reeds and grasses, but leaping up when flushed and fleeing with erratic 'zig-zag' flight, giving alarm call.

207. EURASIAN WOODCOCK *Scolopax rusticola* PLATE 24
(Woodcock)

Description: Large (35 cm), plump, short-legged wader with very long straight bill. Distinguished from snipes by much larger size and transverse bands on top of head. Larger and paler than Rufous Woodcock.

Iris—brown; bill—pinkish at base with black tip; feet—greyish pink.
Voice: Usually silent when flushed.
Range: Eurasia; migrant to SE Asia.
Distribution and status: Occasional winter visitor to Brunei.
Habits: Nocturnal forest bird. It hides by day, sitting tight on the ground, and flies by night to feeding grounds in open areas. Clattering of wings notable when flushed.

208. RUFOUS WOODCOCK *Scolopax saturata* PLATE 24
(Horsfield's/East Indian/Indonesian/Dusky Woodcock)

Description: Largish (30 cm), dark plump, short-legged wader with very long straight bill. Resembles snipes but darker plumage, larger, has transverse bands on top of head, and lives in montane forests. Smaller, darker, and more rufous than Eurasian Woodcock.
Iris—brown; bill—pinkish at base with black tip; feet—greyish.
Voice: Fast, endlessly repeated, harsh cry *do-do-do-do-do*, also a singing *krrr-krrr-krrr*.
Range: Endemic to Sumatra, Java, and New Guinea.
Distribution and status: Mountains of Sumatra, where recorded from Mts. Leuser and Kerinci, from 1900 to 2400 m, and W Java where recorded from Mts. Salak, Tangkubahnprahu, and Pangrango, from 1200 to 2800 m.
Habits: Lives in mountain forests, often near lakes. Hides by day, sitting tight on the ground, and flies by night to feeding grounds in open areas.

209. RED KNOT *Calidris canutus* PLATE 25
(Lesser/The Knot)

Description: Medium-sized (24 cm), thick-set, short-legged, greyish wader with a shortish, thick dark bill and pale eyebrow stripe. Upperparts grey with faint scaling; underparts whitish, lightly buff on neck, breast, and flanks. In flight shows narrow white wing bar and pale grey rump. Underparts rufous in summer.
Iris—dark brown; bill—black; feet—yellowish green.
Voice: Low-pitched throaty *chut chut* and rising *ee-yik*, also a musical chatter when feeding.
Range: Breeds in the Arctic and migrates south in winter as far as Australia.
Distribution and status: In N Borneo, Sumatra, and Java a rare passage migrant. Some non-breeding birds may be seen in summer.
Habits: Frequents sand- and mud-flats, and estuaries. Generally very social, in large flocks. Mixes with other waders; feeds with rapid stabbing of the bill, sometimes completely submerging the head to catch food.

210. GREAT KNOT *Calidris tenuirostris* PLATE 25

Description: Largish (27 cm) long-billed greyish wader. Similar to Red Knot but larger, with longer, thicker bill slightly decurved at tip; upperparts darker with faint streaking; crown streaked; breast and sides spotted black even in non-breeding plumage; rump and wing bar white. Summer birds show blackish breast and rufous wing bar.
Iris—brown; bill—black; feet—greenish grey.
Voice: Low *chucker-chucker-chucker* call, or double noted whistle *nyut-nyut*.
Range: Breeds in E Siberia but migrates south in winter as far as Australia.
Distribution and status: In the Greater Sundas an uncommon visitor. Some non-breeding birds may be seen in summer.
Habits: Frequents tidal mud-flats and sand bars, and sometimes coastal pastures.

211. RUFOUS-NECKED STINT *Calidris ruficollis* PLATE 25
(Red-necked Stint, AG: Sandpiper)

Description: Small (15 cm) greyish brown stint with black legs and pale, streaked upperparts. Upperparts greyish brown, mottled and streaked; eyebrow stripe white; centre of rump and tail dark brown; sides of tail white; underparts white. Distinguished from Long-toed Stint by greyer, more uniform plumage and black legs. In summer plumage neck, crown, and wing coverts are rufous.
Iris—brown; bill—black; feet—black.
Voice: Weak, whistling *chit-chit-chit*.
Range: Breeds in Siberia and Alaska but migrates south in winter as far as Australia.
Distribution and status: A regular and common coastal visitor.
Habits: Frequents coastal mud-flats in large flocks of active birds, walking briskly or running about, picking up tiny food items, and when excited bobbing up and down with a backward throw of the head. More of a coastal bird than Long-toed Stint.

212. TEMMINCK'S STINT *Calidris temminckii* PLATE 25

Description: Small (14 cm) grey wader. Upperparts dull uniform grey. Grey breast grading to whitish belly. Distinguished from other stints by pure white outer tail feathers, most easily seen as birds alight, also by distinct calls and greenish or yellowish legs. Summer plumage: brown breast and rufous wing coverts.
Iris—brown; bill—black; legs and feet—greenish or yellowish.
Voice: Distinctive short rapid trill *titititiii* . . . .

140 SANDPIPERS

Range: Breeds in Eurasia; migrant in winter to Africa, India, SE Asia, Philippines, and Borneo.
Distribution and status: Regular winter visitor to N Borneo in small numbers; recorded once in SE Borneo.
Habits: As other stints, favouring mud-flats and marshy areas. Primarily a freshwater bird but also visits tidal creeks.

213. LONG-TOED STINT *Calidris subminuta*   PLATE 25

Description: Small (14 cm) grey-brown stint with boldly black-streaked upperparts and greenish yellow legs. Crown brown; conspicuous white eyebrow. Breast pale brownish grey; belly white; centre of rump and tail dark brown; outer tail pale brown. Summer birds are browner.
Iris—dark brown; bill—black; feet—greenish yellow.
Voice: Rapid high-pitched *shoo-shoo-shoo* in alarm, and a purring *chrrup*.
Range: Breeds in Siberia but migrates south in winter as far as Australia.
Distribution and status: In the Greater Sundas a regular visitor, sometimes numerous up to 1200 m in the Kelabit highlands. Some birds stay over summer.
Habits: Frequents coastal mud-banks but also found well inland, in paddy fields and other muddy areas. Occurs singly or in flocks, often mingles with other waders. Less shy than other waders, usually being the last to fly off when approached.
Note: May be conspecific with Least Sandpiper *C. minutilla*.

214. SHARP-TAILED SANDPIPER *Calidris acuminata*   PLATE 25
(Siberian Pectoral Sandpiper)

Description: Smallish (19 cm) short-billed wader with rufous cap, pale eyebrow, and buffy breast. Bold black streaks on underparts are characteristic. Belly white; tail centrally black, laterally white; narrow white wing bar. Summer birds more rufous.
Iris—brown; bill—black; legs and feet—yellowish to green.
Voice: Plaintive *chew* or *wheep*, sharp liquid *whit-whit*, *whit-it-it*, and soft grunting.
Range: Breeds in NE Asia but migrates south in winter as far as Australia and New Zealand.
Distribution and status: In N Borneo, Java, and Bali a rather rare visitor.
Habits: Frequents marshes and mud-flats, swamps, lakes, and paddy fields. Mixes freely with other waders.

[215. DUNLIN *Calidris alpina* PLATE 25
(Red-backed Sandpiper)

Description: Small (19 cm), medium-billed, greyish wader with white eyebrow. Bill slightly decurved at tip. Tail black down centre, white on sides. Differs from Curlew Sandpiper by dark rump, shorter legs, and darker breast. Distinguished from Broad-billed Sandpiper by longer legs and more uniform head with single eyebrow. In summer black breast is diagnostic.
Iris—brown; bill—black; feet—greenish grey.
Voice: Harsh nasal whistle *dwee* in flight.
Range: Breeds in Holarctic; rare migrant to SE Asia.
Distribution and status: Doubtfully recorded Borneo and Java. There are no definitive specimens from Indonesia and sight records from Malay Peninsula to Australia require appraisal.
Habits: Frequents coastal and inland mud-flats, singly or in small parties, often mixes with other waders. Feeds busily with a somewhat hunched posture.]

216. CURLEW SANDPIPER *Calidris ferruginea* PLATE 25

Description: Smallish (21 cm) wader with conspicuous white rump and long, decurved, black bill. Upperparts mostly grey; underparts white; eyebrow, wing stripe, and bar across uppertail coverts white. In summer breast and general plumage rufous.
Iris—brown; bill—black; feet—yellowish to green.
Voice: Twittering calls. Plaintive *chew* or *wheep*, sharp *whit-whit*, *whit-it-it*, and soft grunting.
Range: Breeds in NE Asia but migrates south in winter as far as Australia and New Zealand.
Distribution and status: In Java and Bali a rather rare visitor but more common in Borneo and Sumatra; some non-breeding birds can be seen in summer.
Habits: Frequents coastal mud-flats and close to the sea in paddy fields and fish ponds. Usually mixed in with other stints and sandpipers. Runs over mud as the tide falls, probing and picking for food. Rests on sand spits on one leg, and flies in fast, close flocks.

217. SANDERLING *Calidris alba* PLATE 25

Description: Smallish (20 cm) greyish wader with conspicuous black shoulder. Appears whiter than any other sandpiper with prominent white wing bar in flight; tail dark down centre, white on sides. Lack of hind toe distinctive.
Iris—dark brown; bill—black; feet—black.
Voice: Squeaking flight note *cheep cheep cheep.*

Range: Holarctic. Breeds in northern latitudes and migrates south in winter as far as Australia and New Zealand.
Range: Uncommon winter visitor (few oversummer) to the Greater Sundas.
Distribution and status: On Java more frequent on the south coast.
Habits: Frequents sandy beaches along sea coasts, rarely on mud. Generally runs along the water's edge following receding waves, picking at tiny food organisms washed on the beach. Sometimes solitary but normally gregarious.
Note: May be placed in genus *Crocethia*.

218. BROAD-BILLED SANDPIPER *Limicola falcinellus* PLATE 25

Description: Smallish (17 cm) sandpiper with decurved bill, often conspicuous black carpal patch, and diagnostic double white eyebrow. Upperparts streaked grey-brown; underparts white with streaking on breast; rump and tail with black centre and white sides. Distinguished in winter from Dunlin by forked supercilium and short legs.
Iris—brown; bill—black; feet—greenish brown.
Voice: Dry trill *ch-r-r-reep*.
Range: Breeds in Siberia but migrates south in winter as far as Australia and New Zealand.
Distribution and status: In Borneo, Sumatra, Java, and Bali a rare coastal visitor.
Habits: Frequents intertidal mud-flats and sand banks, and marshy areas where it is a quiet solitary bird. Probes with bill held vertically. Crouches when alarmed.

[219. SPOON-BILLED SANDPIPER *Eurynorhynchus pygmaeus*
(Spoonbill Sandpiper) PLATE 25

Description: Small (15 cm), greyish brown, short-legged wader with streaked upperparts and prominent white eyebrow. The diagnostic spatulate bill is not easily seen in the field. Closely resembles Red-necked Stint in winter, but greyer with whiter forehead and breast. In summer upperparts and upper breast are rufous.
Iris—brown; bill—black; feet—black.
Voice: Shrill, quiet *preep preep* on take-off.
Range: Breeds in N Europe and Asia, but migrates to China and Hainan in winter with vagrants reaching SE Asia.
Distribution and status: Single bird seen in Sabah in Borneo in 1967 matches the description of this bird but remains unconfirmed.
Habits: A bird of sand-flats feeding with bill almost vertically downwards, working actively with a diagnostic sideways 'vacuum cleaning' motion.]

STILTS 143

220. RUFF *Philomachus pugnax* PLATE 24
(Reeve (female only))

Description: Largish (male 28 cm, female 23 cm), short-billed, dull brownish, long-legged wader with small head, long neck, and straight bill. Upperparts dark brown with pale scaling; throat pale buff; head and neck buff; underparts white, often lightly barred on flanks. In flight the narrow white wing bar and white oval patches on either side of the dark tail base are distinctive. Female noticeably smaller than male.
Iris—brown; bill—brown with yellowish base; feet—variable, yellow or green to brown-grey.
Voice: Low *chuck-chuck* but generally silent on winter grounds.
Range: Breeds in N Europe and Asia, but migrates south in winter to SE Asia and Indonesia, with uncommon vagrants as far as Australia.
Distribution and status: In the Greater Sundas a rare visitor, but has been recently recorded in Bali and Java. Rare in both Borneo and Sumatra on coasts and inland, up to 900 m.
Habits: Frequents marshy areas and mud-flats where it mingles with other waders.

STILTS—FAMILY RECURVIROSTRIDAE

A WORLDWIDE family of seven species of very long-legged waders. There can be no mistaking the only two members of this family in the Greater Sundas.

221. WHITE-HEADED STILT *Himantopus leucocephalus* PLATE 26

Description: Striking, elongate (37 cm), black and white wader with very long pink legs. Head and body white except for black wings, nape, and hindneck. In young birds the head is greyish and the back washed brownish. Distinguished from Black-winged Stilt by black hindneck patch.
Iris—pink; bill—long, pointed, black; legs and feet—pink.
Voice: High-pitched yelps *kik-kik-kik* and a shrill piping alarm call.
Range: Java and Lesser Sundas, Sulawesi and Moluccas to New Guinea, Australia, and New Zealand.
Distribution and status: In Java and Bali this bird is a rare breeder and uncommon visitor to coastal areas. An uncommon summer visitor to S Sumatra and Borneo.

144 STILTS

Habits: Frequents brackish or freshwater swamps, shallow lakes, and river margins, paddy fields, mud-flats and saltfields; occurs in pairs or small parties.
Note: Some authors treat this species as a subspecies of the Black-winged Stilt.

222. BLACK-WINGED STILT *Himantopus himantopus* PLATE 26
(Pied/Common Stilt)

Description: Tall, elongate (37 cm), pied wader with distinctive thin black bill, black wings, long red legs, and white plumage. Distinguished from White-headed Stilt by lack of black on hindneck.
Iris—pink; bill—black; legs and feet—pinkish red.
Voice: High-pitched piping calls and tern-like *kik-kik-kik*.
Range: India, China, and SE Asia.
Distribution and status: Vagrants reach Philippines and N Borneo.
Habits: Frequents shallow coastal and fresh water swamps.

PHALAROPES—FAMILY PHALAROPIDAE

A FAMILY of three species of specialized pelagic waders. They are slim, dainty birds with narrow pointed bills. They have dense plumage with duck-like down, rendering them highly buoyant. The feet are lobed rather than webbed. Outside the breeding season these birds stay almost permanently at sea in flocks, weaving about and feeding on the sea a short distance apart. All breed in the northern hemisphere and visit the tropics only in the winter. Sibley and Monroe (1990) reassigned the group to the family Scolopacidae.

Two species are found in the Greater Sundas.

223. RED PHALAROPE *Phalaropus fulicaria* PLATE 26
(Grey Phalarope)

Description: Small (21 cm) straight-billed grey wader. Very similar to Red-necked Phalarope but forecrown whiter, upperparts paler and more uniform, and bill deeper and broader, sometimes with yellow base. In the hand the lobes of the feet are yellow.
Iris—brown; bill—black with yellow base; feet—grey.
Voice: Like Red-necked Phalarope.
Range: Breeds in Arctic; winters mainly at sea off W Africa and Chile.
Distribution and status: Recorded once off Sarawak coast.
Habits: As Red-necked Phalarope but not yet recorded inland.

224. RED-NECKED PHALAROPE *Phalaropus lobatus* PLATE 26
(Northern Phalarope)

Description: Very small (18 cm), fine-billed, grey and white wader usually seen swimming on the sea. Crown and eye-patch black; upperparts grey with dark feather centres; underparts whitish; white wing bar conspicuous in flight. Distinguished from stints by fine bill and black eye-patch.
Iris—brown; bill—black; feet—grey.
Voice: Single or repeated *chek*.
Range: Breeds in Holarctic; winters at sea worldwide.
Distribution and status: Not uncommon winter visitor and passage migrant in the seas of the Greater Sundas; on passage not uncommon inland on Borneo.
Habits: Wintering flocks ride out at sea, feeding on surface plankton. They are rather tame, approachable birds. Sometimes they come inland and feed on ponds or mud-flats. Small flocks have been found up to 1200 m in the Kelabit uplands of Borneo.

Thick-knees—FAMILY BURHINIDAE

A SMALL family of specialized waders characterized by long powerful legs, lack of back toe, and enlarged knee-joints—hence the family name. The bill is straight, moderately short, and powerful. The eye is always large and clear yellow, and the wings are often marked black and white. Thick-knees live on sandy open areas of beach or heath wasteland. Only one species occurs in the Greater Sundas.

225. BEACH THICK-KNEE *Burhinus giganteus* PLATE 26

Description: Unmistakable large (55 cm) wader with stout bill. Crown and upperparts greyish brown, side of head boldly marked with black and white stripes; conspicuous grey, black, and white wing markings; inner primaries white, contrasting with black outer primaries in flight; underparts whitish, washed grey on chest.
Iris—yellow; bill—black with yellow base; legs and feet—yellow.
Voice: Mournful, low, whistling notes *wee-loo*.
Range: SE Asia; Philippines, Borneo, Sumatra, Java, Bali, Sulawesi, Moluccas, Lesser Sundas to New Guinea and Australia.
Distribution and status: Regularly seen in small numbers along the coastlines of the Greater Sundas.
Habits: Inhabits sandy and gravel beaches. Generally in pairs, hunting crabs and other prey. Birds are shy of walking humans but will approach inquisitively and

inspect someone sitting quietly. Makes curious head bobbing movements while standing. The nest is a scrape in the sand.

Note: Formerly placed in separate genus *Esacus* in which case name *E. magnirostris* was applicable. May be conspecific with Great Thick-knee *B. recurvirostris*.

Pratincoles—FAMILY GLAREOLIDAE

A SMALL family of curious long-winged birds with powerful, arched, pointed bills, found from Africa to Australia. They are insectivorous, catching food on the wing like a swallow, or running after it on the ground. Most species are migratory.

Two species occur in the Greater Sundas.

226. ORIENTAL PRATINCOLE *Glareola maldivarum*   PLATE 26
(Eastern Collared/Large Indian Pratincole)

Description: Medium-sized (23 cm), long-winged, plover-like bird with a forked tail and black border around the buffy throat (less clear in winter migrants). Upperparts brown with olive gloss; underwing with blackish primaries and chestnut coverts; uppertail coverts white; belly grey; undertail white; forked tail black with white base and outer edges.

Iris—dark brown; bill—black with scarlet base; feet—dark brown.

Voice: Sharp, raucous, grating *tar-rak*.

Range: Breeds in E Asia; migrant south in winter through Indonesia to Australia.

Distribution and status: Uncommon migrant and winter visitor to Sumatra (including islands) and Borneo (where breeding reported); a locally abundant visitor to coastal lowlands of N Java and Bali.

Habits: A graceful wader in small to large noisy flocks, mixing with other waders in open areas, marshes, and rice stubble. Runs well and bobs its head but also flies after insects in the air with graceful swallow-like flight. Often seen on airfields.

227. AUSTRALIAN PRATINCOLE *Stiltia isabella*   PLATE 26
(Long-legged/Isabelline Pratincole/Australian Courser)

Description: Medium-sized (23 cm), long-legged and long-winged, plover-like bird. Upperparts rufous; long black wings extending far beyond tail; upper breast sandy, patches on lower breast and belly chestnut; uppertail coverts, belly, and undertail white; underwing black; short black tail with white edge.

Iris—reddish brown; bill—red with black tip; feet—reddish.

Voice: Shrill thin *kwerree-peet*.

Range: Breeds in Australia; migrates north-west in summer to New Guinea and Indonesia.
Distribution and status: Erratic visitor to the Greater Sundas; recorded once in Sumatra (Belitung, in 1888) and there are a few records scattered through Borneo, Java, and Bali.
Habits: Similar to the Oriental Pratincole but more of a ground runner, living in open flocks. It bobs its head and tips its body. At dusk flocks hawk like terns after flying insects.

Skuas—FAMILY STERCORARIIDAE

A SMALL, worldwide family of dark-backed seabirds, rather similar in appearance to gulls but some are characterized by elongated central tail feathers. They are notorious for their aggressive harrying of other seabirds, forcing them to drop or regurgitate their food. Sibley and Monroe (1990) reassigned the skuas as a tribe under the family Laridae.

Four species occur in the seas of the Greater Sundas; all are rare.

228. POMARINE JAEGER *Stercorarius pomarinus* PLATE 27
(Pomatorhine Jaeger, AG: Skua)

Description: Largish (55 cm) dark seabird with elongated, spatulate central tail feathers. Occurs in two colour phases. Light phase: top of head black, side of head and nape yellowish; underparts white with sooty sides and breast bar; upperparts sooty chocolate with paler whitish grey bases to primaries; central two tail feathers extend 5 cm in blunt, broad, twisted trailers. Dark phase has no white or yellow. Non-breeding adults resemble immature; paler, more mottled and with grey cap.
Iris—dark; bill—black; feet—black.
Voice: Generally silent at sea.
Range: Breeds in Arctic; winter migrant to southern seas.
Distribution and status: The commonest skua in Indonesian waters; regularly reported Sunda Straits, W Sumatra, and SW Java, and presumably occurs throughout on passage.
Habits: As Parasitic Jaeger but keeps more out to sea. Parasitic, pirates food from other seabirds.

SKUAS

229. PARASITIC JAEGER *Stercorarius parasiticus* PLATE 27
(Arctic Jaeger, AG: Skua)

Description: Large (45 cm) dark seabird with elongated central tail feathers. Light phase: top of head black; sides of head and back collar yellow; underparts white with or without grey breast bar; upperparts dark, sooty chocolate except whitish bases to primaries giving conspicuous wing flash in flight. Dark phase: sooty brown all over except pale wing flash. Central tail feathers elongated into sharp, pointed streamers distinct from the shorter, blunt ones of the Pomarine Jaeger. Non-breeding adult paler, more mottled and with grey cap.
Iris—dark; bill—black; feet—black.
Voice: Generally silent at sea.
Range: Breeds in Arctic; winter migrant to southern seas.
Distribution and status: Less common than Pomarine Jaeger; recorded off Bali but most records are tentative.
Habits: Flies low over sea but attacks other feeding seabirds, twisting and turning them until they drop or disgorge food. Sometimes also follows ships, feeding on discarded refuse.

[230. LONG-TAILED JAEGER *Stercorarius longicaudus* PLATE 27
(AG: Skua)

Description: Largish (50 cm) dark seabird with elongated central tail feathers. Similar in both light and dark colour phases to Parasitic Jaeger but smaller, more slender, more buoyant, and with much longer central tail streamers (14–20 cm beyond tip of tail). Light phase does not have grey breast band. Dark phase is rare. Non-breeding adult duller and with tail streamers reduced. Juvenile has bolder black and white barring on vent than other jaegers.
Iris—dark; bill—black; feet—black.
Voice: Generally silent at sea.
Range: Breeds in Arctic; winter migrant to southern seas.
Distribution and status: Rare visitors may be seen in Greater Sundas waters. Sight records only off W Java and Bali.
Habits: As other jaegers.]

231. SOUTH POLAR SKUA *Catharacta maccormicki* PLATE 27
(McCormick's Skua)

Description: Large (53 cm) dark brownish seabird with small pointed projections on central tail feathers. Head, breast, and belly paler brown than wings. White base of primaries conspicuous from below and above. Larger and stockier than other skuas, with broad rounded wings and heavier body. Light phase lacks

black cap. Dark phase has at least some white on face. Non-breeding adults and immature darker and more streaked.
Iris—dark brown; bill—black; feet—black.
Voice: Generally silent at sea.
Range: Breeds in Antarctic.
Distribution and status: Uncommon summer vagrants may range into the Greater Sundas. Probably sometimes overlooked but recorded for Sumatra from time to time.
Habits: As other skuas, pirating regurgitated food from other seabirds. Has powerful hawk-like flight. Follows ships and perches on masts.

Gulls—Family Laridae

A LARGE, worldwide family of fish-eating and scavenging seabirds. Most of the species are white with black wing tips and varying degrees of black, grey, and brown on the head and upperparts. Juveniles are mottled brown and it takes several years to acquire full adult plumage. Gulls are larger, more round-winged, and heavier in flight than terns.

Gulls are most common in temperate latitudes where major sea upwellings support rich pelagic fisheries. There are no resident species in the Greater Sundas but 3 species have been recorded as visitors.

232. COMMON BLACK-HEADED GULL *Larus ridibundus* PLATE 27
(Black-headed Gull)

Description: Medium-sized (40 cm) grey and white gull with black spot behind eye (in winter) and red bill and feet. The leading edge of the wing is white, and the black of wingtip is not extensive nor spotted white. Immature birds have a black subterminal tail band; rear edge of wing black; plumage mottled brown. Distinguished from Brown-headed Gull by smaller size, more conspicuous white leading edge to wing, lack of white spots on black wing tip.

Iris—brown; bill—red (with black tip in immature); feet—red (paler in immature).
Voice: Harsh *kwar* calls.
Range: Breeds in Palaearctic; migrant to India, SE Asia, Philippines.
Distribution and status: Regular winter visitor in small numbers to N Borneo coast.
Habits: In the Greater Sundas it stays mostly out at sea, sitting on water, floating objects, or fish-trap poles, or wheeling in flocks with other gulls over fish shoals.

150 GULLS

233. SABINE'S GULL *Xema sabini* PLATE 27

Description: Small (34 cm), black, white, and grey gull. Wing tricoloured with sharp contrast between black triangle of outer wing and white triangle of secondaries; wing coverts and mantle uniform grey or in immature, brown scalloped buff; rump, underparts and underwing white. Slightly forked tail is white in adult or with black tip in immature. Adult has dark half-collar on nape. Distinguished from Black-headed Gull by forked tail, very distinctive wing pattern, and buoyant tern-like flight.
Iris—brown, red eye-ring; bill—black with yellow tip in adult, black in immature; legs and feet—dark grey in adult, pinkish in immature.
Voice: Harsh and grating, like Arctic Tern.
Range: Breeds in high Arctic; winters in E Atlantic and E Pacific. Vagrants recorded from many parts of the world.
Distribution and status: One record in the Greater Sundas about 10 km off the coast of W Sumatra in 1984.
Habits: Stragglers mix with flocks of other seabirds. More pelagic than other gulls, generally far offshore.

234. BROWN-HEADED GULL *Larus brunnicephalus* PLATE 27

Description: Medium-sized (42 cm) white gull with grey back and large white patch at base of primaries, and diagnostic white spots on black wing tips. Winter birds have dark brown patch behind eye. Summer birds have entire head and neck brown. Distinguished from Black-headed Gull by pale iris, heavier bill, larger size, and different wing tip pattern. Immature lacks white spots of wing tips and has black terminal tail band.
Iris—whitish yellow or grey, red eye-ring; bill—orange-red; feet—orange-red.
Voice: Harsh *gek, gek*, and loud wailing *ko-yek, ko yek* (Smythies).
Range: Breeds in C Asia; in winter to India, China, and SE Asia.
Distribution and status: Recorded once on coast of E Sumatra.
Habits: Joins other gulls in flocks over sea, coasts, and river estuaries.

TERNS AND NODDIES—FAMILY STERNIDAE

A WORLDWIDE family of graceful seabirds. Terns have short legs, long pointed wings, forked tails, and fine pointed bills. They have a buoyant flight and often hover over the water before diving to catch small fish. They congregate in large wheeling flocks wherever the fishing is good and are often found in coastal waters or even on inland lagoons and riverways. Many species are migratory, breeding in the extreme northern and southern latitudes, and only coming into the tropics in their respective winter seasons.

Seven of the 16 species recorded in the Greater Sundas are resident and breed in colonies on beaches or rockpiles. The nest is typically a simple scrape in the sand or gravel.

235. WHISKERED TERN *Chlidonias hybridus* PLATE 28

Description: Smallish (25 cm) pale tern with white forehead (in winter) and slightly forked tail. Adults in non-breeding plumage have white forehead, black-streaked crown, and black hindcrown and nape; underparts white; wings, back, and uppertail coverts grey. Young birds similar to adults but with brown mottling. Distinguished from non-breeding White-winged Tern by blacker crown, grey rump, and lack of separate black cheek patch. In breeding plumage, forehead is black, and breast and belly grey.

Iris—dark brown; bill—red; feet—red.

Voice: Harsh, staccato *kitt* or *ki-kitt.*

Range: Breeds in S Africa, S Europe, Asia, and Australia.

Distribution and status: It makes irregular movements through Indonesia; recorded from Sumatra, Borneo, Java, Bali, Moluccas, and New Guinea. In Sumatra, Java, and Bali a regular winter visitor with some birds also seen in summer.

Habits: Lives in small or occasionally large flocks, often coming up to 20 km inland to feed over flooded land and paddy fields, taking food in shallow plunges or low skimming flights.

236. WHITE-WINGED TERN *Chlidonias leucopterus* PLATE 28
(White-winged Black Tern)

Description: Small (23 cm) tern with slightly forked tail. Non-breeding adult has pale greyish upperparts; back of head mottled greyish black; underparts white. Distinguished from non-breeding Whiskered Tern by more complete white nape collar, crown less black and more mottled, black ear coverts separate from black crown, and paler rump. Breeding adult unmistakable with head, back, and breast black, contrasting sharply with white tail and pale grey wings. Upperwing whitish, underwing coverts distinctively black.

Iris—dark brown; bill—red (breeding), black (non-breeding); feet—orange-red.

Voice: Repeated *kweek* or sharp *kwek-kwek.*

Range: Breeds in S Europe across Asia to C Russia and China; migrates in winter to S Africa and through Indonesia to Australia and occasionally New Zealand.

Distribution and status: Common migrant and winter visitor throughout

Greater Sundas; predominantly coastal but also inland over rice fields, to 400 m. Some birds over summer.

Habits: Frequents coastal areas, estuaries, and river mouths in small flocks; ranges well inland to feed over paddy fields and swamps. Feeds by skimming low over water, working into the wind, and chasing insects. Commonly perches on poles.

237. GULL-BILLED TERN *Sterna nilotica*   PLATE 28

Description: Medium-sized (39 cm) pale tern with slightly forked tail and heavy black bill. Adult in winter has white underparts, grey upperparts, white head with grey mottling on nape, and black patch through eye. In summer entire cap is black.
Iris—brown; bill—black; feet—black.
Voice: Repeated *kuwk-wik* or *kik-hik*, *hik hik hik*.
Range: Almost worldwide, breeding in Americas, Europe, Africa, Asia, and Australia, and passing through Indonesia and New Guinea as a migrant.
Distribution and status: In Sumatra and Borneo waters it is quite common. In Java and Bali an uncommon winter visitor.
Habits: Frequents coastal estuaries, lagoons, and inland fresh waters. Often hovers and generally feeds by skimming over water or mud, rarely plunges into water.

238. CASPIAN TERN *Sterna caspia*   PLATE 28

Description: Very large (49 cm) tern with diagnostic massive red bill. Black cap of summer becomes streaked white in wintering birds. Undersides of primaries black. Immature birds barred brown on upperparts.
Iris—brown; bill—red; feet—black.
Voice: Harsh, rasping *kraaah*.
Range: Cosmopolitan.
Distribution and status: Vagrant in Greater Sundas, where recorded once on coast of E Sumatra and once in Brunei.
Habits: Favours coasts, lakes, mangroves, and estuaries.

239. COMMON TERN *Sterna hirundo*   PLATE 29

Description: Smallish (35 cm) black-naped tern (in winter) with deeply forked tail. Adult non-breeding: upperwing and back grey; uppertail coverts, rump, and tail white; forehead white; crown mottled black and white; black on nape; underparts white. Breeding: black cap, grey breast. In flight, non-breeding adult and immature are characterized by a blackish bar on the forewing and blackish edge to outer tail feathers. Immature has browner upperparts and scaling on mantle.

Iris—brown; bill—black in winter, red base in summer; feet—reddish, darker in winter.
Voice: Harsh *keer-ar* descending, with emphasis on first note.
Range: Breeds in N America, Europe, and Asia but migrates south in winter to S America, Africa, Indonesia, and Australia.
Distribution and status: In the Greater Sundas an irregular winter migrant, occasionally present in very large flocks with some non-breeding birds over summering.
Habits: Frequents coastal waters and occasionally inland fresh waters. Rests on elevated perches such as fishing platforms and rocks. Strong flyer; feeds by plunging steeply into the sea.

240. ROSEATE TERN *Sterna dougallii* PLATE 29

Description: Medium-sized (39 cm) black-crowned tern with very long, deeply forked, white tail. Adult in summer has black crown, pale grey upperwings and back, and white underparts; pinkish on breast. In winter forehead white, crown mottled, pinkish wash absent. Webs of outer primaries blackish. Juvenile: black bill and legs; blackish brown crown, nape, and ear coverts; mantle darker brown than Common Tern; tail white, lacking streamers.
Iris—brown; bill—black with red base (in breeding season); feet—reddish in breeding season, otherwise black.
Voice: Musical *chew-it* while fishing, or harsh *aaak* in alarm.
Range: Cosmopolitan.
Distribution and status: A scarce winter visitor reported throughout the Greater Sundas. Breeding is suspected in Sumatra (islands in Malacca Straits), and known on islets off Brunei and locally around the coasts of Java and islands in Java Sea. Colonies are on coral ridges and sandy beaches of offshore rocks and islets.
Habits: Inhabits coral formations, coral islands, and sandy beaches, but generally uncommon. Often mixes with other terns. Flies in graceful manner, plunging steeply to catch small fish.

241. BLACK-NAPED TERN *Sterna sumatrana* PLATE 29

Description: Smallish (31 cm), very white tern with very long forked tail, and distinctive black nape band and narrow bill. Upperparts pale grey; underparts white; head white except black spot in front of eye and black band over nape. Young birds have brown mottling on crown and blackish mottling on nape. Juvenile: side of head and nape greyish brown; upperparts brownish, scalloped buff and grey; rump whitish, rounded unforked tail.

Iris—brown; bill—black with yellow tip (adult), dirty yellow (juvenile); feet—black (adult) or yellow (juvenile).
Voice: Sharp *tsii-chee-chi-chip* or in alarm *chit-chit-chitrer*.
Range: Tropical islands and coasts of Indian and Pacific Oceans to N Australia.
Distribution and status: This is one of the commonest terns in the Greater Sundas, breeding on offshore rocks and islets.
Habits: Gregarious bird, flocking with other terns along sandy and coral beaches, rarely over mud, and never far inland.

242. BRIDLED TERN *Sterna anaethetus* PLATE 29
(Brown-winged Tern)

Description: Medium-sized (37 cm) dark-backed tern with long, deeply forked tail. Adult: dark brownish grey on upperwings, back, and tail except white leading edge to wing and white outer tail feathers; underparts white. White forehead; narrow supercilium extends beyond eye; white collar. Young birds are browner with mottled brown crown, grey breast, and back barred buff.
Iris—brown; bill—black; feet—black.
Voice: Staccato yapping *wep-wep*, and harsh grating alarm calls *kee-errr-krr*.
Range: Widespread throughout Atlantic, Indian, and Pacific Oceans, as far as Australia.
Distribution and status: An offshore resident in the Greater Sundas, seen inshore mostly during the summer moult.
Habits: Keeps well out to sea, coming inshore only in bad weather or breeding season. Not very social; single or in small parties. The flight is graceful and buoyant. Feeds by scooping insects or fish off the surface; does not dive. Frequently rests on flotsam or at night on the spars of ships. Breeds in mixed colonies with Black-naped Terns.

243. SOOTY TERN *Sterna fuscata* PLATE 29

Description: Medium-sized (44 cm) black-backed tern with deeply forked tail. Similar to Bridled Tern but upperwings and back darker sooty brown, lacks a white collar ring, and white forehead does not extend into supercilium. Immature is sooty brown with white vent and bars of white spots on back and upper wings.
Iris—brown; bill—black; feet—black.
Voice: Nasal *ker-waky-wak*, *wide-a-wake*.
Range: Widespread throughout tropical Atlantic, Indian, and Pacific Oceans.
Distribution and status: Mostly recorded far out to sea in Javan and Sumatran waters.
Habits: Truly oceanic tern staying well out to sea or on small rocky and sandy islets. Follows ships at night. Flight is easy and buoyant, soaring on updraughts.

244. LITTLE TERN *Sterna albifrons*   PLATE 29

Description: Small (24 cm) pale tern with slightly forked tail. Summer: crown, nape, and eye-stripe black; forehead white. Winter: black of crown and nape reduced to crescent. Wing with darker leading edge and white rear edge. Juvenile: similar to non-breeding adult but mottled with brown on crown and mantle; tail white with brown tip; bill dusky.
Iris—brown; bill—yellow with black tip; feet—yellow.
Voice: Rasping, high-pitched shrieks.
Range: Common resident throughout coastal seas of temperate and tropical regions.
Distribution and status: In Sumatra and Borneo a fairly common winter visitor. In Java and Bali there is a small resident population, swelled in winter by a different migrant race.
Habits: Inhabits sandy seashores, mixing with other terns. Fast wing-beat and regular hovering and diving characteristic, with the bird taking off again as soon as it has dived.

245. GREAT CRESTED-TERN *Sterna bergii*   PLATE 30
(Greater Crested-Tern/Swift Tern)

Description: Large (45 cm) crested tern. In summer the crown and crest are black, becoming mottled with white in transition to the winter white crown and grey mottled crest. Upperparts grey, underparts white. Young birds darker grey than adults and mottled with brown and white on upperparts; tail grey. Bill colour is the best character to distinguish from other crested terns.
Iris—brown; bill—yellow; feet—black.
Voice: Sharp, rasping *kirrik*, or clear *chew*.
Range: Distributed throughout coasts and islands of Indian Ocean, Persian Gulf, tropical seas of Pacific, and coasts of Australia and S Africa.
Distribution and status: In the Greater Sundas this is one of the commonest terns in inshore waters and around small rocky islets. Breeds on Karimun Jawa islands and suspected to breed on small islands off Sumatra.
Habits: Fishes in small parties of twos or threes, sometimes with other terns. Rather clumsy plunges. Rests on beaches, buoys, and fishing platforms, or on the water or floating objects. Often ventures quite far out to sea.

246. LESSER CRESTED-TERN *Sterna bengalensis*   PLATE 30
(Crested Tern)

Description: Medium-sized (40 cm) crested tern. Similar to Great Crested-Tern but smaller, with black forehead in breeding plumage, and distinctive orange bill.

156 TERNS AND NODDIES

In winter the forehead and crown become white; the crest remains black. Juvenile similar to non-breeding adult, but brownish mottling on upperparts and dark grey flight feathers.
Iris—brown; bill—orange; feet—black.
Voice: Raucous screams *kirrik*.
Range: Breeds along Red Sea, Persian Gulf, India, SE Asia, Borneo, Philippines, New Guinea, and N Australia.
Distribution and status: In Sumatra and Bornean waters a regular visitor but much less common than the Great Crested-Tern. In Java and Bali a common winter visitor.
Habits: Highly social; in large flocks, often mixed with other species, especially Crested-Terns. Frequents coastal waters and mud, sand, or coral shores, often feeding far out to sea; rests in quarrelsome parties on fishing platforms and buoys. Dives vertically submerging completely.

247. CHINESE CRESTED-TERN *Sterna bernsteini* PLATE 30

Description: Medium-sized (38 cm) crested tern with diagnostic black-tipped yellow bill. Wintering birds have forehead white and black cap with central white streaks leaving 'U'-shaped black patch around nape. Distinguished from Great and Lesser Crested-Terns by black tip to yellow bill. Immature like immature Lesser Crested-Tern but browner, inner part of wing lining paler, two dark bars on inner wing; back and tail mottled whitish and brown.
Iris—brown; bill—yellow with black tip; feet—black.
Voice: Harsh, high-pitched cries.
Range: Breeds in E China migrating south in winter to S China, Philippines, and occasionally N Borneo.
Distribution and status: In the last few years the species has become very rare, close to extinct. In the Greater Sundas it has not been recorded reliably for several years.
Habits: Similar to other crested-terns. Favours open sea and small islets.

248. BROWN NODDY *Anous stolidus* PLATE 30
(Common Noddy/Noddy Tern)

Description: Medium-sized (42 cm), dark sooty brown tern with whitish crown and notched tail. Plumage uniform sooty brown apart from whitish crown and white eye-ring. Juvenile has dark forehead and crown, white eye-ring, and whitish tips to feathers of back and wing coverts.
Iris—brown; bill—black; feet—blackish brown.
Voice: Harsh *karrk* and *kwok-kwok*.

Range: Throughout the tropical and subtropical oceans and N Australia; breeding throughout its range.
Distribution and status: In Greater Sundas waters a not uncommon bird out at sea but uncommon inshore.
Habits: An oceanic bird with a slow, lazy, wheeling flight. Rarely dives like other terns. Takes small fry as they skip out of the water when pursued by predatory fish. Sometimes alights on water to collect food. In courtship, partners nod heads in display, hence the English name.

249. BLACK NODDY *Anous minutus*   PLATE 30

Description: Smallish (32 cm) blackish brown tern with whitish crown and notched tail. Very similar to Brown Noddy but smaller, more slender, and with an almost pure white crown which extends further back to cover nape; bill is longer and more slender. There is a fine white arc below the eye. Juvenile less white on crown; upperwing coverts and secondaries with buff tips.
Iris—brown; bill—black; feet—blackish brown.
Voice: Harsh *kik-kirrik* and rattling *churr.*
Range: Found throughout the tropical and subtropical Atlantic and Pacific Oceans. Replaced in Indian Ocean by closely related Lesser Noddy *A. tenuirostris.*
Distribution and status: A rare vagrant off the west coast of Borneo, a breeding visitor along the eastern coastline of Sumatra, and an uncommon bird along the north coast of Java and Bali.
Habits: As Brown Noddy.
Note: May be conspecific with Lesser Noddy *A. tenuirostris*, in which case name White-capped Noddy becomes applicable.

250. COMMON WHITE TERN *Gygis alba*   PLATE 28
(White/Fairy Tern/ White Noddy)

Description: Small (30 cm) pure white tern with black eye-ring. Entire adult plumage whitish except for black eye-ring. Tail slightly forked with outer feathers shorter than second and third feathers. Bill is unusual, being very slender, sharp, and slightly upturned. Juvenile: dark ear-spot, greyish brown mottled mantle and upper wings, black primary shafts, and more rounded wings.
Iris—brown; bill—blackish with blue base; feet—bluish black with whitish webs.
Voice: Soft buzzing calls.
Range: Widely distributed throughout tropical and subtropical oceans.
Distribution and status: In Sumatran and Javan waters a very rare visitor, the nearest breeding sites being Christmas Island, Cocos Keeling Islands, and Palau.
Habits: Flies with light rising and sinking manner; occasionally dives for food but never submerges.

158 PIGEONS AND DOVES

Pigeons and doves—FAMILY COLUMBIDAE

The large, worldwide family of pigeons feeds predominantly on fruits, seeds, and berries, and all have rather compact, plump bodies and short, stout bills. They nest on flimsy twig platform nests, laying white eggs. Calls are repetitive, melodious coos, and in flight pigeons make a characteristic flapping noise. Thirty species occur in the Greater Sundas, comprising three species groups.

1. Fruit-doves and green-pigeons—(*Treron, Ptilinopus*)—smaller, arboreal birds with generally brightly coloured plumage without metallic colours.
2. Imperial pigeons—(*Ducula, Columba*)—large, arboreal birds with metallic sheen in the plumage and generally with grey or whitish underparts.
3. Ground doves—(*Macropygia, Streptopelia, Geopelia, Chalcophaps* and *Caloenas*)— birds which regularly visit the ground and have either highly iridescent, greenish upperparts, or drab, reddish brown colours.

Fruit doves and green pigeons

251. SUMATRAN GREEN-PIGEON *Treron oxyura* PLATE 31
(Yellow-bellied/Green Spectacled Pigeon)

Description: Medium-sized (34 cm including long, pointed tail) green-pigeon. Male: upperparts dark green with blue-grey collar. Flight feathers blackish without yellow edges; grey tail long and pointed; underparts green with yellowish breast and belly and long, light cinnamon undertail coverts. Female: lacks yellow belly and has buff-streaked green undertail coverts.
Iris—blue-green; bill—blue-green, turquoise base; feet—red.
Voice: A ringing *oo-oowao-oowao* or variations.
Range: Endemic to Sumatra and Java.
Distribution and status: On Sumatra it is uncommon in hill and mountain forests of the Barisan Range between 350 and 1800 m. Uncommon on Java and restricted to the mountains of W Java (east to Papandayan), up to 3000 m.
Habits: Occurs in flocks which wander from place to place but never far from the thick hill forests where they live.

252. WEDGE-TAILED GREEN-PIGEON *Treron sphenura* PLATE 31
(> Korthal's Green-Pigeon, AG: Pigeon)

Description: Medium-sized (30 cm) green-pigeon. Male: head green; breast orange, lacking in Sumatran race; mantle purplish, grey wing coverts, and upper back purplish chestnut; rest of wing and tail dark green; greater wing coverts and darker flight feathers with yellow edges; vent yellowish with dark streaks ; flanks edged yellow; undertail coverts cinnamon. Female: pale yellow undertail coverts and vent with large dark markings; lacks golden and chestnut colouring of male. Iris—pale blue to red; bill—turquoise at base, cream at tip; feet—red.

Voice: A deep whistled note *koo* varied with a curious grunting note.
Range: Himalayas, SW China, SE Asia, Sumatra, Java, and Lombok.
Distribution and status: On Sumatra (in the Barisan Range from Sibayak to Dempu) and Java it is locally common on high mountains between 1400 and 3000 m; usually at higher altitude than Sumatran Green-pigeon. Not recorded for Bali but it could occur.
Habits: Occurs in oak-laurel and montane heath forests and is quite tame and approachable.

253. THICK-BILLED GREEN-PIGEON *Treron curvirostra* PLATE 31
(AG: Pigeon)

Description: Medium-sized (27 cm), thickset green-pigeon. Male: back, mantle, and upper innerwing coverts maroon in male; dark green in female; forehead and crown grey; neck green; underparts yellowish green; wings blackish with yellow edges to feathers and a bold yellow wing bar; central tail feathers green, others grey with medial black bar; flanks green with white bars; undertail coverts cinnamon. Distinguished from Grey-cheeked Green-Pigeon by greenish sides to face, greyer nape, and red cere.
Iris—yellow, bright blue-green eye-ring; bill—green with red (Sumatra and Borneo) or olive base (W Java, Mentawai, and Enggano); feet—crimson.
Voice: Loud, throaty, guttural notes.
Range: NW India and Nepal, SE Asia, Philippines, and Greater Sundas.
Distribution and status: On Sumatra (including endemic races on Sumatra islands) and Borneo a common bird of lowland forests up to 200 m, and locally much higher on Borneo. In Java this species is recorded only from Deli and Tinjil off SW Java.
Habits: Noisy, feeding in flocks, often found thrashing about in low canopy trees.

254. GREY-CHEEKED GREEN-PIGEON *Treron griseicauda* PLATE 31
(AG: Pigeon)

Description: A small (25 cm), yellow-green pigeon. Forehead bluish grey and bare area around eye green. Wing coverts, scapulars, and upper back green in female but maroon in male; flight feathers nearly black with bright yellow edges; orange shoulder patch. Underparts and lower back mostly green with chestnut undertail coverts in male. Tail green with pale grey terminal bar; undertail coverts cinnamon. A rare subspecies on Kangean Island has a yellowish neck and purplish wash over the breast of the male.

Iris—red; bill—yellow with dark green cere in male, all green in female; feet—red.

Voice: Quiet, deep ringing *haw-haw-haw*.

Range: Sulawesi, Java, and Bali.

Distribution and status: Fairly common in Java and Bali up to 1200 m, and in E Java up to 2500 m.

Habits: Frequents lowland forest and gardens. Lives in pairs, but congregates at fruit trees. Occasionally hundreds have been recorded feeding together. Keeps high up in the canopy, occasionally mixing with other species.

Note: May be conspecific with Thick-billed Green-Pigeon.

255. CINNAMON-HEADED GREEN-PIGEON *Treron fulvicollis*
(AG: Pigeon) PLATE 31

Description: Medium-sized (27 cm) green (female) or cinnamon-headed (male) pigeon. Breast ochraceous-orange; belly greyish green with yellow on thighs; undertail coverts cinnamon. In N Borneo race, male has head and breast cinnamon and belly grey, only slightly tinged with green. Cinnamon-headed male with dark green tail is unmistakable. Female is distinguished from Thick-billed Pigeon by yellowish thighs, more slender bill, and greener crown. Distinguished from all other green pigeons by greenish-white bill with red base. In flight the underside of the tail is blackish with a broad, grey terminal bar.

Iris—brown; bill—pale greenish with red base; feet—red.

Voice: No information.

Range: Malay Peninsula, Sumatra and Borneo.

Distribution and status: An uncommon bird around coastal areas up to 200 m. More common in S and C Borneo.

Habits: Prefers coastal forests, mangroves, swamp forest and open scrub. Feeds and nests in small trees.

256. LITTLE GREEN-PIGEON *Treron olax*
(AG: Pigeon) PLATE 31

Description: Smallest (22 cm) green-pigeon. Male: wing coverts, back, and mantle dark maroon; head grey; breast orange; belly green; undertail coverts cinnamon. Female: greyish crown; white chin; green breast and belly; dark green upperparts and yellow undertail coverts.

Iris—white; bill—white to bluish green; feet—red.

Voice: A quiet rising and falling whining whistle (D.A.H.).

Range: Malay Peninsula, Borneo, Sumatra, and W Java.
Distribution and status: In Borneo and Sumatra it is locally common up to 1400 m. In W Java it is a very rare bird of lowland forest.
Habits: Inhabits forest, parks and gardens with a preference for sub-montane habitat. Occurs in pairs and small flocks.

257. PINK-NECKED GREEN-PIGEON *Treron vernans*   PLATE 32
(AG: Pigeon)

Description: A smallish (29 cm) green-pigeon. Male: head blue-grey; sides of neck, lower nape, and band across upper breast pink; lower breast orange; abdomen green, ventrally yellow; white edges on flanks and thighs; undertail coverts cinnamon; back green; uppertail coverts bronze-ochre; wings dark with contrasting yellow edges to greater wing coverts; tail grey with subterminal black bar and pale grey rim. Female green; lacks grey, pink, and orange colouring of male.
Iris—pink; bill—blue-grey with green cere; feet—red.
Voice: A curious chuckling, cooing whistle with characteristic 'winding up' whine at start (D.A.H.). *Ooo-ooo, cheweeo-cheweeoo-cheweeo*, also hoarse *krrak,krrak* rasps when feeding in flocks.
Range: Indochina, Malay Peninsula, Borneo, Philippines, Sulawesi, Sumatra, Java, Bali, and Lesser Sundas.
Distribution and status: A common pigeon of coastal forests, secondary forests, and open country, more rarely found in high forest up to 900 m.
Habits: Small flocks congregate in low forest and descend on fruit trees to feed. Disturbed birds leave in twos and threes with a loud clapping of wings. In the evenings and mornings they give low, soft, cooing calls from communal roosts.

258. ORANGE-BREASTED GREEN-PIGEON *Treron bicincta*
(AG: Pigeon)   PLATE 32

Description: Medium-sized (29 cm) green-pigeon with conspicuous yellow stripes and edges on black wing feathers. Face green, nape and upper back grey. Male: pink breast with deep orange barring on lower breast; female has green breast. Tail dark grey with a black subterminal bar, often broken by central all-grey feathers.
Iris—blue and red; bill—greenish blue; feet—dark red.
Voice: Attractive modulated whistle followed by gurgling notes *ko-wrrrook, ko-wrrroook, ko-wrrroook*, harsh croaking alarm, and a chuckling call *kreeeew-kreeew-kreeew* (Balen, 1991).
Range: India, SE Asia, Java, and Bali.

Distribution and status: Formerly coastal lowlands of mainland W Java, now rare; recent records are from Baluran, E Java and W Bali.
Habits: Typical of the genus. Lives in pairs or sometimes small parties. Feeds in small, fruit-bearing bushes and trees. Has a typical *Treron* tail-flicking display. Prefers wooded lowlands and plantations.

259. LARGE GREEN-PIGEON *Treron capellei*   PLATE 32
(AG: Pigeon)

Description: Large (36 cm), orange-breasted green-pigeon. Greyish green back; wings dark grey with narrow yellow edges on coverts; underparts pale green with prominent yellow-orange breast band (less prominent in female); tail pale green with broken blackish bar and whitish tip concealed by all green central feathers; white edging on flanks and vent; dark cinnamon undertail coverts.
Iris—brown; bill—pale green; feet—yellow.
Voice: Hornbill-like, deep, booming, growled *koo*, and when feeding a gurgling note, also a rasping *kak, kak, kak, kwok-kwok-kwok*.
Range: Malay Peninsula, Borneo, Sumatra, and Java.
Distribution and status: In Sumatra and Borneo it is not uncommon in lowland forests up to 1300 m. In Java it is a rare bird and there are no recent records.
Habits: Lives singly or in pairs but gathers at fruit trees. Prefers primary forest and forest patches.

260. JAMBU FRUIT-DOVE *Ptilinopus jambu*   PLATE 32

Description: Medium-sized (28 cm), green, crimson and white fruit dove. Male: crimson face and black throat patch; nape and upperparts green; underparts white with pink breast patch and chestnut undertail coverts. Female: face dull purple; underparts green with white belly; undertail coverts brownish.
Iris—brown; bill—yellow or orange; feet—dark red.
Voice: A soft *koo*, seldom uttered.
Range: Malay Peninsula, Borneo, Sumatra, and W Java.
Distribution and status: In Borneo and Sumatra a locally common bird up to 1500 m. In W Java it is a rare bird.
Habits: Takes fruit from trees and also fallen fruit on the ground. Lives in coastal districts (including mangroves) and open, wooded areas. Favours small islands and flies from island to island. Often overlooked.

PIGEONS AND DOVES 163

261. PINK-HEADED FRUIT-DOVE *Ptilinopus porphyreus* PLATE 32
(Pink-necked Fruit Dove)

Description: Largish (29 cm), pink-headed fruit-dove. Entire head, neck and throat purplish pink, bordered across the chest by a white band underlined in greenish black; upperparts green; underparts grey with yellow undertail coverts. Female: only face dull pink, breast bands less defined.
Iris—red orange; bill—greenish; feet—pink.
Voice: Soft *hoo*.
Range: Endemic to Sumatra, Java, and Bali.
Distribution and status: Known from the Barisan Range of Sumatra (Mts Kerinci and Dempu) and the mountains of Java and Bali, from 1400 to 2200 m.
Habits: Generally seen singly or in pairs, preferring oak-laurel forest and montane heath forests. Shy and inconspicuous behaviour.

262. BLACK-NAPED FRUIT-DOVE *Ptilinopus melanospila* PLATE 32

Description: Medium-sized (27 cm), green fruit-dove. Male: silvery grey head with black nape and yellow throat patch; lower breast and upperparts green with yellow and red undertail coverts. Female: overall green except for red undertail coverts and yellow edges to flight feathers and lower belly.
Iris—yellow; bill—greenish yellow; feet—red.
Voice: Monotonous, ringing *owook-wook... owook-wook*.
Range: S Philippines, Sulawesi, Moluccas, and Sundas.
Distribution and status: Recorded once Sumatra (Pulau Tegal in Lampung Bay, presumably a stray from Java) and resident on islands off Borneo (Burung, Balambang, Banggi, and Maratuas). On Java (including Matasiri and Kangean) and Bali it is locally common in lowland and hill forests, up to 800 m.
Habits: A shy, pair-living bird more often heard than seen. Sometimes quite large flocks gather at food trees and at roosts.

263. BLACK-BACKED FRUIT-DOVE *Ptilinopus cinctus* PLATE 32
(White-headed/Banded Fruit-Dove)

Description: Medium-sized (34 cm), distinctive black and white fruit-dove. Head and neck greyish white becoming pure white on breast; black bar across lower breast. Upperparts dark greenish black with a grey terminal tail bar. Belly and vent grey.
Iris—brown; bill—grey; feet—red.
Voice: A deep booming call.
Range: Bali and Lesser Sundas to Timor.
Distribution and status: On Bali it is rare; in dry forest.

164 PIGEONS AND DOVES

Habits: Lives singly or in small flocks. Has a fast flight with whistling wings. This pigeon suns on exposed perches.

Imperial-pigeons

264. GREEN IMPERIAL-PIGEON *Ducula aenea*   PLATE 33
(> Enggano Imperial-Pigeon)

Description: Large (45 cm) green and grey pigeon. Head, neck, and underparts pale pinkish grey; undertail coverts chestnut; upperparts dark green with a diagnostic bronze iridescence.
Iris—reddish brown; bill—blue-grey; feet—dark red.
Voice: Loud single *oom*, a reverberant *kruk-kroorr*, and loud full call of several chuckling notes ending in a rolling note (see Appendix 6).
Range: India to S China, SE Asia, Philippines, Sundas, and Sulawesi.
Distribution and status: In the Greater Sundas this is the common large pigeon of lowland rainforest and mangroves, up to 1000 m.
Habits: Pairs and small parties can be seen flying over the forest to roosting trees in the evening, and dispersing again to find food in the morning. A bird of the high treetops. In courtship flight it makes spectacular vertical climbs to stalling point then abruptly dives and levels out again.

265. PIED IMPERIAL-PIGEON *Ducula bicolor*   PLATE 33
(Nutmeg Imperial-Pigeon)

Description: Large (42 cm) white and black pigeon. Entire body creamy white except black flight feathers and tail. Distinguished from Silvery Wood-Pigeon by whiter or more creamy colour.
Iris—brown; bill—grey; feet—blue-grey.
Voice: Deep, loud, resonant, chuckling *hu-hu-hu-hu-hu*.
Range: Malay Peninsula, SE Asia, Borneo, Sumatra, Java, Philippines, Sulawesi, Lesser Sundas, Moluccas to Irian Jaya.
Distribution and status: Still common on islands off Borneo and Sumatra, particularly along coastal mangroves and on small islands, with some populations on the main islands. In Java and Bali it is now rather rare, due to excessive hunting.
Habits: Settles in communal roosts and forages in small parties. Powerful fliers, birds often fly between small islands. Feeds and sits conspicuously in high trees.

266. MOUNTAIN IMPERIAL-PIGEON *Ducula badia*   PLATE 33

Description: Large (45 cm) darkish pigeon. Head, neck, breast, and belly purplish grey; chin and throat white; mantle and wing coverts dark maroon; back and rump dark greyish brown; tail is brownish black with a broad, pale grey terminal band; undertail coverts buff. Distinguished from Dark-backed Imperial-Pigeon by purplish black and buff vent, and from Green Imperial-Pigeon by bicoloured tail.

Iris—white, grey, or red; bill—crimson with white tip; feet—crimson.

Voice: A click followed by two melancholic, booming coos *click-broom-broom* (see Appendix 6).

Range: India, SE Asia, Borneo, Sumatra, and W Java.

Distribution and status: In Borneo and Sumatra this is the commonest large pigeon of montane forests between 400 and 2200 m, but also visits coastal mangroves. In Java a very rare bird of hills and mountains, often confused with the commoner Dark-backed Imperial-Pigeon. There are few recent Javan records, for example Halimun.

Habits: Coastal birds bathe in the mangrove water and make daily trips inland. Montane populations make daily flights to lowland feeding areas. Shier than Green Imperial-Pigeon.

267. DARK-BACKED IMPERIAL-PIGEON *Ducula lacernulata*
(Black-backed Imperial-Pigeon) PLATE 33

Description: Large (45 cm), very dark pigeon. Head, neck, and underparts pinkish grey with variation in head colour—grey in W Java, pink in E Java and Bali. Upperparts dark brownish grey, tail with a broad, light grey terminal bar. Distinguished from Mountain Imperial-Pigeon by darker back and cinnamon undertail coverts.

Iris—reddish brown; bill—dark grey; feet—purplish red.

Voice: Typically very deep, ringing call *hooh-oo* with the emphasis on the second note and some variations (D.A.H.).

Range: Endemic to Java, Bali, Lombok, and Flores.

Distribution and status: In Java and Bali this is the commonest large pigeon of montane forests from 400 to 1500 m.

Habits: Inhabits montane forests. Sits quietly, low in trees through midday. Active in mornings and evenings.

268. GREY IMPERIAL-PIGEON *Ducula pickeringi* PLATE 33
(Malaysian Imperial-Pigeon)

Description: Large (40 cm) brownish grey pigeon with slight greenish gloss on mantle. Similar to Green Imperial-Pigeon but smaller, greyer, and lacks chestnut undertail coverts.

Iris—crimson; bill—bluish grey with dark base; feet—crimson.

Voice: No information.

Range: Found only on small islands around the N and E coast of Borneo, S Philippines, and N Sulawesi.
Distribution and status: Locally common in coastal forest.
Habits: Similar to other imperial-pigeons but confined to small islands.

269. PINK-HEADED IMPERIAL-PIGEON *Ducula rosacea* PLATE 33
(Javan Imperial-Pigeon)

Description: Large (44 cm), pinkish grey pigeon. Head pinkish with white eye-ring and a white band at base of the bill. Underparts grey, washed pink on belly; upperparts dark bluish green with a slight iridescence; undertail coverts chestnut. Iris—brown; bill—blue-grey with dark red or purple cere; feet—purplish red.
Voice: A series of *owoo* notes descending the scale (F.G.R.).
Range: Limited to small islands in the Java Sea, S Sulawesi, Moluccas, and Lesser Sundas.
Distribution and status: Locally common.
Habits: As other imperial-pigeons.

270. METALLIC PIGEON *Columba vitiensis* PLATE 33
(White-throated Pigeon, AG: Wood-Pigeon)

Description: Large (42 cm), dark, glossy pigeon with white throat. Nape, back, wings, and tail dark slaty grey, mottled with glossy, metallic green, iridescent tips to feathers; top of head, mantle, and underparts purplish grey with lilac iridescence; crimson eye-ring.
Iris—brown with yellow ring; bill—yellow with red base; feet—crimson.
Voice: Booming *hooo ooo ooo* with emphasis on last two notes; also soft cooing.
Range: Philippines to Lesser Sundas and W Pacific.
Distribution and status: A bird of small islands, also on some of the islands off the E coast of Borneo—Tiga, Mantanani, and Maratua, where it is locally common.
Habits: Lives in tall coastal forests of small islands.

271. SILVERY WOOD-PIGEON *Columba argentina* PLATE 33
(Grey/Silver Wood-Pigeon)

Description: Large (40 cm), pale grey pigeon with black wings and tail and greyish underparts. Distinguished from Pied Imperial-Pigeon by grey (rather than white) upperparts, completely black terminal half of tail, and red eye-ring.
Iris—brown; bare eye-ring—dark red; bill—pale green with red base; feet—red.
Voice: No information.
Range: E Sumatra, Riau archipelago, and N Borneo.

Distribution and status: A rare bird living on small islands of the Malacca Straits, but occasionally seen on the coast of Sumatra and Borneo below 100 m.
Habits: Lives on small islands where it is rare due to deforestation. Sometimes mixes with Pied Imperial-Pigeons.

272. ROCK PIGEON *Columba livia*   PLATE 33
(Feral Pigeon or Rock Dove)

Description: Medium-sized (32 cm), bluish grey, urban pigeon with black wing bars and terminal tail bar, and a greenish violet gloss on head and breast. This is the feral form of the familiar town and domestic pigeon.
Iris—brown; bill—horn; feet—grey.
Voice: Familiar *oo-roo-coo* of the domestic pigeon.
Range: Palaearctic but introduced almost worldwide and feral populations are now established in many towns and cities.
Distribution and status: In Borneo, colonies are known from Banjarmasin, Kuching, Kota Kinabalu, and Samarinda.
Habits: Originally a bird of cliffs but easily adapted to life in towns. Lives in flocks and frequently perches on buildings or stays on the ground; feeds in gardens, fields, and open spaces. Familiar wheeling flight is characteristic.

Ground doves

273. BARRED CUCKOO-DOVE *Macropygia unchall*   PLATE 34
(Large Cuckoo-Dove)

Description: Large (38 cm), long-tailed, brown dove. Back and tail heavily barred black or brown. Head grey with iridescent blue-green sheen on nape. Breast pinkish, grading into white vent. Female lacks green iridescence. Heavier barring on back and barred tail distinguish it from other Cuckoo-Doves in the region.
Iris—yellow to pale brown; bill—black; feet—red.
Voice: A loud resonant *kro-uum* or *u-wa* with the second note louder and higher than the first, which is only audible at short range.
Range: Widely distributed from the Himalayas to Java and Bali. Absent from Borneo.
Distribution and status: Present in most sub-montane forests of Sumatra, Java, and Bali, between 800 and 3000 m. Generally less common than *M. ruficeps*.
Habits: Small flocks feed in trees in the mountains, occasionally coming to the ground to feed or drink. Quite tame but flight noisy, flapping when alarmed. Flies very swiftly through canopy. When on the ground the bird raises its tail.

274. RUDDY CUCKOO-DOVE *Macropygia emiliana*   PLATE 34
(Indonesian Cuckoo-Dove)

Description: Medium-sized (30 cm), long-tailed, rich reddish brown dove. Breast purplish brown with inconspicuous black barring. Male has pink iridescence on neck and breast. Female has dark barring on mantle. Some variation between different island races.

Iris—bluish inner ring and red outer ring; bill—creamy grey; feet—purplish red.

Voice: Three loud notes *poh wa wao*. (See Appendix 6.)

Range: Greater Sundas, Lombok, Sumbawa, and Flores.

Distribution and status: Status on Sumatra uncertain; endemic races on Simeulue, Nias, Mentawai, and Enggano Islands apparently common; also common on islands in Lampung Bay, and one record from mainland Sumatra probably a stray. On Borneo, Java, and Bali locally common in hill forest; also recorded in the lowlands of Borneo, and recorded to 1500 m on Java.

Habits: Inhabits primary forest or glades and small openings in forest. Does not come out into cultivated land like the Little Cuckoo-Dove. Has a floppy flight through understorey but fast and powerful when above the canopy. Comes to the ground to feed and drink; uses its long tail as a balancing support when feeding in trees.

Note: Some authors include this species in the Red Cuckoo-Dove *M. phasianella*. The endemic race *cinnamomea* on Enggano Island is sometimes treated as a distinct species.

275. LITTLE CUCKOO-DOVE *Macropygia ruficeps*   PLATE 34

Description: Medium-sized (30 cm), long-tailed, reddish dove. Smaller than Ruddy Cuckoo-Dove and with buff breast; black barring on upper parts, and dark subterminal bar on outer tail feathers. The male has a green and lilac iridescent sheen on the nape. Female lacks iridescence and has heavier dark mottling on breast.

Iris—grey white; bill—brown with black tip; feet—coral red.

Voice: Fast *kroo-wuk* with emphasis on the second note, repeated about 30 times. After a short pause the call starts again. (See Appendix 6.)

Range: Widespread and common sub-montane forest bird in SE Asia, Sumatra, Java, Borneo, and Lesser Sundas.

Distribution and status: In Borneo and Sumatra it is generally common in hills and lower montane forest. In Java and Bali it occurs, often in large numbers, on most mountains between 300 and 2000 m.

Habits: Prefers forest edge; flocks often raid adjacent rice fields.

PIGEONS AND DOVES 169

276. ISLAND COLLARED-DOVE *Streptopelia bitorquata* PLATE 34
(Javan/Javanese/Philippine Collared-Dove)

Description: Medium-sized (30 cm), long-tailed, pinkish brown dove. Similar to commoner Spotted Dove but distinguished by greyer head and the white-edged, black nape patch lacking white spots. Central tail feathers brown, outer tail feathers grey with whitish edges.
Iris—orange; bill—black with red base; feet—purplish red.
Voice: A diagnostic deep, throaty *cru-cruuuu* and variations (D.A.H.). (See Appendix 6.)
Range: Philippines, Java, Bali and Lesser Sundas.
Distribution and status: Occasional sightings in Sumatra are probably escapes. Records from N Borneo are probably strays from the Philippines. On Java and Bali it is an occasional lowland bird, rarely found above 600 m.
Habits: Frequents open, wooded country, but found mainly in mangroves. Rests in small trees and feeds in open patches on the ground, in pairs or small flocks.

277. SPOTTED DOVE *Streptopelia chinensis* PLATE 34
(Burmese Spotted/Spot-necked Dove)

Description: Familiar medium-sized (30 cm), pinkish brown dove with a longish tail. Outer tail feathers with broad white tips white. Flight feathers darker than body; has a diagnostic conspicuous black patch on the side of the neck, finely spotted with white.
Iris—orange; bill—black; feet—red.
Voice: A melodious, gently repeated *ter-kuk-kurr* with last note stressed; hence its Indonesian name 'tekukur'. (See Appendix 6.)
Range: Widely distributed and common in SE Asia, through to Lesser Sundas, and introduced elsewhere as far as Australia.
Distribution and status: Commonly found throughout the Greater Sundas in open lowland areas and villages. Often kept as a pet.
Habits: Commensal of man, living around villages and rice fields, feeding on the ground, and frequently encountered in pairs sitting on open roads. When disturbed it flies close to the ground with a distinctive slow wing-beat.

278. ZEBRA DOVE *Geopelia striata* PLATE 34
(Peaceful/Barred Dove, AG: Ground-Dove)

Description: Small (21 cm), slender, brown dove with a long tail. Head grey, nape finely barred; back brown with black edges to feathers. Outer tail feathers blackish with white tips.
Iris—blue-grey; bill—blue-grey; feet—dark pink.

Voice: A melodious, soft, rolling, whistled *croo* repeated 6–8 times in a hurried rattling sequence. (See Appendix 6.)
Range: Philippines, Malay Peninsula, Sumatra, Java, Bali, and Lombok. Introduced in SE Asia, Sulawesi, and elsewhere in Indonesia.
Distribution and status: Common in lowlands of E and S Sumatra, up to 900 m (including Bangka and Belitung), but now quite rare in lowlands of Java and Bali due to excessive trapping. Introduced in Borneo and local populations are well established. Commonly kept as a pet.
Habits: Favours clearings in open, wooded country, often near villages. Pairs or small groups feed on the ground and sometimes gather to drink at water sources.

279. EMERALD DOVE *Chalcophaps indica* PLATE 34
(Green-winged Dove/Common Emerald-Dove)

Description: Medium-sized (25 cm), rather short-tailed, ground dove with reddish-pink underside. Crown grey with white forehead, grey rump, and iridescent green wings. Female lacks grey crown. In flight two conspicuous black and white bars can be seen on the back.
Iris—brown; bill—red with orange tip; feet—red.
Voice: Deep, soft, mournful, drawn-out, two notes *tuk-hoop* with emphasis on second note. (See Appendix 6.)
Range: Common and widespread in lowland and sub-montane, primary, and secondary forests from India to Australia.
Distribution and status: Common in Borneo and Sumatra, but in Java and Bali this bird is now becoming scarce.
Habits: Usually found singly or in pairs, and spends most of its time on the forest floor in thick cover. Flies very fast and low through the forest with a clap of wings on take-off. Drinks at streams and pools.

280. NICOBAR PIGEON *Caloenas nicobarica* PLATE 34
(Hackled Pigeon)

Description: A large (40 cm), long-legged, almost tail-less ground pigeon with a dark grey, iridescent purple mane of long hackles. Back and wings glossy, iridescent green with copper sheen. Short tail is white.
Iris—brown; bill—black with prominent cere; feet—dark purplish red.
Voice: A deep croak, seldom heard.
Range: Andamans, Nicobars, throughout the Indonesian archipelago and Philippines, to New Guinea and N Melanesia.
Distribution and status: Confined to small offshore islands and generally rare. Breeds on Kangean, Karimun Jawa, and other small islands in Java Sea, Lingga

archipelago, most of the Mentaur islands except Mentawai, and small islands off the N and E coasts of Borneo.

Habits: Lives and feeds on the ground on small predator-free islands, flying from island to island over long distances. Rests on low perch for most of the day, becoming active at dusk. Active during day only in dark shady sites.

Parrots—FAMILY PSITTACIDAE

Parrots are a large, diverse family of colourful birds found throughout the world's tropics and Australia. They have large heads, powerful hooked beaks, and strong flexible feet with two toes pointing backwards. They nest in tree holes and feed mostly on fruit, seeds, and pollen. Their flight is fast, and their calls are harsh and piercing.

Nine species are recorded in the Greater Sundas.

281. RED-BREASTED PARAKEET *Psittacula alexandri* PLATE 35
(Moustached Parakeet)

Description: Medium-sized (34 cm) colourful parrot with diagnostic pink breast. Adult: crown and cheeks violet grey with black lores; nape, back, wings, and tail green; pronounced black 'moustache'; thighs and vent pale green. Immature birds have buffy brown head with less prominent black moustache.
Iris—yellow; bill—red; feet—grey.

Voice: Repeated piercing *kekekek*, particularly of young birds, and a raucous trumpet-like scream.

Range: India, S China, SE Asia (except Malay Peninsula), and Greater Sundas.

Distribution and status: Not recorded on mainland Sumatra but there are endemic races on Simeulue, Banyak Islands, and Nias. Still regularly seen in coastal areas of SE Borneo in the Barito region. Formerly common in Java and Bali but the pet trade has reduced its numbers; now usually found only in more remote forested areas, although there is a large wild colony in the Ragunan Zoo grounds in Jakarta.

Habits: A communal bird, travelling, roosting, and nesting in parties. Noisy and conspicuous in flight with birds flashing low over open spaces to settle with a clatter of wings in trees to feed or rest, screeching frequently.

282. LONG-TAILED PARAKEET *Psittacula longicauda* PLATE 35

Description: Largish (40 cm) parrot with very long tapering tail and green breast. Male: green crown, red sides of head, and bold black moustachial stripe; mantle washed pale blue; yellow-tipped tail, and wings bluish. Female: duller, with greenish moustache and no blue on back. In flight the wing lining is yellow. Distinguished from Red-breasted Parakeet by green underparts and red sides of head.

Iris—greenish yellow; bill—red with horn tip; feet—grey.

Voice: Harsh, strident screeches given from treetops and in flight.

Range: Andamans, Nicobars, Malay Peninsula, Sumatra, Riau, Natunas, and Borneo.

Distribution and status: Locally common in coastal and lowland areas in open forests, plantations, forested swamps, and secondary forests, up to 300 m.

Habits: Flies fast in large flocks between feeding sites and roosts. Congregates in huge numbers at coastal roost sites.

283. RAINBOW LORIKEET *Trichoglossus haematodus* PLATE 35
(Coconut Lorikeet)

Description: Medium-sized (24 cm), colourful parrot. Adult: head blackish brown, streaked grey; yellow collar; green back; red breast and underwing; purple-black belly; thighs with green and yellow bands; yellow band under wing conspicuous in flight.

Iris—red; bill—red; feet—grey.

Voice: Harsh, repeated screech during flight *keek, keek, keek, keek*; chattering and twittering calls when settled.

Range: Australia, Pacific, New Guinea, Moluccas, Lesser Sundas to Bali.

Distribution and status: On Bali rare and possibly a recent colonizer from Lombok, but escapes are also occasionally seen in Bali and Java and may sometimes breed. Heavily traded for the pet industry.

Habits: A gregarious bird, flying over forest in noisy screeching parties. Formerly common in sub-montane forests near lake Bratan, Bali, where attracted to coffee plantations and their shading *Erythrina* trees.

284. YELLOW-CRESTED COCKATOO *Cacatua sulphurea* PLATE 35
(Lesser Sulphur-crested Cockatoo)

Description: Large (33 cm), noisy, conspicuous, white parrot with a long, erectile yellow crest and yellow cheeks.

Iris—dark brown; bill—black; feet—dark grey.

Voice: A loud, raucous screeching *kerk-kerk-kerk*, and assorted whistles.

Range: Endemic to Sulawesi and Lesser Sundas. Introduced to Singapore and Hong Kong.

Distribution and status: Occurs on Nusa Penida Island off Bali and a rare subspecies is found on Masalombo Besar, Java Sea. Occasional birds seen on Java and Bali are probably escapes.

Habits: Lives in pairs and congregates in small groups. Conspicuous in flight with heavy, fast flapping interspersed with gliding, screeching at each other. When calling from a perch the crest is erected, then lowered.

285. BLUE-RUMPED PARROT *Psittinus cyanurus* PLATE 35

Description: Smallish (18 cm), green parrot with rounded tail. Distinguished from Hanging-Parrots by much larger size, and from Lorikeets by short tail. Male has diagnostic blue head, blackish back, red shoulder patch, blue lower back and rump, and red bill. Female has brown head, less blue on back, and a brown bill. Immatures are green. In flight underwing is blackish with red underwing coverts and axillaries.
Iris—yellow; bill—red (male), brown (female); feet—blue-grey.
Voice: High-pitched, sharp *chi, chi, chi* or *chew-ee* chittering calls, and melodious trills.
Range: Malay Peninsula, Sumatra, Riau, Lingga, Bangka, Simeulue, Mentawai, and Borneo.
Distribution and status: A common bird of lowland forest, swamp forest, mangroves, and cultivated areas, up to 700 m.
Habits: Flies fast and direct over upper canopy. Lives in small flocks. Prefers open canopy.

[286. GREAT-BILLED PARROT *Tanygnathus megalorhynchos* PLATE 35
(Moluccan/Island Parrot)

Description: Large (40 cm), noisy, green parrot with a massive red bill. General colour green but wings are streaked yellow and washed with blue; back is pale blue; tail slightly pointed, with yellow; underside yellowish green.
Iris—yellow; bill—orange-red; feet—grey.
Voice: A harsh shriek in flight *kaw-kaw*.
Range: Balut (Philippines), Wallacea, and islands off NW New Guinea.
Distribution and status: Doubtfully recorded in Bali and Nusa Penida. Occasional birds, almost invariably escapes, are still seen on Bali and Java.
Habits: Greater Sundas lie outside the species' regular range and only occasional stray birds are seen. Elsewhere, prefers rather open country along the coast and can be locally common. Flies in pairs with characteristic fast, shallow wing-beat.]

287. BLUE-NAPED PARROT *Tanygnathus lucionensis* PLATE 35

Description: Large (30 cm) green parrot with distinctive large red bill, rounded tail, and blue crown. Nape and back washed blue, brownish bar on the shoulders Iris—yellow; bill—red; feet—grey.
Voice: Loud, harsh shrieks.
Range: Palawan and Philippines.
Distribution and status: Local populations have become established on Maratua, Mantanani, and Siamil Islands of NE Borneo, and a feral population is found around Kota Kinabalu.
Habits: Pairs and small flocks feed in trees of coastal forests. Nests in tree holes, sometimes in coconut palms.

288. BLUE-CROWNED HANGING-PARROT *Loriculus galgulus*
(Malaysian Hanging-Parrot, AG: Lorikeet) PLATE 35

Description: Tiny (12 cm), red-rumped parrot, the only Hanging-Parrot in Borneo and Sumatra. Male is green with scarlet rump and tail, red throat patch, blue crown patch, and a golden patch on the mantle. Female lacks scarlet throat. Iris—brown; bill—black ; feet—orange or brown.
Voice: Very high-pitched, whistled *dzi* given in flight.
Range: Malay Peninsula, Sumatra, and Borneo.
Distribution and status: A common inhabitant of lowland forests throughout Sumatra (including islands) and Borneo, up to 500 m. Recorded W coast of Java; probably a stray from Sumatra but populations may become established.
Habits: Flies fast over the forest in small flocks with 'whirring' wings, giving shrill calls as they pass. Settle to feed on flowers, buds, and small fruits; clambering about with comical swagger. Difficult to see because of small size and green colour. Has the strange habit of hanging upside down to sleep. Females carry nesting material tucked into their rump feathers.

289. YELLOW-THROATED HANGING-PARROT *Loriculus pusillus*
(AG: Lorikeet) PLATE 35

Description: A tiny (12 cm), red-rumped, bright green parrot with yellow-green underparts and yellow patch on throat (much reduced in female and immature). Iris—yellow; bill—orange; feet—orange.
Voice: A shrill, ringing *sree-ee* uttered in flight.
Range: Endemic to Java and Bali.

Distribution and status: On Java and Bali a common bird of rainforest from sea level to 2000 m; probably nomadic and easily overlooked.
Habits: As Blue-crowned Hanging-Parrot.
Note: Some authors place under Vernal Hanging-Parrot *L. vernalis*, but colour differences and major break in distribution seem to justify species status.

Cuckoos—FAMILY CUCULIDAE

A LARGE, worldwide family of insectivorous birds with slender bodies, and long wings and tails. The two outer toes point backward and the two inner toes forward. Cuckoos have strong, curved bills which are used for catching large insects. Some species specialize on caterpillars.

Three main groups of cuckoos occur in the Greater Sundas. True Cuckoos are arboreal, have pointed wings, often barred or streaked plumage, and breed parasitically by laying eggs in the nests of other birds that then raise the foster nestling. Malkohas have colourful bills, very long tails, and long, strong legs. They creep about in thick tangles of vines and low bushes and have low, clucking calls. Coucals have black and chestnut plumage with long tails, and live in secondary scrub habitats and long grass. They are poor fliers and hop in a characteristic manner.

Twenty-eight species occur in the Sundas region; some are very difficult to identify. Many species are best recognized by voice.

290. CHESTNUT-WINGED CUCKOO *Clamator coromandus* PLATE 36
(Red-winged Cuckoo, AG: Crested-Cuckoo)

Description: Large (45 cm), black, white, and rufous, long-tailed cuckoo with prominent erectile crest. Crown and crest black; back and tail black with blue gloss; wings chestnut; throat and breast orange-brown; nuchal collar white; belly whitish. Immature: upperparts with rufous scaling; throat and breast whitish. Iris—red brown; bill—black; feet—black.
Voice: Loud harsh screech *chee-ke-kek*, and a hoarse whistle.
Range: Breeds in India, S China, and SE Asia; migrates in winter to the Philippines, Sulawesi, Borneo, Sumatra, Lingga, Bangka, Mentawai, and Java.
Distribution and status: In Borneo and Sumatra an occasional bird in suitable habitat, up to 1500 m. In Java it is very rare.
Habits: A shy bird of thick, low vegetation in scrub forest, mangrove, cultivated land, and gardens. Habits similar to a Malkoha, scrambling about in low vegetation, hunting for insects. In flappy and coucal-like flight the crest is lowered.

176 CUCKOOS

291. LARGE HAWK-CUCKOO *Cuculus sparverioides* PLATE 36

Description: Largish (40 cm), greyish-brown, hawk-like cuckoo with a reddish subterminal bar to white-tipped tail; breast rufous, mottled white and grey; belly barred white and brown, washed rufous; chin black and throat white. Immature: upperparts brown with rufous barring; underparts buff with blackish streaks. Differs from hawk in posture and bill shape.
Iris—orange with large yellow eye-ring; bill—black above, yellow-green below; feet—pale yellow.
Voice: In breeding season calls *pi-peea* or *brain-fever*, with increasing speed and shriller pitch to a frantic climax. (See Appendix 6.)
Range: Resident in Himalayas, S China, Philippines, Borneo, and Sumatra; visits Sulawesi, W Java, and maybe Bali in winter.
Distribution and status: In Sumatra a scarce resident and winter visitor in mountain forests including pine, from 900 to 1600 m. In Borneo a rare winter visitor to lowlands but a resident montane race is common in mountains above 1000 m. In Java it is a rare visitor. A wintering juvenile was reported from Bali in 1990.
Habits: Prefers forested areas. Typical secretive cuckoo of treetops.

292. MOUSTACHED HAWK-CUCKOO *Cuculus vagans* PLATE 36
(Lesser Hawk-Cuckoo)

Description: Medium-sized (30 cm) grey and brown cuckoo; crown and nape grey; back brown; wings and tail greyish brown with broad barring; tip of tail white. Underparts whitish with black streaks. Blackish moustachial stripe diagnostic. Has prominent, bright yellow eye-ring.
Iris—brown; bill—blackish above, greenish below; feet—yellow.
Voice: Repeated, monotonous, dissyllabic phrase *kang koh*, every 2 seconds. Plaintive mellow whistle *peu peu* on ascending scale to frantic sudden stop.
Range: Malay Peninsula, Borneo, and Sumatra.
Distribution and status: In Borneo and Sumatra this seems to be a rare bird. Occasional strays have reached W Java but there are no recent records.
Habits: Prefers forest edge and secondary forest.

293. HODGSON'S HAWK-CUCKOO *Cuculus fugax* PLATE 36
(Fugitive Hawk-Cuckoo)

Description: Medium-sized (29 cm) grey and brown cuckoo. Adult: head and mantle grey, back browner; tail brown with dark bars and narrow, reddish terminal bar. Underparts whitish, strongly streaked reddish on breast. Immature: upperparts entirely brown.

Iris—pale yellow with yellow eye-ring; bill—green and black; feet—yellow.

Voice: Shrill, whistled *pee-weet* at about one second intervals and a rapid series of dissyllabic notes *pee-pee*, repeated faster and faster and rising the scale to a frantic peak, then dropping back to start again. (See Appendix 6.)

Range: E Asia, S China, SE Asia, Philippines, Borneo, Karimatas, Sumatra, and W Java; visitor to Sulawesi and Buru.

Distribution and status: In Borneo and Sumatra an occasional resident and visitor, up to 1400 m. In Java it is rare both as a resident and a winter visitor in the lowlands. Recently recorded on Bali.

Habits: Secretive, poorly known bird of forest, secondary growth, bamboo thickets, and plantations.

294. INDIAN CUCKOO *Cuculus micropterus*   PLATE 36
(Short-winged Cuckoo)

Description: Medium-sized (30 cm) greyish cuckoo. Similar to Common Cuckoo but distinguished by subterminal black tail bar and dull grey eye-ring. Grey head contrasts with brown back. Female is browner than male; immature distinguished by buff barring on head and upper back.

Iris—brown; bill—black above, greenish below; feet—yellow.

Voice: Loud, clear, deliberately enunciated four note whistle *blanda mabok*, persistently repeated, often at night. Fourth note is lower. (See Appendix 6.)

Range: S Asia, SE Asia, Philippines, Borneo, Sumatra with offshore islands, and W Java.

Distribution and status: In Sumatra and Borneo there are both resident and wintering races. An uncommon bird of forests up to 1000 m. In Java there is a small resident race and a larger winter migrant race; both are rare.

Habits: Generally keeps high to the canopy of forest and secondary forest. More frequently heard than seen.

295. COMMON CUCKOO *Cuculus canorus*   PLATE 36
(Eurasian/Grey Cuckoo)

Description: Medium-sized (32 cm) cuckoo. Upperparts grey with blackish tail; black bars on whitish belly. Hepatic female form is rufous, barred black on the back. Distinguished from Indian Cuckoo by yellow eye-ring and lack of subterminal tail bar, and from female of Oriental Cuckoo by unbarred rump. Juvenile has white patch on nape.

Iris—yellow with yellow eye-ring; bill—black above, yellowish below; feet—yellow.

Voice: Classic, loud, clear *kuk-oo* generally heard only in breeding area. (See Appendix 6.)

178 CUCKOOS

Range: Breeds in Eurasia, migrating to Africa and SE Asia.
Distribution and status: Vagrants have reached W Java and maybe Sumatra. There are no recent records.
Habits: Favours open wooded areas.

296. ORIENTAL CUCKOO *Cuculus saturatus* PLATE 36
(Himalayan Cuckoo)

Description: Smallish (26 cm) grey cuckoo with broadly barred belly and flanks. Male and grey female have grey breast and upperparts, blackish-grey unbarred tail, and buffy underparts barred black. Immature and hepatic female have upperparts rufous brown, heavily barred black, and underparts whitish, barred black up to chin. Resident race *lepidus* of Sumatra and Java is small. Bornean resident race *insulindae* is darker and smaller than migrant races. Distinguished from Common and Indian Cuckoos by bolder, broader breast bars and by song. Hepatic female distinguished from Common Cuckoo female by barred rump.
Iris—yellow; eye-ring—yellow; bill—greyish; feet—orange-yellow.
Voice: The Bornean resident race *insulindae* gives a three-hoot call of a grace note followed by three flat, clear hoots *hoop, hoop-hoop* similar to the call of the Golden-naped Barbet. The Sumatran and Javan race also has a three-hoot call. The migrant forms have a four-hoot call lacking the grace note, but are not heard in their winter grounds. (See Appendix 6.)
Range: Migrant races breed in N Eurasia and Himalayas migrating in winter to SE Asia and Greater Sundas. Resident races occur on the Greater Sundas.
Distribution and status: The Bornean resident race is a common bird on higher mountains between 1300 and 2700 m. The Javan and Sumatran race *lepidus* occurs occasionally, at all altitudes. Winter migrants of two other races *saturatus* and *horsfieldi* are not rare at lower altitudes.
Habits: Secretive bird of forest canopy. Rarely seen except in breeding season when calling very frequently (February–March).
Note: The Bornean race was formerly treated as a race of Lesser Cuckoo *C. poliocephalus* but is now accepted as a race of *saturatus* on the basis of its call (Wells and Becking, 1975). All records of *poliocephalus* from the Greater Sundas are erroneous.

297. BANDED BAY CUCKOO *Cacomantis sonneratii* PLATE 37
(Banded Cuckoo)

Description: Small (22 cm), brown, finely barred cuckoo. Adult: upperparts rich brown, underparts whitish, all finely barred black; has a conspicuous pale eyebrow. Immature: brown with black streaks and blotches rather than barring.
Iris—yellow-red; bill—above blackish, below yellowish; feet—grey.

Voice: A shrill, rhythmic four-note call *smoke-yer-pepper*, distinguished from four-note call of Indian Cuckoo by being quicker, more plaintive, and less deliberately enunciated (D.A.H.). In the breeding season, a rising call of four slow notes, followed by three to six faster notes of two or three syllables, rising in pitch to a sudden stop; also *tay-ta-tee* call. (See Appendix 6.)
Range: India, China, Borneo, Sumatra with offshore islands, Java, and Philippines.
Distribution and status: A fairly common lowland bird, up to 900 m and rarely up to 1500 m, in mountains. Recently discovered on Bali.
Habits: Prefers open forests, forest edge, secondary scrub, and cultivated areas. Regularly heard but rarely seen.

298. PLAINTIVE CUCKOO *Cacomantis merulinus* PLATE 37

Description: Small (21 cm) greyish-brown and rufous cuckoo. Adult: head grey; back and tail brown; breast and belly orange-rufous. Similar to Rusty-breasted Cuckoo but paler, and song is distinct. Immature: upperparts brown, barred black; underparts whitish with fine barring; resembles adult Banded Bay Cuckoo but without eye-stripe.
Iris—crimson; bill—blackish above, yellow below; feet—yellow.
Voice: Mournful whistle *tay-ta-tee, tay-ta-tee*, increasing in speed and rising in pitch. Sometimes heard at night. A second call of two or three whistles breaking into a descending series *pwee, pwee, pwee, pee-pee-pee-pee*. (See Appendix 6.)
Range: E India, S China, Borneo, Sumatra, Java, Bali, Sulawesi, and Philippines.
Distribution and status: A common lowland bird up to 1300 m.
Habits: Prefers open woodland, secondary forest, and cultivated areas, including towns and villages. Regularly mobbed by small birds. Has a very familiar call but is difficult to see.

299. RUSTY-BREASTED CUCKOO *Cacomantis sepulcralis* PLATE 37
(Grey-headed Cuckoo)

Description: Small (23 cm) greyish brown cuckoo. Adult: head grey; back, wings, and tail greyish brown; underparts rufous. Similar to Plaintive Cuckoo but darker. Immature is rich brown above and whitish below with rather bold, broad, black barring all over.
Iris—brown; eye-ring—yellow; bill—black with orange patch; feet—grey.
Voice: Mellow whistle *weet* or *peeweet*, repeated 10 to 25 times, slowing and descending by semitones; also a rising call, more rapid and jumbled than the similar call of the Plaintive Cuckoo (D.A.H.). (See Appendix 6.)
Range: Malay Peninsula, Borneo, Sumatra, Belitung, Enggano, Simeulue, Java, Bali, Sulawesi, Moluccas, Lesser Sundas, and Philippines.

180 CUCKOOS

Distribution and status: A widespread and locally common lowland bird of hills up to 1200 m.
Habits: Prefers forest and forest edge, secondary growth, plantations, and village groves.
Note: Sometimes combined with the Brush Cuckoo *C. variolosus* but probably more closely related to *C. merulinus*.

300. ASIAN EMERALD CUCKOO *Chrysococcyx maculatus*   PLATE 37

Description: Small (17 cm) shining green cuckoo. Male: head, upperparts, and breast shining green; belly white with green banding. Female: rufous crown and nape, copper-green upperparts and white underparts barred dark buff. Distinguished from next three species by rufous crown and nape, and buffish barring. Immature has rufous head and streaked crown. In flight a broad white band on underwing at base of flight feathers.
Iris—reddish brown; bare eye-ring—orange; bill—orange-yellow; feet—black.
Voice: Loud whistled twitters.
Range: Breeds in northern SE Asia, migrating south in winter as far as Malay Peninsula and Sumatra.
Distribution and status: An uncommon bird of lowland forests and secondary growth, up to 900 m.
Habits: Recognized by calls but otherwise easily overlooked. Feeds quietly in tree crowns.

301. VIOLET CUCKOO *Chrysococcyx xanthorhynchus*   PLATE 37

Description: Small (16 cm), violet (male) or bronze green (female) cuckoo. Male: head, breast, and upperparts violet; belly white with violet banding. Female: eyebrow, cheeks, and underparts white, banded bronze; crown brownish; upperparts bronze-green. Reddish bars on outer tail feathers distinguishes from Horsfield's and Little Bronze-Cuckoos.
Iris—red; bill—yellow with red base (male), upper mandible black with red base (female); feet—grey.
Voice: High-pitched *kie-vik, kie-vik*, usually given in dipping flight; also a shrill, musical, descending, accelerating trill.
Range: E Asia, SE Asia, Borneo, Sumatra with offshore islands, Java, and Philippines.
Distribution and status: In Borneo and Sumatra it is not uncommon up to 700 m. In Java a rare lowland bird.
Habits: Prefers forest edge, gardens, mangrove, and plantations rather than primary forest. Generally secretive, creeping about branches, catching insects, or perched motionless at the top of a tall tree in an exposed position to call.

CUCKOOS

302. HORSFIELD'S BRONZE-CUCKOO *Chrysococcyx basalis*
(Rufous-tailed/Narrow-billed Bronze-Cuckoo) PLATE 37

Description: Small (15 cm), greenish bronze cuckoo. Head browner than back. Similar to Little and Gould's Bronze-Cuckoos but with conspicuous white eyebrow, brown ear patch and grey eye-ring; outer tail feathers reddish brown at base and underparts barred only on belly and flanks.
Iris—brown; bill—black; feet—grey.
Voice: High-pitched whistle, sliding down scale, repeated endlessly *feeooo-feeooo-feeooo*.
Range: Breeds in Australia; migrates to Malay Peninsula and Greater Sundas.
Distribution and status: Few records Sumatra and Borneo. Uncommon on Java (locally common in NW coastal scrub) and Bali, but probably overlooked.
Habits: A shy bird of open country, especially along the coast.

303. LITTLE BRONZE-CUCKOO *Chrysococcyx minutillus* PLATE 37
(Malay/Malayan/Malaysian/Australian Bronze-Cuckoo)

Description: Small (15 cm), green cuckoo with barred underparts. Upperparts bronze green; underparts white, barred with green. Distinguished from female Violet Cuckoo by all black bill and less rufous on outer tail feathers. Immature has white underwing coverts. Distinguished with difficulty from Gould's Bronze-Cuckoo by darker, bottle green crown, bolder and greener barring on underparts, white of forehead more extensive. Underside of outer tail feathers entirely banded black and white over both webs, and without rufous.
Iris—red (male), brown (female); bill—black; feet—grey.
Voice: Three to five, usually four-note whistle of thin tremulous notes, descending and rather prolonged *teu teu teu teu*; also a high-pitched, drawn-out trill on descending scale.
Range: Malay Peninsula, Greater Sundas, islands in Wallacea and N and E Australia.
Distribution and status: Specimens are known from N Sumatra (sight records in S Sumatra), lowland coastal NW Java (sight records inland from Salak), and N and E Borneo. A rather rare lowland bird up to 800 m.
Habits: Prefers secondary thickets, gardens, plantations. Sits at the top of large trees to call but is otherwise rather inconspicuous. Sometimes joins mixed bird flocks.

304. GOULD'S BRONZE-CUCKOO *Chrysococcyx russatus*   PLATE 37

Description: Small (15 cm), bronze-green cuckoo with barred underparts. Similar to Little Bronze-Cuckoo but has more purplish iridescence, crown not such a deep bottle green, white on forehead less pronounced, and barring on underparts bronze rather than green, and less bold. Female similar but duller. Can be separated from Little Bronze-Cuckoo on basis of pattern of undertail rectrices: outer web of outer tail feathers greenish, washed rufous, not black and white as in Little Bronze-Cuckoo, inner web black and white bands but proximal portions of white compartments rufous.
Iris—red (male), brown (female); bill—blackish; feet—grey.
Voice: Accelerating, descending, high-pitched *see-see-see*.
Range: Malay Peninsula (once), S Philippines, Borneo, Sulawesi, Lesser Sundas, New Guinea, and NE Australia.
Distribution and status: Recorded N and E Borneo.
Habits: Prefers open forests, swamp forests, and mangroves.
Note: Some authors place this form within Little Bronze-Cuckoo but the two forms are sympatric in N Borneo. Species limits in the *Chrysococcyx 'malayanus'* group are complex and not fully resolved. The distribution outlined above is based on specimens and detailed sight records are likely to expand it considerably.

305. DRONGO CUCKOO *Surniculus lugubris*   PLATE 37

Description: Medium-sized (23 cm), black cuckoo. Plumage all over glossy black, except for white thighs and white barring on undertail coverts and underside of outer tail feathers; white nuchal patch is rarely visible. Juveniles are irregularly spotted with white. Tail forked like a drongo.
Iris—brown (male), yellow (female); bill—black; feet—blue grey.
Voice: Loud clear call of four to seven, even-spaced, *pi* notes on steadily ascending scale, preceded by higher introductory note; also a rapidly trilled series of rising notes ending with about three descending notes. (See Appendix 6.)
Range: India, China, Borneo, Sumatra with offshore islands, Java, Bali, Sulawesi, N Moluccas, and Philippines.
Distribution and status: An uncommon lowland bird with resident and migrant populations up to 900 m.
Habits: Inhabits forest, forest edge, and secondary scrub. Secretive. Similar to drongos in appearance but not in posture, movements, or flight.

CUCKOOS 183

306. ASIAN KOEL *Eudynamys scolopacea* PLATE 36
(Indian/Common Koel)

Description: Large (42 cm), entirely black (male) or white-speckled grey-brown (female) cuckoo with green bill.
Iris—red; bill—pale green; feet—blue-grey.
Voice: Loud *kow-wow* with stress on second syllable, repeated up to 12 times with increasing tempo and pitch, by day and at night. Also a shriller, faster *kuil, kuil, kuil, kuil* call. (See Appendix 6.)
Range: India, China, SE Asia, Malay Peninsula, Borneo, Sumatra, Java, Bali, Lesser Sundas, and Moluccas.
Distribution and status: A widespread but uncommon resident and winter migrant in coastal and lowland areas.
Habits: This is a maddening bird. Its loud calls endlessly mock the birdwatcher who rarely gets a glimpse of this shy bird which keeps to dense cover in mangroves, secondary forest, forest, gardens, and plantations. Brood parasite of crows, drongos, and orioles.

307. BLACK-BELLIED MALKOHA *Phaenicophaeus diardi* PLATE 38

Description: Largish (34 cm), grey malkoha. Distinguished from similar Chestnut-bellied Malkoha by dark grey belly. Entire body greyish, wings metallic blue-green. Underside of tail feathers with bold white tips.
Iris—bluish white; bare skin around eye—crimson; bill- green; feet—blue-grey.
Voice: Sharp call *pwew-pwew* (M.M.N.); a single soft *taup* (M. & W.).
Range: Malay Peninsula, Sumatra, and Borneo.
Distribution and status: A common resident up to 900 m.
Habits: Skulks in dense vegetation tangles of middle canopy trees. Prefers primary dryland and swamp forests, and secondary growth.

308. CHESTNUT-BELLIED MALKOHA *Phaenicophaeus sumatranus*
(Rufous-bellied Malkoha) PLATE 38

Description: Large (40 cm) grey malkoha with rufous belly and very long tail. Head, nape, breast, and flanks grey; wings metallic greenish blue; graduated tail bluish grey above, with white tips below.
Iris—pale blue-white; bare skin around eye—red; bill—green; feet—grey.
Voice: *Tok-tok* or *chi-chi* (T.H.).
Range: Malay Peninsula, Sumatra, N Natunas, and Borneo.
Distribution and status: A common resident of lowlands up to 1000 m.

Habits: Sneaks secretively in the dense crowns of smaller trees, singly or in pairs, searching for food. Inhabits primary and secondary forests.

309. GREEN-BILLED MALKOHA *Phaenicophaeus tristis* PLATE 38

Description: Large (55 cm), very long-tailed malkoha. Head and mantle grey; underparts brownish grey with dark shaft lines on throat and breast; back, wings, and tail dark metallic green; tail feathers with white tips.
Iris—brown; bare skin around eye—red; bill—green; feet—black.
Voice: A clucking and croaking call, rather frog-like.
Range: Himalayas, China, SE Asia, Sumatra, and Kangean Islands.
Distribution and status: In Sumatra a frequent bird of hill and lowland montane forests along the Barisan chain, between 500 and 1500 m. On Kangean Island a common lowland bird.
Habits: As other malkohas. Prefers middle canopy of primary and secondary forests, and plantations.
Note: Kangean form is sometimes considered a separate species *P. kangeangensis*.

310. RAFFLES'S MALKOHA *Phaenicophaeus chlorophaeus* PLATE 38

Description: Smallest (30 cm) malkoha with diagnostic chestnut mantle. Male has rufous head and breast with grey vent. Female has grey head, nape and breast with rufous vent. Graduated tail chestnut above, with white tips below.
Iris—dark brown; bill—green with blue base; feet—slaty blue.
Voice: Cat-like *miao*; descending series of soft *miaow* notes (M.M.N.); harsh croaks. (See Appendix 6.)
Range: Malay Peninsula, Sumatra, Bangka, Batu, and Borneo with northern offshore islands.
Distribution and status: A common resident of lowland primary and secondary forests, heath forest, and gardens, up to 900 m.
Habits: Travels in small flocks, flying from small tree to tree, searching for prey.

311. RED-BILLED MALKOHA *Phaenicophaeus javanicus* PLATE 38

Description: Large (46 cm) red-billed malkoha. Upperparts grey, glossy bluish green; chin and throat rufous; breast buffy grey; belly chestnut; tips of tail feathers white.
Iris—brown; bare skin around eye—blue; bill—red; feet—grey.
Voice: Striking, but not loud, whining call.
Range: Malay Peninsula, Borneo, Sumatra, and Java.
Distribution and status: An occasional bird of lowland and hilly areas, up to 1500 m.

Habits: Similar to other malkohas. Frequents drier forests, forest edge, and secondary scrub.

312. CHESTNUT-BREASTED MALKOHA *Phaenicophaeus curvirostris*
PLATE 38

Description: Large (49 cm) long-tailed malkoha with diagnostic rufous-tipped tail. Crown and nape grey; upperparts dull green; facial skin around eye red; underparts rufous; no white in tail which has broad rufous tips. Island races differ. Sumatran form has grey throat and cheeks, and black belly. Bornean race has chestnut throat, cheeks, and belly, and a shorter, squarer tail.
Iris—blue (male), yellow (female); bill—green with red base (male), or brown base (female); feet—grey-brown.
Voice: Deep *tok-tok-tok* call, rather like a clucking chicken; faster, repeated *tok, tok-tok, tok* in flight.
Range: Malay Peninsula, Borneo, Sumatra, Bangka, Mentawai, Java, Bali, and Palawan.
Distribution and status: A locally common lowland bird up to 1100 m; sometimes higher.
Habits: Frequents thickets in forest, sometimes in pairs or small family parties. Sits motionless for long periods in the crowns of small trees, sometimes comes out into *Imperata* grassland.

313. SUNDA GROUND-CUCKOO *Carpococcyx radiceus* PLATE 38
(Malayan/Green-billed Ground-Cuckoo)

Description: Very large (60 cm), terrestrial, forest cuckoo. Adults unmistakable with black head; grey mantle with metallic green wash, tail and wings metallic violet; underparts barred black and white. Young birds have the underparts uniform rufous. Sumatran race is slightly greener and smaller than Bornean.
Iris—brown or grey; bare skin around eye—green; bill—green; feet—green.
Voice: Coughing *heh heh heh*, and loud call of two notes *tock-tor*, with first note rising and second falling, with dove or barbet quality (D.A.H.).
Range: Endemic to Sumatra and Borneo.
Distribution and status: Few records (none recent) from Sumatra, between 300 and 1700 m in the Barisan Range from Singgalang south to Dempu. On Borneo rather rare and patchy in distribution, but recorded from all parts.
Habits: Shy, terrestrial cuckoo.

186 CUCKOOS

314. SHORT-TOED COUCAL *Centropus rectunguis* PLATE 38

Description: Large (30 cm), rufous-winged, black cuckoo. Differs from Greater Coucal only by shorter tail, bluer sheen on head, breast, and mantle, and a different call.
Iris—red; bill—black; feet—black.
Voice: Four to five resonant booming notes *buup* on a descending scale (M. & W.). Like Greater Coucal but slower and more hoarse and resonant (Fogden); also more rapid series of resonant, rising notes uttered at dusk.
Range: Malay Peninsula, Sumatra, and Borneo.
Distribution and status: A scarce bird with patchy distribution probably often overlooked as Greater Coucal with which it sometimes associates. Recorded up to 1700 m in Padang Highlands. More of a forest bird than Greater Coucal.
Habits: Inhabits coastal scrub, grass, and secondary forest.

315. GREATER COUCAL *Centropus sinensis* PLATE 38
(Crow-Pheasant)

Description: Large (52 cm) long-tailed coucal. Plumage entirely black except for uniform chestnut red wings. Some Kangean specimens are pale buff all over with rufous wings.
Iris—red; bill—black; feet—black.
Voice: Series of deep *boop* notes beginning slowly, increasing in tempo, and falling in pitch; then pitch rises and tempo slows to a drawn-out series at constant pitch; or an abbreviated call of four *boop* notes at the same pitch. Also a sudden *plunk* sound. (See Appendix 6.)
Range: India, China, SE Asia, Philippines, Borneo, Sumatra, Nias, Mentawai, Java, and Bali.
Distribution and status: An occasional lowland bird up to 800 m, much less common than Lesser Coucal.
Habits: Frequents forest edge, secondary scrub, reedy riverbanks, and mangroves. Often comes to the ground but also hops about in small bushes and trees. Prefers thicker vegetation than Lesser Coucal.

316. LESSER COUCAL *Centropus bengalensis* PLATE 38

Description: Largish (42 cm), rufous and black, long-tailed cuckoo. Similar to Greater Coucal but smaller and duller, almost dirt coloured. Pale chestnut mantle and wings suffused with black. Immature is streaked brown. Intermediate plumage is common.
Iris—red; bill—black; feet—black.
Voice: Several deep, hollow *hoop* notes, increasing in tempo and descending in

pitch, like water pouring from a bottle; more rapid than Greater Coucal. A second call of three *hup* notes breaking into a series *logokok, logokok, logokok* (D.A.H.).
Range: India, China, SE Asia, Philippines, Borneo, Sumatra with eastern islands, Java, Bali, Sulawesi, Moluccas, and Lesser Sundas.
Distribution and status: A common bird at lower altitudes up to 1000 m, rarely up to 1500 m in mountains.
Habits: Prefers scrub, marshes, and open grassy areas including *Imperata* grassland. Often on the ground or making short, flappy flights low over vegetation.

317. SUNDA COUCAL *Centropus nigrorufus* PLATE 38

Description: Largish (46 cm), black and reddish brown, long-tailed cuckoo. Entire plumage black with purple gloss except for red wings. Differs from other coucals by black back and black wing coverts and inner secondaries.
Iris—red; bill—black; feet—black.
Voice: Not described.
Range: Endemic to Java.
Distribution and status: Restricted to mangrove and associated swamp vegetation in the coastal lowlands of Java. Formerly recorded from inland freshwater swamps but now extremely local, recorded from W (Ujong Kulon), NW (Krawang), NC (Indramayu), SC (Segara Anakan), and probably E (Brantas delta) Java. One record from Sumatra (1902) is thought to be erroneous. Possibly threatened with replacement by Lesser Coucal as habitat is converted to fish ponds.
Habits: Occurs in coastal marshes, *Acrostichum* thickets, and *Imperata* grass near mangroves. Habits as other coucals.

Owls—ORDER STRIGIFORMES

Owls are a familiar, worldwide order of large-eyed, nocturnal birds of prey with haunting calls. They have broad round heads, flat faces, and eyes facing forward. Most species have distinct facial discs around the eyes. There are two families— Barn Owls and True Owls.

All owls lay white eggs and most species nest in tree holes or even holes in buildings. At night when they are active, they are of course difficult to see. The best means of identification is the call.

188 OWLS

Barn Owls—Family Tytonidae

Barn Owls are nocturnal raptors with very round, heart-shaped faces, dark eyes, and broad facial discs which amplify sound to the ears. They hunt largely by ear. The wing feathers are soft for silent flight, and their calls are harsh screeches.
There are 2 Barn Owls in the Greater Sundas both with very wide distributions.

318. BARN OWL *Tyto alba* PLATE 39
(Common Barn Owl)

Description: Unmistakable large (34 cm) white owl. Broad, white heart-shaped face diagnostic. Upperparts buff with fine markings; underside white with fine black spots. General colour variable and immature birds darker buff.
Iris—dark brown; bill—dirty yellow; feet—dirty yellow.
Voice: Harsh, hoarse, high-pitched screech *wheech* or *se-rak*; also a high *ke ke ke ke ke* call. (D.A.H.).
Range: Cosmopolitan.
Distribution and status: An uncommon bird in the lowlands of Sumatra, Java, and Bali, up to 800 m. Not recorded on Borneo but spread in S and C Sumatra is probably a result of deforestation and colonization of S Borneo is likely.
Habits: Hides during the day in dark holes in houses, trees, caves, cliffs, or dense vegetation (including mangroves). Emerges at dusk over open ground, flying low on silent wings. Nests in tree holes or usually in buildings.

319. ORIENTAL BAY OWL *Phodilus badius* PLATE 39
(Asian Bay/Bay Owl)

Description: Medium-sized (27 cm) reddish brown owl, rather similar in shape to Barn Owl, with heart-shaped facial mask and sometimes erect 'ears'. Upperparts reddish brown with black and white spots; underparts pinkish buff with black spots; face pinkish.
Iris—dark; bill—brown; feet—dirty brown.
Voice: A soft hoot and ringing *hooh-weeyoo*; also mournful musical whistles *kwankwit-kwankwit-kek-kek-kek*, given in flight in darkness (see Appendix 6).
Range: India, S China, SE Asia, Philippines, Borneo, Sumatra, Belitung, Java, and Bali.
Distribution and status: A rare forest bird, up to 1500 m.
Habits: Poorly known, shy, nocturnal forest owl. Apparently sits rather horizontally in the daytime, like a frogmouth.

True Owls—Family Strigidae

True Owls are similar to Barn Owls but have generally shorter legs and smaller facial discs. Several species have prominent erectile 'ear' tufts. The plumage of all species is patterned elaborately with grey, brown, white, and black giving them good camouflage when resting during the day.

There are a total of 20 True Owl species in the Greater Sundas, including some insular forms of limited distribution.

320. WHITE-FRONTED SCOPS-OWL *Otus sagittatus* PLATE 40
(Malayan Scops-Owl)

Description: Medium-sized (26 cm), rufous owl with prominent ear tufts, dark eyes, and a white forehead. Larger than other scops owls and distinguished by brown eyes and whitish bill. Lacks neck collar of Collared Scops-Owl. Back tawny rufous; breast cinnamon, scalloped with blackish arrowheads; tail relatively longer than other scops-owls.
Iris—brown; bill—pale grey; feet—grey.
Voice: Low, soft moans.
Range: Malay Peninsula and Sumatra.
Distribution and status: In Sumatra it is a very rare or overlooked bird of lowland forests, recorded only once in the north.
Habits: As other scops-owls.

321. REDDISH SCOPS-OWL *Otus rufescens* PLATE 40

Description: Small (19 cm) reddish owl with conspicuous ear tufts. Upperparts reddish brown, streaked with black and white; underparts reddish buff, streaked black; 'ear' tufts buffish.
Iris—brown; bill—cream; feet—yellowish.
Voice: Hollow, high-pitched whistle *hooee*, repeated at regular intervals (see Appendix 6).
Range: Malay Peninsula, Philippines, Borneo, Sumatra, Bangka, and Java.
Distribution and status: An uncommon lowland bird, probably absent from E Java.
Habits: Frequents understorey of lowland forest; poorly known.

322. MOUNTAIN SCOPS-OWL *Otus spilocephalus* PLATE 40

Description: Small (18 cm), tawny rufous owl with small ear tufts, yellow eyes, and cream bill; lack of bold streaks or bars on breast and row of very large, triangular, white spots on scapulars diagnostic. There is no neck collar in Bornean

190 OWLS

birds but most Sumatran specimens have a collar of white feathers with black tips. Bornean specimens are generally more buff in colour with bolder black markings.
Iris—greenish yellow; bill—cream; feet—whitish grey.
Voice: Soft, far-carrying, double-noted, metallic whistled hoot *plew plew*, given at about 12 second intervals, most of the year. (See Appendix 6.)
Range: Himalayas, S China, SE Asia, Malay Peninsula, Sumatra, and N Borneo.
Distribution and status: An uncommon bird between 1000 and 2500 m, in moist montane forests.
Habits: As other scops-owls. Replies to imitations of its call.
Note: Some authors place the Sumatran form *vandewateri* in the same species as *O. angelinae* of Java.

323. STRESEMANN'S SCOPS-OWL *Otus stresemanni* PLATE 40

Description: Small (18 cm) rufous-brown owl with dark eyes and prominent ear tufts. Paler than Mountain Scops-Owl. Upperparts brownish red, paler, and more fulvous underneath, with black and white spots.
Iris—greenish yellow; bill—white; feet—whitish grey.
Voice: Unknown.
Range: Endemic to Sumatra.
Distribution and status: Known by one specimen taken at 920 m in the Kerinci Valley, C Sumatra.
Habits: As other scops-owls.
Note: May be a local race or light morph of the Mountain Scops-Owl *O. spilocephalus vandewateri*.

324. JAVAN SCOPS-OWL *Otus angelinae* PLATE 40
(Angeline's Scops-Owl)

Description: Small (20 cm) dark owl with prominent ear tufts and white eyebrows. Upperparts greyish brown, heavily streaked and flecked with black. Underparts barred and streaked black on breast, whitish on belly.
Iris—golden yellow; bill—yellow; feet—dirty yellow.
Voice: Fledged young give hard *tch-tsch sch sch*, repeated at six second intervals, reminiscent of young Collared Scops-Owl (Andrew & Milton).
Range: Endemic to Java.
Distribution and status: Known from W Java where only recorded from Pangrango and Tangkubanprahu. There are few field records but the number of specimens and success of mist-netting suggest it is overlooked rather than rare.
Habits: Poorly known; frequents montane forest between 1500 and 2500 m.

325. ORIENTAL SCOPS-OWL *Otus sunia*   PLATE 40
(Asian Scops-Owl)

Description: Small (18 cm), mottled, brown owl with short ear-tufts, yellow eyes, and heavily black-streaked breast. Distinguished from Collared Scops-Owl by lack of pale neck collar; from Mountain and White-fronted Scops-Owls by black-streaked breast. Both grey and rufous forms occur.
Iris—orange-yellow; bill—grey; feet—grey.
Voice: Rough, guttural *toik-toitoink* or *toik toik tatoink* with last note emphasized (see Appendix 6).
Range: E Asia, India, SE Asia.
Distribution and status: Vagrant to N Sumatra.
Habits: Hunts in smaller trees of forest edge, clearings, and secondary growth.

326. MANTANANI SCOPS-OWL *Otus mantananensis*   PLATE 40

Description: Small (18 cm) mottled dark brown owl with short ear tufts, yellow eyes, and pale belly, peppered black. Some birds are paler, greyish brown.
Iris—yellow; bill—grey; feet—grey.
Voice: Single, goose-like, honk note or honk followed by 3 gruff notes of lower pitch.
Range: S Philippines and Mantanani.
Distribution and status: Resident on Mantanani off NW Borneo where common in forest, coconut groves, and casuarinas.
Habits: Hunts in smaller trees of forest edge, clearings, and secondary growth.

327. SIMEULUE SCOPS-OWL *Otus umbra*   PLATE 40
(Simular/Mentaur/Simulu Scops-Owl)

Description: Small (18 cm), dark reddish-brown, blotchy owl with prominent ear tufts and yellow eyes; lacks pale neck collar of Collared Scops-Owl. Underparts rufous, occasionally with thin white and rufous bars with black edges.
Iris—greenish yellow; bill—grey; feet—grey.
Voice: Syncopated duets in flight, and a male territorial song of two steady notes followed quickly by a higher rising, inflected note. The female notes are higher-pitched. Female sometimes gives continuous call of prolonged whines.
Range: Endemic to Simeulue Island off NW Sumatra
Distribution and status: It is not rare.
Habits: As other scops-owls, favours forest edge and patches, also clove plantations.

192 OWLS

328. ENGGANO SCOPS-OWL *Otus enganensis* PLATE 40
(Engano Scops-Owl)

Description: Small (20 cm) brown mottled owl with prominent ear tufts and yellow eyes. Similar to Simeulue Scops-Owl. Dorsal coloration varies from chestnut to brownish olive.
Iris—yellow; bill—grey; feet—grey.
Voice: King reports that the call is distinct from Simeulue Scops-Owl.
Range: Endemic to Enggano Island off SW Sumatra.
Distribution and status: Quite common.
Habits: As other scops-owls.

329. RAJAH SCOPS-OWL *Otus brookii* PLATE 40
(Rajah's Scops-Owl)

Description: Medium (23 cm), brownish owl with conspicuous ear tufts. Similar to Collared Scops-Owl but slightly larger and with broad buff neck collar and yellow eyes.
Iris—yellow; bill—yellowish; feet—dirty yellow.
Voice: Monotonously repeated, explosive, clear note of constant pitch.
Range: Endemic to Greater Sundas.
Distribution and status: On Sumatra probably throughout the Barisan Range, at 1200 to 2400 m. On Borneo known only from Mt Dulit, and on Java known only from the Ijen Plateau.
Habits: Presumed similar to other scops-owls.

330. COLLARED SCOPS-OWL *Otus lempiji* PLATE 40
(>Japanese Scops-Owl)

Description: Small (20 cm), greyish or brownish owl with conspicuous ear tufts and diagnostic pale, sandy neck collar. Upperparts greyish or sandy brown, mottled and blotched with black and buff; underparts buff, streaked black.
Iris—dark brown; bill—yellow; feet—dirty yellow.
Voice: Male gives a soft hoot *woop*, inflected upwards; also a steady series of gruff notes at one second intervals. Female call is higher-pitched, quavering, inflected downward *wheoo* or *pwok*, about five times a minute; also a gentle twitter. Pairs often call in duet. (See Appendix 6.)
Range: SE Asia, Philippines, Borneo, Sumatra, Bangka, Belitung, Java, and Bali.
Distribution and status: A quite common owl up to 1600 m, including tree-lined suburban streets of large towns.
Habits: Sits on a low perch for much of the night, giving its mournful call in season. Hunts from its perch and pounces on prey on the ground.

331. MENTAWAI SCOPS-OWL *Otus mentawi* PLATE 40

Description: Small (20 cm), blotched dark brown owl with prominent ear tufts, yellow eyes, and no neck collar. Most birds are reddish brown but some are dark blackish brown. Underside is streaked with black shafts.
Iris—yellow; bill—horn; feet—grey.
Voice: Irregular, gruff, honking notes of varying inflection and pitch. Duets of female's higher-pitched, quavering notes followed by *po po* of male. Male alone gives series of *po po* calls, ending in descending series of seven or eight single *po* notes (Marshall, 1978).
Range: Endemic to the larger islands of Mentawai, off the W coast of Sumatra.
Distribution and status: An occasional bird of lowland forest and secondary growth.
Habits: Like other scops-owls. Can be attracted by playback of its call.
Note: Some authors treat this as a subspecies of Collared Scops-Owl but the voice is quite different (Marshall, 1978).

332. BARRED EAGLE-OWL *Bubo sumatranus* PLATE 39
(Malaysian/Malay Eagle-Owl)

Description: Large (45 cm), heavily barred, dark grey owl with conspicuous horizontal ear tufts. Upperparts blackish brown, finely barred with buff; eyebrows white; underparts whitish grey, strongly barred black.
Iris—dark brown; bill—yellow; feet—pale yellow.
Voice: In flight, loud deep *whoo* or *hooa-who*, ending in a deep groan. Various other strange loud calls ascribed to demons in local folklore.
Range: Malay Peninsula, Borneo, Sumatra, Bangka, Java, and Bali.
Distribution and status: An occasional, though rarely seen, bird of lowland forest, up to 1000 m, and higher on some Sumatran mountains.
Habits: Fond of bathing in pools or streams. Flies fast and low from its daytime hiding place as dusk falls. Hunts from perches and hops well on the ground.

333. BUFFY FISH-OWL *Ketupa ketupu* PLATE 39
(Malaysian/Malay Fish-Owl)

Description: Large (45 cm) yellowish brown owl with conspicuous ear tufts. Upperparts rich brown, mottled with black streaks and buff edges; underparts rufous buff with bold black streaks.
Iris—bright yellow; bill—grey; feet—yellow.
Voice: Various loud shrill calls *kootookookootook*, a ringing *pof-pof-pof*, and also *hie-ee-ee-eek-heek*.
Range: SE Asia, Borneo, Sumatra with eastern islands, and Nias, Java, and Bali.

Distribution and status: Not uncommon in lowland forests up to 1100 m.
Habits: Mostly nocturnal but partly active by day in shady places. At night prefers open areas outside the forest, woodlands, parks, paddy fields, riversides. Enjoys bathing and stands for long periods in water, catching much of its food in water.

334. COLLARED OWLET *Glaucidium brodiei* PLATE 39
(Collared Pygmy Owl)

Description: Tiny (16 cm) barred owl with yellow eyes, a pale neck collar, and no ear tufts. Upperparts pale brown, barred reddish buff; crown grey with small white or reddish with eye-spots; brown bar across white throat; breast and belly buff, barred with black; thighs and vent white, streaked with brown.
Iris—yellow; bill—horn; feet—grey.
Voice: Mellow, whistled monotone *poop a poop poop*, given by day or night. Imitation of this call is very effective in attracting this owl and also small mobbing songbirds. (See Appendix 6.)
Range: Himalayas, S China, SE Asia, Malay Peninsula, Sumatra, and Borneo.
Distribution and status: In Sumatra and Borneo an occasional owl of montane forests, between 800 and 3500 m.
Habits: Seen during the day in tall trees when calling or mobbed by other birds. At night it hunts from prominent perches, keeping to taller trees. Flies with very fast wing-beat.

335. JAVAN BARRED OWLET *Glaucidium castanopterum* PLATE 39
(Chestnut-winged/Spadiced Owlet, AG: Pygmy-Owl)

Description: Small (24 cm), finely barred, rufous-brown owl without ear tufts. Upperparts rufous chestnut, barred ochre with broken white line at edge of scapulars; underparts mostly brown, barred ochre; underparts whitish with chestnut flanks; bold white chin stripe conspicuous, underlined with brown and buff gorget.
Iris—yellow-brown; bill—greenish with yellow tip; feet—greenish yellow.
Voice: Unlike most owls: a rapid trill, descending in pitch but increasing in volume, given at dawn and dusk. (See Appendix 6.)
Range: Endemic to Java and Bali.
Distribution and status: Not uncommon in fragments of forest in the lowlands and hills of Java and Bali.
Habits: Frequents gardens, villages, primary, and secondary forest. It is principally nocturnal but sometimes active by day. Calls mostly in evening and early morning.
Note: Formerly included in Asian Barred Owlet *G. cuculoides*.

336. BROWN HAWK-OWL *Ninox scutulata* PLATE 39
(Oriental/Philippine Hawk-Owl)

Description: Medium-sized (30 cm), large-eyed, dark hawk-like owl. Lack of facial disc characteristic. Upperparts dark brown; underparts buff, broadly streaked reddish brown; vent, chin, and a spot at the base of the bill, white. Iris—bright yellow; bill—bluish grey with green cere; feet—yellow.

Voice: Mellow, rising falsetto whistle *pung-ok*, the second note short with rising inflection, repeated every one or two seconds, sometimes for long periods, usually at dawn and dusk. (See Appendix 6.)

Range: India, E Asia, SE Asia, Philippines, Sulawesi, Borneo, Sumatra, and W Java.

Distribution and status: Both resident and winter migrant races are found. All are uncommon at low to moderate altitudes, up to 1500 m. The species is more common on Borneo and Sumatra but there are no recent records for Java where it is rare.

Habits: Becomes active shortly before dusk at the edge of the forest or in cultivated areas, and flies after dragonflies and other insects which it catches in mid-air. Sometimes a family group will hunt together around a clearing. Calls irregularly, particularly when the moon is up.

337. SPOTTED WOOD-OWL *Strix seloputo* PLATE 39

Description: Large (47 cm), white-spotted, brown owl without ear tufts. Facial disc pale rufous; upperparts rufous chocolate, boldly marked with black-edged white spots; underparts white, washed with rufous and barred dark brown, with a whitish chin bar.
Iris—dark brown; bill—greenish black; feet—grey.

Voice: Deep, resonant, rising *beloop* or *hoop-hoong*, or *hoo-hoo-hoo-hooo*; also deep growls. (See Appendix 6.)

Range: SE Asia, Palawan, and Java.

Distribution and status: Tentatively added to the Sumatran list (Mt. Kerinci) where possibly overlooked in the past. On Java it is uncommon in mangrove and lowland forests; a pale and smaller form occurs on Bawean.

Habits: Frequents lowland forest and clusters of trees near villages, and even in towns. Habits similar to Brown Wood-owl.

338. BROWN WOOD-OWL *Strix leptogrammica* PLATE 39

Description: Large (47 cm), heavily barred, reddish-brown owl without ear tufts. Conspicuous face disc of rufous 'spectacles', black eye-ring, and white eyebrows;

underparts buff with fine dark brown barring, washed chocolate on breast; upperparts dark brown, strongly barred buff and white.
Iris—dark brown; bill—whitish; feet—bluish grey.
Voice: Distinctive deep *boo-boo*, or four-syllable *goke-galoo*, *huhu-hooo* and others.
Range: S India, China, SE Asia, Borneo, Sumatra, Belitung, Mentawi, Nias, and W and C Java.
Distribution and status: In Borneo and Sumatra an uncommon bird of lowlands. In Java this is a rare montane bird.
Habits: A rarely seen nocturnal forest owl. If disturbed by day it compresses its plumage to look like a dead piece of wood, watching with half closed eyes. Pairs call to each other at dusk prior to hunting.

339. SHORT-EARED OWL *Asio flammeus* PLATE 39
(Pueo)

Description: Medium-sized (37 cm), long-winged, tawny owl. Conspicuous facial disc with short ear tufts, not visible in the field, and piercing bright yellow eyes in dark eye-rings. Upperparts tawny, heavily streaked black and buff; underparts buff, streaked dark brown. Black carpal patch conspicuous in flight.
Iris—yellow; bill—dark grey; feet—whitish.
Voice: Sneezing bark *kee-aw* given in flight.
Range: Holarctic; in SE Asia a winter migrant.
Distribution and status: Vagrant to N Borneo.
Habits: Prefers grassy open areas.

FROGMOUTHS—FAMILY PODARGIDAE

A FAMILY of curious looking nocturnal birds, related to nightjars but adapted to live inside the forest. Frogmouths are aptly named for their enormous wide gape facilitating the capture of insects on the forest floor and off branches. They are found from SE Asia to New Guinea and Australia. All species have mottled camouflaged plumage. They sit very upright by day on a low perch. Frogmouths lay a single egg in a downy, cup-shaped nest, precariously balanced on a horizontal twig.

Six species occur in the Greater Sundas.

340. LARGE FROGMOUTH *Batrachostomus auritus* PLATE 41

Description: Very large (40 cm), chestnut frogmouth with white nape collar, prominent white spots on wing, and smaller spots on breast. Male bright chestnut,

female duller and paler. Throat and breast rufous brown, grading to buff belly (creamy in some Bornean females) and vent. Protruding eyelids orange. Iris—white or brown; bill—horn; feet—yellowish.
Voice: Repeated deep far-carrying tremulous hoot with rising inflection at dusk and dawn.
Range: Malay Peninsula, Sumatra, Natunas, and Borneo.
Distribution and status: An uncommon bird of lowland forests.
Habits: Rarely seen, staying motionless by day in canopy. A nocturnal hunter of insects, generally pounced upon on the ground or plucked from branches. Sometimes found in low bushes along streams.

341. DULIT FROGMOUTH *Batrachostomus harterti* PLATE 41

Description: Large (37 cm) frogmouth. Distinguished from Large Frogmouth by slightly smaller size and purplish-brown coloration of male. Female is chestnut brown with whiter nape collar.
Iris—dark brown; eyelids—yellow; bill—horn; feet—creamy horn.
Voice: No information.
Range: Endemic to Borneo.
Distribution and status: Confined to N mountains of Borneo where it is an uncommon bird, between 300 and 1500 m.
Habits: As other frogmouths.

342. GOULD'S FROGMOUTH *Batrachostomus stellatus* PLATE 41

Description: Smallish (25 cm) reddish brown frogmouth. Males tend to be redder than females but species is dimorphic. Common phase: bright rust, rarer form: dark rufous brown. Rufous-scalloped cream underparts distinctive.
Iris—yellow in male, brown in female; bill—horn; feet—pink in male, yellow in female.
Voice: Male call is two clear notes, higher on second note, with wavering tremolo connecting notes, given at 7-second intervals.
Range: Malay Peninsula, Sumatra, N Natunas, and Borneo.
Distribution and status: An occasional bird below 500 m.
Habits: As other frogmouths, prefers lowland rainforest.

343. SHORT-TAILED FROGMOUTH *Batrachostomus poliolophus*
(Pale-headed Frogmouth) PLATE 41

Description: Medium-sized (30 cm) greyish brown (male) or brownish red (female) frogmouth with long ear tufts and fine black streaking. Similar to

respective grey and rufous phases of Sunda Frogmouth but tail shorter and lacking pale bars.
Iris—yellow; bill—horn; feet—whitish pink.
Voice: No information.
Range: Endemic to Sumatra and Borneo.
Distribution and status: On Sumatra an uncommon resident in the hills of the Barisan Range, recorded from 600 to 1300 m, but there are few records. On Borneo most specimens are from Dulit and Kelabit Highlands but it is probably widespread.
Habits: As other frogmouths.
Note: The Bornean population is sometimes treated as an endemic species *B. mixtus*.

344. JAVAN FROGMOUTH *Batrachostomus javensis* PLATE 41
(Blyth's Frogmouth)

Description: Smallish (25 cm), dark-coloured frogmouth. Male greyish and spotted; female rufous brown, with huge head and gape with a ring of long bristles. Long ear tufts distinctive.
Iris—yellow; bill—brown above, greyish below; feet—brown.
Voice: Hoarse cackling *gwaa* notes, descending in pitch; barks, trills, and a wheezy, rising, plaintive whistle. Different races show variation in calls. (See Appendix 6.)
Range: SE Asia, Palawan, and Greater Sundas.
Distribution and status: Status on Sumatra uncertain; one specimen from Bintang (Riau Archipelago) and one specimen from lowland N Sumatra in 1884. On Borneo recorded in lowlands from most parts, and on Java it is an occasional bird of the wetter lowland and hill forests. Absent from Bali.
Habits: By day it sits very upright with beak pointing upwards and eyes closed, occasionally two birds close together. Generally not far off the ground.
Note: The taxonomy of *javensis* is confused; sometimes treated as two species where *javensis* is endemic on Java and Blyth's *B. affinis* is found in SE Asia, Sumatra, and Borneo. The Palawan population is sometimes put in *affinis*, sometimes in *javensis* and sometimes in *cornutus*.

345. SUNDA FROGMOUTH *Batrachostomus cornutus* PLATE 41
(Horned/Long-tailed Frogmouth)

Description: Medium-sized (28 cm), dimorphic frogmouth. Individuals vary from finely streaked black and white plumage to more or less all brown or dull rufous coloration. The assumption that all males are of the grey form and all females of the red form is correct.
Iris—yellow; bill—dark brown above, greenish below; feet—brown.

Voice: A descending series of *gwaa* notes, all starting on the same pitch.
Range: Endemic to the Greater Sundas.
Distribution and status: Recorded from E Sumatra (including Bangka and Belitung), Borneo, and Kangean. It is an uncommon bird of lowland and coastal areas.
Habits: Prefers secondary forest and forest edge. Rests on low branches by day, sometimes with bill wide open.

Nightjars—FAMILY CAPRIMULGIDAE

NIGHTJARS are short-legged, entirely insectivorous, nocturnal birds which have a net of bristles around the bill for catching insects while on the wing. By day they rest on the ground. They fly in an erratic, slow, flapping manner and emit monotonous calls. The eggs are laid on the ground in a scrape without any nest material.

Two 'eared' and 5 'non-eared' species are found in the Greater Sundas.

346. MALAYSIAN EARED-NIGHTJAR *Eurostopodus temminckii*
PLATE 41

Description: Large (42 cm), dark brown, black-barred nightjar with prominent ear tufts. Has narrow pale collar, barred underparts, and lacks white patches on wings and tail. Uniform crown is same colour as sides of head, without pale eyebrow.
Iris—brown; bill—horn; feet—brown.
Voice: Clear 3-note call. The first starts low, followed by two longer notes, each rising then dipping and broken in the middle, *tap-ti-bau*. Stress is on first and last notes. (See Appendix 6.)
Range: Malay Peninsula, Sumatra, and Borneo.
Distribution and status: A common bird of forest edge and heath forests below 1200 m.
Habits: As other nightjars but flight more rapid and erratic than Great Eared-Nightjar. Prefers open scrub near forest. Calls while feeding high at dusk.

347. GREAT EARED-NIGHTJAR *Eurostopodus macrotis*
PLATE 41
(Giant Eared-Nightjar)

Description: Very large (40 cm), dark brown-barred nightjar with prominent ear tufts. Similar to Malaysian Eared-Nightjar but much larger and distinguished by buffy crown paler than rest of head, and by call and flight. The two species do not overlap in the Greater Sundas.

Iris—brown; bill—horn; feet—brown.
Voice: Loud, hollow, 3-note whistle. The first note is short and sometimes inaudible, the second is longer and dips in pitch, the third note, also long, rises then tails off, *pip piuiu*; also continuous inflected whistle.
Range: India, S China, SE Asia, Philippines, Sulawesi, and Simeulue.
Distribution and status: Common on Simeulue.
Habits: Similar to other nightjars. Glides like a harrier, flight slower and less erratic than Malaysian Eared-Nightjar. Favours forest edge and open scrub. Often seen at dusk over forest.

348. GREY NIGHTJAR *Caprimulgus indicus* PLATE 41
(Jungle/Indian Jungle/Japanese Nightjar)

Description: Largish (28 cm), greyish nightjar. Male: lacks rusty nape collar of Large-tailed Nightjar; white tail markings on outer four pairs of tail feathers; white patches at base of primaries. Female similar to male but white tail patches buffy.
Iris—brown; bill—blackish; feet—chocolate.
Voice: Hard, sharp *chuck*, rapidly repeated at a steady rate of about 6 per second, then ending with a *chrrrr*. Wintering birds rarely call.
Range: Resident in India, China, SE Asia, and Philippines; migrates in winter to Borneo, Sumatra, Java, New Guinea.
Distribution and status: Apparently rare in Sumatra but recorded from both lowlands and uplands of most of Borneo. In Java it is found mostly in mountains. Not recorded from Bali.
Habits: Prefers rather open mountain forest and scrub. Typical nightjar flight, settles on ground or on horizontal branch in daytime.

349. LARGE-TAILED NIGHTJAR *Caprimulgus macrurus* PLATE 41
(White-tailed/Long-tailed Nightjar/Coffinbird)

Description: Largish (30 cm) greyish brown nightjar. Prominent white patch on centre of four outer primaries, and broad white tips of two outer pairs of tail feathers diagnostic. In females these patches are buffy; throat bar white.
Iris—brown; bill—greyish brown; feet—greyish brown.
Voice: Rich, deep *tchoink* like two stones being struck together, given at a steady rate of about 3 per second following a purring warm up; also low growling. (See Appendix 6.)
Range: India, SE Asia, Philippines, Indonesia to New Guinea and Australia.
Distribution and status: A local but common bird of forest edge and wooded country, including mangroves, up to 1200 m. No records from Kalimantan.
Habits: Rests in forest edge or wooded areas during the day in a shady place on

the ground. It calls for 30 min at dusk and dawn, from a perch or on the wing. Hunting interspersed with resting periods on the ground, often on roads where birds are frequently killed by cars.

350. SAVANNAH NIGHTJAR *Caprimulgus affinis* PLATE 41
(Allied Nightjar)

Description: Smallish (22 cm) uniform-coloured nightjar. Male has diagnostic white outer tail feathers. White throat band divided into two side patches. Female is more rufous and lacks white tail markings.
Iris—brown; bill—reddish brown; feet—dull red.
Voice: Penetrating, plaintive *chweep* uttered constantly for 30 min at dusk and dawn, on the wing.
Range: India, S China, SE Asia, Sulawesi, Philippines, Sundas.
Distribution and status: This is the common nightjar of the lowlands, in dry open coastal areas including large cities.
Habits: Typical nightjar sitting on the ground by day, or on the tops of tall, flat buildings in cities. Hawks for insects; attracted by city lights.

351. BONAPARTE'S NIGHTJAR *Caprimulgus concretus* PLATE 41

Description: Smallish (22 cm), rich dark brown nightjar with large white throat patch. Upperparts spotted with black; underparts buffy, narrowly barred with black. Has white patches on outer two tail feathers but lacks white wing patches.
Iris—brown; bill—horn; feet—grey.
Voice: Unknown.
Range: Endemic to Sumatra and Borneo.
Distribution and status: Few records from Sumatra; recorded in lowland forest of mainland and Belitung. On Borneo a rare but widely distributed bird of open lowland forests, including heath forests up to 500 m.
Habits: Poorly known forest nightjar.

352. SALVADORI'S NIGHTJAR *Caprimulgus pulchellus* PLATE 41

Description: Medium-sized (24 cm) nightjar with broad, white throat band; extensive narrow white barring on underparts. White spots on outer two pairs of tail feathers and on two pairs of outer primaries.
Iris—dark brown; bill—brown; feet—brown.
Voice: Five *tock* notes in an irregular rhythm, like the *kwow-kwow* call of Asian noodle vendors.
Range: Endemic to Sumatra and Java.

Distribution and status: On Sumatra known by one specimen taken on Singgalang in 1878; on Java recorded from peaks of W Java and Tengger Highlands in E Java.
Habits: A forest nightjar in sub-montane habitat.

Swifts—FAMILY APODIDAE

A WORLDWIDE family of fast-flying, insectivorous birds; look superficially rather like swallows but are in fact most closely related to hummingbirds.

Swifts have long, pointed backswept wings and either short and squarish or long and pointed tails, and tiny legs. They rarely perch in trees but usually rest by clinging to cliffs with their sharp claws. They nest in caves, hollow trees, and under house roofs in cup-shaped nests made of mud or, in some species, saliva.

They feed on the wing using their wide mouth to catch insects. Some of the cave-nesting swiftlets use a form of sonar echolocation with clicking calls to find their way in the dark.

There are 14 species of swifts in the Greater Sundas and some are very difficult to identify in flight.

353. GIANT SWIFTLET *Hydrochous gigas* PLATE 42

Description: Large (16 cm) swiftlet with sooty black upperparts, dark rump, and dark brown underparts. The tail is slightly forked.
Iris—brown; bill—black; feet—black.
Voice: Sharp, whickering calls.
Range: Malay Peninsula and Greater Sundas.
Distribution and status: On Sumatra known by two specimens collected at 480 m in Padang Highlands. There are sight records from N Borneo. On Java apparently restricted to Salak, Pelabuhanratu, Pangrango, Patuha areas in W Java.
Habits: Generally over forest, in hilly and mountainous terrain. Tends to fly higher and faster than other swiftlets. Does not use echolocation. Nests under waterfalls and in rock crevices. The nest is an inedible cup of roots, moss, and other fibres cemented with saliva.

354. EDIBLE-NEST SWIFTLET *Collocalia fuciphaga* PLATE 42
(> Grey-rumped/Brown-rumped/White-nest/German's Swiftlet)

Description: Smallish (12 cm) swiftlet. Upperparts blackish brown with paler, greyish or brown rump in Java, or dark brown rump in Sumatra and Borneo (sub-

species *vestita*); tail slightly forked; underparts brown. Generally indistinguishable in the field from Mossy-nest, Black-nest, and Volcano Swiftlets unless at nest.
Iris—dark brown; bill—black; feet—black.
Voice: High-pitched *tscheerrr* is commonly uttered near breeding sites.
Range: S China, SE Asia, Philippines, and Sundas.
Distribution and status: Throughout the Greater Sundas; common at all altitudes, up to 2800 m on Sumatra and Borneo. In Java and Bali a rather local species limited by the availability of deep, dark cliff clefts or caves for nesting.
Habits: Generally feeds higher than Glossy Swiftlet and has a more powerful, less fluttery flight on stiffer wings. Uses echolocation in dark caves with a loud 'rattle call'. Often feeds around tall forest trees such as fruiting figs which attract fig wasps; bathes and drinks over fresh water by dipping down with a splash. Breeds in coastal rock crevices or deep in limestone caves. Nests are made entirely of hardened saliva and are the valued 'white birds' nests' which are collected for sale to make birds' nest soup.
Note: Some authors split German's Swiftlet *C. germani* as a distinct species on the basis of sympatric distribution, but ecological isolation of the forms allows them to be treated as the same species. Can be placed in genus *Aerodramus*.

355. BLACK-NEST SWIFTLET *Collocalia maxima* PLATE 42

Description: Smallish (13 cm) blackish brown swiftlet with rump grading from greyish to the same colour as the back, and virtually indistinguishable in the field from the Edible-nest Swiftlet. Legs are well feathered; tail rather square-cut.
Iris—brown; bill—black; feet—black.
Voice: Shrill calls.
Range: Malay Peninsula and Greater Sundas.
Distribution and status: Throughout the Greater Sundas. Breeding mostly on coasts in Sumatra and islands. On Borneo this is the commonest swiftlet in limestone areas. In Java it is an uncommon bird of islands and coastal areas, generally near limestone.
Habits: The nest is made of white, cemented saliva mixed with feathers in limestone caves. These are the so-called 'black nests' which are harvested for sale but are of less value than the cleaner 'white nests' because more labour is needed to remove feathers and grubs. Birds give echolocation 'rattle' call.

356. MOSSY-NEST SWIFTLET *Collocalia salangana*
(Mossy/Sunda/Thunberg's Swiftlet)

Description: Smallish (12 cm) swiftlet, almost indistinguishable in the field from the Edible-nest Swiftlet, but with darker rump and unforked, slightly notched tail.

Can be identified only at the nest which is mossy, or in hand by concealed barbs at the base of feathers of the back.

Iris—dark; bill—black; feet—black.

Voice: Similar to Edible-nest Swiftlet, including 'rattle call' echolocation.

Range: Greater Sundas and Sulawesi.

Distribution and status: On Sumatra breeding only known from Barisan Range (Padang Highlands). Recorded N Borneo (including Natuna). In Java appears to be uncommon, possibly because it is rarely recognized.

Habits: Similar to Edible-nest Swiftlet including echolocation. The nest is diagnostic, being rounder, softer, and mossier than that of the Glossy Swiftlet. The nests are also made deeper in caves as this is a true echolocating swiftlet.

Note: May be conspecific with Uniform Swiftlet *C. vanikorensis* of E Indonesia.

357. VOLCANO SWIFTLET *Collocalia vulcanorum* PLATE 42

Description: Largish (14 cm) blackish swiftlet with long wings and slightly forked tail. The rump varies from greyish to as dark as the back. Legs unfeathered or only slightly feathered.

Iris—dark; bill—black; feet—black.

Voice: Piercing, swift call *teeree-teeree-teeree*.

Range: Endemic to Java.

Distribution and status: Known only from the peaks of Gede, Tangkubanprahu, and Papandayan. Nests in crater crevices and as all the above volcanoes are active, colonies are susceptible to periodic extinction.

Habits: Fast-flying, in flocks around the open peaks and ridges of highest mountains. Most conspicuous in the crater of Gn. Gede, Java. Uses echolocation. Nests in rock crevices and makes an inedible mossy nest.

Note: Sometimes treated as a resident race of Himalayan Swiftlet *A. brevirostris*. The Himalayan Swiftlet, however, is not known to winter south to the Greater Sundas, and *vulcanorum* is completely isolated. Wintering Himalayan Swiftlets could reach N Sumatra but this has never been confirmed.

358. GLOSSY SWIFTLET *Collocalia esculenta* PLATE 42
(> White-bellied Swiftlet)

Description: Small (9 cm), glossy blue-black swiftlet with only slightly notched tail, grey chin, and diagnostic white belly. This is the smallest and commonest swiftlet over the whole Sundas region.

Iris—brown; bill—black; feet—black.

Voice: Shrill, twittering chirrups and burbling calls.

Range: Andamans, Nicobars, and Malay Peninsula through the Philippines, and Indonesia to New Guinea and SW Pacific.

Distribution and status: Common throughout Sumatra and Borneo from sea level to highest peaks.
Habits: Ubiquitous, flying over all types of forest and farmland. Dips to drink over rivers and pools each evening. Nests in cave mouths; does not echolocate.
Note: Some authors include the greener form *C. linchi* of SE Sumatra, N Borneo, Java, and Bali in this species.

359. CAVE-SWIFTLET *Collocalia linchi*

Description: Small (10 cm) swiftlet with dull greenish black upperparts, sooty grey underparts with whitish belly, and slightly notched tail.
Iris—dark brown; bill—black; feet—black.
Voice: High-pitched *cheer-cheer* calls.
Range: Malay Peninsula, Greater Sundas and Lombok.
Distribution and status: Local on Sumatra (probably throughout the Barisan Range but specimens only from Mts Leuser and Lampung) and Borneo (Mt. Kinabalu). On Java (including islands in Java Sea) and Bali this is the commonest swiftlet at all altitudes.
Habits: As Glossy Swiftlet. Nest is an irregular cup of moss, grass, or other vegetable material, cemented with saliva and built in better lit areas near the mouths of caves, rock crevices or in buildings. Flight is weak and fluttery.
Note: Formerly included within Glossy Swiftlet but sympatry in part of range suggests full species status.

360. WHITE-THROATED NEEDLETAIL *Hirundapus caudacutus*
(Spine-tailed Swift/Northern Needletail)   PLATE 42

Description: Large (20 cm) blackish swift with white chin and throat, white undertail coverts, white spot on side of neck, and small white patch on tertials; back brown with silvery whitish saddle. Distinguished from other needletails by white throat.
Iris—dark brown; bill—black; feet—black.
Voice: High-pitched chitters when chasing each other.
Distribution and status: Breeds in N Asia, China, Himalayas, but migrates in winter to Australia and New Zealand.
Distribution and status: In Borneo and Java a rare passage migrant.
Habits: Similar to Brown-backed Needletail.

361. SILVER-BACKED NEEDLETAIL *Hirundapus cochinchinensis*
(White-vented/Grey-throated Needletail, AG: Needle-tailed Swift) PLATE 42

Description: Largish (18 cm) blackish swift with a whitish (Sumatra) or pale brown saddle on back and rump, and a short squarish needle-tail. Chin and throat is greyish, and undertail coverts white. An aberrant all-black form formerly known as *H. c. ernsti* has been recorded in W Java. Distinguished from White-throated Needletail by greyish throat and lack of white on tertials. Lores never white.
Iris—dark brown; bill—black; feet—dark purple.
Voice: No description available.
Range: NW India, SE Asia, Sumatra, and W Java.
Distribution and status: In Sumatra it is an uncommon passage migrant and winter visitor, presumed to come from the Himalayas. In W Java this is a very rare bird known from only a few records. It may breed in Java but this has never been confirmed.
Habits: Similar to Brown-backed Needletail.

362. BROWN-BACKED NEEDLETAIL *Hirundapus giganteus* PLATE 42
(Brown/Giant Needletail, AG: Needle-tailed Swift)

Description: Large (24 cm), glossy black swift with rounded tail and white undertail coverts and flanks. Upperparts dark blue with black gloss, except brown back and rump; underparts brown except white vent and flanks. Distinguished from other needletails by brown back and white lores (absent in resident race).
Iris—dark brown; bill—black; feet—purplish.
Voice: Brittle, chattering squeaks *cheek* or *chirrweet*.
Distribution and status: India, SE Asia, Borneo, Sumatra, Java, Bali, Sulawesi, and Philippines.
Distribution and status: In Borneo and Sumatra an occasional resident. In Java and Bali an uncommon bird at all altitudes. A few winter migrants of mainland race reach Borneo and presumably Sumatra.
Habits: Flies very fast, passing close with an audible whoosh of air. Commonly regarded as the world's fastest bird. Generally hunts in small flocks, frequently over the crest of a high ridge, using the air currents to its advantage. Also flies low over water, making three or four audible wing clicks while dipping to splash water to bathe or drink.

363. SILVER-RUMPED SWIFT *Raphidura leucopygialis* PLATE 42
(Silver-rumped Spinetail)

Description: Small (11 cm) black swift. The rounded tail and silvery grey rump and tail coverts, which reach the tip of the tail, distinguish it from any other swiftlet. In flight the 'butterknife' shape of the broad wings is diagnostic.
Iris—dark brown; bill—black; feet—black.
Voice: High-pitched *tirrr-tirrr* note.
Range: Malay Peninsula, Borneo, Sumatra with offshore islands, and Java.
Distribution and status: In Borneo and Sumatra it is locally common, but in Java an uncommon bird up to 1500 m.
Habits: Flies low over forests or forest clearings, often near streams, with a very fluttery, bat-like flight. Generally in small flocks.

364. FORK-TAILED SWIFT *Apus pacificus* PLATE 42
(White-rumped/Pacific Swift)

Description: Unmistakable, largish (18 cm),˙dusky brown swift with long, deeply forked tail, whitish chin, and white patch on rump. Distinguished from Little House Swift by larger size, paler colour, darker throat, narrower white rump saddle, slimmer shape, and forked tail.
Iris—dark brown; bill—black; feet—purplish.
Voice: Buzzing and twittering sound, and long, high-pitched squeaks *skree-ee-ee*.
Range: Breeds in Siberia and NE Asia, but migrates south in winter through SE Asia and Indonesia to New Guinea and Australia.
Distribution and status: A regular passage migrant, sporadic in occurrence but sometimes numerous, up to 1500 m.
Habits: Generally found in flocks over open places and often mixed with other species of swifts. Flies more slowly than needletail swifts and makes erratic flutters and turns when feeding.

365. LITTLE SWIFT *Apus affinis* PLATE 42
(House Swift)

Description: Medium-sized (15 cm) blackish swift with a white throat and rump, and notched rather than forked tail. Distinguished from the larger Fork-tailed Swift by darker colour, whiter throat and rump, and almost square-cut tail.
Iris—dark brown; bill—black; feet—brown.
Voice: Trill of fast-repeated, loud, shrill, whickering screams given in flight, especially before roosting in the evening.
Range: Africa, Middle East, India, SE Asia, Philippines, Sulawesi, and Greater Sundas.

208 SWIFTS

Distribution and status: Common resident on Sumatra (including Riau and Lingga Archipelagos, and Belitung), locally common on Borneo, Java, and Bali, in towns and coastal areas, occasionally up to 1500 m.
Habits: Lives in large flocks, hunting with a steady, even flight over open areas. Nests under house eaves, cliff overhangs, or cave mouths.

366. ASIAN PALM-SWIFT *Cypsiurus balasiensis* PLATE 42

Description: Small (11 cm), entirely dark brown, slender swift. Distinguished from swiftlets by its larger, narrower wings, and very deeply forked tail.
Iris—dark brown; bill—black; feet—purplish.
Voice: High-pitched chattering *cheereecheet* is regularly uttered.
Range: India, China, SE Asia, Borneo, Sumatra, Java, Bali, Sulawesi, and Philippines.
Distribution and status: Rather local in Borneo but in Sumatra, Java, and Bali a fairly common bird, up to 1500 m, in suitable habitat.
Habits: Distribution is determined by the presence of palms with fan-shaped leaves such as *Livistona*, *Borassus*, *Areca*, or *Corypha*, used as nest sites and resting places. The nest is cemented under a palm leaf.

TREESWIFTS—FAMILY HEMIPROCNIDAE

TREESWIFTS are a small family confined to SE Asia. Very similar to true swifts but members of this family perch in trees and are distinguished by more elongated wings and tails. Treeswifts hawk in wide, circling flights from high vantage points on tree perches; often in shrill-calling flocks. Their tiny nests are also attached to branches where they glue a single white egg. There are only two species in the Greater Sundas.

367. GREY-RUMPED TREESWIFT *Hemiprocne longipennis* PLATE 42

Description: Largish (20 cm) tree perching swift with very long tail and wings, and grey patch on tertials. Cheeks chestnut in male, green in female; otherwise sexes similar. Short crest on forecrown; crown, nape, back, and wing coverts glossy greenish grey; rump grey; wings and tail black; throat, breast, and flanks grey; belly and undertail coverts white. Immature is brown, scaled and mottled with white.
Iris—dark brown; bill—black; feet—black.
Voice: Loud, harsh, high-pitched cries *cher tee too—cher tee too—cher tee too*, and variations.
Range: Malay Peninsula, Borneo, Sumatra, Java, Bali, and Sulawesi.

Distribution and status: Common in Borneo and Sumatra, widespread in Java and Bali, though rarely numerous, up to 1500 m.
Habits: Prefers forest edge or open forest with tall emergent trees with bare branches from which treeswifts make aerial hawking sallies after flying insects. The wide circular flight is more like a bee-eater or wood-swallow than other swifts.

368. WHISKERED TREESWIFT *Hemiprocne comata* PLATE 42
(Lesser Treeswift)

Description: Small (15 cm) tree perching swift with extremely long wings and forked tail, conspicuous white facial whiskers, and white patch on tertials. Plumage of head, wings, and tail is dark bluish black; back, rump, and breast are bronze-brown with a greenish gloss. Male has chestnut cheeks. Immature is brown, mottled with white.
Iris—brown; bill—black; feet—black.
Voice: Loud, clear, high-pitched cries *cheer-ter, cheer-ter*.
Range: India to SE Asia, Malay Peninsula, Sumatra, islands off W Sumatra, Borneo, and Philippines.
Distribution and status: In Sumatra and Borneo a common bird, up to 1500 m.
Habits: Generally in small parties, perched on exposed branches of tall trees from which they make short dashing flights after passing insects.

Trogons—FAMILY TROGONIDAE

TROGONS are medium-sized, brightly coloured birds with short bills, short legs and feet, short wings, long broad tails, and soft, fluffy plumage. The family is pantropical. Two toes point backwards. Buff coloured eggs are laid in nests in tree holes. Trogons are insectivorous, hunting from sometimes quite low perches in thick forest. They have harsh, distinct calls.
Eight species occur in the Greater Sundas.

369. BLUE-TAILED TROGON *Harpactes reinwardtii* PLATE 43
(Blue-billed Trogon)

Description: Medium-sized (34 cm), green and yellow trogon. Adult: upperparts iridescent bluish green; naked skin around eye blue. Tail iridescent greenish blue with three outer feathers with white edges and broad white tips. Primaries black with white edges; wing coverts green, finely barred yellow (male) or brown (female). Underparts yellow with green-grey band across upper breast. Sumatran race is smaller, has maroon rump in male, and smaller yellow bib. Immature: generally brownish with some blue-green on back.

Iris—brown; bill—orange-red in adult, brown in immature; feet—orange.

Voice: Penetrating, hoarse *chierr*, *chierr*, or loud *turrr* while the tail is flicked back and forth, spreading and folding.

Range: Endemic to Sumatra (throughout the Barisan Range) and W Java (east to Papandayan).

Distribution and status: Confined to montane rainforest between 1000 and 2500 m where it is uncommon.

Habits: Sits upright on a horizontal branch in a shady spot in the forest, on the look out for insects or giving its curious loud calls. Flies from perch to perch with a noisy flapping of wings.

370. RED-NAPED TROGON *Harpactes kasumba* PLATE 43

Description: Large (33 cm) black-headed trogon. Broad red neck collar of male distinctive. Male has broad white crescent on breast. Female differs from other female trogons by having brownish grey throat and breast, and tawny belly.

Iris—brown; bare skin around eye—blue; bill—bluish; feet—pink.

Voice: Male gives short, melancholic sequence *kar*, *kar*, *kar*, *kar*, harsh and descending, trailing away at end. Female gives a quiet whirring rattle (M. & W.). (See Appendix 6.)

Range: Malay Peninsula, Sumatra, and Borneo.

Distribution and status: An occasional bird of lowland primary and logged forests, up to 600 m.

Habits: Hunts from low perch in dense forest.

371. DIARD'S TROGON *Harpactes diardii* PLATE 43

Description: Large (30 cm) black-headed trogon. Male is distinguished by narrow pinkish crescent on breast and pink neck collar. The maroon crown patch is rarely visible in the field. Female has brown breast and pinkish belly, and is distinguished from female Scarlet-rumped Trogon by brown rather than pink rump.

Iris—brown; bare skin around eye—mauve blue; bill—blue; feet—grey.

Voice: Series of 10 to 12 *kau* notes, with second note on higher pitch then descending scale and slowing down at end. (See Appendix 6.)

Range: Malay Peninsula, Sumatra, and Borneo.

Distribution and status: An occasional bird of lowland primary and logged forests, below 1000 m.

Habits: Typical trogon behaviour.

TROGONS 211

372. WHITEHEAD'S TROGON *Harpactes whiteheadi* PLATE 43

Description: Large (33 cm) reddish or cinnamon trogon with black throat. Male: similar to Red-naped Trogon but light grey breast distinctive. Forehead, crown, nape, and ear coverts are scarlet. Female: cinnamon instead of red but shares male's grey breast; differs from female Orange-breasted Trogon by cinnamon belly and crown.
Iris—brown; bare skin around eye—blue; bill—bluish with black tip; feet—pale grey.
Voice: Growling notes.
Range: Endemic to N Borneo.
Distribution and status: Confined to high mountains over 1000 m.
Habits: Prefers dark, wet patches of old forest, keeping to higher branches and hawking insects from a prominent perch.

373. CINNAMON-RUMPED TROGON *Harpactes orrhophaeus*
PLATE 43

Description: Medium-size (25 cm) red and black trogon. Male has black head, scarlet breast, and cinnamon back without any neck collar or breast crescent. Distinguished from male Scarlet-rumped Trogon by cinnamon rump. Female differs from other female trogons by rusty area around eyes and lack of pink on belly.
Iris—brown; bare skin around eye—blue; bill—blue; feet—grey.
Voice: A harsh, explosive *purrr*, or three or four falling *taup* notes in descending sequence.
Range: Malay Peninsula, Sumatra, and Borneo.
Distribution and status: A rare montane resident apparently absent from much of Kalimantan. Recorded between 1000 and 1500 m. Sometimes found in lowlands, e.g. Danum Valley.
Habits: As other trogons. Keeps to tall primary montane forest.

374. SCARLET-RUMPED TROGON *Harpactes duvaucelii* PLATE 43
(Red-rumped Trogon)

Description: Smallish (23 cm) red and black trogon. Male has black head, scarlet belly, and cinnamon back with diagnostic scarlet rump. It has no neck collar or breast crescent. The female has brownish breast and pinkish belly but is distinguished from larger female Diard's and Cinnamon-rumped Trogons by pinkish rump.
Iris—brown; bare skin around eye—blue; bill—blue; feet—bluish.
Voice: Male song is quiet, descending 12 notes *yau, yau, yau* ... accelerating

rhythm, repeated many times. Female gives a quiet, whirring rattle (M. & W.). (See Appendix 6.)

Range: Malay Peninsula, Sumatra, and Borneo.

Distribution and status: A common bird of lowland primary and logged forests below 600 m.

Habits: As other trogons.

375. ORANGE-BREASTED TROGON *Harpactes oreskios* PLATE 43

Description: Medium-sized (25 cm) brown and orange trogon. Head, neck, and breast greenish grey (more grey in female, brown in immature); back and tail reddish brown; primaries black and wing coverts barred black; yellowish to orange lower breast and belly. Edges and underside of graduated tail white.

Iris—olive; bill—bluish black; feet—grey.

Voice: Male song is a five note cadence *kek tau-tau-tau-tau* (M. & W.). Repeated harsh *kek-kek*. (See Appendix 6.)

Range: S China, SE Asia, Borneo, Sumatra, and Java.

Distribution and status: In Borneo and Sumatra a not uncommon bird between 300 and 1500 m. In Java it is an uncommon bird of lowland and hill forests, up to about 1200 m.

Habits: A solitary but noisy and quite conspicuous bird of the lower levels of the forest. Hunts from a perch and not shy of humans.

376. RED-HEADED TROGON *Harpactes erythrocephalus* PLATE 43

Description: Large (30 cm) red-headed trogon. Red head of male diagnostic. Lacks neck collar and has narrow white crescent on red breast. Female differs from other female trogons by having red belly and white breast crescent, and differs from all male trogons in having cinnamon brown head.

Iris—brown; bare skin around eye—blue; bill—bluish; feet—pinkish.

Voice: Mellow, repeated *tiaup*, and also a rattled note *tewirrr* (M. & W.).

Range: Himalayas to S China, SE Asia, Malay Peninsula, and Sumatra.

Distribution and status: In Sumatra it is a rather rare bird of hill forests above 700 m.

Habits: Hunts from low perch in dense forest.

KINGFISHERS—FAMILY ALCEDINIDAE

KINGFISHERS are a group of brightly coloured birds (many with metallic blue feathers), with short legs and tail, big heads, and long powerful beaks. They eat insects, small vertebrates, and some species take fish. They nest in burrows in the ground, in tree trunks, river banks, or termite mounds. The eggs are whitish and almost spherical. The family has a worldwide distribution. Some kingfishers give rather loud, harsh calls. The three front toes are partly fused at the base. There are 15 species recorded in the Greater Sundas.

377. COMMON KINGFISHER *Alcedo atthis* PLATE 44
(River/European/The Kingfisher)

Description: Small (15 cm), electric blue and rufous kingfisher. Upperparts shimmering, pale greenish blue; underparts rufous orange with white chin. White spot on the side of the neck; orange stripe running through the eye and covering the ear coverts is diagnostic and distinguishes this species from the darker Blue-eared Kingfisher.
Iris—brown; bill—black; feet—red.
Voice: High-pitched squeak *peep-peep* when in flight, and a softer click when perched.
Range: Widespread in Eurasia, SE Asia, Indonesia to New Guinea.
Distribution and status: In Sumatra a common visitor up to 1500 m. In Borneo, Java, and Bali it occurs only as a rare winter visitor.
Habits: Frequents freshwater habitats and mangrove in open country. Perches on rocks or overhanging branches and plunges into the water to catch fish.

378. BLUE-EARED KINGFISHER *Alcedo meninting* PLATE 44
(Deep-Blue Kingfisher)

Description: Small (15 cm) bright blue-backed kingfisher. Back darker metallic blue than Common Kingfisher; underparts brighter orange-red; blue ear coverts distinctive.
Iris—brown; bill—blackish; feet—red.
Voice: A high-pitched *chiet*, usually uttered in flight; rapid chatter when perched.
Range: India to China and SE Asia, Philippines, Sulawesi, Sumatra, Borneo, Java, Bali, and Lombok.
Distribution and status: It is regularly seen by freshwater streams, rivers, and lakes and also sometimes over brackish water from sea-level to 1000 m. Prefers more wooded terrain than Common Kingfisher.
Habits: Flies very fast from perch to perch; makes curious head-bobbing movements, watching for food. Makes lightning dives into the water to catch prey which is then carried to a perch, killed, and eaten.

379. BLUE-BANDED KINGFISHER *Alcedo euryzona* PLATE 44

Description: Medium-sized (18 cm) dark blue and white kingfisher. Crown, sides of head, and wings dark bluish black; breast band, back, and tail light blue. Lores, ear coverts, throat, and belly whitish, washed rufous. Female has rufous orange belly and creamy throat.
Iris—brown; bill—black (with reddish lower mandible in female); feet—vermilion.
Voice: A squeak similar to Common Kingfisher.
Range: Malay Peninsula and Greater Sundas.
Distribution and status: Resident on mainland Sumatra (uncommon), Borneo (not uncommon), and Java (rare). A bird of small forested streams in lowlands and hills, up to 1500 m; probably predominantly sub-montane.
Habits: A shy forest bird generally found close to streams. Hunts from a low perch; habits similar to Blue-eared Kingfisher.

380. SMALL BLUE KINGFISHER *Alcedo coerulescens* PLATE 44

Description: Very small (14 cm) blue and white kingfisher. Upperparts and breast band shimmering, greenish blue; crown and wing coverts barred with bluish black; lores, throat, and belly white.
Iris—brown; bill—black; feet—red.
Voice: Rather high-pitched, two-note squeak *tew-tew*, uttered in flight.
Range: Sumatra, Java, Bali, Lombok, and Sumbawa.
Distribution and status: Resident S Sumatra (where probably a recent colonizer from Java), Java, and Bali. Common in coastal swamps, mangroves, and estuaries.
Habits: Perches on trees beside small streams or brackish fish ponds, and in mangroves.

381. BLACK-BACKED KINGFISHER *Ceyx erithacus* PLATE 44
(Oriental/Three-toed/Malay Kingfisher, AG: Pygmy-Kingfisher)

Description: Very small (14 cm) red and yellow kingfisher. Bright yellow underparts, and bluish-black back and wing coverts diagnostic. Blue lores and ear coverts.
Iris—brown; bill—red; feet—red.
Voice: During flight a high-pitched whistle *tsriet-siet* or *tsie-tsie*.
Range: Malay Peninsula, Philippines, Greater Sundas.
Distribution and status: An occasional bird of lowland forests, up to 1500 m.
Habits: Confined to forest, usually but not always close to rivers or swamps. Flies at great speed from one low perch to another, then hunts from the perch for insects or other prey. Spiders are taken from their webs on the wing.

382. RUFOUS-BACKED KINGFISHER *Ceyx rufidorsa* PLATE 44
(Red-backed Kingfisher, AG: Pygmy-Kingfisher)

Description: Very small (14 cm) reddish kingfisher with yellow underparts. Upperparts dark rufous with lilac tinge and lilac stripe down back and uppertail coverts. Distinguished from Black-backed Kingfisher by rufous rather than bluish black mantle and lack of blue spot on forehead and behind eye.
Iris—brown; bill—red; feet—red.
Voice: High-pitched squeaky whistle given in flight.
Range: Malay Peninsula, Sumatra, W islands of Sumatra, Borneo, Java, and Bali.
Distribution and status: An occasional bird of lowland primary and secondary forest, and mangroves.
Habits: Lives in lower storey of dark forest close to smallest streams. Flies at great speed while calling. A shy bird.
Note: Intermediates between Black-backed and Rufous-backed Kingfishers are common, so some authors follow Sims (1959) in putting them both in the same polymorphic species *C. erithacus*.

383. STORK-BILLED KINGFISHER *Pelargopsis capensis* PLATE 44

Description: Very large (35 cm) blue-backed kingfisher with diagnostic massive red bill; crown, sides of face, and nape grey-brown. Underparts pinkish orange.
Iris—brown; bill—red; feet—red.
Voice: Contact calls between members of a pair; alarm call when disturbed is a very loud, sharp, laughing cry *wiak-wiak*. Also gives a laughing *kak, kak, kak, kak, kak* . . . beginning very loud and becoming softer (D.A.H.).
Range: India, SE Asia, Philippines, Borneo, Sumatra, Java, and W Lesser Sundas.
Distribution and status: A common bird on the large rivers of Borneo and less common in Sumatra. Occasional bird along Java's coastline and on large, lowland rivers. It is now rare in E Java, and there are no recent records from Bali where it was formerly reported.
Habits: Lives in pairs but hunts alone. Frequents large rivers, mangroves, and coastline. Sits on dead branches overlooking water and makes spectacular dives to catch fish. Flies off with noisy alarm calls when disturbed.

384. BANDED KINGFISHER *Lacedo pulchella* PLATE 44

Description: Medium-sized (20 cm), conspicuously barred, forest kingfisher. Male: crown and upperparts blue with black and white bars; underparts whitish with reddish breast flanks. Nominate race has rufous forehead; cheeks, and malar stripe. Black in Bornean race (also on Bangka). Female: rufous

and black-barred upperparts; underparts white with black bars on breast and flanks.
Iris—purplish grey; bill—red; feet—pale green.
Voice: Drawn-out series of high-pitched *taweo* notes given at constant speed, beginning with one or two loud notes, becoming softer (D.A.H.). (See Appendix 6.)
Range: SE Asia, Borneo, Sumatra with eastern islands, and Java.
Distribution and status: In Sumatra it is locally common up to 1000 m. In Borneo it is an uncommon bird of headwater streams up to 1300 m. In Java a rare bird, and it is absent from Bali.
Habits: A rather active, forest-living kingfisher of hill forest and even submontane forest. It hunts from high and low perches.

385. RUDDY KINGFISHER *Halcyon coromanda* PLATE 45

Description: Medium-sized (25 cm) violet and rufous kingfisher. Upperparts bright violet rufous, except for contrasting pale blue rump; underparts rufous.
Iris—brown; bill—red-orange; feet—red-orange.
Voice: Rapid, mellow, dissyllabic or trisyllabic note, slowed down, most often heard just after dusk or before dawn (D.A.H.).
Range: Widespread from India to Japan, China, SE Asia, Philippines, Sulawesi, Malay Peninsula, Sumatra, Borneo, and Java.
Distribution and status: In the Sundas, the resident race is an uncommon coastal bird. It is commoner in NE Borneo than elsewhere on the island, and the only recent records from Java are from the west. Absent from Bali. A winter migrant race visits the E coast of Sumatra and occasionally N Borneo.
Habits: Inhabits coastal forest, swamp forest, and mangrove. Rarely found far from the sea where it catches most of its food.

386. WHITE-THROATED KINGFISHER *Halcyon smyrnensis*
(White-breasted/Smyrna Kingfisher) PLATE 45

Description: Largish (27 cm) blue and brown kingfisher with a white chin, throat, and breast. Head, neck, and rest of underparts brown; mantle, wings and tail bright iridescent blue; upper wing coverts and wing tips dark brown.
Iris—dark brown; bill—deep red; feet—red.
Voice: Loud screaming, tittering call *kee kee kee kee* given in flight or from perch, and harsh *chewer chewer chewer*.
Range: Middle East, India, China, SE Asia, Philippines, Andamans, Malay Peninsula, and Sumatra.
Distribution and status: In Sumatra a frequent bird of open land, near water, up

to 900 m. In W Java it has only been recorded a few times, including one nesting record. In Sumatra it is replacing the Collared Kingfisher as the common kingfisher of cultivated areas.

Habits: Active, noisy hunter of open fields, rivers, ponds, and coast.

387. JAVAN KINGFISHER *Halcyon cyanoventris* PLATE 45

Description: Medium-sized (25 cm) very dark kingfisher. Adult has dark brown head; brown throat and collar; purple blue belly and back; black wing coverts and bright blue flight feathers. White patch on wing visible in flight. Immature has whitish throat.
Iris—dark brown; bill—red; feet—red.
Voice: A clear ringing call *chee-ree-ree-ree-ree* or *cheree-cheree-cheree*.
Range: Endemic to Java and Bali.
Distribution and status: A widespread and not uncommon bird of open spaces, near clean water, up to 1000 m on Java and Bali. It has disappeared from some former haunts such as Bogor.
Habits: Sits on a low branch on an isolated tree or on a pole in open grassy areas, diving on insects and other prey. More rarely hunts over water. A quieter bird than the Collared Kingfisher but its striking call is often heard.

388. BLACK-CAPPED KINGFISHER *Halcyon pileata* PLATE 45

Description: Large (30 cm) blue, white, and black kingfisher. Black head diagnostic. Wing coverts black; rest of upperparts bright metallic blue. Flanks and vent washed rufous. White wing patch conspicuous in flight.
Iris—dark brown; bill—red; feet—red.
Voice: Shrill, loud call when alarmed.
Range: Breeds in China and Korea but migrates south in winter.
Distribution and status: A common visitor in N Borneo but rare in the south. It is uncommon in Sumatra, and in Java it is a very rare visitor of coastal regions, up to 500 m. Not recorded on Bali.
Habits: Prefers mangroves, estuaries, and the banks of large rivers. Perches over rivers on overhanging branches. Sometimes hunts over swampy grassland.

389. COLLARED KINGFISHER *Todirhamphus chloris* PLATE 45
(White-Collared/Mangrove Kingfisher)

Description: Medium-sized (24 cm) blue and white kingfisher. Crown, wings, back, and tail bright iridescent greenish blue; black stripe through eye; white lores; clean white collar and underparts distinguish it from the dirtier white of the Sacred Kingfisher.

Iris—brown; bill—dark grey above, paler below; feet—grey.

Voice: Harsh scream *chew-chew-chew-chew-chew* with downward inflection, or double notes *chek-chek*, *chek-chek*, *chek-chek*; a variety of other calls during breeding.

Range: S and SE Asia, Indonesia to New Guinea and Australia.

Distribution and status: The commonest kingfisher in Sumatra, Java, and Bali from sea level to 1200 m. In Borneo it is common around the coasts but rare inland.

Habits: This is a regular bird of open country, especially in coastal areas. It perches on a rock or tree and hunts along the beach or in any open space near water, including gardens, towns, and plantations. Large prey are bashed repeatedly on the perch prior to eating. A very noisy bird whose harsh call can be heard all day.

390. SACRED KINGFISHER *Todirhamphus sanctus* PLATE 45

Description: Medium-sized (22 cm) blue and white kingfisher; similar to Collared Kingfisher but slightly smaller, blue parts are more greenish, and breast is washed with buff or rufous instead of being clean white. Looks like a dirty version of Collared Kingfisher.

Iris—brown; bill—black; feet—light grey.

Voice: Similar to Collared Kingfisher; typical shrill, four-note phrases *kee-kee-kee-kee—kee-kee-kee-kee*.

Range: Resident in Australia but a regular visitor to New Guinea and Indonesia in the southern winter.

Distribution and status: An occasional visitor to Borneo and Sumatra, more common in the south. In Java and Bali most common near the sea, and more common in the east.

Habits: Sits on a pole, mangrove tree, or even down on the sand or mud-flats, hunting along the coasts, pouncing on insects and crustaceans on the ground. Tamer and less obtrusive than Collared Kingfisher.

391. RUFOUS-COLLARED KINGFISHER *Actenoides concretus*
(Chestnut-collared Kingfisher) PLATE 45

Description: Medium-size (23 cm), rufous and blue, forest kingfisher with diagnostic greenish crown. Male has neck collar and underparts rufous orange, black eye-stripe, and dark blue malar stripe and upperparts. Female has mantle and wings dark green, spotted with buff.

Iris—brown; bill—yellow with dark culmen; feet—pale yellow.

Voice: A loud, rising whistle *kweee, kweee* ..., of about 1 note per second, or single calls at a rate of about 1 per minute.
Range: Malay Peninsula, Sumatra, and Borneo.
Distribution and status: An uncommon bird of lowland and hill forests up to 1500 m, on Sumatra (including Mentawai, Bangka, and Belitung) and Borneo.
Habits: Lives in forest, generally far from water. Hunts from low perch, taking most of its food on the ground. Rather shy and unobtrusive.

Bee-eaters—FAMILY MEROPIDAE

BEE-EATERS are a small family found throughout the Old World. They are colourful birds with predominantly green plumage. They have short legs; graceful outlines with long, slender, curved bills; long pointed wings, and in many species, elongated streamer-like central tail feathers. Most species are gregarious and prefer open country. Parties of birds sit on bare perches and make sweeping flights after insects which they carry back to the perch. The prey is then hit against a hard surface to break it and soften it before eating. The three front toes are partly fused. Bee-eaters nest in burrows in the ground where they lay white eggs.

Five species occur in the Greater Sundas region, one of which is a northern winter visitor.

392. CHESTNUT-HEADED BEE-EATER *Merops leschenaulti*
(Bay-headed Bee-Eater)   PLATE 46

Description: Smallish (20 cm) green and brown bee-eater lacking elongated central tail streamers. Crown, nape, and mantle bright chestnut; wings and tail green; rump bright blue; throat yellow, bordered with chestnut; narrow black bar on upper breast; belly pale green; eye-stripe black. Orange underwing visible in flight.
Iris—reddish brown; bill—black; feet—dark brown.
Voice: A liquid-ringing trill given in flight, *kree-kree-weet-weet-weet* and variations.
Range: S and SE Asia, Malay Peninsula, Sumatra, Java, and Bali.
Distribution and status: In Sumatra, Java, and Bali it is widespread and locally common in open and wooded areas, up to 1200 m. It is absent from Borneo.
Habits: Typical of the group. Parties of bee-eaters seem to wander from place to place.

393. BLUE-TAILED BEE-EATER *Merops philippinus* PLATE 46

Description: Largish (30 cm, including elongated tail streamers), elegant bee-eater. Black stripe through eye is outlined with blue above and below. Head and mantle green, rump and tail blue. Chin yellow, throat chestnut, and breast and belly pale green. Underwing in flight orange.
Voice: Plaintive trill *kwink-kwink, kwink-kwink, kwink-kwink-kwink* given in flight.
Range: Breeds in S Asia, Philippines, Sulawesi, and New Guinea. Visits the Sunda Islands in winter.
Distribution and status: In Sumatra, Java, and Bali it is common in open habitats, up to 1200 m. Common in S Borneo, rarer in the north.
Habits: Gregarious parties gather in open areas to hunt. They settle on bare twigs and telegraph wires and make lazy, circular, swallow-like, gliding flights after insects. The bill snaps audibly as the bird catches its prey. More of an aerial feeder than other bee-eaters. Calling groups sometimes pass high overhead.

394. BLUE-THROATED BEE-EATER *Merops viridis* PLATE 46

Description: Medium-sized (28 cm, including elongated central tail streamers) bluish bee-eater. Adult: crown and mantle chocolate; eye-stripe black; wings bluish green; rump and tail with streamer pale blue; underparts pale green with diagnostic blue throat. Immature lacks elongated tail feathers, and head and mantle are green.
Iris—red or brown; bill—black; feet—grey or brown.
Voice: Fast trilling notes given in flight *kerik-kerik-kerik*.
Range: S China, SE Asia, Philippines, Malay Peninsula, Sumatra, Borneo, North Natunas, Karimata, and W Java. Not known from Bali.
Distribution and status: In Sumatra and Borneo a locally common resident but it is uncommon in Java. Regional migrations are suspected.
Habits: Favours open country and woodlands in low-lying areas, usually close to the sea. Birds congregate at breeding sites in sandy areas. Does less gliding and flying than the Blue-tailed Bee-Eater, preferring to hunt flying insects by waiting on a perch. Occasionally picks insects off the water surface or off the ground.

395. RAINBOW BEE-EATER *Merops ornatus* PLATE 46
(Rainbow-Bird)

Description: Medium-sized (25 cm, including elongated central tail streamers) greenish bee-eater with black line through the eye outlined by blue above and below. Similar to Blue-tailed Bee-Eater but distinguished by black throat bar, black tail, and orange underwing which is prominent in flight. Immature lacks black gorget and tail streamers.

Iris—red; bill—black; feet—grey.
Voice: Light, rolling chitter *pirr pirr pirr*, usually given in flight.
Range: Breeds in Australia; migrant to New Guinea and E Indonesia.
Distribution and status: A rare summer vagrant to Bali.
Habits: Typical of the group, hunting from tree perches in open grassy areas.

396. RED-BEARDED BEE-EATER *Nyctyornis amictus*   PLATE 46

Description: Medium-sized (30 cm) green forest bee-eater with unmistakable strawberry-pink, puffy breast. Adult has lilac crown and red breast, but immature birds are all green.
Iris—orange; bill—blackish; feet—dull green.
Voice: Attractive musical duet; normally gruff, descending *qua-qua-qua-qua* chuckle calls, and growling alarm *krerkrer*.
Range: Malay Peninsula, Sumatra, and Borneo.
Distribution and status: A regular bird of primary and logged forests, up to 1200 m.
Habits: Lives in upper and middle canopy of tall forest. Hunts quietly from high perches. Fans and flicks tail at intervals. More of a forest bird than other bee-eaters.

ROLLERS—FAMILY CORACIIDAE

ROLLERS are medium-sized, brightly coloured, long-winged birds found in Europe, Asia, Africa, and Australia. They have powerful, sharp beaks and feed mostly on large insects. Like the kingfishers and bee-eaters they have short legs, with the front three toes fused at the base, lay roundish white eggs in burrows and tree holes, and the young retain their feather sheaths till they are nearly fully grown.

Only one species, the dollarbird, occurs in the Greater Sundas and it is somewhat atypical of the family, having an unusually broad bill.

397. DOLLARBIRD *Eurystomus orientalis*   PLATE 46
(Broad-billed Roller)

Description: Medium-sized (30 cm) dark roller with broad red bill (black in immatures). Overall colour is dark bluish grey but throat bright blue. In flight there are contrasting, circular, light blue patches in the centre of each wing which gives the bird its name.
Iris—brown; bill—red with black tip; feet—orange-red.
Voice: Hoarse, rasping croaks *kreck-kreck*, in flight or from perch.

Range: Widely distributed from E Asia, SE Asia, Japan, Philippines, Indonesia to New Guinea and Australia.
Distribution and status: Resident and migrant races occur throughout the Greater Sundas. Widespread, but never very common, in semi-open country at the edge of forest, up to 1200 m.
Habits: Usually seen sitting in a dead tree in open country, making occasional flights after passing insects or diving onto insects on the ground. Has a curious, heavy, flapping flight similar to a nightjar. Two or three sometimes fly and dive together at dusk, especially during courtship. The Dollarbird is sometimes mobbed by small birds because its head and beak give it a predatory appearance.

Hoopoes—FAMILY UPUPIDAE

HOOPOES are a small family of only two species distributed through Africa, Europe, Madagascar, and Asia. Hoopoes are characterized by colourful plumage, an erectile crest, and long curved bills. One species is found in the Greater Sundas.

398. EURASIAN HOOPOE *Upupa epops*   PLATE 46
(Common/The Hoopoe)

Description: Unmistakable, medium-sized (30 cm), brightly coloured bird with long erectile crest of black-tipped, pinkish rufous plumes. Head, mantle, shoulders, and underparts pinkish rufous with wings and tail striped black and white. Long and decurved bill.
Iris—brown; bill—black; feet—black.
Voice: Low, soft *hoop—hoop hoop*, on a monotone accompanied by head bobbing display. (See Appendix 6.)
Range: Africa, Eurasia, Indochina.
Distribution and status: A rare vagrant to Sumatra and N Borneo.
Habits: An active bird of open, often moist ground where it feeds by probing the ground with its long bill. Erects crest when alarmed and on alighting.
Note: Some authors include the African Hoopoe *U. africana* in this species.

Hornbills—FAMILY BUCEROTIDAE

HORNBILLS are large, black or brown, and white, mainly arboreal birds, with long, heavy bills. Many species have large protuberant casques on top of the bill which may be gaudily coloured. Hornbills are found throughout Africa and tropical Asia, and throughout Indonesia to New Guinea. They eat fruit and insects and have harsh, penetrating calls.

The nesting habits of the family are interesting. The incubating females are usually sealed into tree hole nests with mud, leaving only a small aperture through which food can be passed by the male. When the young are hatched the female breaks out but reseals the nest entrance again until the young are ready to leave. Ten species of hornbill occur in Sumatra, eight in Borneo, but only three are found in Java.

399. BUSHY-CRESTED HORNBILL *Anorrhinus galeritus* PLATE 47

Description: Smallish (70 cm) black hornbill with floppy crest. Lack of white in plumage diagnostic. Tail is greyish brown with wide terminal black bar. Bare skin around eye and on throat blue.

Iris—red in male, black in female, blue in immature; bill—black in male, whitish in female; feet—black.

Voice: High-pitched strident, chattering yelps given by several birds together.
Range: Malay Peninsula, Sumatra, N Natuna, Borneo.
Distribution and status: Abundant hornbill of lowland and sub-montane forests up to Kelabit highlands in Borneo, and up to 1800 m on Mt. Kerinci in Sumatra.
Habits: Living in noisy flocks of 5–15, generally feeding in middle canopy of closed forest.

400. WHITE-CROWNED HORNBILL *Aceros comatus* PLATE 47
(White-crested/Long-crested Hornbill)

Description: Large (85 cm) long-tailed, black and white hornbill with distinctive fuzzy white crown. Both sexes have white crown, tail, and trailing edge of wing. Throat and neck are white in male, black in female.

Iris—yellow; bill—grey; feet—black.

Voice: Lively hollow, pigeon-like *kuk kuk*, *kuk kuk kuk*, or single soft *hao* call. First note is longest.
Range: NE India, coastal Burma, Malay Peninsula, Sumatra, and Borneo.
Distribution and status: A local bird of hill forests in the northern two thirds of Sumatra and northern half of Borneo.
Habits: A bird of the middle and lower forest storeys.

401. WRINKLED HORNBILL *Aceros corrugatus* PLATE 47

Description: Medium-sized (75 cm) black and white hornbill with short, red, buckled casque. Male is black with sides of head, neck, and terminal two thirds of tail white. Female like male but head and neck black, bare skin of throat bluish.

Iris—red; orbital skin—blue; bill—yellow and red; feet—horn.
Voice: Deep, echoing calls *rowwow* or *wakowwakowkow* given from treetops or in flight, and harsh *kak kak* contact calls.
Range: Malay Peninsula, Sumatra, Batu Islands, and Borneo.
Distribution and status: An occasional bird of lowland forests and swamp forests, up to 1000 m.
Habits: Lives singly or in groups, feeding largely on figs in upper canopy. Flies high over forest to evening roost trees. Rather shy.

402. WREATHED HORNBILL *Aceros undulatus* PLATE 47

Description: Large (100 cm) white-tailed hornbill. Both sexes have black back, wings, and belly, but male has creamy head with reddish plume from the nape, and naked yellow gular pouch with a distinct black stripe. Female has black head and neck, and blue gular pouch.
Iris—red; bill—yellow with small corrugated casque; feet—black.
Voice: A repeated, short, hoarse, dog-like double yelp *koe-guk*.
Range: E India, SW China, SE Asia, Malay Peninsula, Borneo, Sumatra, Java, and Bali.
Distribution and status: In Borneo and Sumatra this is a fairly common bird of lowland and hill forests, up to 2000 m. In Java and Bali it is rather local.
Habits: Flies in pairs or small flocks over the forest with heavy wing-beats, seeking fruiting trees. Often mixes at feeding trees with other hornbills.

403. PLAIN-POUCHED HORNBILL *Aceros subruficollis* PLATE 47

Description: Large (90 cm) white-tailed hornbill. Both sexes similar to Wreathed Hornbill but differ in smaller size, lack of dark bar on gular pouch and lack of corrugations on lower bill.
Iris—red; bill—creamy tinged brown at base, with flat corrugated casque; feet—blackish.
Voice: Harsh *keh-kek-kek*, higher pitched than Wreathed Hornbill.
Range: Malay Peninsula and Sumatra, but poorly known and could occur in Borneo.
Distribution and status: Status unclear due to confusion with Wreathed Hornbill but appears to be rare.
Habits: Typical habits of a lowland hornbill.
Note: Authors have placed this species in Blyth's Hornbill *A. plicatus* or treated it as the juvenile of Wreathed Hornbill. Now generally recognized as a distinct species.

HORNBILLS 225

404. ASIAN BLACK HORNBILL *Anthracoceros malayanus* PLATE 47
(Malaysian Black/Black Hornbill)

Description: Medium-sized (75 cm) black hornbill with white-tipped outer tail feathers and proportionally large casque. Male sometimes has white stripe over eye to nape.
Iris—reddish brown; bill and casque—white in male, blackish in female; feet—black.
Voice: Harsh creaking cries and repetitive piping brays, similar to Oriental Pied Hornbill.
Range: Malay Peninsula, Sumatra, Lingga, Bangka, Belitung, and Borneo.
Distribution and status: A rather uncommon hornbill of lowland primary and logged forests, and swamp forests, usually below 500 m.
Habits: Forages in upper and middle canopy of closed forests.

405. ORIENTAL PIED HORNBILL *Anthracoceros albirostris* PLATE 47
(Malaysian Pied Hornbill > Southern Pied Hornbill)

Description: Small (75 cm), black and white hornbill with a large yellow-white casque. Plumage black except for a white patch under the eye, white lower belly, thighs, and undertail coverts, white tips to flight feathers and white outer tail feathers.
Iris—dark brown; naked skin around eye and gular skin—white; bill and casque—yellow-white with black spots on base of lower mandible and front of casque; feet—black.
Voice: Incessant, strident cackle *ayak-yak-yak-yak-yak*.
Range: N India, S China, SE Asia, Malay Peninsula, and Greater Sundas.
Distribution and status: A conspicuous bird of lowland primary and secondary forests throughout the Greater Sundas.
Habits: Prefers more open habitat such as forest edge, clearings and secondary forest than other hornbills. Found in pairs or noisy parties, flapping or gliding from tree to tree. Eats insects more than fruit.
Note: Includes both Northern (*albirostris*) and Southern (*convexus*) Pied Hornbills, but not the Malabar Hornbill (*A. coronatus*). *A. coronatus* becomes the correct name if lumped together.

406. RHINOCEROS HORNBILL *Buceros rhinoceros* PLATE 47

Description: Very large (110 cm), black and white hornbill with large yellow and red bill and casque, and diagnostic white tail with a broad black band. Head, back, wings, and breast black; belly and thighs white.
Iris—white to blue in females, red in males; skin around eyes—dark grey; bill—

yellow with red base and surmounted by upturned spiralling casque; feet—greenish grey.

Voice: Loud, harsh roar *kronnk*, repeated by either sex and often given in chorus with one partner slightly later than the other. Sharper *gak* note given just before flight.

Range: SE Asia, Malay Peninsula, Sumatra, Borneo, and Java. Absent from Bali.

Distribution and status: This species is found in low densities in most large blocks of lowland and hill forest. It is very conspicuous because of its size, habits, and call but in fact is generally present at low density.

Habits: Pairs inhabit the crowns of the tallest trees. A regular visitor to fruiting giant strangling figs. Gives a dramatic whooshing sound of wing-beats in flight.

407. GREAT HORNBILL *Buceros bicornis*   PLATE 47
(Great Indian Hornbill)

Description: Very large (125 cm), black and cream hornbill with black subterminal bar on white tail and broad, yellow-stained white bar on black wing. Bill and casque yellow; casque flattened and concave on top. Face black; white plumage of neck and breast often stained yellow. A rare form with black neck and enlarged casque, of unknown provenance, (shown below) turns up occasionally in captive collections.
Iris—red in male, whitish in female; bill—yellow; feet—black.

Voice: Loud, barked *gok* or *wer gok*, harsher than those of Rhinoceros Hornbill.

Rare form of Great Hornbill, *Buceros Bicornis* (407), with black hind neck and enlarged casque.

Range: India, SE Asia, Malay Peninsula, and Sumatra.
Distribution and status: In Sumatra this is an uncommon bird of lowland and sub-montane forests.
Habits: Generally in pairs. Flies noisily over forest. Feeds and roosts in upper canopy of primary forests, logged forests, and swamp forests.

408. HELMETED HORNBILL *Buceros vigil* PLATE 47

Description: Unmistakable, very large (120 cm, plus 50 cm elongated central tail streamers), dark brown and white hornbill with elongated central tail feathers;

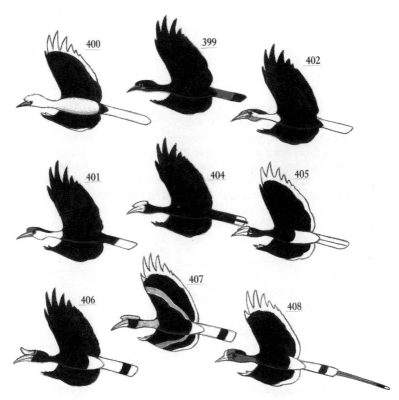

Male hornbills in flight. 399, Bushy-crested (*Anorrhinus galeritus*); 400, White-crowned (*Aceros comatus*); 401, Wrinkled (*A. corrugatus*); 402, Wreathed (*A. undulutus*); 404, Asian Black (*Anthracoceros malayanus*); 405, Oriental Pied (*A. albirostris*); 406, Rhinoceros (*Buceros rhinoceros*); 407, Great (*B. bicornis*); 408, Helmeted (*B. vigil*).

white tail with black band and broad white bar at rear of wing diagnostic. High, yellow and scarlet, box-like casque which is used for carving as 'hornbill ivory'. Neck: red bare skin in male, pale blue in female.
Iris—red; bill—yellow and red; feet—brown.
Voice: A series of identical, loud, hollow *took* notes, gaining in speed before drawing to an amazing climax of maniacal laughter, *tee poop* notes.
Range: Malay Peninsula, Sumatra, and Borneo.
Distribution and status: This is a familiar but uncommon bird of tall, lowland forests, up to 1500 m.
Habits: Pairs live in upper canopy, mixing with other hornbills, pigeons, and primates in major fruiting trees.

Barbets—FAMILY CAPITONIDAE

BARBETS are small, colourful birds with large, powerful bills. They are closely related to woodpeckers and share their habits of excavating tree holes for nesting. They also have the same unusual arrangement of toes, with two pointing forward and two backwards, for clinging to vertical tree trunks. However, their diet is comprised of fruit, seeds, and flowers, and they are particularly fond of small figs. Almost all species have the habit of sitting motionless in a treetop for long periods, emitting monotonous, loud, repetitive calls. Since the general colour of most species is bright green, they are very inconspicuous but species may be identified by call.

Sixteen species occur in the Greater Sundas.

409. FIRE-TUFTED BARBET *Psilopogon pyrolophus* PLATE 48

Description: Medium-sized (26 cm) green barbet with creamy bill and yellow band on breast outlined black below. Tuft of amazingly bright orange bristles above bill. Head gaudily marked black, grey, green, and mauve. Immature is duller with an olive crown.

Iris—brown; bill—creamy green with black central bar; feet—yellowish green.
Voice: Harsh, cicada-like, buzzing call rising in pitch and speeding up before ending abruptly (see Appendix 6).
Range: Malay Peninsula and Sumatra.
Distribution and status: A common forest resident, between 500 and 1500 m.
Habits: Forages among crowns of canopy trees, preferring tall forests.

410. LINEATED BARBET *Megalaima lineata* PLATE 48

Description: Largish (29 cm), pale-headed, streaky barbet. Plumage generally green with a pale yellowish brown head and neck, and diagnostic white streaking on head and underparts.
Iris—buff; bill—pale pink; feet—yellow.
Voice: A low-pitched, irregular ringing *bul-tok* .. *bul-tok*, at about 1 second intervals; also a loud *kuerr-kuerr*, and a rare 'counting call' consisting of a long trill followed by a series of four-syllable trilled notes (D.A.H.). (See Appendix 6.)
Range: W Himalayas, S E Asia, Java, and Bali.
Distribution and status: A fairly common bird of open forest, forest edge, and secondary forests in the remoter areas.
Habits: Similar to other barbets but most common in relatively dry, open, wooded habitats and coastal forest.

411. BROWN-THROATED BARBET *Megalaima corvina* PLATE 49
(Javan Brown-throated Barbet)

Description: Largish (26 cm) dull-coloured barbet. Body uniformly green with dark brown head, neck, and upper breast, with golden flecks.
Iris—brown; bill—black; feet—grey.
Voice: A ringing call *hoo-too too-too too* or *boo*, *toot-boo*, *toot-bootootoot*. Also a rapid rattling alarm note. (See Appendix 6.)
Range: Endemic to Java. Absent from Bali.
Distribution and status: Confined to moist montane forests of W Java. Can be quite common within its limited habitat.
Habits: Usually seen fairly high in the canopy, singing or looking for food. Usually solitary but sometimes found in small parties or mixing with other species.

412. GOLD-WHISKERED BARBET *Megalaima chrysopogon* PLATE 48

Description: Large (30 cm) green barbet with red forehead and hind-crown, dirty yellow-brown fore-crown and golden yellow cheek patch. Bornean birds have black cheeks and eyebrow. Sumatran birds have dirty brown cheeks and eyebrow. Chin is dirty grey, edged blue and violet. Immature is duller.
Iris—brown; bill—black; feet—grey.
Voice: Loud, mellow *kootuk-ootuk-ootuk* about 70 notes per minute, and low-pitched trill of a single note, repeated for 20 to 40 seconds, ever more slowly then speeding at end. (See Appendix 6.)
Range: Malay Peninsula, Sumatra, and Borneo.
Distribution and status: This is a common bird of tall forests up to 1500 m, but more often heard than seen.

230   BARBETS

Habits: Typical barbet of upper canopy in hill forests of Borneo and lowlands of Sumatra.

413. RED-CROWNED BARBET   *Megalaima rafflesii*   PLATE 48
(Many-coloured Barbet)

Description: Medium-sized (25 cm) green barbet with colourful blue, red, black, and yellow-marked head with entire crown red, blue throat, and yellow cheek patch diagnostic. Immature is duller.
Iris—brown; bill—black; feet—grey.
Voice: Two deep-toned *tuk* notes followed by a pause, then about 10 to 20 fast *tuk* notes at about 3 per second. (See Appendix 6.)
Range: Malay Peninsula, Sumatra, and Borneo.
Distribution and status: A common bird of lowland forests below 800 m.
Habits: Typical barbet of upper canopy.

414. RED-THROATED BARBET   *Megalaima mystacophanos*   PLATE 48
(Gaudy Barbet)

Description: Medium-sized (23 cm) green barbet with gaudily marked red, yellow, blue, and black head. Male differs from other barbets by yellow forehead and red throat; female by red lores and hind-crown, with no black on head. Immature is like female but duller.
Iris—brown; bill—black; feet—grey.
Voice: Irregular series of *tok* notes uttered in phrases of 1–4 notes, at rate of 1 per second; also a high-pitched trill, shortening on repetition. (See Appendix 6.)
Range: Malay Peninsula, Tana Islands, Sumatra, and Borneo.
Distribution and status: A common bird of lowland forests below 800 m, rarer in swamp and peat forests.
Habits: Typical barbet of upper and middle canopy, preferring tall primary and logged forests.

415. BLACK-BANDED BARBET.   *Megalaima javensis*   PLATE 49
(Javan Barbet)

Description: Largish (26 cm) colourful barbet. Adult plumage generally uniformly green with yellow crown and yellow spot under eye, red throat and red patch on side of breast, and broad black collar across upper breast and side of head to eye. A second black stripe runs through the eye.
Iris—brown; bill—black; feet—dull olive.
Voice: A repeated ringing call *tooloong-tumpook*. (See Appendix 6.)
Range: Endemic to Java and Bali.

Distribution and status: Not uncommon in lowland and hill forest to about 1500 m, throughout mainland Java and Bali.
Habits: Similar to other barbets. Generally found in rather light, open parts of the forest.

416. BLACK-BROWED BARBET *Megalaima oorti* PLATE 48
(Mueller's Barbet)

Description: Smallish (20 cm) green barbet with gaudy blue, red, yellow, and black head. Differs from Gold-whiskered Barbet by smaller size, black eyebrow, blue cheeks, yellow throat, and red spot above shoulder. Immature is duller.
Iris—brown; bill—black; feet—greyish green.
Voice: Hollow *tok-tr-trrrrrt* call, about 20 times per minute, with emphasis on third syllable. (See Appendix 6.)
Range: S China, SE Asia, Malay Peninsula, and Sumatra.
Distribution and status: In Sumatra it is a common bird in montane and submontane forests of the Barisan chain, between 1000 and 2000 m.
Habits: Typical barbet of upper and middle canopy.

417. MOUNTAIN BARBET *Megalaima monticola* PLATE 49

Description: Smallish (20 cm) green barbet with blue, yellow, and red head markings. Forehead green; crown streaked brown and pale yellow; nape blue with a red patch; ear coverts bluish green; throat yellowish buff, mottled with blue feathers with grey edges; two red spots on sides of breast. Immature is duller.
Iris—brown; bill—black; feet—grey.
Voice: Hollow, fast, repeated *took-took-took-took-* (*hiccup*) *took-took-took-took-* (*hiccup*). (See Appendix 6.)
Range: Endemic to Borneo.
Distribution and status: Confined to the mountains of N Borneo; recorded from most peaks, from Kinabalu south to upper Kapuas and upper Kayan. A frequent forest bird between 600 and 1200 m.
Habits: As other barbets. Favours forest and village fruit groves.

418. YELLOW-CROWNED BARBET *Megalaima henricii* PLATE 48

Description: Medium-sized (21 cm) green barbet with distinctive yellow forehead and eyebrow, and blue throat. Hindcrown blue; two red spots on nape and sides of neck. Immature is duller.
Iris—brown; bill—black; feet—grey.
Voice: Hollow, rattled trill *trrrt—tok-tok-tok-tok*, about 30 phrases per minute with either 4 or 6 *tok* notes. (See Appendix 6.)

Range: Malay Peninsula, Sumatra, and Borneo.
Distribution and status: An uncommon bird of lowland forests below 700 m.
Habits: Typical barbet of upper canopy in primary and swamp forests.

419. ORANGE-FRONTED BARBET *Megalaima armillaris* PLATE 49
(Blue-crowned/Flame-fronted Barbet)

Description: Medium-sized (20 cm) barbet. Plumage entirely green except for orange-yellow bar across the breast, orange-yellow forehead, and blue on hindcrown. Sometimes the breast bar is reduced to two spots.
Iris—brown; bill—black; feet—blue.
Voice: Monotonous, repeated *trrk trrk trrk trrrrk*, and variations. (See Appendix 6.)
Range: Endemic to Java and Bali.
Distribution and status: A not uncommon resident of primary forest and the forest edge on mainland Java and Bali; occurs from sea level up to 2500 m, but more common above 900 m.
Habits: Mixes with other birds at fruit trees. This is probably Java's commonest barbet.

420. GOLDEN-NAPED BARBET *Megalaima pulcherrima* PLATE 49
(Kinabalu Barbet)

Description: Smallish (20 cm) green barbet with bluish head. Forehead, crown, upper nape, and throat blue; sides of head yellowish green, and orange-yellow nape collar. Immature birds are duller.
Iris—brown; bill—black; feet—grey.
Voice: Repeated hollow call *took-took-tarrrook*; also a rolling trill *rrrrrr, rrrrr, rrrr, rrrr* . . ., gradually decreasing in note length with each repetition. (See Appendix 6.)
Range: Endemic to Borneo.
Distribution and status: Confined to the mountains of N Borneo; from Mts Kinabalu and Trus Madi, south to Murud and Mulu. A rather local bird of forests between 1500 and 3100 m, occasionally down to 1000 m.
Habits: As other barbets.

421. BLUE-EARED BARBET *Megalaima australis* PLATE 49

Description: Small (18 cm) barbet with blue crown and chin and black malar stripe and throat bar. Javan race *australis* has yellow cheeks and upper breast; Borneo and Sumatra form *duvauceli* lacks yellow but has red cheeks and side of head.
Iris—brown; bill—black; feet—greenish grey.

Voice: Fast, repeated, endless rattle *tu-trruk*, about 100 times a minute with continuous head turning; or shrill, trilled notes, repeated more slowly (like a pea whistle) with the head held very still. (See Appendix 6.)
Range: E India to SW China, Borneo, Sumatra, Java, and Bali.
Distribution and status: A common bird of primary forests, plantations, and secondary forest, from sea level to about 2000 m.
Habits: A familiar call in wooded countryside though the bird itself is not so easily seen. Singly or in pairs, quietly joins doves and other birds feeding in fig trees. Nests in a tiny tree hole, often under a branch.

422. BORNEAN BARBET *Megalaima eximia* PLATE 48
(Black-throated/Crimson-crowned Barbet)

Description: Small (15 cm) green barbet with forehead and throat black or dark blue. Blue eyebrow and two small blue patches on the ear coverts. Cheek patch yellow and upper breast, crown, and two patches below cheeks crimson. Specimens ascribed to a Kinabalu race *cyanea* have a blue throat, but typical birds can also be found on Kinabalu.
Iris—brown; bill—black; feet—olive green.
Voice: Hollow *took took-took* call repeated; also a trill. (See Appendix 6.)
Range: Endemic to Borneo.
Distribution and status: Found in the mountains of N Borneo; from Kinabalu south to the upper Kapuas. An uncommon bird, between 400 and 1200 m.
Habits: As other barbets but prefers montane forests.

423. COPPERSMITH BARBET *Megalaima haemacephala* PLATE 49
(Crimson-breasted Barbet)

Description: Small (15 cm) red-crowned barbet. Races vary. Sumatran adult *delica* has crown and breast red; throat, cheek, and eyebrow yellow; black stripe across crown separates yellow and red face from bluish-green nape. Javan and Bali race *rosea* has crown, eyebrows, cheek, throat, and upper breast crimson. Back, wings, and tail bluish green; underparts dirty white with heavy black streaks. Immature lacks red and black on head but has yellow patch under eye and under chin.
Iris—brown; bill—black; feet—red.
Voice: A resonant, monotonous, metallic *took-took-took*, continued for several minutes at a steady rate of about 110 notes per minute. Tail flicks forwards with each *took*. Another slower, less regular call when head bobs but tail is held still. (See Appendix 6.)
Range: W Pakistan to SW China, Philippines, Sumatra, Java and Bali.

Distribution and status: A widespread resident of open lowland forests, up to 1000 m.
Habits: Similar to other barbets but prefers open habitat such as woods, gardens, city parks, and plantations. In the early morning several birds may congregate to call from the top of a bare branch.

424. BROWN BARBET *Calorhamphus fuliginosus*   PLATE 49

Description: Small (17 cm) barbet with dark brown upperparts and rusty buff underparts with distinctive orange-red legs. Bornean races have orange throat. Bill is very large.
Iris—red; bill—dark brown in male, paler in female; feet—orange-red.
Voice: Thin, wheezy, squeaky whistle, and a 6–note phrase unlike other barbets.
Range: Malay Peninsula, Sumatra, and Borneo.
Distribution and status: This is a common bird of tall lowland forests and swamp forests, up to 1000 m.
Habits: An aberrant barbet in appearance and behaviour. Lives in groups working actively through the crown of figs and other fruiting trees.

HONEYGUIDES—FAMILY INDICATORIDAE

THIS small family is largely African in distribution with only two rare species in Asia, one living in the Greater Sundas.

Honeyguides have two toes pointing backwards like barbets but they resemble finches in colour and shape, and have short, strong bills and no bristles. They nest in tree holes and feed largely on bees and wasps. Their name is derived from some African species that have learned to lead humans and ratels to wild beehives to encourage them to open the hive.

425. MALAYSIAN HONEYGUIDE *Indicator archipelagicus*   PLATE 49
(Malayan Honeyguide)

Description: Smallish (16 cm), dull, brownish grey finch-like bird with a small bright yellow shoulder patch. Underparts whitish with grey wash on breast. Female lacks the yellow shoulder patch. Immature has streaked underparts.
Iris—brown; bill—black; feet—black.
Voice: Male gives harsh, cat-like cry followed by churring *miaow-krrruuu* or *miaw-miaw-krruuu*. The second syllable is a revving rattle that rises in pitch like a toy airplane. (See Appendix 6.) Alarm is agitated trill.
Range: Malay Peninsula, Sumatra, and Borneo.

Distribution and status: This is a rare bird of lowland forests, up to 1000 m.
Habits: Little known.

WOODPECKERS—FAMILY PICIDAE

WOODPECKERS are a large, well-known family of medium-sized birds with long, powerful beaks for boring into wood. They are found worldwide except in Australia. All species work over the trunks and branches of trees, drilling or poking under bark for insects and grubs which are flicked up by the long, protrusive, sticky tongue. The feet are adapted for clinging to trees with only two forward toes and one or two toes pointing backwards. The stiffened tail feathers are used as a stabilizing prop when drilling. Woodpeckers excavate tree holes for nesting, fly with an erratic dipping flight, and have harsh, discordant calls.

Twenty three species occur in the Greater Sundas, of which only six reach Bali. They range in size from the tiny (10 cm), Rufous Piculet to the large (50 cm), Great Slaty Woodpecker.

426. SPECKLED PICULET *Picumnus innominatus*   PLATE 50
Description: Tiny (10 cm), olive-backed, short-tailed, tit-like woodpecker with diagnostic boldly spotted underparts and black and white striped face and tail. Forehead orange in male.
Iris—red; bill—black above; yellow below; feet—grey.
Voice: Sharp noted *tsit* repeated; or alarm rattle. (See Appendix 6.)
Range: SE Asia, Borneo, and Sumatra.
Distribution and status: In Borneo this is a poorly known bird recorded from only a few localities but not uncommon in lowlands of NE. In Sumatra this is an uncommon resident between 100 and 1800 m, and maybe occasionally down to sea level.
Habits: Found in mixed lower montane forest, on dead trees or branches, particularly in areas of bamboo. Makes a slight, persistent tapping noise when looking for food.

427. RUFOUS PICULET *Sasia abnormis*   PLATE 50
Description: Tiny (10 cm), green and orange, short-tailed, tit-like woodpecker. Forehead golden in male, colour as breast in female; upperparts olive-green; underparts dark reddish yellow with orange breast; has only three toes.
Iris—red; bill—black above; yellow below; feet—yellow.
Voice: Single, sharp noted *tsit*, repeated several times; or in alarm a rapid, insistent *kih-kih-kih-kih-kih*.

Range: SE Asia, Borneo, Sumatra, Belitung, Nias, and Java.
Distribution and status: In Borneo and Sumatra this is a fairly common bird of lowlands and hills, up to 800 m. In Java it is confined to west and central lowlands.
Habits: Found in lower and middle storeys of secondary forests, and to a lesser extent primary forest, on dead trees or branches, particularly in areas of bamboo. Makes a slight, persistent tapping noise when looking for food on tree trunks and fine stems.

428. RUFOUS WOODPECKER *Celeus brachyurus* PLATE 50

Description: Medium-sized (21 cm), dark reddish brown woodpecker. Entire body reddish brown with black barring on wings and upperparts and to a lesser extent on underparts. Male has red cheek patch.
Iris—red; bill—black; feet—brown.
Voice: Short, hurried, high-pitched laugh *kwee-kwee-kwee- kwee* ... of 5–10 notes on a descending scale. Drums in short, decelerating bursts. (See Appendix 6.)
Range: S Asia, SE Asia, Borneo, Sumatra, Bangka, Belitung, Nias, and Java.
Distribution and status: In Borneo and Sumatra this is a common bird up to 1500 m. In Java it is confined to lowland forest of W and C Java where it is less common.
Habits: Prefers rather open forest, secondary forest, forest edge, gardens, and plantations at low altitudes. Pecks rarely audible.

429. LACED WOODPECKER *Picus vittatus* PLATE 51
(Laced Green Woodpecker)

Description: Medium-sized (30 cm) green woodpecker. Crown red in male, black in female; back green; rump yellow; tail black; primaries black with white stripes; throat buff; breast buff with bold green lacing of dark feather edges; black eyestripe and malar stripe flecked with white; cheeks bluish.
Iris—red; bill—black; feet—greenish.
Voice: Plaintive, shrill call *kweep* with falling tone.
Range: Bangladesh, SE Asia, Sumatra, Lingga Archipelago, Java and Bali.
Distribution and status: Locally common in suitable habitat up to 200 m. Sumatran records all from the east where it now seems rare.
Habits: Inhabits rather open, coastal forest, including mangroves and plantations. Feeds on the ground as well as on fallen trees or among bamboo and coconuts.

430. GREY-HEADED WOODPECKER *Picus canus*   PLATE 51
(Grey-faced/Black-naped Green Woodpecker)

Description: Largish (32 cm), reddish woodpecker. Crown scarlet in male, black in female. Black nape, tail, and malar stripe; distinctive grey cheeks and throat; upperparts reddish; underparts plain buffy rufous; rump orange. Primaries are black finely barred with white. This is an aberrant red form of the green races found on the Asian mainland.
Iris—brown; bill—dark slaty green; feet—grey.
Voice: In Malaya, a clear, musical whistle *pew, pew, pew* . . . of 4–8 descending notes. Sumatran race may differ. Drums frequently.
Range: Europe, India, S China, SE Asia, Malay Peninsula, and Sumatra.
Distribution and status: In Sumatra this is a rare bird of upper montane forests, from 1000 to 2000 m.
Habits: Typical woodpecker behaviour. Sometimes comes to ground to find ants.

431. GREATER YELLOWNAPE *Picus flavinucha*   PLATE 51
(Greater Yellow-naped Woodpecker)

Description: Large (34 cm) green woodpecker with yellow throat and long yellow crest. Tail black; flight feathers barred black and brown; rest of plumage green. Female has rufous-brown throat. Distinguished from Lesser Yellownape by lack of red on head.
Iris—reddish; bill—greenish grey; feet—greenish grey.
Voice: Slow *chup* or *chup-chup*, followed by staccato roll (Smythies).
Range: Himalayas, S China, SE Asia, Malay Peninsula, and Sumatra.
Distribution and status: In Sumatra it is quite common in mixed lower montane forests, pine forests, and secondary growth, between 800 and 2000 m.
Habits: A flamboyant and noisy woodpecker often seen in small family groups.

432. CRIMSON-WINGED WOODPECKER *Picus puniceus*   PLATE 51

Description: Medium-sized (25 cm) green and red woodpecker. Adult: long red crest with yellow tip; green body with red wings and black tail; yellow throat, and in male a red malar stripe; creamy barring on flanks and belly, heavier in female. Inner web of primaries is spotted white. Immature lacks red on crown.
Iris—reddish brown; bill—black above, yellow below; feet—green.
Voice: A plaintive call note *tiuik*, unlike other woodpeckers. Five to 7 *kwee* notes in a descending scale. (See Appendix 6.)
Range: Malay Peninsula, Borneo, Sumatra, Bangka, Nias, and Java. Absent from Bali.
Distribution and status: A relatively common bird up to 900 m.

238 WOODPECKERS

Habits: Occurs in the canopy of primary and secondary forest, gardens, and coastal scrub. Taps in short bursts.

433. LESSER YELLOWNAPE *Picus chlorolophus* PLATE 51
(Lesser Yellow-naped Woodpecker)

Description: Medium-sized (26 cm), bright green woodpecker with fluffy yellow edged crest and red markings on face. Male has distinctive red eyebrow and malar stripe with white upper edge. Female has red only on side of crown. Flanks barred with white; flight feathers black. Distinguished from Greater Yellownape by red on head and white edge to malar stripe. Differs from Checker-throated and Crimson-winged Woodpeckers by lack of red wings and by white edge to malar stripe.
Iris—red; bill—grey; feet—greenish grey.
Voice: Loud, shrill, descending alarm *kwee-kwee-kwee*, or single *pee-ui* call. (See Appendix 6.)
Range: Himalayas, S China, SE Asia, Malay Peninsula, and Sumatra.
Distribution and status: In Sumatra it is not uncommon in lower montane forests, between 800 and 1400 m, throughout the Barisan range.
Habits: A noisy and conspicuous woodpecker, sometimes travelling in small groups or following mixed-species bird flocks.

434. CHECKER-THROATED WOODPECKER *Picus mentalis*
(Chequer-throated Woodpecker) PLATE 51

Description: Medium-sized (27 cm) green woodpecker with red wings and black tail. Long crest with yellow tip but lacks red. Orange collar, and black and white checkered throat diagnostic. Malar area checkered in male, chestnut in female.
Iris—red; bill—black above, grey below; feet—green.
Voice: A quiet, upward-inflected *kwee* or *chok*, unlike other woodpecker calls.
Range: Malay Peninsula, Borneo, Sumatra, Bangka and Western Java.
Distribution and status: In Sumatra and Borneo it is fairly common in lowland and montane forests, up to 1600 m. In W and C Java it is a relatively uncommon bird.
Habits: Prefers the middle strata of forest but sometimes also seen at the edge of the forest or coming into cultivated lands.

435. BANDED WOODPECKER *Picus miniaceus* PLATE 51

Description: Medium-sized (23 cm) long-crested woodpecker. Adult: red crest with yellow tip; red wings; upperparts banded green and yellow with yellow rump;

underparts buff, banded brown and washed reddish on breast; tail black. Cheeks of male red and of female, spotted white. Immature lacks red on crest and upper body but has red flecking on crown.

Iris—reddish brown; bill—black above, grey below; feet—greenish grey.

Voice: Plaintive mewing *chewerk-chewerk-chewerk* on rising scale, and harsh screaming *kwee-kwee-kwee* on same note. (See Appendix 6.)

Range: Malay Peninsula, Borneo, Sumatra, Bangka, Belitung, Nias, and Java. Absent from Bali.

Distribution and status: Common in Borneo and Sumatra, in lowland and montane forests to 1400 m, but rare in Java.

Habits: Prefers secondary forest, forest edge, and open woods, including gardens. Forages among vines and rotten logs of lower storeys, singly or in small parties.

436. COMMON GOLDENBACK *Dinopium javanense* PLATE 50
(Common Golden-backed (Three-toed) Woodpecker/Common Flameback)

Description: Medium-sized (30 cm) colourful woodpecker. Face striped black and white. Male has red crown and crest; crown of female black streaked with white; back and rump red; mantle and wing coverts golden; breast scaly appearance, white feathers with black edges. Differs from Greater Goldenback by having only one broad black malar stripe and only one hind toe.

Iris—red; bill—black; feet—black with only three toes.

Voice: A harsh prolonged trill between partners *churrrr*, and a small *chee chee* call or a hard *kluuk-kluuk-kluuk* in flight.

Range: India, SE Asia, Philippines, Borneo, Sumatra, Riau Archipelago, Java, and Bali.

Distribution and status: This is a fairly common woodpecker in rather open lowland forest and cultivated areas, up to 1000 m.

Habits: Lives in pairs which call to each other regularly; prefers secondary and open forests, mangroves, plantations, and gardens.

437. OLIVE-BACKED WOODPECKER *Dinopium rafflesii* PLATE 50

Description: Medium-sized (25 cm), black-tailed, green woodpecker with long crest, and black and white striped face. Male's crest is red, female's is black. Flight feathers are black and flanks have conspicuous white spots. Distinguished from goldenbacks by olive-green back and lack of red rump. Differs from all green-backed woodpeckers by black and white striped face.

Iris—red; bill—black; feet—grey.

Voice: Rapid, loud, descending, staccato laugh *chakchakchak-chak*, sometimes ending with one or two spaced notes; also single *chak* note.

Range: Malay Peninsula, Sumatra, and Borneo.

Distribution and status: Uncommon resident on Sumatra (including Bangka) and Borneo. Found in mangroves, lowland and hill forests to 1100 m, locally to 1600 m on Borneo.
Habits: Typical woodpecker, preferring low and medium storeys of wet, swampy forests, including mangroves. Generally not in secondary forest. Forages mostly on rotten wood.

438. BUFF-RUMPED WOODPECKER *Meiglyptes tristis* PLATE 50
(Fulvous-rumped Woodpecker)

Description: Small (15 cm), heavily barred, black and white woodpecker. Entire body barred black with white, except buffy white rump. Head washed fulvous and malar area reddish in male. Crown feathers sometimes raised to form a crest. Iris—red brown; bill—black; feet—greenish.
Voice: A chittering call *chit-chit-chit-tee*.
Range: Malay Peninsula, Borneo, Sumatra, and Java.
Distribution and status: Common in parts of its range in Borneo and Sumatra, but rare in W Java.
Habits: Prefers rather open coastal habitats. Forages quietly in canopy and small branches of primary and secondary forest, and forest edge. Joins mixed-species flocks.

439. BUFF-NECKED WOODPECKER *Meiglyptes tukki* PLATE 50

Description: Smallish (21 cm) dark brown woodpecker with distinctive broad, buff neck patch and buffy barring on back. Adult male has red malar stripe and blackish throat bar. Immature like adult but with bolder buff barring. Iris—crimson; bill—blackish; feet—greyish green.
Voice: Rolling *kirr-r-r*, and harsh, high-pitched churring. Also loud drumming. (See Appendix 6.)
Range: Malay Peninsula, Sumatra with most offshore islands, N Natuna, Bangka, and Borneo.
Distribution and status: This is a common bird of lowland primary and secondary forests, below 1000 m.
Habits: Prefers middle and lower forest strata. Sometimes joins mixed-species flocks.

440. GREAT SLATY WOODPECKER *Mulleripicus pulverulentus* PLATE 51

Description: Unmistakable, very large (50 cm), lanky, grey woodpecker. Entire plumage grey with buff throat; male has red cheek patch and reddish wash on throat and neck.

Iris—dark brown; bill—dirty white with grey base and tip; feet dark grey.
Voice: A loud braying or whinnying cackle, typically *woik woik*. (See Appendix 6.)
Range: Malay Peninsula, Borneo, Natunas, Sumatra, Riau, and Java, in lowland forest below 1000 m.
Distribution and status: In Sumatra and Java it is now a rare bird, but in Borneo it is not so rare and is also found in uplands.
Habits: A noisy and conspicuous bird where present. Prefers semi-open, low-lying habitats. Sometimes travels in noisy family parties. Birds forage in emergent trees and sometimes drum loudly.

441. WHITE-BELLIED WOODPECKER *Dryocopus javensis* PLATE 50
(Great Black Woodpecker)

Description: Unmistakable, large (42 cm), black and white woodpecker. Upperparts and breast black, belly white. Male has red crest and cheek patches; female is all black with white belly.
Iris—yellow; bill—black; feet—grey-blue.
Voice: A loud, sharp, rising yelp *kiyow*; also a loud laugh in flight *kiau kiau kiau* . . . (M. & W.). Hammers loudly. (See Appendix 6.)
Range: India, China, SE Asia, Philippines, Borneo, N Natunas, Sumatra, Java, and Bali.
Distribution and status: In Borneo and Sumatra it is uncommon up to 1000 m. In Java and Bali it is rare.
Habits: Prefers open lowland forest, including mangroves. Where present it is a noisy and conspicuous bird, usually solitary. It forages at all levels.

442. FULVOUS-BREASTED WOODPECKER *Dendrocopus macei*
PLATE 50
(Streak-bellied Woodpecker)

Description: Rather small (18 cm), black and white barred woodpecker. Crown red in male, black in female; side of face white with black malar stripe and collar; upperparts barred black and white; underparts buff with black streaks, red undertail coverts.
Iris—brown; bill—bluish black above, bluish grey below; feet—olive.
Voice: Ringing call *tuk-tuk*, and trill *tirri-tierrier-tierrierie*.
Range: Himalayas, India, SE Asia (except Malay Peninsula), Sumatra, Java, and Bali.
Distribution and status: Status on Sumatra uncertain; one specimen and recent sight records from S Sumatra. On Java and Bali a common bird with a wide altitude range, up to 2000 m.

242 WOODPECKERS

Habits: Prefers open forest, secondary forest, plantations, and gardens. A fairly tame, approachable bird.

443. GREY-CAPPED WOODPECKER *Dendrocopus canicapillus*
PLATE 50
(Grey-headed Pygmy-Woodpecker)

Description: Small (15 cm), black and white-striped woodpecker with no red on underparts and grey crown; male has a red streak above and behind the eye. Orange-buff wash on breast and belly is streaked blackish.
Iris—whitish brown; bill—grey; feet—greenish grey.
Voice: Shrill, squeaking trill *ki-ki ki ki rrr* . . .
Range: Pakistan, China, SE Asia, Borneo, and Sumatra.
Distribution and status: On Sumatra (including Nias and Riau Archipelago), found in the Barisan Range from 1000 to 2800 m; on Borneo occurs in lowland forest.
Habits: As other small woodpeckers.

444. SUNDA WOODPECKER *Picoides moluccensis*
PLATE 50
(Brown-capped/Malaysian/Malaysian Pygmy-Woodpecker)

Description: Small (13 cm) black and white woodpecker with dark brown cap; upperparts dark brown with white spots; underparts dirty white, streaked black; side of face white with grey cheek patch and broad black malar stripe. Male has thin red line behind eye.
Iris—red; bill—black above, grey below; feet—green.
Voice: Short, sharp wheezy trill *kikikikikiki* or whirring churr *trrrrr-i-i*.
Range: India, SE Asia, Borneo, Sumatra, Java, and Lesser Sundas.
Distribution and status: In Sumatra it was formerly rather rare, but it is more common and widespread with opening up of forests. In Borneo it is mostly a coastal species. In Java and Bali it is quite common at low elevations.
Habits: Typical small woodpecker, working slowly over dead branches and dead trees for food; usually solitary. A bird of secondary forest, open country, and mangroves.

445. GREY-AND-BUFF WOODPECKER *Hemicircus concretus*
PLATE 50
(Grey-breasted Woodpecker)

Description: Small (14 cm), scaly-backed, crested woodpecker. Head and breast grey; long crest and red forehead in male, grey in female and juvenile; belly

buffish; rump white; back and wing coverts black with buff-white edges to feathers.
Iris—brown; bill—dark grey; feet—grey.
Voice: Various typical woodpecker calls. Vibrating chitter in flight; a loud *wikawik* call and trilling *chee-chee-chee-chee-chee*.
Range: Malay Peninsula, Borneo, Sumatra, and Java.
Distribution and status: This species can be locally abundant in Borneo and Sumatra but is uncommon in Java.
Habits: Prefers open country, secondary forest, gardens, and plantations. Uses canopy layers where it sometimes hangs upside down gleaning insects in a tit-like manner.

446. MAROON WOODPECKER *Blythipicus rubiginosus* PLATE 50

Description: Smallish (23 cm) reddish brown woodpecker with ivory-yellow bill. Upperparts maroon without black barring. Male has red nape. Female has black base to bill and pale greyish barring on wings. Underparts brownish black.
Iris—brownish red; bill—yellowish white; feet—blue-black.
Voice: Single sharp *chikick* notes, and high-pitched, harsh, cackling *chai-chai-chai-chai* . . . call. Both are regarded as omen calls by Bornean Dyaks.
Range: India, SE Asia, Malay Peninsula, Sumatra, and Borneo.
Distribution and status: This is a common bird of primary and secondary forests and rubber plantations, up to 2200 m.
Habits: Prefers lower forest strata and forages quietly on rotted wood.

447. ORANGE-BACKED WOODPECKER *Reinwardtipicus validus*
PLATE 51

Description: Largish (30 cm) colourful woodpecker with lanky, long-necked appearance. Black wings and tail contrast with pale body. Male has a red crest, red throat, orange wash on back and rump, and reddish underparts. Female has dark brown crest, white back, and grey underparts. Both sexes have chestnut bars on primary and secondary wing feathers.
Iris—orange yellow; bill—yellowish; feet—reddish grey.
Voice: Various loud, ringing, typical woodpecker calls. Chattering *cha-cha-cha*, loud rapid *wheet-wheet-wheet-wheet- wheow*, repeated high-pitched *pit* or *pelleet* and alarm cry *toweetit-toweetit*.
Range: Malay Peninsula, Borneo, Sumatra, and Java; absent from Bali.
Distribution and status: In Borneo and Sumatra it is quite common. In Java it is an uncommon bird of lowland and occasionally montane, primary or secondary rain forest, up to 2200 m, largely confined to W Java.
Habits: Typical woodpecker behaviour. Lives in pairs or small parties. Noisy calling and resonant hammering when foraging at all levels.

448. GREATER GOLDENBACK *Chrysocolaptes lucidus* PLATE 51
(Greater Flameback/Greater Golden-backed Woodpecker)

Description: Largish (31 cm) colourful woodpecker. Very similar to Common Goldenback but slightly larger, has two black malar stripes which fuse on the cheek, and four toes instead of three. Crown of female is yellowish in E Java and Bali form *C. l. strictus*, elsewhere black with white spots.
Iris—pale yellow; bill—grey; feet—black.
Voice: Harsh, loud, strident, explosive burst of stuttered shrieks, similar to a large cicada.
Range: India, China, Philippines, NE Borneo, Sumatra with eastern islands, Java, and Bali.
Distribution and status: It is not uncommon in some lowland areas. In Sumatra it is mostly confined to mangroves.
Habits: Prefers rather open forest, forest edge, and mangroves. Lives in pairs and sometimes drums loudly.

BROADBILLS—FAMILY EURYLAIMIDAE

BROADBILLS are a small African and Asian family, with large heads, heavy, broad bills, short legs, and generally elongated tails. Most species are rather colourful.

They are forest birds which catch flying insects from a perch in mid-air with a loud snap of the large bill. Some species also eat fruits. They nest in elaborate, hanging, purse-shaped nests.

Nine species occur in the Greater Sundas.

449. DUSKY BROADBILL *Corydon sumatranus* PLATE 52

Description: Large (27 cm) broadbill with distinctive, very large, pink bill, dusky plumage, and a narrow, white subterminal tail bar. Throat and upper breast paler, mottled reddish. In flight it shows white stripes on wings and orange-yellow patch on back. Immature has blackish throat and lacks pale back patch.
Iris—brown; bill—pink; feet—brown.
Voice: *Ki-ip, ki-ip* (T.H.). A sequence of about 8 screaming notes, climbing a scale *hi-chu-ui, chu-ui, chu-ui* . . . accented on the *chu* note; also a rattling laugh, and clear, high-pitched *tsiu* (M. & W.). (See Appendix 6.)
Range: SE Asia, Malay Peninsula, Sumatra, and Borneo.
Distribution and status: Not uncommon in forest on Sumatra and Borneo (including Natunas), in lowlands and hills up to 800 m, locally to 1200 m on Borneo.

Habits: Sits in higher canopy, catching insects in flight and among foliage, often in small flocks.

450. BLACK-AND-RED BROADBILL *Cymbirhynchus macrorhynchos*
PLATE 52

Description: Large (23 cm) broadbill with dark crimson underparts and black upper breast band, large blue and yellow bill, black upperparts, and white wing stripe diagnostic. Crimson rump distinguishes it from Black-and-yellow Broadbill. Immature has greyer throat and buffish belly, washed red.
Iris—green; bill—blue above, yellow below; feet—purplish blue.
Voice: Monotonous rasping *wiark*, and rising trill, softer and briefer than Black-and-yellow Broadbill.
Range: SE Asia, Malay Peninsula, Sumatra, and Borneo.
Distribution and status: Common in lowland forest to 900 m, on Sumatra (including Bangka and Belitung) and Borneo.
Habits: Hawks after insects from exposed perch in upper or lower canopy. Generally found at edge of forest along rivers, streams, and roads.

451. BANDED BROADBILL *Eurylaimus javanicus*
PLATE 52

Description: Medium-sized (21 cm), large-headed, purplish bird with solid, very broad, blue bill. Head and throat greyish pink; narrow breast bar, back and wings blackish, heavily marked by yellow plumes and a transverse yellow wing bar; white spot on the leading edge of the wing; base of tail obscured by long yellow plumes; underparts purplish with, in male, a dark grey bar across the chest; vent yellow.
Iris—blue; bill—greenish blue with black tip; feet—pinkish.
Voice: Plaintive, shrill, slurred whistle. Song is a loud run of cicada-like buzzing notes, rising and then falling in pitch, and beginning with a single *yeow* note. (See Appendix 6.)
Range: Malay Peninsula and Greater Sundas.
Distribution and status: In Sumatra (including islands) and Borneo (including Natunas) this is an occasional bird, up to 1000 m. On Java it is a local bird occurring up to 1500 m. Not recorded from Bali.
Habits: Frequents upper and middle storeys of primary and secondary forest, heath forests, and plantations. Sits quietly in the forest, often in a small clearing, and hunts from a tree perch.

452. BLACK-AND-YELLOW BROADBILL *Eurylaimus ochromalus*
PLATE 52

Description: Small (15 cm), gaudily coloured broadbill with blue bill, black head, and complete white collar distinctive. Upperparts generally black with yellow rump and wing streaks, and white terminal spots on tail. Underparts pinkish grading into yellow towards the vent, black band across upper breast (partial in female) and black thighs. Immature is duller with yellow forehead.
Iris—yellow; bill—bluish; feet—pink.

Voice: Very notable and often heard call of accelerating, monotonous trill of about 7 seconds without the introductory whistle or fade out of the Banded Broadbill; also a plaintive peeping cry. (See Appendix 6.)

Range: Malay Peninsula, Sumatra, and Borneo.

Distribution and status: Common in primary and secondary forests up to 900 m on Sumatra (including islands) and Borneo (including Natunas).

Habits: Hunts insects from low perch in forest. Keeps to lower and middle canopy.

453. SILVER-BREASTED BROADBILL *Serilophus lunatus* PLATE 52
(Gould's/Hodgson's Broadbill)

Description: Small (15 cm), gracile-billed, pinkish-grey broadbill with arched black eyebrow and blue wing patch. Scapulars, back, and rump chestnut; tail black with narrow white tip. Female has narrow white band across greyish breast.
Iris—brown and green; bill—yellow, or blue with yellow base; feet- yellowish green.

Voice: Clear, whistled *piu*.

Range: Nepal to SW China, SE Asia, Malay Peninsula, and Sumatra.

Distribution and status: On Sumatra this is an occasional bird of hill forests between 800 and 1500 m.

Habits: Lives in small flocks in the sub-canopy and understorey of open forests, along streams and river banks. Flocks with other species. Catches insects in the foliage and on the wing.

454. LONG-TAILED BROADBILL *Psarisomus dalhousiae* PLATE 52

Description: Unmistakable, elongate (25 cm) green broadbill with long, blue graduated tail, yellow throat and face with black cap and nape. Wings black with a prominent blue patch. Has small blue spot on the top of the head and a yellow spot behind the eye. Immature is mostly green.
Iris—green and grey; bill—green with blue tip above, yellow beneath; feet—green.

Voice: Loud, sweet, whistling calls of 5–8 notes on the same pitch.

Range: Himalayas, S China, SE Asia, Malay Peninsula, Sumatra, and Borneo.
Distribution and status: On Sumatra it is an occasional bird of primary and mature secondary hill forests, usually 700 to 1500 m, but locally to 2000 m (e.g. Mt. Kerinci). On Borneo it is a confined to the mountain range from Mts Kinabalu and Trus Madi to Tama Abo, and the outliers Dulit and Mulu.
Habits: Lives in noisy flocks working through the middle canopy, sometimes mixing with other species.

455. GREEN BROADBILL *Calyptomena viridis* PLATE 52
(Lesser Green Broadbill)

Description: Small (18 cm), curiously spherical, bright green broadbill. Male is iridescent grass green with black wing bars and two black spots above and behind eye. Female is duller and lacks black markings. Frontal crest conceals shape of head.
Iris—brown; bill—green and black; feet—green.
Voice: One or two soft notes followed by brief, rising, low-pitched trill *tu-tut-trrrr*, also loud *oink*. (See Appendix 6.)
Range: Malay Peninsula, Sumatra, and Borneo.
Distribution and status: A common but often overlooked bird of primary and mature secondary lowland and hill forests, up to 1000 m but rarely above 1700 m on Sumatra (including islands), and to 1200 m on Borneo (including Natunas).
Habits: Generally solitary; found in middle and high canopy where it is almost invisible when stationary. Best recognized by voice.

456. HOSE'S BROADBILL *Calyptomena hosii* PLATE 52
(Blue-bellied/Magnificent/Magnificent Green Broadbill)

Description: Medium-sized (20 cm), rounded, green broadbill with diagnostic blue breast. Similar to smaller Green Broadbill, but breast bright blue in male and light blue in female. Has round black spots instead of bars on the wing.
Iris—dark brown; bill—blackish green; feet—olive green.
Voice: Soft, dove-like cooing accompanied by head bobbing (T.H.).
Range: Endemic to Borneo.
Distribution and status: A local bird of hill forest, up to 1000 m, with a patchy distribution: quite common in Baram drainage (where recorded at sea-level on Mt. Dulit), very common on Liang Kubung, and recorded upper Telen, but apparently absent or rare on other intervening mountains.
Habits: Similar to Green Broadbill but mostly at higher altitudes.

248 BROADBILLS

457. WHITEHEAD'S BROADBILL *Calyptomena whiteheadi* PLATE 52
(Black-throated Green/Black-throated Broadbill)

Description: Large (25 cm), curiously rounded, green broadbill with black throat. Similar to smaller Green Broadbill but distinguished by black throat and more speckled appearance caused by black bases of body feathers.
Iris—brown; bill—green; feet—green.
Voice: Generally silent but occasionally gives shrill *saat* call, and snore-like wheezes.
Range: Endemic to Borneo.
Distribution and status: Restricted to N Borneo, where occurs usually between 1000 and 1500 m, throughout the range from Mts Kinabalu to Batu Tibang, and the outliers Mulu and Dulit.
Habits: As other green broadbills but living in the lower canopy of dark ravine forests, at higher altitude.

PITTAS—FAMILY PITTIDAE

PITTAS are a family of colourful, ground-living birds found from Africa to Australia. They are plump, short-tailed, long-legged birds that hop around on the forest floor or in low vegetation, in search of invertebrates. They give simple, plaintive calls or whistles, some species calling from high in canopy.

Pittas nest in hollow, ball-like structures made of vegetation, often close to the ground. They fly with a rapid wing-beat when disturbed, keeping close to the forest floor, and in some species revealing conspicuous white wing flashes. Some species are migratory. Almost all species are beautifully coloured with rich blue, gold, red or green patterns.

Nine species are resident in the Greater Sundas. Two additional species visit as migrants.

458. SCHNEIDER'S PITTA *Pitta schneideri* PLATE 53

Description: Largish (22 cm), brown and blue, large-headed pitta. Male has chestnut crown and nape, bright blue upperparts, and blackish-brown flight feathers. Underparts orange-buff, with whitish throat and narrow broken black line across upper breast. The female is darker, duller blue than the male with paler, less orange crown and browner flight feathers. The belly is more rufous than the male's and the black breast band is more pronounced. Immature is mottled brown with white spots.
Iris—red-brown; bill—brown; feet—pinkish grey.

Voice: Low, soft, prolonged tremulous whistle rising on first note and falling on second, repeated at 5–6 second intervals (Hurrell).
Range: Endemic to Sumatra.
Distribution and status: Confined to the higher mountains of the Barisan Range (Mts Sibayak, Kerinci, Kaba, and Dempu). Inhabits the forest floor of mountain forests, from 900 to 2400 m. Formerly common in the Kerinci valley, the only recent records are at higher elevations on the peak.
Habits: As other pittas but prefers high mountain forests. It works steadily along the forest floor turning over leaves in search of insects and snails.

Juvenile Schneider's Pitta, *Pitta schneideri* (458).

459. GIANT PITTA *Pitta caerulea* PLATE 53

Description: Very large (27 cm), thick-billed, blue-backed or reddish brown pitta. Sexes differ. Male has buffish grey head with black eye-stripe, crown, nape, and fine collar, blue upperparts, and grey-buff underparts. Female has reddish brown upperparts with blue only on rump and tail. Immature is dark brown, mottled with buff scaling.
Iris—dark grey; bill—black; feet—pink.
Voice: Slow, mournful whistle with descending pitch and downward inflection given by males in breeding season, similar calls given by females and juveniles at fledgling stage.
Range: Malay Peninsula, Sumatra, and Borneo.
Distribution and status: On Sumatra and Borneo a rare bird of lowland forests, recorded to 1200 m. In Sumatra not recorded this century.
Habits: Lives singly or in pairs, on the floor of dark damp forests, forest edge, and secondary forests. Birds break snails on rock anvil. Shyer than other pittas, flying off longer distances when disturbed.

460. BLUE-BANDED PITTA *Pitta arquata* PLATE 53

Description: Small (17 cm), gaudy, red, green, and blue pitta. Head is chestnut with narrow, glistening blue eye-stripe. Upperparts dull bluish green and underparts bright red with a band of pointed, shiny blue feathers across the breast. Immature is brown and mottled.
Iris—grey; bill—reddish black; feet—slaty grey-brown.
Voice: 2–3 sec monotonous soft whistle with rising inflection.
Range: Endemic to Borneo.
Distribution and status: Not uncommon in primary hill forests, between 700 and 1600 m, but also known from the lowlands.
Habits: As other small pittas.
Note: Gould's original name *arquata* was unjustifiably amended to *arcuata* by Salvadori in 1874. The latter name still persists *in lit.*

461. BLUE-HEADED PITTA *Pitta baudii* PLATE 53

Description: Small (17 cm) rufous-backed pitta with conspicuous white bar on black wings. Sexes differ. Male has maroon-red back, bright blue crown and tail, black upper breast and eye-stripe, and contrasting white throat. Dark purple-blue lower breast and belly look black in the field. Female is duller with rufous upperparts, blue tail, pale brown underparts and pale grey throat.
Iris—grey; bill—black; feet—horn.
Voice: Disyllabic *wu-wooo* with a distinct tremulous quality; second note fades.
Range: Endemic to Borneo.
Distribution and status: Occurs throughout the island in lowland forests, up to 600 m, and is locally common.
Habits: As other pittas; searches for insects under fallen leaves on forest floor.

462. BLUE-WINGED PITTA *Pitta moluccensis* PLATE 53
(Moluccan Pitta)

Description: Medium-sized (18 cm), plump, colourful pitta with rufous chest. Head black with pale brown eyebrow; back green; wings bright blue with white wing patch; throat white; vent red.
Iris—brown; bill—blackish; feet—pale brown.
Voice: Loud whistle *pu-wiu, pu-wiu* with second note highest. In Borneo the call is said to warn of rain. (See Appendix 6.)
Range: E India, SW China, SE Asia; wintering to the Malay Peninsula, Sumatra, and Borneo.
Distribution and status: In Sumatra (including islands) and Borneo (including Natunas) it is a locally common migrant and winter visitor, up to 1000 m;

recorded from most woodland habitats, including gardens. Vagrants have reached Sulawesi and the Philippines, but records from Java and Australia are dubious.
Habits: Frequents scrub, secondary forest, and mangrove. Never far from the coast. Typically skulking pitta, hopping along the ground like a giant thrush. Roosts at night only about 1 m off the ground in a small bush. Often hunts along stream beds.
Note: May be conspecific with Indian Pitta *P. brachyura*.

463. FAIRY PITTA *Pitta nympha* PLATE 53

Description: Medium-sized (20 cm), colourful pitta. As Blue-winged Pitta but with paler and greyer underparts, and sky blue wing and rump patches.
Iris—brown; bill—blackish; feet—blackish.
Voice: Similar to Blue-winged Pitta.
Range: Breeds in Japan, Korea, and E China and winters south to S China, Indochina, and Borneo.
Distribution and status: Recorded from October to March south through Kalimantan. Occasionally locally common up to 1000 m, particularly in north.
Habits: As Blue-winged Pitta.
Note: May be conspecific with Indian Pitta *P. brachyura*.

464. MANGROVE PITTA *Pitta megarhyncha* PLATE 53
(Long-billed Pitta)

Description: Medium-sized (20 cm) colourful pitta. As Blue-winged Pitta but with crown almost pure brown, and rump and wing patches cobalt blue. The bill is larger.
Iris—brown; bill—blackish; feet—blackish.
Voice: As Blue-winged Pitta but more slurred and with shorter intervals between notes. (See Appendix 6.)
Range: W coast of Malay Peninsula, from Sunderbans to Singapore.
Distribution and status: Resident in mangroves, south through Riau Archipelago, eastern lowlands of Sumatra, and Bangka. Once collected in N Borneo (Baram district).
Habits: As Blue-winged Pitta but prefers mangrove and beach forests.
Note: May be conspecific with Indian Pitta *P. brachyura*.

465. ELEGANT PITTA *Pitta elegans* PLATE 53

Description: Medium-sized (20 cm) colourful pitta. Crown black, buff eyebrow extending from bill to nape where it becomes blue, black throat; upperparea green with glittering light blue rump and wing coverts; primaries black with white band;

underparts mainly buff with black and crimson patch on belly and vent.
Iris—brown; **bill**—blackish; **feet**—brown.
Voice: No information.
Range: Islands in Flores and Banda Seas, and Lesser Sundas.
Distribution and status: Found once Nusa Penida, maybe migrant from Lombok.
Habits: Favours forest edge, woodlands, and thickets.

466. GARNET PITTA *Pitta granatina*   PLATE 53

Description: Small (17 cm) purplish pitta with red (or black) crown and red belly. Bright blue patch on wing; narrow, shiny, pale blue brow line behind the eye. Immature birds are brown with a rufous crown and underparts, and blackish sides to head. Races vary. The E Sumatran form *coccinea* has black spots on red belly and black forehead. In N Borneo the red-capped form *granatina* is replaced by the black-capped form *ussheri* between the Lawas and Merapok Rivers.
Iris—brown; **bill**—black; **feet**—pinkish grey.
Voice: Prolonged, rising, monotonous whistle. Birds can be called up by playback or whistled imitation. (See Appendix 6.)
Range: Malay Peninsula, Sumatra, and Borneo.
Distribution and status: In Sumatra it is an occasional bird of lowland forests on the E coast south to around Medan. In Borneo it is a frequent bird of lowland forest.
Habits: Hops along the floor of dark, wet, often swampy forests, sometimes on fallen logs and brush piles. Quite tame if one remains motionless.
Note: The Bornean black-crowned form *P. g. ussheri* is sometimes included in *P. venusta*, and sometimes treated as a separate species *P. ussheri*, but *granatina* x *ussheri* hybrids are so well documented that it must be included in *granatina*.

467. BLACK-CROWNED PITTA *Pitta venusta*   PLATE 53
(Graceful/Black-and-scarlet Pitta)

Description: Small (16 cm) dark pitta with black head and red belly. There is a pale blue patch on the wing and a short blue browline behind the eye. Immature is entirely dark brown with blue brow stripe.
Iris—brown; **bill**—black; **feet**—pinkish.
Voice: Soft, melodious whistles.
Range: Endemic to Sumatra.
Distribution and status: Found in the hills of the Barisan Range, from Ophir south to Dempu. A rare bird of hill forests between 400 and 1400 m; the last published record was in 1918.
Habits: As Garnet Pitta.

Note: Some authors treat this species as a form of the Garnet Pitta. Others lump it with the Bornean black-capped race of Garnet Pitta. The latter treatment is incorrect (see Garnet Pitta).

468. HOODED PITTA *Pitta sordida*   PLATE 53
(Black-headed Pitta)

Description: Unmistakable, medium-sized (18 cm), plump, long-legged, short-tailed, green bird with black head. Upperparts green; wing coverts blue with white wing patches; head black; breast and belly green; vent bright red. The migrant race has a brownish cap; the resident's cap is all black, but both forms look black in the forest.
Iris—brown; bill—black; feet—flesh-coloured.
Voice: Double whistle *pih-pih*, repeated at short intervals. (See Appendix 6.)
Range: India to SW China, SE Asia, Philippines, Sulawesi, Greater Sundas, and New Guinea.
Distribution and status: Northern populations winter to Malay Peninsula, Sumatra (including Nias, islands in Malacca Strait, and Bangka), and Java. Resident Sumatra (including Bangka and Belitung), Borneo, and W Java. Frequents wetter forest, mostly lowlands, up to 800 m. Not recorded in Bali.
Habits: Hops about on forest floor, turning over leaves, and probing dead wood in search of invertebrates. Conspicuous in flight with large white patches on dark wings. Sleeps at night on a low bare branch or vine; completely tame and approachable in torchlight.

469. BANDED PITTA *Pitta guajana*   PLATE 53
(Blue-tailed Pitta)

Description: Unmistakable, medium-sized (22 cm), plump, golden-barred pitta. Black head with conspicuous broad yellow eyebrow diagnostic; back and wings brown with white wing bar; tail blue; chin white. Island races differ. Breast and flanks barred black and yellow in Borneo and Java; blue and orange in Sumatra. Java male has blue upper breast bar; Bornean race has blue lower breast patch; Sumatran race has blue belly and orange nape, also broader white wing bars. Females are all duller and more rufous.
Iris—brown; bill—black; feet—black.
Voice: Loudish *pa-oh*, a high *pieuw*, or a soft, strident *purrrr*. (See Appendix 6.)
Range: Malay Peninsula and Greater Sundas.
Distribution and status: In Sumatra mostly lowlands, to 500 m; in Borneo recorded in lowlands but mostly hills, from 500 to 1500 m; on both islands rather local. In Java and Bali this bird is uncommon in forest up to 1500 m; but also

locally abundant and populations in remnant patches of forest withstand massive trapping pressure, at least in the short-term.

Habits: Favours shady, primary or secondary forests where it hops quickly along the floor or along dead logs. Sometimes found in low bushes, salak, and rattan thickets. It is a shy bird sometimes seen dashing across a forest path but more often only heard. Sleeps in low vegetation, only 1-3 m above the ground.

Larks—Family Alaudidae

Larks are a family of moderate-sized birds of worldwide distribution. They are short-legged, terrestrial birds of open spaces, looking superficially like pipits (see Wagtails) but have weaker flight, shorter tails, and thicker bills, and several have short, erectile crests.

Larks sing while on the wing and several species hover in a fluttering manner giving their beautiful, melodious songs. They feed and nest on the ground.

In the Greater Sundas there are only two species of larks.

470. AUSTRALIAN LARK *Mirafra javanica* PLATE 54
(Singing/Horsfield's/Javan/Eastern Lark, AG: Bushlark)

Description: Small (14 cm) reddish brown ground bird, heavily mottled black. Underparts pale buff, streaked with black; outer tail feathers white. Superficially resembles a pipit but has a thicker bill and shorter tail and legs. Distinguished from Oriental Skylark by rufous on wings.

Iris—dark brown; bill—brown above; yellowish below; feet—pinkish with very elongated hind claw.

Voice: Sweet, clear song of shrill, trilling notes; a *chirrup* in alarm.

Range: Africa, India, SE China, SE Asia, Philippines, S Borneo, Java, Bali, and through eastern Indonesia to New Guinea and Australia.

Distribution and status: In Borneo known from Banjarmasin area; on Java and Bali quite common in open habitat, up to 800 m.

Habits: Singly or in flocks of dispersed individuals; frequents open areas of short grass, padangs, and paddy stubble. Usually walking on the ground or making a weak, fluttery, undulating flight. Sings on the ground or in the air, during flight or hovering while slowly descending vertically. Sometimes calls at night. Rests on telegraph wires or bushes. Takes regular dust baths.

471. ORIENTAL SKYLARK *Alauda gulgula* PLATE 54
(Eastern/Small/Indian/Lesser Skylark)

Description: Small (19 cm), mottled-brown, pipit-like bird with faint pale eyebrow and a slight crest. Distinguished from pipits by heavier bill, weaker flight, by posture, and from Bushlark by lack of rufous on wings and by behaviour.
Iris—brown; bill—horn; feet—grey.
Voice: Beautiful, liquid, bubbling song given in flight, often at great height but not given by vagrants.
Range: Breeds in Palaearctic and migrates south in winter.
Distribution and status: Rare vagrants reach N Borneo.
Habits: Prefers open land with short grass. Differs from bushlarks by never perching in trees.

Swallows—FAMILY HIRUNDINIDAE

FAMILIAR, worldwide family of graceful birds with slender bodies and long, pointed wings. Swallows are gregarious and catch insects in mid-air, hawking to and fro along waterways or circling high in the sky. They superficially resemble swifts but the flight is not as fast; swallows glide with wings half closed, unlike swifts which glide with sickle wings fully extended. The sexes look alike.

Unlike swifts, swallows frequently perch in trees, on telegraph wires, television aerials, poles, or houses and will settle on the ground to drink at pools, collect mud for nests, and occasionally to catch ants and other insects. Swallows nest in cup-shaped nests of mud, built under house roofs or on cliff underhangs; some species burrow in banks. Swallows are famous for their migrating abilities.

There are six species of swallows in the Greater Sundas of which four are only visitors.

472. SAND-MARTIN *Riparia riparia* PLATE 54
(Common/Gorgeted Sand-Martin/Bank Swallow)

Description: Small (12 cm) brown swallow with white underparts and a diagnostic brown breast band. Immature has buff throat.
Iris—brown; bill—black; feet—black.
Voice: Shrill twitters.
Range: Cosmopolitan (except Australia). Eurasian birds winter south to SE Asia and Philippines.
Distribution and status: Small numbers regularly reach N Borneo. Recorded once in Kalimantan (Kapuas).

Habits: Lives over marshes and rivers, making sweeping flights over water or perching on prominent dead branches.

473. BARN SWALLOW *Hirundo rustica* PLATE 54
(Common/House/Rustic/The Swallow)

Description: Medium-sized (20 cm including elongated tail feathers) glossy blue and white swallow. Upperparts steely blue; throat reddish edged with a blue upper breast bar, belly white; tail very elongated, with white spots near tips of feathers. Distinguished from Pacific Swallow by cleaner, white belly, more elongated tail, and blue breast bar. Immature have duller plumage, lack tail streamers, and are more difficult to distinguish from Pacific Swallow.
Iris—brown; bill—black; feet—black.
Voice: High-pitched *twit* and twittering calls.
Range: Nearly worldwide. Breeds in northern latitudes and migrates south in winter through Africa, Asia, SE Asia, Philippines, and Indonesia to New Guinea and Australia.
Distribution and status: In the Greater Sundas (including islands) this is a common winter visitor at all altitudes.
Habits: Glides and circles high in the sky, or low over land or water to catch small insects. Alights on dead branches, poles and telegraph wires to perch. Feeds independently but with large numbers feeding at same site. Sometimes congregates in large roosting flocks, even in cities.

474. PACIFIC SWALLOW *Hirundo tahitica* PLATE 54
(Eastern House/Small House Swallow/Least Swallow)

Description: Smallish (14 cm) blue, red and buff swallow. Upperparts steely blue; forehead chestnut. Distinguished from Barn Swallow by dirty white underparts, less elongated tail without streamers, lack of dark blue chest bar, and slightly smaller and less smart appearance.
Iris—brown; bill—black; feet—brown.
Voice: Pleasant twittering calls and high-pitched *tweet* in alarm.
Range: S India, SE Asia, Philippines, Malay Peninsula, and Sundas to New Guinea and Tahiti.
Distribution and status: In the Greater Sundas (including islands) this is a common bird of open spaces especially over water, up to about 1500 m.
Habits: Generally found in small, loose parties, working independently in circles or gliding low over water. In winter they mix freely with swiftlets, but do not congregate in large roosting flocks. The nest is a cup of mud pellets, stuck under a roof, bridge, or rock overhang. It has an open entrance in the rim.

[475. RED-RUMPED SWALLOW *Hirundo daurica* PLATE 54
(Lesser Striated/Ceylon Swallow)

Description: Large (18 cm) swallow with pale chestnut rump contrasting with steely dark blue upperparts; underparts white, finely streaked with black. The tail is long and deeply forked. Cannot be reliably distinguished in field from the Striated Swallow.
Iris—brown; bill and feet—black.
Voice: Shrill calls in flight.
Range: Breeds in Eurasia, migrating south in winter to Africa, India, and SE Asia.
Distribution and status: Recorded in N Sumatra and probably N Borneo. There are no specimens and confusion with Striated Swallow is possible.
Habits: Like Barn Swallow.
Note: Some authors place the Striated Swallow in this species but see Vaurie (1951).]

476. STRIATED SWALLOW *Hirundo striolata* PLATE 54
(Oriental Mosque/Greater Striated Swallow)

Description: Large (20 cm) swallow with streaky chest and red rump. Upperparts steely blue; underparts dirty white with black streaks; tail deeply forked.
Iris—brown; bill—black; feet—greyish.
Voice: Usually silent but sometimes gives loud *chew-chew* or vibrating *schwirrr*.
Range: NE India, SE Asia, Philippines, Malay Peninsula, Java, Bali, and Lesser Sundas.
Distribution and status: Reported from S Sumatra (presumably non-breeding birds from Java) and N Borneo (possibly resident or from Philippines). On Java and Bali this is an uncommon swallow of local occurrence, up to 1500 m.
Habits: Similar to other swallows but keeps more to lowlands near cultivated areas. Lives in pairs or small flocks and flies with slower beat and more soaring than the other swallows. The nest is a cup made of mud pellets with a tunnel entrance plastered on the under-surface of a ceiling or overhang.

477. ASIAN HOUSE-MARTIN *Delichon dasypus* PLATE 54
(Asiatic House-Swallow)

Description: Small (13 cm) plump black and white swallow. White rump and shallow forked tail diagnostic. Upperparts steely blue; rump white; breast greyish white.
Iris—brown; bill black; feet—pinkish.
Voice: Excited whickering calls.

258 SWALLOWS

Range: Breeds N India to Japan and winters south to SE Asia, Philippines (rarely) and Greater Sundas.
Distribution and status: Not uncommon N Sumatra (including Riau Archipelago) but sporadic in Borneo, Java and Bali, recorded mostly in hills to 1500 m.
Habits: Single birds mix with other swallows or swiftlets. More aerial than other swallows and mainly seen in soaring flight.

Cuckoo-shrikes — FAMILY CAMPEPHAGIDAE

An Old World family which, despite its name, is closely related neither to the cuckoos nor the shrikes. Some species resemble cuckoos in shape and plumage while all are somewhat shrike-like in having strong, hooked bills for catching insects.

Cuckoo-shrikes have soft, fluffy feathers and short legs. Most species are noisy, conspicuous, flocking birds of the forest canopy. Most are dull coloured, black, white, and grey, but minivets are colourful with mostly bright red and yellow plumage. All eat insects, some also take fruit. They do not alight on the ground. Cuckoo-shrikes make cup-shaped nests in the canopy. On the basis of DNA-hybridization studies, Sibley and Monroe (1990) now reassign Cuckoo-Shrikes together with Orioles as a tribe of the Corvidae (crows).

There are 15 species of Cuckoo-Shrikes in the Greater Sundas.

478. BAR-WINGED FLYCATCHER-SHRIKE *Hemipus picatus*
(Pied Flycatcher-Shrike/Bar-winged Pygmy-Triller) PLATE 55

Description: Small (15 cm) pied cuckoo-shrike. Resembles Black-winged Flycatcher-Shrike but has a broad white wing bar. Distinguished from Ashy Minivet by smaller size and whitish rump. Distinguished from Pied Triller and Little Pied Flycatcher by lack of white eyebrow.
Iris—brown; bill—black; feet—black.
Voice: Noisy, high-pitched *chir-rup, chir-rup* (Smythies).
Range: India, SW China, SE Asia, Malay Peninsula, Sumatra, and Borneo.
Distribution and status: On Sumatra and Borneo this is a common bird of hill and mountain forests, between 700 and 2000 m, occasionally lower in Borneo.
Habits: Lives in flocks, often mixed with other species, working through forest canopy, crouching to uncover hiding or disturbed insects then swooping on them like a shrike.

CUCKOO-SHRIKES 259

479. BLACK-WINGED FLYCATCHER-SHRIKE *Hemipus hirundinaceus* PLATE 55
(AG: Pygmy-Triller)

Description: Small (15 cm) pied cuckoo-shrike. Male: black upperparts with white rump, and edges of outer tail feathers; underparts white. Female: similar but black replaced by brown. Distinguished from Bar-winged Flycatcher-Shrike by lack of white wing bar; from Ashy Minivet by dark upperparts and white rump and from Pied Triller and Little Pied Flycatcher by lack of white eyebrow.
Iris—brown; bill—black; feet—black.
Voice: Coarse *tu-tu-tu-tu*, *hee-tee-tee-teet*, or *hee-too-weet*, alternating with high-pitched *cheet-weet-weet-weet*.
Range: Malay Peninsula and Greater Sundas.
Distribution and status: A fairly common bird of lowlands and hills, up to 1100 m on Sumatra (including islands) and Borneo (including N Bornean Islands), and up to at least 1500 m on Java and Bali.
Habits: Frequents forest and particularly forest edge. Lives in pairs or small parties, often mixing with other species, travelling through the crowns of small trees.

480. LARGE WOODSHRIKE *Tephrodornis gularis* PLATE 55
(Brown-tailed/Hook-billed Woodshrike)

Description: Smallish (18 cm) grey and white bird with a white rump. Upperparts grey in male, brown in female; underparts white with greyish wash over breast. Dark mask; bill hooked at tip.
Iris—brown; bill—black; feet—black.
Voice: Repetitive *wit wit wit*.... (D.A.H.). Loud *chew-chew*, incessant noisy *kee-a*, *keea*, or a harsh *chreek-chreek chee-ree* or *chee-ree-ree—che ree reeoo-reeoo*. (See Appendix 6.)
Range: India, China, SE Asia, Malay Peninsula, and Greater Sundas.
Distribution and status: In Sumatra and Borneo this is a frequent bird of lowland forests, up to 1000 m; on Sumatra most common near the coast. In Java this is an uncommon bird of lowland rainforest, up to about 1500 m. Not recorded on Bali.
Habits: Occurs in pairs or small noisy flocks, travelling among tree crowns. Pursues disturbed insects and often hunts from a perch. Also takes insects off water surfaces; favours forest edge and clearings.
Note: DNA studies indicate that *Tephrodornis* is not orioline like other Cuckoo-shrikes but related to Helmet-Shrikes.

481. MALAYSIAN CUCKOO-SHRIKE *Coracina javensis* PLATE 55

Description: Large (28 cm), grey, shrike-like bird. Male: upperparts grey, wings with whitish edges to feathers; tail with two central feathers grey, others grading paler outwards; belly whitish; lores and eye-ring black and throat dark grey. Female paler, with grey barring on lower breast and flanks. Immature similar to female but browner and with bolder barring on underparts and rump.
Iris—brown; bill—black; feet—dark brown.
Voice: Loud, piercing whistle *pee-eeo-pee-eeo*, *tweer*, or *twee-eet*.
Range: Malay Peninsula, Java, and Bali.
Distribution and status: On Java and Bali this is a widespread and locally fairly common bird in coastal and low-lying areas, up to 1500 m.
Habits: Generally single or in pairs. Keeps to the tops of the tallest trees, often at the edge of forest clearings.
Note: Treated by some authors as a race of Large Cuckoo-Shrike *C. novaehollandiae* = *caledonica*.

482. SUNDA CUCKOO-SHRIKE *Coracina larvata* PLATE 55
(Black-faced Cuckoo-Shrike)

Description: Medium-sized (23 cm), greyish black, shrike-like bird. Adult ashy grey with blackish flight feathers and tail, and a black face mask (more extensive in male). Female unbarred. Immature: like adult but with white edges to feathers.
Iris—dark brown; bill—black; feet—black.
Voice: Harsh, ringing whistles *eeooo-eeooo-eeoooo* or *shreeok*, and a curious, loud, wheezy song.
Range: Endemic to Greater Sundas.
Distribution and status: Endemic to the mountains of the Greater Sundas. In Sumatra this is a common bird in the N and W ranges (but not recorded in the Barisan Range south of Mt. Kerinci), between 850 and 2200 m; in Borneo common in the higher ranges from Mt. Kinabalu to at least the Mueller range of Kalimantan. In Java this is a very local bird of the higher mountains. Not recorded on Bali.
Habits: Frequents treetops of montane forest and pine forests, generally singly or in pairs, sometimes mixes with flocks of other birds. Eats fruit as well as insects.

483. BAR-BELLIED CUCKOO-SHRIKE *Coracina striata* PLATE 55
(Barred Cuckoo-Shrike)

Description: Medium-sized (28 cm), grey, shrike-like bird. Male: grey; tips of wing and tail blackish; rump and vent paler with dark grey barring. Female

barring is bolder, more black and white, and extends further up belly.
Iris—pale yellow; bill—black; feet—brown.

Voice: A clear, mellow, double whistle or four-noted *kliu kliu*.

Range: Philippines, Andamans, Malay Peninsula, Sumatra, Borneo, and Kangean.

Distribution and status: On Sumatra (including islands) and Borneo (including islands) frequent bird in lowland forests (up to 900 m on Sumatra). On Kangean Islands the endemic subspecies *vordermanni* is not uncommon in coastal areas.

Habits: Frequents the tallest treetops and flies very high, in small parties.

484. LESSER CUCKOO-SHRIKE *Coracina fimbriata* PLATE 55

Description: Smallish (20 cm), dark, shrike-like bird. Male: dark grey, paler below with black tail and wings, and blackish head. Female: paler with pale barring on underparts. Immature: browner with grey and white barring over white body.
Iris—brown; bill—black; feet—black.

Voice: Clear, deliberate whistle on one pitch *chee-wee . . . chee-wee* or *chweet-weet-weet-weet-weet-weet*, given while searching for food.

Range: Malay Peninsula and Greater Sundas.

Distribution and status: In the Greater Sundas an occasional bird of lowland and hill forests (up to 1000 m on Sumatra; to 1500 m on Java).

Habits: Prefers primary forest but also visits surrounding cultivated areas and plantations. Keeps to the treetops, singly or in pairs, and often in mixed flocks.

485. PIED TRILLER *Lalage nigra* PLATE 55

Description: Small (16 cm) pied bird. Male: black upperparts with white wing bar, and white edges to wing coverts and outer tail feathers; broad white eyebrow and black eye-stripe; rump grey; underparts white. Female: similar but brown instead of black, and breast finely barred with black. Immature: as female but with upperparts scaled buff, and white underparts streaked with grey.
Iris—brown; bill—grey with black tip; feet—black.

Voice: Double croak *chook-chook* or a rolling, descending *tre-tre-tre-tre*, more musical than White-shouldered Triller.

Range: Philippines, Malay Peninsula, and Greater Sundas.

Distribution and status: In the Greater Sundas (including islands) this is a fairly common bird of open and cultivated lowland areas, up to 1000 m. On mainland Java it is replaced in the east by the White-shouldered Triller. Not recorded on Bali.

Habits: Prefers open habitats, coastal mangroves, and *Casuarina* groves where it searches for insects among the foliage of small trees. Flies from tree to tree with a sluggish, undulating flight. Sometimes comes down onto the ground. Generally rather shy and retiring, keeping well hidden in the tree foliage. Solitary, in pairs, or sometimes in small flocks.

486. WHITE-SHOULDERED TRILLER *Lalage sueurii*   PLATE 55
(White-winged/Sueur's Triller)

Description: Small (17 cm) pied bird. Similar to Pied Triller but slightly larger with narrower eyebrow and less white on wing (despite its name!). Male sometimes has yellowish chin; female has rufous edges to wing feathers.
Iris—brown; bill—grey with black tip; feet—black.
Voice: A rich, vigorous, metallic whistle; song sometimes given in flight *chi-chi-chi-joey-joey-joey*.
Range: Java, Bali, Lesser Sundas, and Sulawesi.
Distribution and status: In E Java and Bali this is a fairly common bird of open and cultivated lowland areas. Co-occurrence or intergradation with the Pied Triller is to be expected in the Kediri–Malang area.
Habits: As Pied Triller but prefers drier habitats. More obtrusive in habits.

487. ASHY MINIVET *Pericrocotus divaricatus*   PLATE 56

Description: Large (20 cm) minivet with diagnostic black, grey, and white plumage. Distinguished from trillers by larger size and lack of wing bar, and from cuckoo-shrikes by white underparts and grey rump. Male: cap, eye-stripe, and flight feathers black; otherwise grey above and white below. Female: paler and greyer.
Iris—brown; bill—black; feet—black.
Voice: Metallic, jingling trill given in flight.
Range: NE Asia and E China. Winters S to SE Asia, Philippines, and Greater Sundas.
Distribution and status: Recorded in Sumatra and N Borneo. An occasional visitor in coastal lowlands and rarely beyond 900 m.
Habits: Hunts insects in tree canopy. Less showy in flight than brightly coloured minivets. Forms flocks of up to 15 birds.

488. SMALL MINIVET *Pericrocotus cinnamomeus*   PLATE 56
(Little/Lesser Minivet)

Description: Small (15 cm), red-grey, and black minivet. Distinguished from other minivets by grey head and mantle of male, and whitish underparts of duller female.

Iris—brown; bill—black; feet—black.
Voice: High-pitched, ringing *tsch-tsch-tsch-tsch* calls between flock members.
Range: India, SE Asia (except Malay Peninsula), Borneo, Java, and Bali.
Distribution and status: Status on Borneo unknown; one specimen collected S Borneo in last century was probably a vagrant from Java. Resident in Java and Bali, where it is a widespread and not uncommon bird of lowland areas. Replaced on Sumatra and Borneo by Fiery Minivet.
Habits: Prefers open forests, mangroves, cultivated lands, villages. Travels in small, restless, noisy flocks, keeping to the tops of taller trees.

489. FIERY MINIVET *Pericrocotus igneus* PLATE 56

Description: Small (15 cm) scarlet and black minivet. Male: vermilion with glossy black head, back, wings, and centre of tail; has orange tinge on belly and edge of tail. Female: head and back grey; face and underparts yellow, becoming orange on undertail coverts and rump.
Iris—brown; bill—black; feet—black.
Voice: Rising, musical *swee-eet*.
Range: Palawan, Malay Peninsula, Sumatra, and Borneo.
Distribution and status: Resident in Sumatra (including islands) and Borneo. Found in coastal mangroves, forests, and woodlands, up to 200 m on Sumatra.
Habits: As other minivets.
Note: Some authors treat this bird as a subspecies of the Small Minivet.

490. GREY-CHINNED MINIVET *Pericrocotus solaris* PLATE 56
(Mountain/Grey-throated/Yellow-throated Minivet)

Description: Medium-sized (17 cm) red or yellow minivet. Red male is distinguished from other minivets by dull, dark grey throat and ear coverts. Yellow female is distinguished by lack of yellow on forehead, ear coverts, and throat.
Iris—dark brown; bill—black; feet—black.
Voice: Soft, slightly rasping *tsee-sip*.
Range: Himalayas, S China, SE Asia, Malay Peninsula, Sumatra, and Borneo.
Distribution and status: Common resident in mountains of Sumatra and N Borneo (from Mt. Kinabalu south to Liang Kubung and Penrissen), in forests between 1200 and 2000 m.
Habits: As other minivets.

491. SUNDA MINIVET *Pericrocotus miniatus* PLATE 56

Description: Large (19 cm), black and red, long-tailed minivet. Male distinguished by combination of black head, very long tail, and absence of red spot on

secondaries. Female is unique in being black and red like male; red covers throat, chin, and forehead, and mantle is also reddish.
Iris—brown; bill—black; feet—black.
Voice: A hard, shrill *chee-chee-chee*, or a protracted loud *tsree-ee*.
Range: Endemic to Sumatra and Java.
Distribution and status: Mountains of Sumatra (Leuser and all ranges to Dempu) and Java. A common bird of mountain forests between 1200 and 2400 m. Not recorded on Bali.
Habits: Lives in large flocks of up to 30 birds. Frequents treetops in and close to primary forest and pine plantations, but flocks sometimes come into cultivated lands.

492. SCARLET MINIVET *Pericrocotus flammeus* PLATE 56
(Indian/Flame Minivet)

Description: Large (19 cm) colourful minivet. Male is blue-black with red breast and belly, rump, outer edge of tail feathers, and 2 wing patches. The female is greyer on the back, has yellow in place of red, and yellow extends to the throat, chin, ear coverts, and forehead.
Iris—brown; bill—black; feet—black.
Voice: Soft *kroo-oo-oo-tu-tup*, *tu-turr*, or repeated *hurr*, also a higher-pitched *sigit, sigit, sigit*.
Range: India, S China, SE Asia, Philippines, Malay Peninsula, Greater Sundas, and Lombok.
Distribution and status: Resident in Sumatra (including islands), Borneo (patchy distribution but recorded all districts), Java, and Bali. A locally common bird in the lowlands and hills up to 1500 m, higher on Java.
Habits: Prefers primary forest where it flits among treetops of the finer leaved trees, in pairs or small groups.

LEAFBIRDS—FAMILY CHLOROPSEIDAE

A SMALL Oriental family of small to medium-sized, green-coloured, sweet-voiced birds. Leafbirds have short, thick, legs and long, slightly curved bills. The plumage is long, thick and fluffy, especially on the rump. Most species eat fruit and/or insects and make neat, cup-shaped nests in the leafy terminal branches of trees and bushes. They do not migrate.

Recent DNA studies suggest that Ioras and Leafbirds should be separated into two families.

Seven species occur in the Greater Sundas.

LEAFBIRDS 265

493. GREEN IORA *Aegithina viridissima* PLATE 57

Description: Small (13 cm) dark green leafbird with white wing bars. Male is similar to Common Iora but with green breast and darker upperparts. Female like female Common Iora but slightly darker green, with green lores and ear coverts, and yellowish rather than white wing bars.
Iris—grey; bill—black; feet—black.
Voice: Similar to Common Iora.
Range: Malay Peninsula, Sumatra, and Borneo.
Distribution and status: A common resident in the lowlands of Sumatra (including islands) and Borneo (including Natunas), up to 600 m.
Habits: Similar to Common Iora, living in canopy of primary and tall secondary forests.

494. COMMON IORA *Aegithina tiphia* PLATE 57
(Black-winged/Small Iora)

Description: Small (14 cm) green and yellow leafbird with two conspicuous whitish wing bars. Upperparts olive green; wing blackish but feathers with white edges; eye-ring yellow; underparts yellow. Island races vary in greenness. Distinguished from Green Iora by yellow lores and breast.
Iris—greyish white; bill—bluish black; feet—bluish black.
Voice: Several calls including a musical, monotonous trill or whistled *cheeepow* or *cheeepow cheeepow*, with the explosive *pow* dropping with whiplash ending. (See Appendix 6.)
Range: India, SW China, SE Asia, Palawan, Malay Peninsula, and Greater Sundas.
Distribution and status: Resident in Sumatra (including islands), Borneo (including N Bornean islands and Maratuas), Java, and Bali. Widespread and locally common in coastal lowlands, up to 1000 m.
Habits: Inhabits gardens, mangroves, open woodlands, and secondary forest. Generally single or in pairs, hopping about in the leafy branches of small trees and keeping well concealed.

495. LESSER GREEN LEAFBIRD *Chloropsis cyanopogon* PLATE 57
(Lesser Leafbird)

Description: Smallish (17 cm) bright green leafbird. Difficult to distinguish in field from Greater Green Leafbird but smaller, with proportionally smaller bill, and lacks blue patch on wing. Male has narrow yellow edge to black throat patch. Female has green instead of yellow throat.
Iris—dark brown; bill—black; feet—blackish.

Voice: Rich-toned song with distinct pauses between phrases of 2 or 3 notes; each phrase usually repeated several times.
Range: Malay Peninsula, Sumatra, and Borneo.
Distribution and status: Resident in lowland forest of Sumatra (up to 700 m) and Borneo (including N Bornean islands). Common on Borneo, but either overlooked or truly local on Sumatra.
Habits: As other leafbirds; favours primary and tall secondary forests.

496. GREATER GREEN LEAFBIRD *Chloropsis sonnerati* PLATE 57
(Greater/Great Green Leafbird)

Description: Large (22 cm) bright green leafbird with a black (male) or yellow (female) throat, blue malar stripe, bluish spot on shoulder, but no blue on wing; female has yellow eye-ring. Immature: similar to female but more yellow.
Iris—dark brown; bill—bluish grey; feet—bluish grey.
Voice: Short bursts of rich, liquid musical whistles interspersed by brief chatter (M. & W.). (See Appendix 6.)
Range: Malay Peninsula and Greater Sundas.
Distribution and status: Resident on Sumatra (including islands), Borneo (including Natunas), Java, and Bali. A widespread but uncommon bird of lowland and hill forests, up to 1000 m.
Habits: A bird of tall treetops in primary and secondary forest, and mangroves. It keeps to the dense leafy canopy where it is generally found singly or in pairs, sometimes in mixed flocks.

497. GOLDEN-FRONTED LEAFBIRD *Chloropsis aurifrons* PLATE 57

Description: Largish (19 cm) bright green leafbird with yellowish forehead (male) and shiny blue shoulder patch. Male has black chin and throat. Female has pale yellowish-green crown, blue malar stripe, and shoulder stripe and green throat. Immature like female but has green crown.
Iris—dark brown; bill—black; feet—blackish.
Voice: Musical song; and mimics other birds.
Range: India, SW China, SE Asia (except Malay Peninsula), and Sumatra.
Distribution and status: Common in the hill forests of Sumatra between 750 and 1500 m.
Habits: Actively searches out insects from upper and middle canopy by patrolling branches systematically. Often joins other species in mixed flocks.

498. BLUE-WINGED LEAFBIRD *Chloropsis cochinchinensis* PLATE 57
(Golden-hooded/Yellow-headed Leafbird)

Description: Medium-sized (17 cm) bright green leafbird with blue wings and black throat (male). Distinguished from other leafbirds by blue wings and edge of tail. Female lacks yellow eye-ring. Male has yellowish ring around black throat patch. Both sexes have blue malar stripe. Several races vary. Female of Kinabalu race *flavocincta* has black throat while other races have green throat. Bornean males of *flavocincta* and *viridinucha* have yellow forehead; Javan male *nigricollis* has green crown but golden-yellow upper breast, and Sumatran *icterocephala* has yellow crown and nape. All have bluer wings than Golden-fronted Leafbird.
Iris—dark brown; bill—black; feet—bluish grey.

Voice: Clear, liquid, musical *chee, chee, cheeweet* or *chee, cheeweet*, a twittering call and a sweet song. (See Appendix 6.)

Range: India, SW China, SE Asia, Malay Peninsula, and Greater Sundas.

Distribution and status: A locally common bird of lowland and hill forests, up to 1000 m on Sumatra (including islands), Borneo (including Natunas), and Java (where it occurs up to 1500 m). Not recorded on Bali.

Habits: Inhabits woodland, primary, and tall secondary forests, keeping to the tops of larger trees. Occurs singly, in pairs, or sometimes in small parties, and mixes readily with other species.

Note: The race *flavocincta*, occurring from Mts Kinabalu to Usun Apau and Dulit, with throat black in both sexes might be better treated as a separate species.

499. BLUE-MASKED LEAFBIRD *Chloropsis venusta* PLATE 57
(Masked Leafbird)

Description: Small (14 cm), pretty, bright green leafbird. Male has diagnostic blue forehead and sides of head, purple malar stripe with black throat patch, golden upper breast, blue shoulder patches, and blue tail. Female is lighter with diagnostic blue throat and sides of head. Tail bluish green.
Iris—dark brown; bill—black; feet—black.

Voice: Unknown.

Range: Endemic to Sumatra.

Distribution and status: Local but not uncommon in hill forest on ranges of Sumatra, usually between 600 and 1500 m.

Habits: As other leafbirds.

Bulbuls—Family Pycnonotidae

A LARGE African and Asian family of short-necked and short-winged birds with longish tails and rather slender bills. They have fluffy, soft plumage and several have an erectile crest. Plumage of males and females is similar and most bulbuls are rather dull in colour with, at most, yellow, orange, black, or white patterns.

Bulbuls are primarily frugivorous though they also eat many insects. They are confident birds with lively, and in some species, very musical songs. They tend to be arboreal and make untidy, cup-shaped nests in trees. None is migratory.

There are 29 species of bulbuls in the Greater Sundas.

500. STRAW-HEADED BULBUL *Pycnonotus zeylanicus* PLATE 58
(Straw-crowned/Yellow-crowned Bulbul)

Description: Large (28 cm) pale-headed bulbul with conspicuous black moustache. Crown and ear coverts straw-orange; back olive-brown, streaked with white; wings and tail greenish brown; chin and throat white; chest grey, streaked with white; belly grey; vent yellow.
Iris—reddish; bill—black; feet—dark brown.
Voice: Strong, clear, ringing, melodious, warbled but stereotyped song given antiphonally or in chorus.
Range: Malay Peninsula and Greater Sundas.
Distribution and status: In the lowlands and hills of Sumatra (including Nias) and Borneo, it remains a widespread bird, though formerly more common. On Java restricted to the west and now very rare, up to 800 m. A popular cage bird which faces constant pressure from trappers; birds are now imported from Sumatra and Borneo for the Javan markets. Not recorded on Bali.
Habits: Frequents secondary forest and forest edge, often in marshy reed bed areas, near rivers or swamps. A rather shy, inconspicuous bird, more often heard than seen.

501. CREAM-STRIPED BULBUL *Pycnonotus leucogrammicus* PLATE 58
(Streaked/Striated/Sumatran Bulbul)

Description: Medium-sized (18 cm) olive-brown bulbul with plumage prominently streaked with white. Wings and tail olive-green. Full, rather rounded crest. Sides of head and breast are olive-brown, streaked with white. Throat and belly white with yellow undertail coverts.
Iris—yellow; bill—blackish; feet—blackish.
Voice: Lively, bubbling calls.

Range: Endemic to Sumatra.
Distribution and status: Confined to the hill and mountain forests of the Barisan Range of Sumatra where it is locally common between 800 and 1200 m; recorded to 1900 m (Mt. Kerinci).
Habits: As other forest bulbuls. Favours lower montane and dense secondary forests, and coffee plantations.
Note: Placed by some authors with the Striated Bulbul *P. striatus* of SE Asia.

502. SPOT-NECKED BULBUL *Pycnonotus tympanistrigus* PLATE 59
(Olive-crowned Bulbul)

Description: Small (16 cm), crestless, olive-brown bulbul. Crown dark olive-brown; rump, wings, and tail greenish; lores and throat whitish; skin around eye black; ear coverts yellow. Underparts are mottled brown and white with barred olive-buff undertail coverts.
Iris—brown; bill—black; feet—brown.
Voice: No description.
Range: Endemic to Sumatra.
Distribution and status: Uncommon in hill forest on Sumatra, between 600 and 900 m. Recorded in Barisan Range south to Mt. Kerinci, but likely to occur S to Dempu.
Habits: Poorly known. Prefers secondary forest and lower storeys at edges of forest.

503. BLACK-AND-WHITE BULBUL *Pycnonotus melanoleucos* PLATE 58

Description: Medium-sized (18 cm), black, crestless bulbul with diagnostic white wing coverts. Immature birds are mottled with brown.
Iris—red to brown; bill—black; feet—black.
Voice: Tuneless dissyllable *pet-it* (P.R.).
Range: Malay Peninsula, Sumatra, and Borneo.
Distribution and status: Forest in lowlands and hills of Sumatra (including Mentawai) and Borneo. Few records from Sumatra but occurs at least up to 1200 m; on Borneo probably nomadic.
Habits: Prefers heath forest and swamp forest.

504. BLACK-HEADED BULBUL *Pycnonotus atriceps* PLATE 58

Description: Medium-sized (17 cm) yellowish bulbul with diagnostic glossy black head and black throat. Upperparts yellowish olive; wings blackish; tail blackish with conspicuous yellow tip; underparts greenish yellow. A rare colour form is grey with a white tip to the tail. A dull greenish form is confined to Bawean Island.

Iris—pale blue; bill—black; feet—brown.
Voice: Lively, sharp, musical chirps.
Range: NE India, SE Asia, Palawan, Malay Peninsula, and Greater Sundas.
Distribution and status: On Sumatra (including islands), Borneo (including Maratuas), and Java (including Bawean) fairly common bird lowland areas up to 900 m. In Bali it appears rare.
Habits: Frequents forest edge, secondary forest, and coastal scrub. Single or in small flocks, often mixing with other species.

505. BLACK-CRESTED BULBUL *Pycnonotus melanicterus* PLATE 58
(Black-headed/Black-crested Yellow Bulbul)

Description: Largish (18 cm) yellowish bulbul with black head and crest. Sumatran and Javan birds have bright red throat; on Borneo throat is yellow. Upperparts brownish olive; underparts yellow.
Iris—cream or reddish; bill—black; feet—black.
Voice: Noisy, penetrating *hee-tee-hee-tee-weet* with last note falling.
Range: India, S China, SE Asia, Malay Peninsula, and Greater Sundas.
Distribution and status: Not uncommon in lowland and hill forests of Sumatra, up to 1200 m. A common mountain resident on Borneo, found from Mt. Kinabalu to the upper Kayan and Liang Kubung. In Java region it is more common in S and W Java and rare on Bali.
Habits: Rather shy, favouring densely leaved, taller trees of the forest edge and secondary forests. Occasionally chases flying insects but generally searches actively for fruits. Erects crest when excited.
Note: Red-throated race *dispar* on Sumatra, Java, and Bali; yellow throated *montis* on Borneo. Asian races have black throat and species limits probably need revision. Some authors treat red-throated form as a separate species Ruby-throated Bulbul.

506. SCALY-BREASTED BULBUL *Pycnonotus squamatus* PLATE 58

Description: Small (15 cm) black-headed bulbul with white chin and throat and diagnostic black and white scaled breast and flanks. Back olive; rump and vent bright yellow; wing tips black; tail black with white tips to outer feathers; belly white.
Iris—red; bill—black; feet—black.
Voice: Cheerful, whistling song and babbling calls.
Range: Malay Peninsula and Greater Sundas.
Distribution and status: A rather rare bird of the hill forests of Sumatra, Borneo,

and W Java. The few records from Sumatra are from 400 to 700 m; there are no recent records from Java. Not recorded on Bali.

Habits: Sociable flocking bird inhabiting the tops of small bushes.

507. GREY-BELLIED BULBUL *Pycnonotus cyaniventris* PLATE 58

Description: Smallish (16 cm), dark-headed, olive bulbul with diagnostic grey underparts. Crown and nape dark grey; wing edges and central tail feathers blackish; mantle, back, and wing coverts olive-green with lighter secondaries; vent golden-yellow.
Iris—dark brown; bill—black; feet—black.

Voice: Sharp, lively *chirrup* calls.

Range: Malay Peninsula, Sumatra, and Borneo.

Distribution and status: On Sumatra (including Mentawai) and Borneo uncommon in lowland and hill forest, up to 1000 m.

Habits: Dainty bulbul, living in small flocks in open canopy of lowland forests.

508. RED-WHISKERED BULBUL *Pycnonotus jocosus* PLATE 58

Description: Medium-sized (20 cm) bulbul with long, narrow, forward-pointing, black crest and diagnostic red ear patch on black and white patterned head. Rest of upperparts brownish and underparts buff with red vent. Tip of tail with white edges. Immature lacks red ear patch and has pink vent.
Iris—brown; bill—black; feet—black.

Voice: Loud, incessant chattering and a sweet, short, whistled 2- or 3-note song.

Range: India, S China, and SE Asia. Introduced to Australia and other regions.

Distribution and status: Records from Java can be attributed to cage bird escapes but populations formerly established on the E coast of Sumatra may have been natural colonizations from the Asian continent. No populations known.

Habits: Lives in noisy, active flocks. Perky birds sit on prominent perches, often the very highest point of small trees, to sing and chatter. Favours open wooded areas, forest edge and secondary growth, and villages.

509. SOOTY-HEADED BULBUL *Pycnonotus aurigaster* PLATE 58
(Black-capped/Golden-vented Bulbul)

Description: Medium-sized (20 cm) black-capped bulbul with whitish rump and yellow-orange vent; chin and top of head black; collar, rump, breast, and belly white; wings black; tail brown.
Iris—red; bill—black; feet—black.

Voice: Melodious calls and loud notes *chook, chook.*

Range: S China, SE Asia (except Malay Peninsula), and Java. Introduced to Sumatra and S Sulawesi. Recently reached S Borneo.
Distribution and status: Established on Sumatra; in S Sumatra colonization might have been from Java. First record on Borneo (Palangkarya) in 1984. On Java and Bali this is one of the most widespread and common species, up to about 1600 m.
Habits: Lives in noisy, active flocks, often mixing with other bulbuls. Prefers open wooded or bushy habitats, forest edge, secondary growth, parks, and gardens, even in large towns.

510. PUFF-BACKED BULBUL *Pycnonotus eutilotus* PLATE 59

Description: Medium-sized (20 cm) brown bulbul with full crest. Upperparts brown; underparts whitish, washed grey and with buff vent. Usually has distinctive white tips to tail feathers. Bornean race has only very short crest. Sumatran form is paler underneath.
Iris—red; bill—black; feet—grey.
Voice: Whistled call of a single note followed by a triplet with last note descending.
Range: Malay Peninsula, Sumatra, and Borneo.
Distribution and status: On Sumatra (including Bangka) and Borneo uncommon and local in forests and secondary growth, up to 400 m.
Habits: Typical bulbul of scrub and forest edge, keeping to lower and middle storeys.

511. BLUE-WATTLED BULBUL *Pycnonotus nieuwenhuisi* PLATE 59
(Malaysian Wattled Bulbul)

Description: Smallish (18 cm) olive-green bulbul with blackish head, short crest, blue fleshy eyelids, and white-tipped tail.
Iris—brown; bill—black; feet—grey.
Voice: Unknown.
Range: Endemic to Sumatra and Borneo.
Distribution and status: Very rare. Known only from one specimen from N Sumatra (Lesten, Aceh at 700 m) and one specimen from S Borneo (upper Kayan at 600 m).
Habits: Very poorly known. Recorded from open scrubby forest at moderate altitudes.

512. ORANGE-SPOTTED BULBUL *Pycnonotus bimaculatus*   PLATE 59

Description: Medium-sized (20 cm) brown and white bulbul with a yellow vent and diagnostic orange lores and spot above eye. Upperparts olive brown; throat and upper breast blackish brown; lower breast mottled brown and white; belly white or dusky. Specimens from W Java have yellowish ear coverts.
Iris—brown; bill—black; feet—black.
Voice: Hard, loud calls *tuk-tuk-tuk-turuk*.
Range: Endemic to Sumatra, Java, and Bali.
Distribution and status: Common in the mountains of Sumatra, Java, and Bali, from 800 to 3000 m.
Habits: Favours forest edge and open glades in the montane forests right up to the *Vaccinium* zones of the highest peaks. A loud active bird, single or in small parties.

513. FLAVESCENT BULBUL *Pycnonotus flavescens*   PLATE 59
(Blyth's/Pale-faced Bulbul)

Description: Medium-sized (20 cm) olive-green bulbul with bright yellow vent, white lores, face, and throat, and greyish, streaky breast. There is a short full crest. Immature has duller lores. Bornean race *leucops* is paler with white lores.
Iris—brown; bill—black; feet—grey.
Voice: A short song of five notes and a very harsh chatter (Smythies).
Range: NE India, SW China, SE Asia (except Malay Peninsula), and Borneo.
Distribution and status: A common mountain bird in N Borneo between 1000 and 3000 m, recorded from Mts Kinabalu to Murud and Mulu.
Habits: Typical flock-living bulbul of open forests, forest edge, and secondary growth of mid-mountains.

514. YELLOW-VENTED BULBUL *Pycnonotus goiavier*   PLATE 59

Description: Medium-sized (20 cm) brown and white bulbul with distinct yellow vent. Crown dark brown; eyebrow white; lores black; upperparts brown; throat, breast, and belly white with light brown streaks on flanks.
Iris—brown; bill—black; feet—pinkish grey.
Voice: Repetitive, cheerful, rich, non-musical *tidli-tidli*.
Range: SE Asia, Philippines, Malay Peninsula, Greater Sundas, and Lombok. Introduced to S Sulawesi.
Distribution and status: On Sumatra (including eastern islands), Borneo (including Batambangan and Maratuas), Java, and Bali this is a common bird, up to 1500 m.
Habits: Forms flocks, often mixing with other bulbuls. Gathers in communal

roosts for the night. Prefers open habitat, secondary growth, roadsides, and gardens. Spends more time feeding on the ground than other bulbuls.

515. OLIVE-WINGED BULBUL *Pycnonotus plumosus* PLATE 59
(Olive-brown/Large Olive Bulbul)

Description: Medium-sized (20 cm), dull, greyish-brown bulbul with red eyes and olive wings. Upperparts washed greenish; chin and throat whitish; ear coverts streaked with whitish; underparts faintly streaked ochraceous; undertail buffy brown. Distinguished from Cream-vented Bulbul by larger size, more greenish appearance, and red eye (brown in immature).
Iris—red; bill—black; feet—brown.
Voice: Rather infrequent, melodious, quiet, chattering song; similar to Yellow-vented Bulbul but more clipped and sharply phrased (C.H.).
Range: Malay Peninsula, Palawan, and Greater Sundas.
Distribution and status: On Sumatra (including islands) and Borneo (including islands) a common bird in lowland forest, up to 300 m. In Java occurs principally in W and C Java, up to 800 m; also occurs on Bawean. Not recorded on Bali.
Habits: Frequents forest edge, plantations, and lightly wooded areas. Generally solitary or in pairs, keeping to the middle and upper canopy.

516. CREAM-VENTED BULBUL *Pycnonotus simplex* PLATE 59
(White-eyed Brown/White-eyed Bulbul)

Description: Smallish (17 cm), dull, brownish-grey bulbul. Similar to Olive-winged Bulbul but smaller, less green, with a whitish chin and throat, and white belly. Γ stinguished in Sumatra by pale whitish eyes (brown in juveniles). Bornean adults have red eyes though a few white-eyed individuals have been found. Javan birds have reddish eyes. Bornean specimens distinguished from the Red-eyed Bulbul by paler creamy vent.
Iris—white or red; bill—black; feet—brown.
Voice: Soft chattering calls *chirriup*, rapidly repeated hysterically.
Range: Malay Peninsula and Greater Sundas.
Distribution and status: On Sumatra (including islands) and Borneo (including Natunas and Anambas) not uncommon up to 600 m (locally to 1300 m on Borneo). On Java it is common only in a few areas, mostly at altitudes below 500 m along the S coast. Not recorded on Bali.
Habits: Lives in primary forest, open country with secondary growth, or deserted clearings. Generally keeps to the middle and tops of trees; sometimes mixes with other bulbuls.

517. RED-EYED BULBUL *Pycnonotus brunneus* PLATE 59
(Brown/Red-eyed Brown Bulbul)

Description: Smallish (17 cm) plain brown bulbul with red eyes. Lacks orange eye-ring of Spectacled Bulbul, and smaller and less green than Olive-winged Bulbul and without white streaks on ear coverts. Similar to Bornean Cream-vented Bulbul but vent darker brownish buff.
Iris—red (brown in immature); bill—brown; feet—brown.
Voice: Typical bulbul chatters.
Range: Malay Peninsula and Greater Sundas.
Distribution and status: Common in lowland forest of Sumatra (including islands) and Borneo (including islands). In Java known only from the Mata Siri Islands of the Java Sea. Not recorded on Bali.
Habits: Prefers secondary forest, forest edge, and scrub.

518. SPECTACLED BULBUL *Pycnonotus erythrophthalmos* PLATE 59
(Lesser Brown Bulbul)

Description: Smallish (18 cm) brown bulbul with red eyes. Distinguished from Red-eyed Bulbul and Olive-winged Bulbul by a fleshy orange eye-ring. Underparts are creamy with whitish throat and grey wash on breast.
Iris—red; bare eye-ring—orange; bill—black; feet—grey.
Voice: High-pitched, metallic tinkling (P.R.).
Range: Malay Peninsula, Sumatra, and Borneo.
Distribution and status: On Sumatra (including islands) and Borneo a rather common bird of primary, secondary, and swamp forests at low altitudes.
Habits: As other brown bulbuls with which it sometimes mixes at feeding sites.

519. FINSCH'S BULBUL *Criniger finschii* PLATE 60

Description: Small (16 cm), stocky, brown bulbul with yellow underparts and reddish brown eye. Distinguished from other red-eyed brown bulbuls by yellow underparts (washed grey-green on breast); from other *Criniger* bulbuls by yellow throat, brown sides of face, and lack of eyebrow or crest; distinguished from Hairy-backed Bulbul by yellower throat and lack of eye-ring.
Iris—red-brown; bill—black; feet—grey.
Voice: Musical song and loud calls *bierchef-bierchef*, *pek-pek-pek-pek*, and others.
Range: Malay Peninsula, Sumatra, and Borneo.
Distribution and status: Status on Sumatra uncertain; presumably resident but known by only one specimen from Tasik river, N Sumatra. On Borneo sparsely distributed in lowland and hill forest.
Habits: Puffs out throat like other *Criniger* bulbuls.

520. OCHRACEOUS BULBUL *Alophoixus ochraceus* PLATE 60
(Brown White-throated Bulbul)

Description: Large (23 cm), crested, brown bulbul with white puffy throat and buffy underparts with cinnamon vent. Bornean birds are browner with dark chestnut wings and tail; Sumatran birds are greenish yellow underneath with ochraceous vent.
Iris—reddish; bill—horn; feet—pale horn.
Voice: Noisy with various rich, grating cackles and chatters *chi-wau, chi-wau*, also musical whistling.
Range: SE Asia, Malay Peninsula, Sumatra, and Borneo.
Distribution and status: On Sumatra and Borneo not uncommon in hill forests, from 300 to 1600 m.
Habits: Keeps more to the forest interior than other bulbuls and often takes insects off the forest floor. A noisy, active bird. Puffs out throat feathers conspicuously.

521. GREY-CHEEKED BULBUL *Alophoixus bres* PLATE 60
(Olive White-throated/Scrub Bulbul)

Description: Largish (22 cm) crested brownish bulbul with yellow underparts and conspicuous white chin and throat. Upperparts olive brown, more rufous on tail; cheeks grey. White throat is often puffed out prominently.
Iris—reddish; bill—black and heavy; feet—greyish brown.
Voice: Harsh, unmelodious call *tru-tu-tu-tu-tu-tu-tu-tu-tu-chok-chok*, and others; also harsh, chattering alarm call.
Range: Malay Peninsula, Palawan, and Greater Sundas.
Distribution and status: Local in lowland forest on Sumatra and Borneo, widespread and locally common on Java and Bali, up to 1500 m.
Habits: Favours primary or secondary forest with thickets of low bushes. Generally single or in pairs. A noisy, active bird of the lower canopy. Occasionally joins mixed species flocks.

522. YELLOW-BELLIED BULBUL *Alophoixus phaeocephalus* PLATE 60
(Crestless White-throated/Grey-headed Bulbul)

Description: Medium-sized (20 cm), grey-headed, olive-brown bulbul with yellow underparts and white throat. Distinguished from Grey-cheeked Bulbul by grey crown and lack of crest. Most races have yellow tip on tail but Sabah race *connectens* has plain tail.
Iris—reddish brown; bill—black; feet—pinkish brown.
Voice: Grating chattering.

Range: Malay Peninsula, Sumatra, and Borneo.
Distribution and status: Common in lowland forest on Sumatra (including islands) and Borneo (including Natunas), up to at least 800 m.
Habits: Lives in noisy, active flocks in forest, forest edge, and secondary scrub. Puffs out throat feathers like Grey-cheeked Bulbul.

523. HOOK-BILLED BULBUL *Setornis criniger* PLATE 60
(Long-billed Bulbul)

Description: Medium-sized (20 cm), crestless, dark bulbul with flattened and hooked bill. Upperparts mostly brown with crown, tail, and wings darker than back. Whitish eyebrow, black eye-stripe and malar stripe; cheeks mottled grey. Underparts white, washed grey on sides of breast and flanks. White spots at tip of tail conspicuous in flight.
Iris—brown; bill—black; feet—black.
Voice: Harsh alarm call *currrk*.
Range: Endemic to Sumatra and Borneo.
Distribution and status: Few records from lowlands of E Sumatra (including Bangka) where restricted to peat swamp and heath forests. On Borneo apparently more widespread and commoner, and locally recorded up to 1000 m.
Habits: Rarely found in primary forest but prefers peat swamp and heath forests where it lives in the middle and lower canopy.

524. HAIRY-BACKED BULBUL *Tricholestes criniger* PLATE 60

Description: Smallish (16 cm), crestless, olive-brown bulbul with pale eye-ring, yellowish sides of head, and yellowish grey underparts. In the hand, the bird shows long hairs on the back which give it its common name. Throat white; breast mottled grey; belly and vent yellow. Small size, lack of crest, and yellow ear coverts separate it from other similarly coloured bulbuls.
Iris—brown or grey; bill—black; feet—black.
Voice: Short chattering notes followed by low, then high whistle and rising final note of middle pitch.
Range: Malay Peninsula, Sumatra, and Borneo.
Distribution and status: Common in lowland forest, up to 1000 m on Sumatra (including Batu Islands and Lingga Archipelago) and Borneo (including Natunas).
Habits: Lives in small flocks in thickets, brush, and lower storey of scrubby forests and secondary growth.

525. BUFF-VENTED BULBUL *Iole olivacea*   PLATE 60
(Crested Olive/Finsch's Olive/Dull-brown/Charlotte's Bulbul)

Description: Medium-sized (19 cm) brown bulbul with white throat, underparts with greyish wash on breast and flanks, and buff vent. Slight crest and indistinct pale eyebrow. Distinguished from most other plain brown bulbuls by white or grey eyes. Birds from E Borneo are paler with a yellow wash on underparts.
Iris—grey or white; bill—brown; feet—brown.
Voice: Various harsh notes.
Range: Malay Peninsula, Sumatra, and Borneo.
Distribution and status: On Sumatra (including islands) and Borneo (including Natunas and Anambas) locally abundant in forest and secondary growth from sealevel to 600 m.
Habits: Noisy, flock-living bulbul of all forest layers.

526. SUNDA BULBUL *Iole virescens*   PLATE 60
(Green Mountain/Green-winged Bulbul)

Description: Medium-sized (20 cm) dull bulbul with streaked underparts and slight crest. Crown grey; back, wings, and tail greenish olive (Java) or brownish olive (Sumatra); cheeks, throat, breast, and flanks greenish grey, heavily streaked yellowish white; vent yellowish white. Javan birds are greener than the Sumatran race.
Iris—red; bill—black; feet—blue-grey.
Voice: Loud, clear ringing call *chet-chet-chet*.
Range: Endemic to Sumatra and Java.
Distribution and status: Locally common in the mountains of Sumatra and Java, between 850 and 2400 m.
Habits: Frequents montane and sub-montane forests. A social flocking bird, sometimes mixing with other species. Generally high in the canopy. Occasional in alpine heath zone.
Note: Treated by some authors as conspecific with the Streaked Bulbul.

527. STREAKED BULBUL *Ixos malaccensis*   PLATE 60
(Green-backed Bulbul)

Description: Medium-sized (22 cm), dark olive, crestless bulbul with grey breast streaked with white, and white belly and vent. Differs from Sunda Bulbul by lack of crest, lack of streaking on crown, and white not yellow vent.
Iris—reddish brown; bill—horn; feet—pink.
Voice: Loud song *kajie ka-jie ka-jie juee*, running into delicate down-scale trills; also rasping *jizeer jizeer* in flight (Wells).

Range: Malay Peninsula, Sumatra, and Borneo.
Distribution and status: An occasional bird of lowland forests up to 900 m in Sumatra (including islands), and up to 1200 m on Borneo.
Habits: Lives in small flocks in canopy of tall forests.

528. ASHY BULBUL *Hypsipetes flavala* PLATE 60
(Brown-eared Bulbul)

Description: Medium-sized (20 cm) bulbul with slight crest, dark brown crown, brown upperparts, and white throat. Two Sundaic races differ greatly in appearance: in Borneo wings and undertail coverts are yellowish green; Sumatra birds lack green in plumage and vent is white.
Iris—reddish brown; bill—black; feet—black.
Voice: Harsh *trrk*, low reedy *li-deet di-deet*, and low throaty *to to to*.
Range: Himalayas, SW China, SE Asia, Malay Peninsula, Sumatra, and Borneo.
Distribution and status: Common in the hills and mountains of Sumatra, from 500 to 1000 m; on Borneo locally abundant in the mountains and recorded in lowlands to over 2400 m.
Habits: Typical forest bulbul living in small parties and keeping to middle and lower storeys of open, sub-montane forests and scrub. Puffs out throat feathers like *Criniger* bulbuls.

Drongos—FAMILY DICRURIDAE

A SMALL family of blackish, insectivorous birds found from Africa to Asia, Australia, and the Solomons. Most species have glossy black plumage, powerful beaks, and long forked tails. They hunt large insects in the air from a tree perch. Their calls are loud and sometimes melodious, but usually harsh with discordant screeching. They are also excellent mimics of other birds. Drongos boldly attack hawks and cuckoos. Their nests are neatly woven cups, placed in low branch forks.

Seven resident species and one northern visitor occur in the Greater Sundas. Two of these species have extraordinary elongated outer tail feathers with terminal racket webs.

529. BLACK DRONGO *Dicrurus macrocercus* PLATE 61

Description: Medium-sized (29 cm) dull black drongo. Beak relatively small, tail very long and very deeply forked, often held at curious angle in wind. Immature has whitish barring on lower underparts.
Iris—red; bill—black; feet—black.

Voice: Varied ringing calls, *hee-luu-luu*, *eluu-wee-weet*, or *hoke-chok-wak-we-wak*.
Range: Iran to India, China, SE Asia (except Malay Peninsula), Java, and Bali.
Distribution and status: Migrants from SE Asia may reach Sumatra. Resident in Java and Bali where it is a common bird of low-lying open country, occasionally up to 1600 m.
Habits: A bird of open spaces, often sitting on small trees or telegraph wires.

530. ASHY DRONGO *Dicrurus leucophaeus* PLATE 61
(Pale/Grey Drongo)

Description: Medium-sized (29 cm) grey drongo with long, deeply forked tail. Races vary in paleness. Bornean race *stigmatops* has whitish patch around eye. Iris—orange-red; bill—grey black; feet—black.
Voice: Clear, loud song *huur-uur-cheluu* or *wee-peet*, *wee-peet*. Mewing and mimics the calls of other birds; is reported to sometimes call at night.
Range: Afghanistan to China, SE Asia, Palawan, and Greater Sundas.
Distribution and status: In Sumatra (including Mentawai and Simeulue), Borneo, Java, and Bali it is a common bird of open woodlands and forest edge, in hills and mountains, from 600 to 2500 m.
Habits: Lives in pairs and sits on a bare branch or vines in open clearings or gaps in the forest, hawking after passing insects, climbing after rising moths, or diving after flying prey.

531. CROW-BILLED DRONGO *Dicrurus annectans* PLATE 61

Description: Medium-sized (26 cm) black drongo with heavy crow-like bill and deeply forked tail with upturned outer feathers. Distinguished from Black Drongo by heavier bill and broader, less deeply forked tail. Immature has whitish barring on lower underparts.
Iris—red brown ; bill—black; feet—black.
Voice: Typical drongo calls with loud musical whistles and harsh churs; also characteristic descending series of harp-like notes (M. & W.).
Range: Resident in the Himalayas and S China; migrant south to SE Asia and the Greater Sundas.
Distribution and status: Uncommon migrant to Sumatra, NW (mostly) Borneo and W Java, mostly in coastal forests and mangroves. Not recorded on Bali.
Habits: Prefers open woodland, coastal scrub, and low mangrove forest. Typical drongo hunting behaviour.

532. BRONZED DRONGO *Dicrurus aeneus* PLATE 61

Description: Small (23 cm), glossy blue-black drongo. Distinguished from Black Drongo by smaller size, glossy plumage, only moderately forked tail, and by habits and habitat; from Crow-billed Drongo by shorter tail, and from Drongo-cuckoo by lack of white undertail barring.
Iris—brown; bill—black; feet—black.
Voice: Loud calls including clear notes and harsh discords.
Range: India, S China, SE Asia, Malay Peninsula, Sumatra, and Borneo.
Distribution and status: Common resident on Sumatra and Borneo, in primary and secondary lowland forests, up to 1400 m.
Habits: Sits on prominent perches, sallying after insects in hawking flights through upper and middle canopy. Mobs raptors and cuckoos boldly. Several birds may chase each other noisily. Favours gaps in canopy.

533. LESSER RACKET-TAILED DRONGO *Dicrurus remifer* PLATE 61

Description: Medium-sized (26 cm without rackets) glossy black drongo with amazing elongated outer tail feathers with terminal rackets. Tuft of short feathers forms ridge above bill. Smaller than Greater Racket-tailed Drongo and lacks frontal crest; most easily distinguished by square-cut tail. Generally separated from Greater Racket-tailed by altitude as the two species overlap only from 1000 to 1500 m. Moulting birds may lack rackets.
Iris—red; bill—black; feet—black.
Voice: Varied and melodious whistling notes *weet-weet-weet-weet-chewee-chewee* and occasional harsh screeches. Mimics the calls of other birds.
Range: India, S China, SE Asia, Malay Peninsula, Sumatra, and Java.
Distribution and status: Found in mountains of Sumatra and W Java where locally common between 1000 and 2500 m. Not recorded on Bali.
Habits: Inhabits dense rainforest, secondary forest, and forest edge. Reported to follow grass fires to catch fleeing grasshoppers and other disturbed prey. Boldly mobs raptors and crows.

534. HAIR-CRESTED DRONGO *Dicrurus hottentottus* PLATE 61
(Spangled Drongo)

Description: Largish (32 cm) glossy black drongo. Plumage spangled with light iridescent spots (spangles). Tail long and forked with blunt, strongly upcurved outer feathers giving it a lyre shape. Some races have crest of long hair-like feathers on crown. Eastern races have white eyes.
Iris—red or white; bill—black; feet—black.

Voice: Melodious loud singing with occasional harsh screeches.
Range: India, China, SE Asia (except Malay Peninsula), Mentawai Islands off W Sumatra, Borneo, Java, and Bali.
Distribution and status: In Mentawai, Borneo (including Maratuas), E Java (including Masalombo Besar and Matasiri in Java Sea, and Kangean), and Bali it is quite a common bird of lowland and sub-montane forest, especially in drier areas.
Habits: Prefers rather open parts of the forest and sometimes gathers in noisy parties, singing and chasing insects in the sky, especially at dawn and dusk. Hawks after insects from low perch; mixes with other species and also follows monkeys and squirrels, catching insects disturbed by their movements. Sometimes settles on dead trees to eat emerging beetles or termites like a woodpecker.
Note: If the Sumatran form *sumatranus* is treated (as it is here) as a distinct species, the name Spangled Drongo should then be reserved for the Wallacean forms *D. bracteatus*. This leaves the Mentawai race *viridinitens* as a very remote relative of *hottentottus* and it may be better to put it in *sumatranus* or to consider it as a distinct species.

535. SUMATRAN DRONGO *Dicrurus sumatranus*　　　PLATE 61

Description: Largish (29 cm) glossy black drongo with slightly forked, wide tail. Similar to Hair-crested Drongo which it replaces in Sumatra, but smaller, lacks hair crest, has shorter tail with outer feathers less upturned, shorter bill and fewer spangles.
Iris—red; bill—black; feet—black.
Voice: Melodious loud singing with occasional harsh screeches.
Distribution and status: Endemic to Sumatra and offshore islands. It is a common bird of lowland forest, and secondary forest especially in drier parts of the region.
Habits: As Hair-crested Drongo.
Note: Some authors place it within the larger species Hair-crested Drongo, together with Wallacean races, which can then all be referred to as Spangled Drongo.

536. GREATER RACKET-TAILED DRONGO *Dicrurus paradiseus*
PLATE 61

Description: Large (30 cm without rackets) glossy black drongo with amazing elongated outer tail feathers with terminal rackets. Rackets are webbed only on outer side of feather shafts and are twisted. Distinguished from Lesser Racket-tailed Drongo by forked tail. Crest of elongated feathers on crown of adults not readily visible in the forest.

Iris—red; bill—black; feet—black.

Voice: Glorious variety of lusty, resonant melodies of warbles, whistles, and bell-like notes with typical drongo harsh churrs. Often mimics other songbirds.

Range: India to China, SE Asia, Malay Peninsula, and Greater Sundas.

Distribution and status: In Sumatra (including islands) and Borneo it remains common in forests up to 700 m. In Java and Bali it was a widespread and common bird of lowland forests up to 1400 m; as this forest is vanishing the bird is increasingly rare.

Habits: Occupies mangroves, swamp forest, primary and secondary forests. Pair-living bird, sometimes gathering in groups to display, giving noisy, lusty song, and hawking for insects from low exposed perches in the forest.

Orioles—Family Oriolidae

A SMALL family of robust birds with often colourful plumage and powerful straight beaks. Orioles feed on fruits and insects. Nests are woven cups of roots and fibres twisted around supporting twigs and suspended from tree forks. They have clear, loud, melodious songs. The flight is leisurely and undulating. The family is reassigned by Sibley and Monroe (1990) as a tribe of the enlarged crow family Corvidae.

Six species occur in the Greater Sundas.

537. DARK-THROATED ORIOLE *Oriolus xanthonotus* PLATE 62
(Black-throated/Malaysian Oriole)

Description: Smallish (18 cm) black and yellow oriole. Male: head, neck, and throat black. Flight feathers black. Breast whitish, streaked with black, distinguishes it from Black-hooded Oriole. Otherwise bright yellow. Female and immature: greenish upperparts, yellow vent, rest of underparts white with bold black streaks.

Iris—red; bill—pink; feet—black.

Voice: Repeated, long, descending, ringing whistle of *pee-pheeu*, liquid *ti-ti-lu-i*, and a creaky screech. Call is a weaker, less melodious version of Black-naped Oriole. (See Appendix 6.)

Range: Philippines, Malay Peninsula, and Greater Sundas.

Distribution and status: In Sumatra (including Mentawai and Bangka), Borneo (including islands), and Java it is an occasional bird of lowland forest, locally up to 1000 m. Not recorded on Bali.

Habits: Similar to Black-naped Oriole but generally found in coastal areas, preferring the canopy of primary and tall secondary forests, forest edge, and swamp forest.

538. BLACK-NAPED ORIOLE *Oriolus chinensis* PLATE 62

Description: Medium-sized (26 cm) yellow and black oriole with black stripe through eye and nape; flight feathers largely black. Male otherwise bright yellow; female duller with olive-yellow back. Immature olive instead of black; underparts whitish with black streaks.
Iris—red; bill—pink; feet—black.
Voice: Clear, liquid, flute-like whistle *leeuw*, *klee-lee-tee- leeuw*, or *u-dli-u*, and variations. Also a very harsh scolding call note and steady, plaintive, quiet whistle. (See Appendix 6.)
Range: India, China, SE Asia, Philippines, Sulawesi, Malay Peninsula, and Sundas.
Distribution and status: In Sumatra (including islands), Java (including Kangean), and Bali this is a locally common bird, up to 1600 m. On Borneo apparently rare and known by a few specimens from Sarawak and Kalimantan.
Habits: Inhabits open forests, plantations, park land, villages, mangroves, and beach forests. Lives in pairs or family parties. Keeps to the trees but will come down in search of insects. Has a conspicuous, slow, powerful wing-beat and an undulating flight.

539. BLACK-HOODED ORIOLE *Oriolus xanthornus* PLATE 62
(Asian/Indian/Oriental Black-headed Oriole)

Description: Medium-sized (20 cm), black-hooded, yellow oriole with yellow underparts. Wings and tail are black. Immature is similar to adult but has yellow forehead, whitish bill, whitish eye-ring, and blackish stripes on dirty white throat.
Iris—red; bill—red in adult; feet—black.
Voice: Similar to Black-naped Oriole, liquid, whistled call of 4 notes *yiu-hu-a-yu* with middle notes stressed; answered by 3 note *tu huee* or *tee-heh* (Smythies). Also mimics other bird calls.
Range: India, S China, SE Asia, Sumatra, and Borneo.
Distribution and status: A winter visitor to N and E Sumatra. A resident race in NE Borneo from Tawau south to Maratua Islands is not uncommon in forest and wooded country, up to 1000 m.
Habits: As Black-naped Oriole but prefers forest edge, open forests, cultivated areas, and secondary forest.

540. BLACK ORIOLE *Oriolus hosii* PLATE 62

Description: Medium-sized (21 cm) black oriole with diagnostic chestnut vent. Distinguished from Black-and-crimson Oriole by black breast and from black crows by small size, chestnut vent, and pale bill.
Iris—red; bill—dull pink; feet—grey.
Voice: Clear whistles with downward inflection.
Range: Endemic to Borneo.
Distribution and status: Confined to the northern mountains between 1200 to 2000 m where known from the outlier Mt. Dulit, Ulu Sabai south to Batu Tibang, but absent from Mt. Kinabalu; at Ulu Sabai replaced by the Black-and-crimson Oriole above 1000 m.
Habits: A quiet but active bird of the forest canopy, generally found in small parties.

541. BLACK-AND-CRIMSON ORIOLE *Oriolus cruentus* PLATE 62
(Crimson-breasted Oriole)

Description: Medium-sized (22 cm) black and red oriole. Male: black with crimson lower breast and wing patch. Female: black with streaky chestnut breast.
Iris—brown; bill—pale blue-grey; feet—black.
Voice: Cat-like *meow*, *kek kreo*, *kreo kek*, or *hee-hee-hee-hiew*.
Range: Malay Peninsula and Greater Sundas.
Distribution and status: On Sumatra it is an occasional bird between 500 and 2400 m. On Borneo it is a locally common montane bird on Mt. Mulu and from Mt. Kinabalu south to Batu Tibang. On Java it is an uncommon bird of hill forest and montane forests, mostly 1200 to 1800 m. Not recorded on Bali.
Habits: Inhabits montane and moss forest where it lives in the tree canopy, rarely coming to the ground. Travels singly or in pairs.

542. ASIAN FAIRY-BLUEBIRD *Irena puella* PLATE 62
(Fairy Bluebird/Blue-mantled/Blue-backed Fairy-Bluebird)

Description: Medium-sized (25 cm) blue and black bird. Male: unmistakable with crown, nape, back, upperwing coverts, rump, uppertail coverts, and vent rich, shining blue; rest of plumage black. Female: greenish blue all over with brighter rump.
Iris—red; bill—black; feet—black.
Voice: Loud, ringing, drawn-out, liquid whistles *whee-eet* on rising scale, sometimes preceded by several introductory notes on a descending scale. Often calls in flight. (See Appendix 6.)

Range: India, SW China, SE Asia, Malay Peninsula, Palawan, and Greater Sundas.

Distribution and status: In Sumatra (including islands) and Borneo this is a common lowland bird in undisturbed forests, up to 1100 m. On Java it is an occasional bird of the lowland forest. Not recorded on Bali.

Habits: Seen singly or in small flocks. Keeps to the tops of tall trees and is most frequently seen when visiting fruiting figs where it mixes with other birds. Flies with undulating beat. Frequents swamp forest, primary forest, and tall secondary forests.

Note: Placed by some authors in its own family Irenidae. DNA evidence suggests it should be placed with leafbirds.

Crows and jays—FAMILY CORVIDAE

Crows and jays are a family of generally large birds with powerful, straight bills and strong feet. They occur almost worldwide. They are intelligent wily birds and several species have learned to live as commensals with man. Most species have a lot of black in their plumage though some of the jays and magpies are colourful, with bright blue, green, and brown feathers. They have harsh calls, make large, untidy, stick nests, and feed on a mixture of fruit and animal material. Some are scavengers.

Eleven species occur in the Greater Sundas.

543. CRESTED JAY *Platylophus galericulatus* PLATE 63
(Malay/Crested Malay Jay)

Description: Unmistakable, medium-sized (28 cm), dark brown or blackish jay with a white neck patch and a long, vertically erectile crest. Sumatran and Javan birds are blackish grey; Bornean birds are rich dark brown.
Iris—brownish red; bill—black; feet—blackish blue.

Voice: Harsh, fast, reeling note and a chirpy, fast, rattling, squirrel-like chatter *tut-ut-ut-ut-ut*.

Range: Malay Peninsula and Greater Sundas.

Distribution and status: On Sumatra, Borneo, and Java not uncommon in lowland forest, up to 1200 m. Not recorded on Bali.

Habits: Travels in small, noisy parties and generally makes quite a fuss when it meets people, erecting its crest repeatedly, flitting from bush to bush, calling, and making weaving body movements. Feeds on large insects.

544. SHORT-TAILED MAGPIE *Cissa thalassina* PLATE 63
(Bornean/Short-tailed Green Magpie)

Description: Medium-sized (32 cm) green jay with black eye-stripe, red bill, and chestnut wing. Pale tertials on wing. Distinguished from Green Magpie by shorter tail, lack of black barring on tertials, and lack of yellow on head.
Iris—brown (Java) or white (Borneo); bill—bright red; feet—red.
Voice: Loud, penetrating shriek *tee-tee-tee-ter* with the first three notes fast and on a similar pitch, and the fourth note prolonged and descending.
Range: SE China, Indochina, Borneo, and Java.
Distribution and status: Not uncommon in Borneo forest between 900 and 2400 m, from Mt. Kinabalu south to Usun Apau and Dulit. Absent from E Java and Bali, and rather a rare bird of sub-montane rainforest in W Java.
Habits: Travels in small groups, often calling noisily but surprisingly hard to see despite colourful plumage. Hunts for insects in the lower storeys of the forest.

545. GREEN MAGPIE *Cissa chinensis* PLATE 63
(Hunting Cissa/Chinese Green Magpie)

Description: Large (34 cm), bright, green magpie with long tail, red bill, and chestnut wing. Black eye-stripe and graduated green tail feathers with black and white tips. Distinguished from Short-tailed Magpie by longer tail, more yellow on head, tertials with black tips, and red eye.
Iris—red; bill—red; feet—red.
Voice: A series of shrill, vibrant whistles *keep keep*, *ton-ka-kis*, harsh trumpet notes, and rapid chattering.
Range: Himalayas, S China, SE Asia, Malay Peninsula, Sumatra, and Borneo.
Distribution and status: On Sumatra an uncommon resident of the Barisan range, between 700 and 2100 m. On Borneo a sub-montane species found from Mt. Kinabalu south to Kelabit highlands, between 300 and 1200 m.
Habits: A shy bird of dense forest, more often heard than seen. Lives in small, noisy families in primary and disturbed montane and sub-montane forests.

546. SUMATRAN TREEPIE *Dendrocitta occipitalis* PLATE 63
(Malaysian/Mountain Treepie)

Description: Large (41 cm) brownish magpie with whitish nape and very long graduated tail. Underparts and back light brown; tail grey with blackish tip; rump and lower back pale grey, and wings black with a white patch at the base of the primaries.
Iris—red; bill—black with grey base; feet—dark grey.
Voice: Bell-like calls.

Range: Endemic to Sumatra.

Distribution and status: Rather common in tall forest at moderate to high altitudes, from 400 to 2300 m.

Habits: A shy but noisy bird that sits on low perches, waiting for prey which it catches on the ground or in foliage, or working steadily and jerkily through crowns of middle and upper canopy like a Malkoha. Sometimes travels in noisy groups. Occurs in primary and secondary, montane and sub-montane forests, plantations, pine forest, and bamboo.

Note: Some authors include the mainland Grey Treepie *Dendrocitta formosae* and Bornean Treepie with this species.

547. BORNEAN TREEPIE *Dendrocitta cinerascens* PLATE 63

Description: Large (40 cm) tawny brown magpie with very long graduated tail and white patches on black wings. Forehead and supercilium dark brown; crown silvery grey; mantle grey, washed with brown.

Iris—reddish brown; bill—black with grey base; feet—dark grey.

Voice: Bell-like, 3–note whistle, guttural chattering, and other calls.

Range: Endemic to Borneo.

Distribution and status: Not uncommon throughout the N and C mountains but absent from Penrissen, the Poi range, and the Kapuas and Mahakam drainages; from 300 to 2600 m locally down to almost sea-level in valleys.

Habits: As other treepies; inhabits forest, scrub, forest edge, and cultivated lands.

Note: Treated by some authors as a subspecies of Sumatran Treepie.

548. RACKET-TAILED TREEPIE *Crypsirina temia* PLATE 63
(Black Racket-tailed/Bronzed/Black Treepie)

Description: Medium-sized (35 cm, including 18 cm long tail), blackish magpie with a very long, terminally flared tail. Entire plumage glossy dark grey with a bronze-green sheen.

Iris—blue; bill—black and thick; feet—black.

Voice: Harsh, unmusical, whining, metallic call of 2 or 3 syllables.

Range: SE Asia, N Malay Peninsula, and Greater Sundas.

Distribution and status: Doubtfully recorded on Sumatra, but possibly a vagrant from Java. On Borneo known by two specimens probably collected in SE Borneo; probably resident given the Javan element in this area. Resident on Java and Bali in forest, secondary growth, and cultivation, up to 1500 m, but becoming increasingly rare.

Habits: Travels singly or in pairs through secondary forest, bamboo, scrub forest, and gardens.

CROWS AND JAYS 289

549. BLACK MAGPIE *Platysmurus leucopterus* PLATE 63
(White-winged/Black-crested Magpie)

Description: Large (38 cm), black magpie with short erect crest. Sumatran birds have white wing bar. Immature birds lack crest.
Iris—red; bill—black; feet—black.
Voice: Noisy with strange rattling churrs, squawks, and a repeated double bell-like *keting ka-longk* or single *keting*. (See Appendix 6.)
Range: Malay Peninsula, Sumatra (including Bangka), and Borneo.
Distribution and status: Uncommon in primary forests, up to 800 m.
Habits: A shy but noisy bird of the canopy in tall forest.

550. SLENDER-BILLED CROW *Corvus enca* PLATE 63
(Little Crow)

Description: Large (45 cm) black crow. Less glossy than Large-billed Crow and with a greyish sheen; bill much less massive. Distinguished in flight by shallow wing-beats.
Iris—brown; bill—black; feet—black.
Voice: Hoarse *kak-kak*.
Range: Philippines, Sulawesi, Malay Peninsula, and Greater Sundas.
Distribution and status: In Sumatra (including islands), Borneo, Java, and Bali a common forest crow, especially along the coast; rarely up to 1000 m.
Habits: Lives in pairs or occasionally small parties. Frequents coastline and forest edge, generally shy.

551. HOUSE CROW *Corvus splendens* PLATE 63
(Indian/Grey-headed/Colombo Crow)

Description: Large (43 cm) black crow. Similar to Slender-billed Crow but with grey collar and shorter bill.
Iris—brown; bill—black; feet—black.
Voice: High-pitched, rasping *ka*.
Range: Iran to SW China, Burma, and Thailand. Introduced elsewhere.
Distribution and status: Recorded on Ujong Kulon and Krakatau in the Sunda Strait. Probably ship-assisted.
Habits: A crow of open and wooded coastal areas and towns.

CROWS AND JAYS

552. LARGE-BILLED CROW   *Corvus macrorhynchos*   PLATE 63
(Thick-billed Crow)
Description: Large (48 cm) glossy black crow with a very heavy bill.
Iris—brown; bill—black; feet—black.
Voice: Harsh, throaty *kaw* and higher-pitched *awa, awa, awa*; also low gurgling.
Range: Iran to China, SE Asia, Philippines, Sulawesi, Malay Peninsula, and Sundas.
Distribution and status: Fairly common around villages and open plains throughout Sumatra, Java, and Bali. Apparently much rarer on Borneo where known by a few specimens from widely scattered localities.
Habits: Pair-living crow frequently found around villages.
Note: Formerly included in Jungle Crow *C. levaillantii*.

553. BORNEAN BRISTLEHEAD   *Pityriasis gymnocephala*   PLATE 63
(Bristled Shrike/Bald-headed Crow/Bald-headed Wood-Shrike)
Description: Smallish (26 cm), very unusual, red and black bird with massive black, hook-tipped bill, black wings, lower back, and tail; bare red head with short straw-coloured bristles on crown and long, thick, streaky, grey-brown bristles over ear. Upper breast, upper back, and thighs scarlet and lower breast covered in bristle-like brown and red feathers. Female has red patches on flanks. Tail short and feet rather small giving a top-heavy look.
Iris—red-brown; bill—black; feet—yellow.
Voice: Loud, curious honks and chortles.
Range: Endemic to Borneo.
Distribution and status: Scarce but well distributed in lowland forests, up to 1000 m.
Habits: As strange as its appearance. This species has unusual calls and curious behaviour, crouching and peering around like a tailless coucal, looking for insects. Lives in noisy groups. Flies with fast shallow wing-beat.
Diet: Large insects and small vertebrates.
Note: Authors have disagreed as to which family this bird belongs—crows, butcherbirds, shrikes, starlings, or in its own family. DNA evidence indicates it is related to the wood-swallows which are in turn related to crows.

LONG-TAILED TITS—FAMILY AEGITHALIDAE

LONG-TAILED TITS are small, agile, perching birds with small, sharp, conical beaks and longish to very long tails. They forage actively on insects and seeds, and usually live in small flocks. They make nests in suspended pouch-like nests.

There is only a single species in the Greater Sundas.

554. PYGMY TIT *Psaltria exilis*   PLATE 64

Description: Tiny (8 cm), nondescript, long-tailed tit with brown upperparts and off-white underparts. Identified by size as 'Java's smallest bird'.
Voice: A soft *trrr-trrr-trrr* or a high, soft *tee-tee-tee-tee* or *sisisirr*.
Range: Endemic to W Java.
Distribution and status: Confined to montane forests and plantations, mostly above 1000 m; locally common, e.g. in Cibodas.
Habits: An active bird which travels in small groups, visiting conifers, *Casuarina*, and other open trees, often at the edge of the forest. Regularly feeds low near the ground where it can be seen easily.

TITS—FAMILY PARIDAE

TRUE tits are small perching birds. They are agile, active, and acrobatic with sharp little beaks with which they hammer furiously on insects or seeds to crack them open. They are aggressive towards other birds. Tits nest in tree holes.

The tits are well represented in N America and Eurasia but only two species occur in the Greater Sundas. Of these the Sultan Tit remains of doubtful status and is certainly not extant in the region at present.

555. GREAT TIT *Parus major*   PLATE 64
(Grey/Cineraceous/Japanese Tit)

Description: Small (13 cm), black, grey, and white tit. Head and throat black, except for large contrasting white patch on side of face. Distinguished from Java Sparrow by small black bill.
Voice: Noisy chirps *chee-weet* or *chee-chee-chee*.
Range: Palaearctic, India, China, SE Asia, Malay Peninsula, and Greater Sundas.
Distribution and status: In Sumatra, Java, and Bali a not uncommon but local resident found in coastal mangroves, up to over 2000 m. In Borneo it is much rarer, largely confined to mangroves and coastal forests, though locally abundant around Banjarmasin.
Habits: Frequents mangroves, gardens, and open forests. Active, versatile little bird which may be active in treetops or down at ground level. Eats a wide range of food but mostly insects collected on trees. Hunts in pairs or family parties.

[556. **SULTAN TIT** *Melanochlora sultanea* PLATE 64

Description: Unmistakable, smallish (20 cm), yellow and black tit with spectacular long, fluffy, yellow crest. Female similar to male, but throat and breast are dark olive-yellow and upperparts washed olive.
Iris—brown; bill—black; feet—grey.
Voice: Repeated, loud, squeaky whistle *tcheery-tcheery-tcheery*; and shrill, chattering alarm calls.
Range: Himalayas, S China, SE Asia, and Malay Peninsula.
Distribution and status: Status unknown; former listing for Sumatra is based on specimens of uncertain provenance and one sight record (1938) of a party in the forest canopy at 100 m in N Sumatra.
Habits: Lives in canopy of primary and secondary forest, in mixed flocks, actively chasing large insects.]

Nuthatches—FAMILY SITTIDAE

NUTHATCHES are small, non-migratory, insectivorous forest birds found in Europe, Asia, and Australia. They specialize in foraging on tree trunks and branches. Two species occur in the Greater Sundas; both easily recognized.

557. **VELVET-FRONTED NUTHATCH** *Sitta frontalis* PLATE 64

Description: Colourful small (12 cm) nuthatch with red bill. Forehead velvety black; nape, back, and tail violet with a bright blue patch on the primaries. Male has black supercilium. Underparts pinkish with a whitish chin.
Iris—yellow or brown; bill—red; feet—reddish brown.
Voice: A shrill, insistent *chih-chih*, or sharp chittering call. In flight a *seep-seep-seep* call is recorded.
Range: India, S China, SE Asia, Philippines, Malay Peninsula, and Greater Sundas.
Distribution and status: In Sumatra (including islands), Borneo (including Maratuas), and Java a fairly common inhabitant of lowland and hill forests, up to 1500 m. Absent from Bali.
Habits: Travels in pairs or family parties, searching for insects on the trunks and branches of forest trees, often from top to bottom, head downwards. Shows an active, jerky motion, always appearing to hurry before flying off to the next tree. Frequents middle storeys of forest, swamp forests, plantations, and pines at lower altitudes than Blue Nuthatch.

558. BLUE NUTHATCH *Sitta azurea* PLATE 64

Description: A small (13 cm) blue and white nuthatch. Crown, nape, and sides of head black; back, wings, and tail glossy blue appearing black in poor light; throat and breast white, belly and vent black (Sumatra and W Java) or blue black (E Java).
Iris—white; bill—yellow; feet—blue grey.
Voice: High, shrill cheeps and twitters, similar to Velvet-fronted Nuthatch.
Range: Malay Peninsula, Sumatra, and Java.
Distribution and status: On Sumatra and Java a common inhabitant of submontane and montane forests between 900 and 2400 m. Absent from Bali.
Habits: As Velvet-fronted Nuthatch but prefers sub-montane forests.

BABBLERS—FAMILY TIMALIIDAE

THIS is a large, poorly defined family which encompasses many, rather diverse, bird groups. Babblers are generally gregarious and noisy and most have rather harsh, chattering calls. Many species tend to be active on, or close to the ground. They have short wings and are not strong fliers. None is migratory. They make cup-shaped nests in trees and bushes.

DNA studies show that most babblers are related to the warblers. Several tribes are recognized and it is convenient to break down the family into five distinct groups.

1. Jungle-Babblers: inconspicuous, rather quiet babblers living on or close to the ground in thickets. Fifteen species occur in the Greater Sundas (*Pellorneum*, *Trichastoma*, *Malacocincla*, *Malacopteron*).
2. Scimitar and Wren-Babblers: feed mostly on the ground in dense forest. The Wren-Babblers have characteristically very short, almost invisible, tails; the Scimitar Babbler is distinct with a strong decurved bill. Eleven species occur in the Greater Sundas (*Pomatorhinus*, *Rimator*, *Ptilocichla*, *Kenopia*, *Napothera*, *Pnoepyga*).
3. Tree-Babblers and Tit-Babblers: small agile birds found in bushes, grass, and bamboo, but only rarely on the ground. They have small tit-like beaks, short wings, long strong legs, and long, soft, fluffy feathers. There are sixteen species in the Greater Sundas (*Stachyris*, *Macronous*, *Timalia*).
4. Song-Babblers: small to larger, often colourful birds often with loud songs. They are mostly arboreal though will come to the ground. They hop with a characteristic jerky motion and make short, fluttery flights, calling and flicking to and fro. There are fourteen species in the Greater Sundas (*Garrulax*, *Leiothrix*, *Pteruthius*, *Alcippe*, *Crocias*, *Heterophasia*, *Yuhina*).

5. Ground-Babblers: primitive, ground-living, insectivorous group of Australo-Papuan region. Only one species occurs in the Greater Sundas (Eupetes).

Jungle-Babblers

559. BLACK-CAPPED BABBLER *Pellorneum capistratum* PLATE 65

Description: Medium-sized (17 cm) brown babbler with blackish crown and distinct supercilium, which is rufous buff anteriorly and white posteriorly. Upperparts rufous brown; underparts rufous buff with whitish throat. N Borneo birds have paler, yellower underparts and grey ear coverts.
Iris—brown; bill—black above, whitish below; feet—brown.
Voice: High-pitched, penetrating *weet* inflected upwards, and variants. (See Appendix 6.)
Range: Malay Peninsula and Greater Sundas.
Distribution and status: In Sumatra (including Bangka and Belitung) and Borneo (including Natunas and N Bornean islands) this is a common lowland bird up to 700 m, locally higher on Borneo. In W Java common in lowland forest; in E Java more a bird of sub-montane forest. Not recorded on Bali.
Habits: Inhabits undergrowth of primary or secondary forest, and bamboo or palm thickets. Usually solitary, occasionally in pairs or small flocks. Shy, on or close to the ground.

560. TEMMINCK'S BABBLER *Pellorneum pyrrogenys* PLATE 65
(Buff-breasted Babbler, AG: Jungle-Babbler)

Description: Small (15 cm) reddish brown babbler. Crown greyish brown; upperparts rufous brown with conspicuous pale shafts. Inconspicuous reddish brown supercilium. Underparts white with a rufous chest bar and tawny flanks.
Iris—red brown; bill—blackish; feet—brown.
Voice: Dissyllabic whistles *pityu* or *pii-tyu* (Wells).
Range: Assam to SW China, SE Asia, Malay Peninsula, Borneo, and Java.
Distribution and status: On Borneo (including N Bornean islands) a common sub-montane resident found throughout the mountains. On Java local and uncommon in lowlands and hills, up to 1200 m.
Habits: Generally keeps to the undergrowth of primary forest and forest edge but sometimes works up vine-covered trees to the canopy, hunting insects.
Note: Some authors include with this species the endemic Sumatran species *buettikoferi*. The form *concreta* of Belitung Island has been incorrectly ascribed to this species and belongs in Abbott's Babbler. Inclusion of the Javan form *pyrrogenys* means this name precedes *tickelli*. May be placed in genus *Trichastoma*.

561. BUETTIKOFER'S BABBLER *Pellorneum buettikoferi* PLATE 65
(AG: Jungle-Babbler)

Description: Small (15 cm) olive brown babbler similar to Temminck's Babbler but lacking rufescence, and cap does not contrast with back.
Iris—red brown; bill—blackish; feet—brown.
Voice: Whistled song with basic phrase *piyu pii biyo* (Wells).
Range: Endemic to Sumatra.
Distribution and status: A local resident throughout mainland Sumatra in lowlands and hills, up to 900 m.
Habits: As Temminck's Babbler.
Note: May be placed in *Trichastoma*. The arrangement here is after D. R. Wells and is based mostly on song. Other arrangements include: (1) *P. tickelli* of Assam to SW China and SE Asia and *P. pyrrogenys* of the Greater Sundas; (2) *P. tickelli* of Assam to SW China and SE Asia and Sumatra, and *P. pyrrogenys* of Borneo and Java.

562. WHITE-CHESTED BABBLER *Trichastoma rostratum* PLATE 65
(Mangrove Brown/Blyth's Babbler, AG: Jungle-Babbler)

Description: Small (15 cm), plain dark brown babbler with whitish underparts. Distinguished from Moustached and Sooty-capped Babblers by brown sides of head, lack of supercilium or dark moustache, and mantle same colour as back. Underparts white with greyish sides and brownish flanks.
Iris—brown; bill—black; feet—pink.
Voice: Male song is variable with 3—6 notes starting with a harsh whistle or short buzz with the last note either rising or falling, and usually one note rising, falling, and levelling. The female response is a loud, higher-pitched single *peew* note with falling inflection, repeated 2—4 times at about 6 second intervals. In Borneo described as a 3-note call *minta duit*. (See Appendix 6.)
Range: Malay Peninsula, Sumatra, and Borneo.
Distribution and status: On Sumatra (including islands) and Borneo (including N Bornean Islands) a less common bird of swamp forest, mangroves, and riverine forests, up to 500 m.
Habits: Lives in pairs or family flocks in dense moist forests, in middle and lower storeys.

563. FERRUGINOUS BABBLER *Trichastoma bicolor* PLATE 65
(AG: Jungle-Babbler)

Description: Smallish (17 cm) rufous babbler with longish tail. Upperparts rufous, lores pale; underparts creamy white, washed with rufous on the sides.

Iris—brown; bill—dark above, whitish below; feet—pale creamy horn.
Voice: Simple loud whistles *hweet* rising sharply in pitch, a repeated *piow* falling in pitch, and a trill in addition to churrs and chatterings of flock. (See Appendix 6.)
Range: Malay Peninsula, Sumatra, and Borneo.
Distribution and status: On Sumatra (including Bangka) and Borneo a fairly common bird at low altitudes, up to 600 m, locally higher on Borneo.
Habits: Lives in small flocks in undergrowth of dense forest. Keeps close to the ground but can be raised by 'pishing'.

564. SHORT-TAILED BABBLER *Malacocincla malaccensis* PLATE 65
(AG: Jungle-Babbler)

Description: Small (14 cm), short-tailed, brown babbler with grey or brown head and black moustache, contrasting strongly with white throat. Underparts white with tawny flanks.
Iris—chestnut; bill—black above, bluish with black tip below; feet—pink.
Voice: Female gives clear whistled song of 6 or 7 notes each, falling in pitch and lengthening in duration. The male call is a short trill, then a series of whistles, followed by about 6 level-pitched, drawn-out whistles. The male call may immediately follow the female or be given separately (Nash and Nash, 1987). (See Appendix 6.)
Range: Malay Peninsula, Sumatra, and Borneo.
Distribution and status: In Sumatra (including islands) and Borneo (including Natunas and Anambas) this is a common bird of the undergrowth of primary and secondary forests, including swamp forests, up to 1000 m.
Habits: Mousy, skulking, almost terrestrial habits. Lives in small groups. Can be attracted by 'pishing'.
Note: All *Malacocincla* may be placed in *Trichastoma*.

565. HORSFIELD'S BABBLER *Malacocincla sepiarium* PLATE 65
(Rusty Brown Babbler, AG: Jungle-Babbler)

Description: Smallish (14 cm) rich rufous brown, rufous-vented babbler with heavy bill. Crown greyish, lores whitish; upperparts brown becoming rufous on rump; throat white; breast grey; centre of belly white with flanks buff; thighs brown; undertail coverts rufous. Distinguished from Abbott's Babbler by darker and greyer crown and lack of pale supercilium.
Iris—red-brown; bill—black above, bluish below; feet—pink.
Voice: A harsh, penetrating, monotonous *pee-oo-weet* or *oo-weet* call, given continuously at dawn and dusk. The *oo* note is lower pitched and less stressed. (See Appendix 6.)
Range: Malay Peninsula and Greater Sundas.

Distribution and status: On Sumatra, Borneo, Java, and Bali locally common in suitable forest habitat, from 300 m up to 1400 m.
Habits: Favours the undergrowth of dense hill and lower montane forests, and forest edge where it clambers about low vegetation, singly, in pairs, or in small flocks. An inquisitive and noisy bird.

566. ABBOTT'S BABBLER *Malacocincla abbotti* PLATE 65
(Common Brown Babbler, AG: Jungle-Babbler)

Description: Smallish (16 cm) dull brown babbler with a heavy bill. Upperparts brownish olive, more rufous on rump; uppertail coverts cinnamon. Sides of face buffish brown with inconspicuous grey lores and supercilium. Underparts whitish on chin and throat, greenish greyish on breast, and buffy to brown on belly.
Iris—brown; bill—black above, pale below; feet—buff.
Voice: On Borneo a simple call of 3 or 4 notes, wavering, loud, rather monotonous (D.A.H.). In Sumatra, Nash recorded duets of 2 to 3 notes by the female accompanying a 6–7 note male song of slurred whistles, ending in an upward inflected whistle. (See Appendix 6.)
Range: Nepal to Assam, SE Asia, Malay Peninsula, Sumatra, Borneo, and Bawean.
Distribution and status: On Sumatra (including Belitung) and Borneo locally common bird of lowland forests, up to 400 m. In Java known from Bawean and Matasiri Islands in the Java Sea.
Habits: Inhabits light forest, scrub, thickets, and undergrowth of teak plantations. Lives singly, in pairs, or small parties, skulking close to the ground, and climbing into the canopy to sing at dawn.

567. BLACK-BROWED BABBLER *Malacocincla perspicillata* PLATE 65

Description: Smallish (16 cm) brown babbler with streaked grey belly. Differs from Horsfield's Babbler in larger size, black forehead and supercilium.
Iris—yellow; bill—black; feet—pink.
Voice: Unknown.
Range: Endemic to Borneo.
Distribution and status: Known only from a type specimen collected somewhere in S Borneo.
Habits: Unknown.
Note: Treated by some authors as conspecific with Vanderbilt's Babbler.

568. VANDERBILT'S BABBLER *Malacocincla vanderbilti* PLATE 65
(Koengke Babbler)

Description: Small (15 cm), short-tailed, heavy-billed babbler with grey underparts, washed with brown on sides. Resembles Horsfield's Babbler but found at higher altitudes.
Iris—brown; bill—black; feet—pink.
Voice: Unknown.
Range: Endemic to Sumatra.
Distribution and status: Presumably confined to northern mountains of the Barisan chain. A rare bird between 800 and 1200 m, known from only one specimen from the upper Alas Valley.
Habits: A skulker of undergrowth in montane forests.
Note: Should probably be treated as a montane race of Horsfield's Babbler. It was formerly treated as a race of Black-browed Babbler *M. perspicillata*.

569. MOUSTACHED BABBLER *Malacopteron magnirostre* PLATE 65
(Brown-headed Babbler, AG: Tree-Babbler)

Description: Small (16 cm) brown babbler with rufous tail, dull brown crown, and broad, dark grey moustache (looks black in field), contrasting with white throat. Borneo race has sooty head. Lores and sides of head grey; underparts greyish white with brownish grey streaks on breast. Immature has less distinct moustache.
Iris—dark brown; bill—creamy grey; feet—bluish.
Voice: Maddening call of 6 loud, deliberate notes descending the scale, repeated again and again. (See Appendix 6.)
Range: Malay Peninsula, Sumatra, and Borneo.
Distribution and status: On Sumatra (including Lingga Archipelago) and Borneo (including Anambas) a familiar bird of lowland forests up to 800 m, locally higher on Borneo.
Habits: Raises crown feathers and puffs out throat when excited. Lives in middle and lower canopy of swamp forest, forest edge, and scrub wasteland.

570. SOOTY-CAPPED BABBLER *Malacopteron affine* PLATE 65
(Sooty-headed/Plain Babbler, AG: Tree-Babbler)

Description: Small (16 cm) dark brown babbler with black crown, pale supercilium, and grey underparts. Resembles Moustached Babbler but crown much darker, tail dark brown, and lacks dark moustache. Borneo race has brown head and rufous tail. Immature has paler crown and rusty flight feathers.
Iris—dark brown; bill—creamy grey; feet—pink.

Voice: A song of about 8 hesitant, distinct, unbird-like whistles, rising and falling in pitch; flock contact calls of 4 or 5 rhythmical notes, varying in pitch but mostly descending; also monotonous, chattering, double notes *tiu tiu*. (See Appendix 6.)
Range: Malay Peninsula, Sumatra, and Borneo.
Distribution and status: On Sumatra (including Banyak and Bangka) and Borneo a common bird of lowlands, up to 700 m.
Habits: Lives in flocks, in crowns of small trees and bushes, in forests, forest edge, and secondary scrub.

571. SCALY-CROWNED BABBLER *Malacopteron cinereum* PLATE 65
(Lesser Red-headed Babbler, AG: Tree-Babbler)

Description: Small (15 cm), red-crowned, brown babbler. Crown rufous or rufous scaled black; nape rufous in Java, black in Borneo and Sumatra; underparts white, greyish on sides; tail rufous brown.
Iris—brown; bill black above, pink below; feet—pinkish.
Voice: Four to 6 thin notes rising up the scale in a minor key. (See Appendix 6.)
Range: SE Asia, Malay Peninsula, and Greater Sundas.
Distribution and status: On Sumatra (including islands) and Borneo (including Natunas) a common bird of lowland forests, locally to 1200 m. On Java locally distributed along the south coast. Not recorded on Bali.
Habits: Frequents open coastal forests and low-lying, primary and hill forests. Travels in small flocks through the lower levels of the forest where it is active, agile, and noisy.

572. RUFOUS-CROWNED BABBLER *Malacopteron magnum* PLATE 65
(Red-headed/Greater Red-headed Babbler, AG: Tree-Babbler)

Description: Smallish (17 cm) olive-brown babbler with rufous crown, black nape (reduced in N Borneo), chestnut tail, and whitish underparts with grey streaks on breast and throat. Similar to but slightly larger than some races (Borneo) of the Scaly-crowned Babbler but red feathers of crown not scaled with black, tail more rufous and throat streaked.
Iris—red; bill—creamy grey; feet—pink.
Voice: Varied song of several clear, dissyllabic notes, generally descending a scale but less regular than Sooty-capped Babbler and usually rising at end of phrase. Sometimes a rising call. (See Appendix 6.)
Range: Palawan, Malay Peninsula, Sumatra, and Borneo.
Distribution and status: In Sumatra and Borneo (including Natunas) a common bird of lowland forest, up to 800 m.
Habits: Lives in small flocks in smaller trees and bushes of primary forests; rarely visits ground.

573. GREY-BREASTED BABBLER *Malacopteron albogulare* PLATE 65
(White-throated Babbler, AG: Tree-Babbler)

Description: Small (15 cm) dark brown babbler with dark grey crown and diagnostic grey breast contrasting with white throat and belly; flanks and vent washed with rufous brown. Cheeks blackish. Short, bright white eyebrow distinguishes it from similarly coloured *Rhinomyias* flycatchers.
Iris—brown; bill—black above, pale below; feet—bluish.
Voice: Churring alarm calls.
Range: Malay Peninsula, Sumatra, and Borneo.
Distribution and status: Rather uncommon and local in lowland forest in Sumatra (including Batu Islands and Lingga Archipelago) and W Borneo. Apparently absent from E Borneo.
Habits: Lives in small trees and bushes of peat swamp and heath forests.

Scimitar and Wren-Babblers

574. CHESTNUT-BACKED SCIMITAR-BABBLER *Pomatorhinus montanus* PLATE 66
(Yellow-billed Scimitar-Babbler)

Description: Medium-sized (20 cm), rufous-backed, long-tailed babbler with conspicuous white supercilium and long, decurved, yellow bill. Crown greyish black; supercilium white; back chestnut; wings and tail brown; chin, throat, breast, and upper belly white; flanks and undertail coverts rufous.
Iris—yellow; bill—yellow with black base; feet—grey.
Voice: Ringing *tur-du-du-weet-weet*, *he-hoi-hoi*, and many variants. (See Appendix 6.)
Range: Malay Peninsula and Greater Sundas.
Distribution and status: On Sumatra (including Bangka) and Borneo a bird of lowland and hill forest upto 1200 m, locally higher on Borneo. In Java and Bali not uncommon in montane forest above 1200 m.
Habits: Lives singly or in small parties, in lower and middle storeys; often mixes with other species, especially laughing-thrushes. Generally feeds on or close to the ground.

575. LONG-BILLED WREN-BABBLER *Rimator malacoptilus*
(White-throated/Sumatran Wren-Babbler) PLATE 66

Description: Small (13 cm), brown, short-tailed, fluffy, buff-streaked brown babbler with long decurved bill. Distinguished from Eye-browed Wren-Babbler

by longer bill and lack of supercilium. Rump and tail rufous; underparts dark brown with white throat and whitish streaks on breast.
Iris—brown; bill—black; feet—pink.
Voice: In Burma reported as a beautiful whistled call; also soft chatters.
Range: Himalayas and Sumatra.
Distribution and status: On Sumatra recorded from the Barisan Range south to Mt. Kerinci where it is an uncommon bird of mountain forest, from 1500 to 2500 m.
Habits: Lives on or close to the ground, hopping among dense ground vegetation.

576. BORNEAN WREN-BABBLER *Ptilocichla leucogrammica*   PLATE 66
(Bornean Ground-Babbler)

Description: Small (17 cm), short-tailed, rich brown babbler with black underparts broadly streaked with white. Crown darker than back; face white with fine black streaks.
Iris—brown; bill—black above, pale grey below; feet—pink.
Voice: Two pure-tone notes, clear double whistle.
Range: Endemic to Borneo.
Distribution and status: Confined to Borneo where it is a rare bird of lowland forest.
Habits: Feeds on ground, hopping about in dense undergrowth of lowland forests.

577. STRIPED WREN-BABBLER *Kenopia striata*   PLATE 66

Description: Small (14 cm) brown babbler with conspicuous white streaking on dark crown, and rufous brown upperparts; creamy white face, supercilium, throat, and breast; orange-buff lores and flanks mottled rufous. Breast finely scalloped brown.
Iris—brown; bill—black; feet—pink.
Voice: Quiet *pee-pee-pee* whistle, soft frog-like croaking, and whistled song.
Range: Malay Peninsula, Sumatra, and Borneo.
Distribution and status: On Sumatra and Borneo an uncommon bird of lowland forest, up to 1000 m.
Habits: A mostly terrestrial skulker of spiny salak palm thickets and swampy forest undergrowth.

302 BABBLERS

578. RUSTY-BREASTED WREN BABBLER *Napothera rufipectus*
(Rufous-chested/Red-breasted/Sumatran Wren-Babbler)   PLATE 66

Description: Largish (18 cm), short-tailed, dark brown wren-babbler with white throat and chestnut-brown underparts. Upperparts scaled with black. Similar to Large Wren-Babbler but with chestnut underparts.
Iris—brown; bill—blackish above, horn below; feet—pinkish.
Voice: Various loud, sharp whistles.
Range: Endemic to Sumatra.
Distribution and status: Locally common in the Barisan Range and isolated mountains between 900 and 2500 m.
Habits: Lives on the forest floor under dense vegetation cover, favouring lower and upper montane forests.
Note: Treated by some authors as a subspecies of Large Wren-Babbler.

579. BLACK-THROATED WREN-BABBLER *Napothera atrigularis*
PLATE 66

Description: Largish (18 cm), short-tailed, dark brown babbler with black throat and ear coverts, buff breast mottled with black, and whitish centre of belly. Upperparts and underparts broadly scaled with black. There is a pale area of bare skin behind the eye.
Iris—brown; bill—blackish above, horn below; feet—pinkish.
Voice: Loud, sharp whistles.
Range: Endemic to Borneo.
Distribution and status: A rare bird of lowland and hill forests, up to 1500 m.
Habits: Lives on the ground in dense cover, favouring sub-montane forests.
Note: Treated by some authors as a subspecies of the Large Wren-Babbler.

580. LARGE WREN-BABBLER *Napothera macrodactyla*   PLATE 66
(Large-footed Wren-Babbler)

Description: Largish (19 cm), plump, short-tailed, mottled-brown babbler. Upperparts dark brown, each feather with pale centre, white shaft, and black edge; underparts whitish with grey edges to feathers; rufous on flanks, thighs, and undertail coverts. There is a pale area of bare skin behind the eye.
Iris—brown; bill—black; feet—pink.
Voice: Variable, loud, musical song of ventriloquial, vibrating whistles, often incorporating sliding notes (see Appendix 6).
Range: Malay Peninsula, Sumatra, and Java.

Distribution and status: Few records from Sumatra and Java; an uncommon bird of lowland forest, up to 1200 m.
Habits: An active noisy babbler, living singly or in small flocks, in dense forest. Keeps on or close to the ground, noisily turning over leaves like a thrush or pitta.

581. MARBLED WREN-BABBLER *Napothera marmorata* PLATE 66
(Mueller's Wren-Babbler)

Description: Largish (20 cm), short-tailed, rufous babbler with diagnostic white lores, white throat, and blackish breast and belly. Heavily scaled with black on white throat, and with white on black breast. Sides of head rufous; flanks and vent rufous-brown. Immature has rufous feather shafts on head and crown.
Iris—brown; bill—black; feet—pink.
Voice: No information.
Range: Malay Peninsula and Sumatra.
Distribution and status: On Sumatra recorded on slopes of the Barisan Range south to Kaba. A rather rare bird of sub-montane forest, between 1000 and 1800 m.
Habits: Lives quietly on the ground among rock piles of dark forested hillsides.

582. MOUNTAIN WREN-BABBLER *Napothera crassa* PLATE 66
(Bornean Streaked Wren-Babbler)

Description: Small (15 cm), short-tailed, scaly brown babbler with uniform pale grey throat and rufous belly and vent. Brown feathers of upperparts with broad black edges.
Iris—brown; bill—black; feet—pinkish.
Voice: Low, churring, chattering alarm notes and a loud call of shrill whistles, descending at end (Smythies).
Range: Endemic to Borneo.
Distribution and status: Confined to the mountains of N Borneo where recorded from Mt. Kinabalu south to Tama Abo and Mulu; also on Niut. An occasional bird of sub-montane and montane forests, between 1000 and 2500 m.
Habits: Lives on the ground in dark montane forests, often under the branches of fallen trees or in other dense cover.
Note: Treated by some authors as a subspecies of Streaked Wren-Babbler *N brevicauda*.

583. EYE-BROWED WREN-BABBLER *Napothera epilepidota* PLATE 66

Description: Very small (11 cm), short-tailed, fluffy, brown babbler with a short, white, pronounced supercilium. General colour reddish brown but mottled with

black and with conspicuous pale feather shafts; chin and throat buffy white; centre of belly white; underparts streaked buff. Dimorphic: either white or rufous buff scaling on underparts.
Iris—brown; bill—brown; feet—pale brown.
Voice: Loud, prolonged, 1-second, almost flat *peeeow* and churrs. (See Appendix 6.)
Range: Assam to SW China, SE Asia, Malay Peninsula, and Greater Sundas.
Distribution and status: On Sumatra not uncommon through the Barisan Range, from 900 to 2200 m. Apparently a rare sub-montane resident in the mountains of N Borneo where recorded from Mt. Kinabalu south to Kapuas and Mahakam drainages; also on Niut. On Java locally common in mountain forest between 1000 and 2000 m. Not recorded on Bali.
Habits: A shy, inconspicuous bird of dense undergrowth.

584. PYGMY WREN-BABBLER *Pnoepyga pusilla*   PLATE 66
(Lesser Scaly-breasted/Brown Wren-Babbler)

Description: Tiny (9 cm), almost tailless, dark rufous brown babbler. Small size, tailless appearance, and lack of supercilium diagnostic. Underparts paler than upperparts; adults have a scalloped pattern of dark markings below.
Iris—dark brown; bill—black; feet—pink.
Voice: Loud, piercing whistle of 2 or 3 well spaced notes of decreasing pitch. High-pitched squeak followed by short chatter. In alarm a quick *zeek-zeek*. (See Appendix 6.)
Range: Nepal to S China, SE Asia, Malay Peninsula, Sumatra, Java, Flores, and Timor.
Distribution and status: On Sumatra and Java quite common in mountain forest, from 900 to 3000 m. Not recorded for Bali but could occur there.
Habits: Scuttles about the forest floor in thickets in a mouse-like manner. Shy and secretive except when calling.

Tree-Babblers and Tit-Babblers

585. RUFOUS-FRONTED BABBLER *Stachyris rufifrons*   PLATE 67
(Red-fronted/Hume's Babbler)

Description: Small (12 cm), pale olive-brown babbler with rufous crown, whitish throat with fine black streaks, orange-buff breast, and grey belly.
Iris—brown; bill—black; feet—brown.
Voice: Plaintive whistled call of 5–7 mellow notes on same pitch, with a short break after first note, *per, pe-pe-pe-pe-pe-pe* given at various speeds. (See Appendix 6.)
Range: Burma, Thailand, Malay Peninsula, Sumatra, and Borneo.

Distribution and status: Rare on Sumatra and Borneo. On Sumatra recorded south to Mt. Kerinci in lowland and hill forest, up to 900 m. On Borneo this a rare but widely distributed sub-montane resident recorded up to 1200 m.
Habits: Skulks near ground in undergrowth, lower and middle storeys of primary and secondary forests, and bamboo thickets.

586. GOLDEN BABBLER *Stachyris chrysaea* PLATE 67
(Golden-headed Babbler)

Description: Tiny (10 cm) yellowish-olive babbler with distinctive black lores and yellow throat. Crown yellow, streaked with black. The Sumatran race *frigida* is less yellow than in mainland Asia.
Iris—reddish; bill—black; feet—black.
Voice: Surprisingly loud, low-pitched whistle of 4–8 notes with first note or 2 emphasized; also flock chatters. (See Appendix 6.)
Range: Nepal to SW China, SE Asia, Malay Peninsula, and Sumatra.
Distribution and status: On Sumatra common in hill and montane forest between 800 and 3000 m.
Habits: Lives in small parties, often mixed with other species, in foliage of low bushes in primary, secondary, and pine forests.

587. WHITE-BREASTED BABBLER *Stachyris grammiceps* PLATE 67
(Javan Babbler, AG: Tree-Babbler)

Description: Small (15 cm) grey and chestnut-brown babbler. Crown black with white edges to feathers; cheeks and flanks grey; moustache of black spots; wings and tail bright rufous brown; chin, throat, breast, and centre of belly white.
Iris—red; bill—black; feet—black.
Voice: Loud rattling call *krrreee-krrrreereeree*.
Range: Endemic to Java.
Distribution and status: Local in remnant patches of lowland and hill forest, up to 900 m. Most records are from W Java but recorded from Lawu and east to Malang.
Habits: Keeps to undergrowth of primary forest; lives in small flocks.

588. GREY-THROATED BABBLER *Stachyris nigriceps* PLATE 67
(Black-throated/Grey Babbler, AG: Tree-Babbler)

Description: Small (13 cm) brown babbler with blackish crown and nape streaked with white, white supercilium, white moustache stripe, and dark grey chin and throat. Several races vary slightly.
Iris—light brown; bill—blackish; feet—pink.

Voice: Rattling *prrreee-prrreee* calls (Smythies).
Range: Himalayas, SW China, SE Asia, Malay Peninsula, Sumatra, and Borneo.
Distribution and status: On Sumatra (including Lingga Archipelago) and Borneo (including Natunas) a common bird of forest undergrowth, in hills and mountains between 500 and 2300 m, presumably lower on islands.
Habits: Lives in small parties, skulking close to the ground in moist undergrowth of hill and montane forest.

589. GREY-HEADED BABBLER *Stachyris poliocephala* PLATE 67
(AG: Tree-Babbler)

Description: Small (15 cm) chestnut-brown babbler with dark grey head with fine white streaks on crown and throat. Underparts rufous. Distinguished from Chestnut-winged Babbler by lack of blue eye-ring, pale eye, white streaks on throat, and brown breast. Immature paler and lacking white streaks.
Iris—cream; bill—black; feet—grey.
Voice: A simple song of 2–4 warbled, wavering, descending or rising notes. Alarm and contact calls are fast chatters, descending series of churrs. (See Appendix 6.)
Range: Malay Peninsula, Sumatra, and Borneo.
Distribution and status: On Sumatra (including Lingga Archipelago) and Borneo not uncommon in secondary forest undergrowth and scrub, up to 900 m.
Habits: Flocking skulker of dense undergrowth, favouring secondary forests at low altitudes.

590. SPOT-NECKED BABBLER *Stachyris striolata* PLATE 67
(Spotted Babbler, AG: Tree-Babbler)

Description: Smallish (17 cm) brown babbler with white throat and supercilium, and black sides of neck spotted with white. Underparts rusty brown. Distinguished from other *Stachyris* babblers by white throat.
Iris—red; bill—black; feet—greenish black.
Voice: Simple song and chattering alarm.
Range: S China, SE Asia (except Malay Peninsula), and Sumatra.
Distribution and status: On Sumatra this is a common bird in hill and mountain forest and bamboo breaks, between 250 and 2200 m.
Habits: Flock-living skulker of forest floor and undergrowth in dense, moist montane forests.

591. CHESTNUT-RUMPED BABBLER *Stachyris maculata*   PLATE 67
(Red-rumped Babbler, AG: Tree-Babbler)

Description: Smallish (17 cm) olive-brown babbler with bare blue lores, bright chestnut rump, and black throat. Forehead black with fine white streaks; breast black scaled with white, grading to white belly streaked with black; vent pale brown. Several races vary slightly.
Iris—pale yellow; bill—black; feet—brown.
Voice: In Malaya, gives strident, loud bursts of 5–10 notes given at a rate of about 2 per second, and throaty, quavering, piping contact trill. (See Appendix 6.)
Range: Malay Peninsula, Sumatra, and Borneo.
Distribution and status: On Sumatra (including islands) and Borneo rather common in lowland forest below 800 m.
Habits: Typical skulking babbler of forest undergrowth. Can be raised by 'pishing'.

592. WHITE-NECKED BABBLER *Stachyris leucotis*   PLATE 67
(White-eared Babbler, AG: Tree-Babbler)

Description: Small (15 cm) rufous brown babbler with grey underparts, black throat, and grey head with white forehead, lores, and supercilium; slight white streaking on crown and line of white spots on side of neck. Similar to Black-throated Babbler but lacks white malar patch.
Iris—red; bill—blackish; feet—black.
Voice: Varied song of 4 or 5 whistled notes of different pitch. Brisk *choi-oi, chu-u, choi-oi* (M. & W.).
Range: Malay Peninsula, Sumatra, and Borneo.
Distribution and status: Few records from N and C Sumatra in lowland and hill forest, up to 800 m. On Borneo a local bird of tall forests between 500 and 1000 m, locally lower.
Habits: Skulker of forest undergrowth, favouring primary forest.

593. BLACK-THROATED BABBLER *Stachyris nigricollis*   PLATE 67
(Black-necked Babbler, AG: Tree-Babbler)

Description: Smallish (16 cm) chestnut babbler with blackish head and conspicuous white malar stripe and line behind eye. Dark grey crown streaked with white; throat outlined by white scaling, belly grey. Distinguished from White-necked Babbler by malar stripe and lack of whitish forehead.
Iris—dark red; bill—black with pale base; feet—blackish.
Voice: Contact and alarm calls consist of harsh, throaty, scolding chatters; a

simple song of 1 or 2 slow, low-piped notes, followed by 10 to 26 faster notes; also 4–7 evenly spaced mellow notes *hu-hu-hu* . . . of same pitch.
Range: Malay Peninsula, Sumatra, and Borneo.
Distribution and status: On Sumatra and Borneo locally common in undergrowth of damp lowland forests, up to 1000 m, rarely up to 1400 m.
Habits: Lives in skulking family flocks close to the ground in dense undergrowth of dark forests, often close to streams.

594. WHITE-BIBBED TREE-BABBLER *Stachyris thoracica* PLATE 67
(White-collared Babbler, AG: Tree-Babbler)

Description: Medium-sized (18 cm) dark chestnut babbler with a diagnostic broad white breast band, sometimes with black edge. Cheeks dark grey. E Javan birds have grey crown.
Iris—red brown; bill—black above, creamy grey below; feet—black.
Voice: Loud, harsh, ringing *churrr-churrr*. Slightly liquid rattle (D.A.H.).
Range: Endemic to Java.
Distribution and status: Uncommon in sub-montane and montane forest from 600 to 1800 m. One specimen labelled 'Sumatra' in the British Museum may be erroneous.
Habits: A shy babbler keeping to the undergrowth and dense thickets; more often heard than seen. Lives in small groups, sometimes climbs up vine-covered trees.

595. CHESTNUT-WINGED BABBLER *Stachyris erythroptera* PLATE 67
(Red-winged Babbler, AG: Tree-Babbler)

Description: Small (12 cm) rufous brown babbler with grey face and conspicuous blue eye-ring. Tail and wings chestnut. Several races differ in the extent of grey on forehead and redness of upperparts. Patch of pale blue skin on side of neck revealed when throat puffed out.
Iris—reddish brown; bare eye-ring—blue; bill—blackish; feet—grey or creamy grey.
Voice: Loud, rapid series of about 7–12 hollow poop notes, dropping in pitch after 4 notes; also scolding churrs. (See Appendix 6.)
Range: Malay Peninsula, Sumatra, and Borneo.
Distribution and status: On Sumatra (including islands) and Borneo (including Natunas) a common skulker of the undergrowth in forest and secondary scrub below 800 m.
Habits: Lives in small flocks, skulking close to forest floor. Can be raised by 'pishing'. Prefers drier secondary forest and scrub.

596. CRESCENT-CHESTED BABBLER *Stachyris melanothorax*
(Pearly-cheeked/Pearl-chested Babbler, AG: Tree-Babbler) PLATE 67

Description: Small (13 cm) rufous babbler with white throat patch with black edges. Supercilium whitish, crown and wings reddish brown; back and tail olive-brown; white throat patch with irregular black outline; cheeks, breast, and belly pearly grey; vent washed with rufous.
Iris—brown; bill—dark brown; feet—brown.
Voice: Short phrases of a very rapid trill at various speeds, and variations (D.A.H.).
Range: Endemic to Java and Bali.
Distribution and status: On Java (including Bawean) and Bali an occasional bird of hill forest, forest edge, and gardens between 500 and 1500 m, locally at sea-level.
Habits: A shy secretive bird remaining hidden in dense undergrowth. Lives in small parties.

597. GREY-CHEEKED TIT-BABBLER *Macronous flavicollis* PLATE 68
(Grey-faced/Javan/Yellow-throated Tit-Babbler)

Description: Small (14 cm) brown babbler. Crown brown (Java) or grey (Kangean). Upperparts pale brown; wings and tail chestnut; face grey; underparts buff washed with olive-grey on sides; throat and upper breast finely streaked with black. The Kangean race has pale yellow underparts and lacks breast streaks.
Iris—light yellow; bill—black with pale tip; feet—olive-green.
Voice: Typical churring babbler call.
Range: Endemic to Java and Kangean Island.
Distribution and status: Locally common in lowland scrub, particularly dry coastal scrub.
Habits: Favours dense undergrowth and tangles of vines. Generally found in small flocks. Less shy but less noisy than Striped Tit-Babbler.

598. STRIPED TIT-BABBLER *Macronous gularis* PLATE 68
(Yellow-breasted/Striated/Stripe-throated Tit-Babbler)

Description: Small (13 cm) rufous babbler with streaked underparts. Crown, back, wings, and tail chestnut; cheeks, face grey; underparts pale greenish yellow to white, with conspicuous black streaks, especially on breast. Underparts variable: whitish with bold dark streaks (Borneo); grey, more finely streaked (Java), or yellowish, finely streaked (Sumatra).
Iris—pale yellow; bill—dark brown, paler below; feet—bluish.

Voice: A monotonous *chunk chunk chunk* in phases of 3 to 10 or more notes, repeated incessantly throughout the day. (See Appendix 6.)
Range: E India to SW China, SE Asia, Palawan, Malay Peninsula, and Greater Sundas.
Distribution and status: On Sumatra (including islands) and Borneo (including islands) a common bird of lowland areas in suitable habitat, up to 1000 m. On Java very local and only doubtfully recorded east of Indramayu; abundant on offshore islets of Ujung Kulon, otherwise there are few records. Not recorded on Bali.
Habits: Occurs in pairs or small flocks in the dense secondary growth of abandoned fields and bamboo thickets. The birds spend most of their time within a few metres of the ground but sometimes climb higher in vine-laden trees.

599. FLUFFY-BACKED TIT-BABBLER *Macronous ptilosus*   PLATE 68
(Plume-backed Tit-Babbler)

Description: Small (15 cm), dark, skulking babbler. Rufous brown plumage, chestnut cap, black throat, and blue eye-ring and lores distinctive. The long, pale-shafted plumes of back are rarely visible in the field.
Iris—red brown; bill—blackish; feet—blackish.
Voice: Harsh creaks, scolding call *cher cher cher cherung*, rising on last note; also low hollow song *poop-poop poop poop poop poop*.
Range: Malay Peninsula, Sumatra, and Borneo.
Distribution and status: On Sumatra (including islands) and Borneo a common bird of lowland forest, up to 700 m.
Habits: Lives in small flocks in dense lower storey of forest and forest edge, generally in damper gullies. Responds well to 'pishing'.

600. CHESTNUT-CAPPED BABBLER *Timalia pileata*   PLATE 68
(Red-capped Babbler)

Description: Medium-sized (17 cm) red-crowned babbler with creamy supercilium. Upperparts reddish brown; crown bright chestnut; white supercilium separated from crown by fine black line; cheeks white to grey; lores black; breast white with black shaft streaks; belly grey washed with fulvous brown on flanks and vent.
Iris—red; bill—black; feet—greenish.
Voice: Wide variety of clear, loud, whistled notes, metallic trills, slurred notes, warbles, rising and falling in pitch; and loud, sharp, dissyllabic whistles rising in pitch (D.A.H.).
Range: Nepal to S China, SE Asia (except Malay Peninsula), and Java.
Distribution and status: On Java it is quite common in lowland areas, up to 1500 m.

Habits: Skulks in thick undergrowth, thick grass, and dense shrub layer in more open scrub areas. Generally in pairs, sometimes in small flocks, keeping close to the ground and calling from thick cover. Extremely difficult to see.

Song-Babblers

601. SUNDA LAUGHINGTHRUSH *Garrulax palliatus* PLATE 69
(Grey and Brown/Catbird Laughingthrush)

Description: Large (27 cm), grey-headed, chestnut-brown babbler with black lores distinctive. Bare orbital skin pale blue.
Iris—brown; bill—black; feet—black.
Voice: Noisy whistles and cat-like mews.
Range: Endemic to Sumatra and Borneo.
Distribution and status: On Sumatra found throughout the Barisan range, in mountain forest from 850 to 2200 m. On Borneo restricted to the mountains of N Borneo where found from Mt. Kinabalu south to Usun Apau and Dulit, and the outlier Mt. Mulu, from 300 to 2000 m.
Habits: Lives in small flocks, mixing with other species, in the lower and middle canopy of montane forests. Lively and conspicuous.

602. RUFOUS-FRONTED LAUGHINGTHRUSH *Garrulax rufifrons*
(Red-fronted Laughingthrush) PLATE 69

Description: Large (27 cm) noisy babbler. Plumage olive-brown above and fulvous brown below, forehead and chin rufous, cheeks greyish, and throat pale brown. Sometimes entire bird with a rufous wash.
Iris—yellowish orange; bill—black; feet—brownish green.
Voice: Noisy chuckling calls between party members, *kee-tee-tee-tee* and *huur-too-too-too*. Captive bird gave long series of whistled notes at the same pitch (D.A.H.).
Range: Endemic to W and C Java.
Distribution and status: Recorded east to Slamat. Within this very restricted distribution it is locally not uncommon in mountain forests, from 1000 to 2400 m.
Habits: Travels in small noisy parties in the understorey of primary mountain forests, sometimes foraging for insects higher. Flocks with other species.

603. WHITE-CRESTED LAUGHINGTHRUSH *Garrulax leucolophus*
PLATE 69

Description: Large (30 cm) brownish-black babbler with unmistakable white head with slight erectile crest, and black forehead, lores, and drooping eye-stripe.

Iris—brown; bill—brown; feet—brownish.

Voice: A very noisy bird. A lead-up of chattering notes followed by loud cackling, melodious 'laughter'.

Range: Himalayas, SE Asia (except Malay Peninsula), and Sumatra.

Distribution and status: In the mountains of Sumatra an occasional bird of primary and secondary forests at moderate altitudes, between 750 and 2000 m.

Habits: Flock-living, in middle and lower storeys of forest. Sometimes comes down to the ground. Flies with typical laughingthrush glides.

604. BLACK LAUGHINGTHRUSH *Garrulax lugubris* PLATE 69

Description: Large (26 cm), dark greyish-brown babbler. Bornean and Sumatran races are quite different. Sumatran birds have orange-yellow bill and blackish plumage. Bornean adults have red bill, browner plumage, and bald yellow head. Immature birds are duller and have fully feathered crown.

Iris—brown; bare skin round eye and on cheek—blue in Sumatra, yellowish in Borneo; bill—orange in Sumatra, red in Borneo; feet—brownish yellow.

Voice: Like White-crested Laughingthrush. Discordant cackling alarm cries and a call of several loud, clear, hollow hoop notes, followed by maniacal 'laughter', also double *hoopoop* notes.

Range: Malay Peninsula, Sumatra, and Borneo.

Distribution and status: On Sumatra not uncommon in mountain forest, from 600 to 1600 m. On Borneo restricted to the mountains of N Borneo from Mt. Kinabalu south to Usun Apau and Dulit, and Mt. Mulu, where it is an uncommon resident of montane forests between 1000 and 2000 m.

Habits: Lives in small flocks, keeping to lower and middle storeys of forest.

Note: The Bornean subspecies *G. l. calvus* is sometimes treated as a separate species. In view of the difference in skin colour, plumage colour, degree of baldness, and bill shape this seems appropriate.

605. CHESTNUT-CAPPED LAUGHINGTHRUSH *Garrulax mitratus* PLATE 69

Description: Large (23 cm) grey babbler with chestnut cap, orange bill, pale eye-ring, and distinctive white wing stripe. Forehead streaked with white. Eye-ring is yellow in Borneo, white in Sumatra.

Iris—brown; bill—orange; feet—yellow.

Voice: Repeated, slurred whistles *ker-keweet*, sliding up and down in pitch. (See Appendix 6.)

Range: Malay Peninsula, Sumatra, and Borneo.

Distribution and status: On Sumatra it is a common bird of mountain forest, forest edge, and plantations, between 700 and 2000 m. Common in the mountains

of Borneo from Mt. Kinabalu south to Batu Tibang, occurring as low as 300 m in valleys and recorded to 3000 m.
Habits: Noisy, conspicuous flocks forage in small trees.

606. SILVER-EARED MESIA *Leiothrix argentauris* PLATE 69
(Silver-eared Leothrix)

Description: Medium-sized (18 cm), colourful, reddish babbler. Black head with silvery white cheeks and red forehead distinctive. Tail, back, and wing coverts olive; throat and breast reddish orange; wings red and yellow; rump and undertail coverts red.
Iris—red; bill—yellow; feet—yellow.
Voice: Hollow, rattled chatter and cheery whistled song *chi-uwi, chi-uwi, chi-uwi,* or *chi-uwi-chiu*. (See Appendix 6.)
Range: Himalayas, S China, SE Asia, Malay Peninsula, and Sumatra.
Distribution and status: On Sumatra this is a locally common bird in the Barisan Range south to at least Ranau, between 600 and 2200 m.
Habits: A restless bird of dense thickets in lower and middle storeys of montane forests.

607. WHITE-BROWED SHRIKE-BABBLER *Pteruthius flaviscapis*
(Red-winged/Black-crowned/Greater Shrike-Babbler) PLATE 69

Description: Small (13 cm) pied babbler. Male: head black with white supercilium; mantle and back grey; tail black; wings black with white tips on primaries, and gold and chestnut tertiaries; underparts white. Female: duller with buff underparts; head olive-green; coloured wing tips less bright.
Iris—greyish; bill—bluish black; feet—whitish.
Voice: Loud and piercing monotone *too-too-too*, *klip klip*, or *chip chip chap chip chap*. (See Appendix 6.)
Range: Pakistan to China, SE Asia, Malay Peninsula, and Greater Sundas.
Distribution and status: An occasional bird of montane forest, between 1000 and 2800 m, throughout Sumatra, Borneo, and Java. Not recorded on Bali.
Habits: Lives in pairs or in mixed-species flocks, moving through the lower and upper canopy, catching insects. Shuffles sideways along small twigs searching keenly for food.

314 BABBLERS

608. CHESTNUT-FRONTED SHRIKE-BABBLER *Pteruthius aenobarbus*   PLATE 69

Description: Very small (11 cm) brightly coloured babbler. Male: forehead, chin, and throat chestnut; upperparts olive-green with 2 bold white bars on black upperwing coverts; white eye-ring and greyish-white supercilium; underparts yellow. Female: whitish underparts, chestnut only on forehead.
Voice: Thin, high-pitched, piercing call *too-weet-weet-weet*.
Range: Assam to S China, SE Asia (except Malay Peninsula), and Java.
Distribution and status: Confined to W Java, recorded east to Papandayan where this is an uncommon bird of higher mountains, from 1000 to 3000 m. A specimen labelled 'Lampung' exists but there is no other indication that this species occurs on Sumatra.
Habits: Lives in the top of low trees and bushes in montane forest, sometimes mixing in flocks with other species.

609. BROWN FULVETTA *Alcippe brunneicauda*   PLATE 68
(Malaysian Fulvetta, AG: Nun-Babbler)

Description: Small (14 cm), grey-headed brown babbler. Dark brown above, paler below with grey head and greyish breast. Distinguished from similarly coloured *Rhinomyias* flycatchers by gregarious habits and active behaviour.
Iris—grey; bill—black; feet—brown.
Voice: A quick descending scale of 5–7 thin, whistled notes and chattering alarm calls with penultimate note sometimes higher (M. & W.). (See Appendix 6.)
Range: Malay Peninsula, Sumatra, and Borneo.
Distribution and status: On Sumatra (including Batu islands) and Borneo (including Natunas) an uncommon bird of lowland and hill forests, and secondary growth, up to 1000 m.
Habits: Noisy, active flocks, flit through undergrowth and small trees and bushes.

610. JAVAN FULVETTA *Alcippe pyrrhoptera*   PLATE 68
(Javanese Fulvetta, AG: Nun-Babbler)

Description: Small (14 cm) reddish-brown babbler. Upperparts rufous brown, slightly greyer on head; rump and uppertail coverts rufous; chin and throat ashy white, washed with buff; breast and belly whitish buff in centre, and ochre buff on flanks and undertail coverts.
Iris—brown; bill—brown; feet—brown.
Voice: Rather loud, melodious, piercing calls *chee-chee-chee-cheweeweet* or *boo-rey-chet-chet*.
Range: Endemic to Java.

Distribution and status: Confined to W Java, recorded east to Papandayan and Cerimay, and found only on a few of the higher mountains above 1000 m; sometimes locally abundant.
Habits: Lives in forest and forest edge, in small groups, rarely mixing with other species.

611. SPOTTED CROCIAS *Crocias albonotatus* PLATE 68
(Spotted Sibia)

Description: Medium-sized (20 cm) rufous-backed, long-tailed babbler. Crown and sides of head, wings and tail blackish grey; tips of tail feathers and edges of flight feathers white; back chestnut, streaked with white; underparts white.
Iris—brown; bill—grey; feet—yellow.
Voice: Harsh, piercing *breeoow-breeoow-breeoow-cheeoow-cheeoow-cheeoo-chee-chee-cheeoow*, similar to a laughingthrush.
Range: Endemic to Java.
Distribution and status: Uncommon, confined to W Java, recorded east to Papandayan and Cerimay, where it occurs only on a few of the highest mountains.
Habits: A shy bird living in leafy tree crowns, more often heard than seen. Single or joining in mixed flocks.
Note: Some authors place it in genus *Minla*.

612. LONG-TAILED SIBIA *Heterophasia picaoides* PLATE 68

Description: Large (32 cm), grey and white, arboreal babbler with very long, pointed tail. Plumage dull grey with darker crown, whitish vent, and white wing patch conspicuous in flight. Tail feathers with paler grey tips.
Iris—brown; bill—black; feet—black.
Voice: A noisy bird, constantly uttering shrill, twittering calls *tsip-tsip-tsip-tsip*.
Range: Himalayas, S China, SE Asia, Malay Peninsula, and Sumatra.
Distribution and status: On Sumatra this is a common bird of the higher mountains between 600 and 3000 m.
Habits: Lives in small flocks, keeping to the tops of taller trees. Flies powerfully while calling.

613. CHESTNUT-CRESTED YUHINA *Yuhina everetti* PLATE 68
(Chestnut-headed Yuhina)

Description: Small (14 cm) greyish-brown babbler of treetops with erectile chestnut crest. General appearance tit-like. Underparts whitish; outer tail feathers mostly white.
Iris—brown; bill—black; feet—brown.

316 BABBLERS

Voice: Constant, harsh, low *chit-chit-chit* calls.
Range: Endemic to Borneo.
Distribution and status: A common sub-montane species recorded up to 1800 m and locally in the lowlands; found from Mt. Kinabalu south to Mueller range and Penrissen.
Habits: Lives in small flocks, feeding in the tops of small to large-sized trees in montane forests.
Note: Several authors treat this form as a subspecies of Striated Yuhina *Yuhina castaniceps*, though it is widely different from the mainland form and has a pronounced crest.

614. WHITE-BELLIED YUHINA *Yuhina zantholeuca* PLATE 68
(Erpornis)

Description: Small (12 cm), olive-green, treetop babbler with greyish white underparts, yellow undertail coverts, and prominent crest. Distinguished from similarly coloured warblers by its crest.
Iris—brown; bill—horn; feet—horn.
Voice: Wheezy, squeaked 3–note calls.
Range: Himalayas, S China, SE Asia, Malay Peninsula, Sumatra, and Borneo.
Distribution and status: On Sumatra it is known by 1 specimen from N Sumatra, taken in 1937 in primary forest between 550 and 800 m. On Borneo a rather uncommon resident of primary and secondary forests, from sea level to 1500 m.
Habits: Lives in flocks and feeds in the crowns of tall trees, often mixing with warblers and other species.

Ground-Babblers

615. MALAYSIAN RAIL-BABBLER *Eupetes macrocerus* PLATE 68
(Rail-Babbler/Malay Rail-Babbler/Eupetes, AG: Jewel-Babbler/Scrub-Robin)

Description: Large (29 cm), rufous, terrestrial babbler with long slender head, neck, and tail. Face is masked by broad, black eye-stripe and white supercilium. Immature has grey forehead, white throat, and dark brownish-grey underparts.
Iris—brown; bare skin on side of neck—blue; bill—black; feet—black.
Voice: Low-pitched, monotonous whistle similar to Garnet Pitta; rich frog-like *kok* calls in irregular series, about 4 per second, audible only at close range. (See Appendix 6.)
Range: Malay Peninsula, Sumatra, and Borneo.
Distribution and status: On Sumatra and Borneo (including Natunas) an occasional bird of tall lowland forests, up to 900 m.

Habits: A shy bird often heard but rarely seen. Lives on the floor of tall primary forest and logged forests. Rail-like in habits, running fast when disturbed.
Note: Some authors place this species in the log-runner family *Orthonychidae*. DNA evidence suggests this is in fact an aberrant crow!

Thrushes—Family Turdidae

A VERY large, worldwide group, divided into true thrushes, whistling thrushes, robins, chats, forktails, and other groups. The birds vary greatly in coloration but are mostly medium-sized, round-headed birds with longish legs, sharp, slender bills, and broad wings. The tail varies from short to very long but in all species, it shows some tendency to be cocked periodically. Food consists of insects, other invertebrates, and berries. Most species feed at least partly on or close to the ground. Thrushes make solid, cup-shaped, fibrous nests, often reinforced with mud and decorated with moss. Many species have melodious songs.

Twenty nine species occur in the Greater Sundas, of which six are only winter visitors.

616. LESSER SHORTWING *Brachypteryx leucophrys* PLATE 64
(Mrs. La Touche's Shortwing)

Description: Very small (11 cm), short-tailed, long-legged thrush (looks like a babbler) with inconspicuous pale supercilium, buffy eye-ring, and heavy bill. Adults: upperparts rufous brown; underparts whitish with buffy brown sides and flanks and mottled buff brown breast. Female underparts more buff than male. Races vary in rufescence. Immature birds are streaked and spotted.
Iris—light brown; bill—brownish black; feet—pinkish purple.
Voice: A ringing *turrr, turrr*, and a high, piercing alarm call. Also fast, sweet, high, warbled song preceded by 2 or 3 emphatic notes (D.A.H.), ending in a jingle. (See Appendix 6.)
Range: Himalayas, S China, SE Asia, Malay Peninsula, Sumatra, Java, and Lesser Sundas.
Distribution and status: On Sumatra, Java and Bali this is a common bird of submontane and montane forests, from 900 to 1900 m.
Habits: Shy, keeping to undergrowth and forest floor, generally at lower altitude than White-browed Shortwing.

617. WHITE-BROWED SHORTWING *Brachypteryx montana* PLATE 64
(Blue/Himalayan Blue/Indigo-Blue Shortwing)

Description: Small (15 cm) bluish (male) or blue and rufous (female), long-legged, short-winged thrush. Male is dark blue all over with a conspicuous white supercilium. Female varies with race being all blue in Sumatra, blue head and nape with rufous wings, back, and tail in Java, and all rufous with only mantle blue in Borneo. Female has smaller, concealed white supercilium. Immature is mottled with brown.
Iris—brown; bill—black; feet—black.
Voice: Less vocal than Lesser Shortwing. Gives one burst of song from each station of its territory boundary. Song commences slowly with several single notes, quickens to a plaintive babble, then stops abruptly (D.A.H.). (See Appendix 6.)
Range: Nepal to S China, Philippines, SE Asia, Malay Peninsula, Greater Sundas, and Flores.
Distribution and status: On Sumatra recorded from Mts Leuser, Singgalang, Kerinci, and Dempu, and on Borneo from Mts Kinabalu, Trus Madi, Mulu, and Kelabits; it is locally common between 1400 and 3000 m. In W Java it is recorded east to Papandayan and is locally common above 1500 m. Not recorded on Bali.
Habits: Shy, stays in dense thickets close to the ground, often near streams. Comes out into open clearings and even on the bare rocky slopes of mountain tops. Variable in its habits according to availability of suitable foods.

618. SIBERIAN RUBYTHROAT *Luscinia calliope* PLATE 64

Description: Medium-sized (16 cm), plump, brown robin with bold white supercilium and malar stripe. Upperparts: brown without rufous in tail; underparts: buff on flanks; buffy white on belly. Female has brownish breast band. Black and white striped head pattern distinctive. Adult male has diagnostic red throat.
Iris—brown; bill—dark brown; feet—pinkish brown.
Voice: Loud, falling-tone whistle; soft deep *tschuck*.
Distribution and status: Breeds in NE Asia. Migrates in winter to India, S China, SE Asia. One record from N Borneo.
Habits: A skulker of thickets in forest and secondary growth; generally near streams.

619. SIBERIAN BLUE ROBIN *Erithacus cyane* PLATE 64
(Siberian Bluechat)

Description: Medium-sized (14 cm), blue and white or brown robin. Male unmistakable with slaty-blue upperparts and broad black band through eye and down side of neck; underparts white. Female olive-brown upperparts with throat and breast brown, scaled with buff. Immature birds and some females have some blue on tail and rump.

Iris—brown; bill—black; feet—pinkish white.

Voice: In winter a hard, low *tak*; also loud *se-ic*.

Range: Breeds in NE Asia; migrant in winter to India, S China, SE Asia, Malay Peninsula, Sumatra, and Borneo.

Distribution and status: A winter visitor in small numbers, recorded throughout Sumatra and Borneo in forests up to 1800 m.

Habits: Stays on or near the ground in dense forest.

620. ORANGE-FLANKED BUSH-ROBIN *Tarsiger cyanurus* PLATE 64
(Siberian/Red-flanked Bush-Robin/Blue-tailed Robin, AG: Bluetail)

Description: Smallish (15 cm) white-throated robin with diagnostic orange flanks contrasting with white belly and vent. Male upperparts blue; immature and female brown with blue tail. Female is distinguished from female Siberian Blue Robin by white mesial stripe on brown throat rather than entire throat white, also by orange rather than buff flanks.

Iris—brown; bill—black; feet—grey.

Voice: Full song not heard in wintering birds; quiet croaking *chuck* (P.R.).

Distribution and status: Breeds in NE Asia. Migrates in winter to India, S China, SE Asia. One record from N Borneo.

Habits: Keeps low in undergrowth of montane forest and secondary growth.

621. MAGPIE ROBIN *Copsychus saularis* PLATE 70

Description: Medium-sized (20 cm), cocky, black and white robin. Male: head, breast, and back shiny blue black. Sumatra, W Java, and W Borneo: wings and central tail feathers black; outer tail feathers and stripe across the wing coverts white; belly and vent white. E Java, N and E Borneo: belly and vent black. Female as male but dull grey instead of black. Immature similar to female but mottled.

Iris—brown; bill—black; feet—black.

Voice: Varied lusty singing, including imitations of other birds but lacking the rich tone of White-rumped Shama.

Range: India, S China, Philippines, SE Asia, Malay Peninsula, and Greater Sundas.

Distribution and status: On Sumatra (including islands) and Borneo it is common in the lowlands, up to 1500 m. On Java and Bali this is a fairly common bird of the lowlands but getting scarce as a result of trapping.

Habits: A familiar bird of gardens, villages, secondary forests, open forests, and mangroves. Conspicuous in flight and perches conspicuously to sing or display.

Feeds mostly on the ground where it constantly lowers and fans its tail before jerking it shut and upright again.

622. WHITE-RUMPED SHAMA *Copsychus malabaricus* PLATE 70

Description: Largish (27 cm), long-tailed, black, white, and rufous robin. Head, neck, and back black with blue gloss; wings and central tail feathers dull black; rump and outer tail feathers white; belly orange-rufous.
Iris—dark brown; bill—black; feet—grey brown.
Voice: Beautiful, complex, melodious songs including imitations of other birds.
Range: India to SW China, SE Asia, Malay Peninsula, and Greater Sundas.
Distribution and status: On Sumatra (including islands) and Borneo (including islands) it remains quite common in the lowlands, locally up to 1500 m. On Java (including Kangean) this is now a rare bird of lowland forests as a result of excessive trapping. Absent from Bali. A black-tailed form occurs on Kangean Island and some islands of the Sumatran west coast.
Habits: Shy, keeps to thickets in denser forest. Sings lustily in morning and evening from a low perch with wings drooped and tail held high. Hops on the ground or makes short flights through undergrowth, flicking its long tail on landing.
Note: White-browed Shama may be included in this species but overlap in distribution suggests that it is a valid species.

623. WHITE-BROWED SHAMA *Copsychus stricklandi* PLATE 70

Description: Largish (27 cm), long-tailed, black, white, and rufous robin. Similar to White-rumped Shama but with white crown.
Iris—dark brown; bill—black; feet—grey brown.
Voice: As White-rumped Shama.
Range: Endemic to NE Borneo.
Distribution and status: Limited from Darvel Bay to Padas river. Replaces White-rumped Shama with some overlap. A common bird up to 1200 m.
Habits: As White-rumped Shama.
Note: May be included in White-rumped Shama but overlap zone suggests that this is a valid species.

624. RUFOUS-TAILED SHAMA *Trichixos pyrrhopygus* PLATE 70

Description: Large (21 cm), long-tailed, black and orange robin. Male is like White-rumped Shama but with much shorter rufous tail, dark grey instead of black, short white supercilium, and rufous rump. Female is browner and lacks white supercilium. Immature is browner, spotted with rufous-buff.

Iris—brown; bill—black; feet—black.
Voice: Song less musical than White-rumped Shama. Long series of slurred, melodious whistles *pi-uuu*, rising and falling irregularly.
Range: SE Asia, Malay Peninsula, Sumatra, and Borneo.
Distribution and status: On Sumatra and Borneo an occasional bird of dense primary and secondary forest, up to 1200 m.
Habits: Similar to White-rumped Shama. Prefers dense moist forests including swamp forests.
Note: Included by some authors in *Copsychus* as *C. pyrrhopygus*.

625. SUNDA BLUE ROBIN *Cinclidium diana* PLATE 70

Description: Small (15 cm) indigo blue (male) or reddish-brown (female) robin. Male has silvery white forehead. Female is chestnut on upperparts and breast, greyish white on throat, belly, and vent. Immature: reddish brown, mottled and spotted with black.
Iris—brown; bill—black; feet—black.
Voice: Quiet. Song is a trivial warble of 2 to 5 sweet melancholy notes.
Range: Endemic to Sumatra and Java.
Distribution and status: On Sumatra it is scarce throughout the mountains, between 1100 and 1500 m. On Java it is an uncommon local resident of high mountains, recorded east to Ciremay, between 1000 and 2400 m.
Habits: A shy bird keeping to the undergrowth of high montane forests.

626. LESSER FORKTAIL *Enicurus velatus* PLATE 70

Description: Smallish (16 cm) white and slaty forktail. Forehead and short supercilium white; crown and nape grey in male, chestnut in female; rest of head, neck, wings, and back dark grey; breast, belly, and rump white, white wing bar, and white tips to feathers of long, forked, graduated black tail. Two outer tail feathers entirely white.
Iris—brown; bill—black; feet—whitish.
Voice: A hard shrill *hie-tie-tie*, or just *chee*.
Range: Endemic to Sumatra and Java.
Distribution and status: On Sumatra common in hill and mountain forest, between 600 and 2000 m. On Java an occasional bird of forest streams, mostly from 600 to 1800 m but locally near sea-level; less common than the White-crowned Forktail.
Habits: Similar to White-crowned Forktail. An active bird of fast flowing streams, taking insects from under the water.

627. CHESTNUT-NAPED FORKTAIL *Enicurus ruficapillus* PLATE 70

Description: Medium-sized (20 cm) black and white forktail with diagnostic chestnut crown and nape. Male has chestnut crown, nape, and upper back, white forehead, and black scaling on white breast. Female is similar but entire back is chestnut.
Iris—brown; bill—black; feet—black.
Voice: High-pitched *shweet-shweet* calls in flight. Shrill whistled alarm call.
Range: Malay Peninsula, Sumatra, and Borneo.
Distribution and status: On Sumatra and Borneo an uncommon bird along smaller forested streams in the lowlands and hills, up to 1300 m.
Habits: Keeps to stream beds. Hunts for food along stream edges and on boulders in streams. Flicks tail on landing after fluttering flight.

628. WHITE-CROWNED FORKTAIL *Enicurus leschenaulti* PLATE 70

Description: Medium-sized (25 cm) black and white forktail. Forehead and forecrown white (feathers sometimes raised into small crest); rest of head, nape, and breast black; belly, lower back, and rump white; white bar in black wing, tail feathers black except for white tips on feathers of very long, forked and graduated tail; two outermost tail feathers entirely white.
Iris—brown; bill—black; feet—pinkish.
Voice: A loud, thin, shrill double whistle, extremely sharp to the human ear, *tsee-eet*.
Range: N India, S China, SE Asia, Malay Peninsula, and Greater Sundas.
Distribution and status: On Sumatra (including Nias and Batu Islands) and Borneo an uncommon bird mostly confined to lowland and hill streams, up to 1400 m. On Java and Bali fairly common along rocky streams at all altitudes where there is forest cover.
Habits: Active, restless bird of fast-flowing rocky streams and rivers. Settles on rocks or walks along the water's edge, pecking left and right at food and constantly spreading its long forked tail. Flies with undulating flight close to the ground while calling.

629. BLACK-BREASTED FRUIT-HUNTER *Chlamydochaera jefferyi*
(AG: Triller/Thrush) PLATE 55

Description: Medium-sized (21 cm), greyish, thrush-like bird with diagnostic bold black breast patch and buffish head with black eye-stripe. Sexes differ. Male: chin and crown reddish buff becoming grey on nape; eye-stripe extending to nape, primaries and edges of tail black; rest of plumage mostly grey with outer tail feathers tipped white. Female is reddish brown where male is grey. Immature: browner on back and speckled with black on head, throat and breast.

Iris—reddish; bill—black; feet—black.

Voice: No information.

Range: Endemic to Borneo.

Distribution and status: Confined to mountains of N Borneo, recorded from Mts Kinabalu and Trus Madi south to Niut and Schwaner range, but uncommon and local. Absent from Kelabit highlands but common on Mt. Dulit.

Habits: Behaves like a pigeon (T.H.) and eats fruits.

Note: Some authors have treated this as a triller or oriole. DNA-hybridization data show it is related to thrushes.

630. JAVAN COCHOA *Cochoa azurea* PLATE 71

Description: Medium-sized (23 cm) dark iridescent blue bird. Male: upperparts shiny dark blue with paler, shiny blue crown and edges to greater wing coverts and flight feathers. Underparts deep purplish blue. Female: brown with blue on forehead and edges of flight feathers. Immature: like female but chest spotted with buffy brown.

Iris—dark brown; bill—black; feet—black.

Range: Endemic to W Java.

Distribution and status: Recorded east to Ciremay, from forest on the higher mountains, from 1000 to 3000 m.

Habits: Completely arboreal and generally in the tallest trees, searching for fruits. Its serrated beak is used to tear flesh off fruits.

Note: Some authors include the Sumatran Cochoa in this species.

631. SUMATRAN COCHOA *Cochoa beccarii* PLATE 71

Description: Large (28 cm) black and iridescent blue thrush. Forehead and crown pale blue; median wing coverts and wing patch greyish blue; median tail feathers blue, others with bluish outer webs; outer tail feathers black; rest of plumage glossy black.

Iris—dark brown; bill black; feet—black.

Voice: Unknown.

Range: Endemic to Sumatra

Distribution and status: Rare; known by only 4 specimens, all males, collected on Singgalang and Kerinci, from 1200 to 1600 m.

Habits: Completely arboreal; forages for fruit in canopy.

Note: Treated by some authors as conspecific with Javan Cochoa, but much brighter coloured and much larger.

632. STONECHAT *Saxicola torquata* PLATE 71

Description: Medium-sized (14 cm), black, white, and rufous chat. Male has black head and flight feathers, dark brown back, bold white patches on neck and wing, whitish rump, and orange breast. Female is duller without black, buffy underparts, and white patch only on wing. Distinguished from female Pied Bushchat by paler colour and white wing patch.
Iris—dark brown; bill—black; feet—blackish.
Voice: Scolding *tsack-tsack*, like striking of two stones together.
Range: Breeds in Palaearctic: migrant in winter to Africa, India and SE Asia.
Distribution and status: A few birds reach Sumatra (including Nias) and N Borneo.
Habits: Prefers open habitat such as farmland, gardens, and secondary scrub. Uses prominent low perches from which to pounce on prey on ground.

633. PIED BUSHCHAT *Saxicola caprata* PLATE 71
(AG: Chat)

Description: Small (13 cm) black and white chat. Male is entirely sooty black except for conspicuous white wing bar and white rump. Female is streaked with brown and has pale brown rump. Immature birds are brown and spotted.
Iris—dark brown; bill—black; feet—black.
Voice: A scolding alarm *chuh*, and a pretty little whistling song *chip-chepee-cheweechu*.
Range: Iran to SW China, SE Asia, Philippines, Sulawesi, Borneo, Java, Bali, Lesser Sundas, and New Guinea.
Distribution and status: Recorded N Borneo, presumably a migrant from Asia. Resident on Java and Bali where it is a common bird of open country, especially in the drier, eastern parts. Usually a lowland bird but occasionally as high as 2400 m.
Habits: A bird of dry, open grassy country. Perches prominently on the top of a bush, rock, post, or wire, and flutters down onto small insect prey. When singing or excited the male cocks his tail.

634. WHEATEAR *Oenanthe oenanthe* PLATE 71

Description: Small (14 cm) sandy brown chat with dark wings and white rump. Wintering male has dark eye-stripe and white supercilium, buffy brown crown and back, blackish wings and centre and tip of tail, and rufous breast. Rump and sides of tail white. Female similar but duller.
Iris—brown; bill—black; feet—black.
Voice: Hard *chack* or *chack-weet* calls.
Range: Breeds in Palaearctic; migrant to India.
Distribution and status: Vagrant recorded in N Borneo.

THRUSHES 325

Habits: A bird of open country where it stays mostly on the ground with a characteristic upright posture, perched on rocks or raised mound.

635. BLUE ROCK-THRUSH *Monticola solitarius* PLATE 71
(Red-bellied Rock-Thrush)

Description: Medium-sized (23 cm) slaty thrush. Male is dull bluish grey with faint black and whitish scaling. Belly sometimes rufous. Female has bluish wash on grey upperparts with black underparts, heavily scaled with buff. Immature birds are like female but with black and white scaling on upperparts.
Iris—brown; bill—black; feet—black.
Voice: Quiet croaks, harsh grating cries, and a short sweet whistled song (Smythies).
Range: Widespread resident and migrant in Eurasia, China, Philippines, SE Asia, Malay Peninsula, Sumatra, and Borneo.
Distribution and status: An uncommon winter visitor at moderate altitudes to N Sumatra and N Borneo (including Natunas).
Habits: Uses prominent open perches such as rocks, houses, poles, and dead trees from which to pounce on insects on the ground.

636. SHINY WHISTLING-THRUSH *Myiophoneus melanurus* PLATE 71

Description: Smallish (22 cm), spangled, blackish-blue thrush. Distinguished from Blue Whistling-Thrush by black bill and smaller size. Distinguished from blue races of Sunda Whistling-Thrush by spangled plumage and shorter bill.
Iris—brown; bill—black; feet—black.
Voice: High-pitched ringing screech.
Range: Endemic to Sumatra.
Distribution and status: On Sumatra common, confined to the slopes of the higher mountains, between 800 and 3300 m.
Habits: A shy bird, keeping on or close to the floor of mossy montane forests, usually near water in primary hill and montane forests.
Note: Sibley and Monroe (1990) use genus spelling *Myiophonus* incorrectly.

637. SUNDA WHISTLING-THRUSH *Myiophoneus glaucinus* PLATE 71
(> Sumatran/Brown-winged Whistling-Thrush)

Description: Largish (25 cm) whistling-thrush with a black bill and no spangles. Three races vary considerably. Bornean *borneensis* is large; male: dark purplish blue all over; female: dark brown; immature: streaked with white underneath. Sumatran *castaneus* is large; male: crown, nape, chin, throat, and breast dark purplish blue, grading into chestnut belly and undertail coverts; tail, mantle, and wings chestnut with bright blue shoulders; female and immature: chestnut (duller than male) with greyish wash on face and underparts; bright blue shoulder patch; blackish wash on crown and nape. Javan *glaucinus* is small; male: dark blue all over, duller and blacker underneath; female: duller. Distinguished from other whistling-thrushes by lack of spangles, and from Blue Whistling-Thrush by black bill.

Iris—brown; bill—black; feet—dark brown.

Voice: A variety of loud ringing calls *ooweet-oweet-tee-teet*, followed by *truuu-truuu*, or a raucous cheet or *tee-ee-eet* . . . *tee-ee-ee-eet* in alarm. Typically a trisyllabic, squirrel-like chortle.

Range: Endemic to the Greater Sundas.

Distribution and status: Throughout the Greater Sundas, in hill and mountain forest, mostly from 400 to 1500 m on Sumatra; mostly sub-montane but locally in the lowlands, and to 2400 m on Borneo; and a montane species on Java and Bali, recorded from 800 to 2400 m, locally lower on Bali.

Habits: Similar to Blue Whistling-Thrush but at higher altitudes. Likes dark caves and crevices as refuges. Calls from tree branch with a pleasant whistling note.

Note: The Sumatran brown-winged race *M. g. castaneus* is almost certainly better treated as a distinct species. Indeed, the group might be better treated as allospecies *M. castaneus* of Sumatra, *M. borneensis* of Borneo, and *M. glaucinus* of Java.

638. BLUE WHISTLING-THRUSH *Myiophoneus caeruleus* PLATE 71

Description: Large (32 cm) black thrush with yellow beak. Plumage black all over with a few white flecks on wing coverts. Wings and tail have iridescent purplish wash. Feathers of head and neck have small reflective spangles at tips. Distinguished from Sunda Whistling-Thrush by yellow bill.

Iris—brown; bill—yellow (Sumatran race sometimes with black culmen); feet—black.

Voice: A whistling song and imitations of other birds. In alarm it gives a high-pitched screech *eer-ee-ee*, similar to a forktail.

Range: Turkestan to India and China, SE Asia, Malay Peninsula, Sumatra, and Java.

Distribution and status: On Sumatra and Java this is an uncommon bird of lowland and hill forest at moderate altitudes up to 1250 m. Replaced at higher elevations by the Sunda Whistling-Thrush.

Habits: Lives close to large rivers or among rocky outcrops in dense forests. Feeds on the ground, coming out in the open but fleeing into thick cover with alarm shrieks when disturbed.

Note: Sundaic birds are placed by some authors in a separate species Large Whistling-Thrush *M. flavirostris*.

639. CHESTNUT-CAPPED THRUSH *Zoothera interpres* PLATE 72
(Kuhl's/Chestnut-headed Thrush > Enggano Thrush)

Description: Small (16 cm) pied and chestnut thrush. Crown and nape chestnut; mantle and back slaty grey; breast blackish; wings and tail blackish with two conspicuous white wing bands; cheeks grey with white markings; belly white with black spots on flanks.
Iris—brown; bill—black; feet—pinkish.
Voice: Melodious song not unlike White-rumped Shama. Alarm call is harsh, ringing *turrrr-turrrr*.
Range: Malay Peninsula, Philippines, Greater Sundas, Lombok, Sumba, and Flores.
Distribution and status: On Sumatra known by one specimen taken on Mt. Kerinci but common on Enggano Island. On Borneo and Java (including Krakatau) it is an occasional bird of lowland forest. May occur on Bali.
Habits: A shy bird, generally seen skulking about with a hopping gait on the forest floor or in fruiting trees.

640. ORANGE-HEADED THRUSH *Zoothera citrina* PLATE 72
(White-throated Thrush)

Description: Medium-sized (21 cm) orange-headed ground thrush. Male: head, nape, and underparts rich orange; vent white; upperparts bluish grey with white bar on the upper wing. Female has upperparts olive-brown. Young like female but with streaks and scaling on back.
Iris—brown; bill—black; feet—brown.
Voice: One of the region's best song birds with a fine clear song. Also a loud, screeching alarm whistle *teer-teer-teerrr*.
Range: Pakistan to S China, SE Asia, Malay Peninsula, and Greater Sundas.
Distribution and status: The few records from Sumatra are of migrants from Asia. On Borneo this is a rare sub-montane resident, between 1000 and 1500 m, known only from Mts Kinabalu and Trus Madi. On Java and Bali this is an occasional bird of lowland and hill forest, up to 1500 m.
Habits: A shy bird which prefers shady forest where it skulks in thick cover on the ground. It sings from tree perches.

641. EVERETT'S THRUSH *Zoothera everetti* PLATE 72
(AG: Ground-Thrush)

Description: Medium-sized (23 cm) dark thrush. Upperparts dark brown with pale throat and sides of head spotted with black and rufous; breast rufous; belly white.

Iris—brown; bill—black; feet—horn.
Voice: Not recorded.
Range: Endemic to Borneo.
Distribution and status: Confined to the high mountains of N Borneo, between 1400 and 2200 m, where it is rare, recorded from Mt. Kinabalu south to Usun Apau and Dulit.
Habits: Ground living and solitary, preferring tall forests at moderate altitudes.

642. SUNDA THRUSH *Zoothera andromedae*   PLATE 72
(AG: Ground-Thrush)

Description: Large (25 cm), short-tailed, dark ground-thrush. Upperparts dark grey laced with black scaling; face and throat speckled black and whitish grey; breast light bluish grey; belly white, heavily laced at sides with black scaling.
Iris—brown; bill—grey; feet—blackish brown.
Voice: Not recorded.
Range: Philippines, Sumatra, Java, Bali, and Lesser Sundas.
Distribution and status: On Sumatra (including Enggano Island) it is locally not uncommon in forest, between 1200 and 2200 m, lower on Enggano Island. On Java restricted to Pangrango but apparently more widespread on Bali.
Habits: A shy bird, keeping to shady cover on the forest floor.

643. SIBERIAN THRUSH *Zoothera sibirica*   PLATE 72

Description: Medium-sized (22 cm), blackish (male) or brown (female) thrush with conspicuous supercilium. Male slaty black with white supercilium and white tips to feathers of tail and vent. Female olive-brown with buffy white and rufous underparts, and buffy white supercilium.
Iris—brown; bill—black; feet—yellow.
Voice: On winter grounds gives only a quiet, whistled contact note *chit*.
Range: Breeds in N Asia; migrant in winter through SE Asia to the Greater Sundas.
Distribution and status: On Sumatra (including islands) and W Java this is a regular visitor to hill and mountain forest. A rare visitor to N Borneo and Bali.
Habits: An active bird of forest floor and canopy, sometimes in flocks.

644. SCALY THRUSH *Zoothera dauma*   PLATE 72
(White's/White's Scaly/Golden/Small-billed/Tiger Thrush)

Description: Large (28 cm) scaly brown thrush. Upperparts brown and underparts white, entirely laced with black and golden buff feather edges.
Iris—brown; bill—dark-brown; feet—pinkish.

Voice: Soft monotonous whistle and thin short *tzeet*.
Range: Widely distributed from Europe and India to China, SE Asia, Philippines, Sumatra, Java, Bali, and Lombok.
Distribution and status: A rare resident of the mountains of N Sumatra recorded south to Mt. Kerinci, between 2000 and 3000 m. On Java and Bali an uncommon resident of montane forests. Vagrant Borneo presumably from N Asia.
Habits: Inhabits dense forest where it feeds on the ground.

645. EYEBROWED THRUSH *Turdus obscurus* PLATE 72
(White-browed/Dark/Grey-headed Thrush)

Description: Medium-sized (23 cm) brownish thrush with a conspicuous white eyebrow. Upperparts olive-brown with darker, greyish head and white supercilium; breast orange; belly white with rufous wash on the sides.
Iris—brown; bill—yellow at base, black at tip; feet—yellowish.
Voice: Thin *zip-zip* contact call.
Range: Breeds in N Asia; migrant in winter to Philippines, Sulawesi, and Greater Sundas.
Distribution and status: A regular winter visitor to hill and mountain forest of Sumatra (including islands) and N Borneo, between 1000 and 2000 m; less common south to Java and Bali.
Habits: Prefers open and secondary forests; moves through low bushes and trees in active noisy flocks. Quite tame and inquisitive.

646. ISLAND THRUSH *Turdus poliocephalus* PLATE 72

Description: Medium-sized (20 cm) blackish and chestnut bird. Plumage overall dusky, from blackish to greyish brown depending on race, except for chestnut belly and sometimes white vent. Yellow eye-ring.
Iris—brown; bill—yellow; feet—yellow.
Voice: A rattling alarm call and a loud, clear, melodious song, which commences very slowly with alternating high and low notes, then speeds gradually in tempo to a gloriously lusty song.
Range: Philippines, Sulawesi, Greater Sundas, Ceram, Timor, New Guinea and Pacific islands to Samoa.
Distribution and status: In the Greater Sundas it is confined to the tops of the highest mountains where it may be locally common between 2000 and 3450 m. On Sumatra known from Mt. Leuser and a few peaks south to Mt. Kerinci; on Borneo known only from Mt. Kinabalu and Trus Madi, but on Java it is common on most of the high mountains above 2000 m. Not recorded Bali.
Habits: Feeds on berries and invertebrates on ground and in bushes. Keeps to dense cover but comes out into open patches when all is quiet. Sings from tree perches.

Old World Warblers—FAMILY SYLVIIDAE

WARBLERS are a very large Old World family of small, very active, insectivorous birds with narrow, pointed bills. Most are drab in colour and difficult to identify in the field. They have generally clear, pretty songs and make neat cup-shaped or domed nests. Tailorbirds make elaborate nests of leaves stitched together with plugs of spider's web. Several species are migratory.
The family can be divided into six groups.

1. Australian Wren-Warblers: one representative in the Greater Sundas (*Gerygone*). Sibley and Monroe (1990) reassigned these to the family Pardalotidae.
2. Leaf-Warblers: small, canopy-feeding birds, including several winter migrants (*Seicercus, Abroscopus, Phylloscopus*). There are 8 species in the Greater Sundas.
3. Grass-Warblers and Bush-Warblers: drab brownish birds inhabiting scrub, swamps, and grassland (*Urosphena, Cettia, Acrocephalus, Locustella, Megalurus, Bradypterus*).
4. Tailorbirds: small warblers with cocky tails and reddish-coloured heads (*Orthotomus*).
5. Wren-Warblers and Fantail-Warblers: medium-sized, drab warblers with long cocky tails inhabiting scrub and secondary vegetation (*Prinia, Cisticola*).
6. Ground-Warblers: almost tailless ground skulkers, of which only one species occurs in the Greater Sundas (*Tesia*).

Australian Wren-Warblers

647. GOLDEN-BELLIED GERYGONE *Gerygone sulphurea* PLATE 73
(Yellow-breasted Gerygone/Flyeater)

Description: Very small (9 cm) yellow-bellied warbler with distinctive white lores. Upperparts greyish brown; chin and throat white; underparts bright yellow; tail has subterminal row of white spots. In immature underparts white, washed with yellow.
Iris—brown; bill—black; feet—dark olive-green.
Voice: A reedy whistled song of 3 to 5 thin notes, sliding from note to note in various descending phrases; ventriloquial and difficult to locate. (See Appendix 6.)
Range: Philippines, Sulawesi, Malay Peninsula, and Greater Sundas.
Distribution and status: On Sumatra (including Enggano and Bangka Islands), Borneo (including Maratuas), Java (inc. Karimunjawa), and Bali it is a locally common bird, up to 1500 m.

OLD WORLD WARBLERS 331

Habits: Frequents coastal scrub, mangroves, rubber plantations, and open forest, especially bamboo and conifer clumps. Generally solitary or in pairs. An inconspicuous little bird betrayed by its constant sweet song.

Leaf-Warblers

648. CHESTNUT-CROWNED WARBLER *Seicercus castaniceps*
(Chestnut/Chestnut-headed Warbler)   PLATE 73

Description: Very small (9 cm) olive warbler with rufous-brown cap, black lateral crown stripe, and black eye-stripe, white eye-ring, grey cheeks, yellow wing bars, yellow rump and flanks, and white centre of belly. Differs from Sunda Warbler by grey cheeks and breast.
Iris—brown; bill—black above, pale below; feet—pinkish.
Voice: Song is high-pitched, metallic, and glissading (P.R.); also double call note *chi-chi*, and wren-like *tsik*.
Range: Himalayas to S China, SE Asia, Malay Peninsula, and Sumatra.
Distribution and status: On Sumatra known from few localities in Sibayak and Kerinci areas of the Barisan Range, from 1200 to 1400 m.
Habits: Actively searches the canopy of small trees in montane forest. Forms mixed flocks with other species.

649. SUNDA WARBLER *Seicercus grammiceps*   PLATE 73
(AG: Flycatcher-Warbler)

Description: Small (10 cm) olive warbler with chestnut head, dark supercilium, and narrow white eye-ring. Upperparts greyish olive with rump whitish in Java and Bali, or grey in Sumatra. Yellow edges of wing coverts form two bars across wing; underparts whitish. Sumatran race has grey rump and back.
Iris—reddish brown; bill—black; feet—orange.
Voice: High-pitched, ringing *chee-chee-chechee* and a buzzing *turrr*.
Range: Endemic to Sumatra, Java, and Bali.
Distribution and status: On Sumatra known from Mts Talamau and Kerinci where it is a common bird between 1400 and 2200 m. On Java and Bali a locally common bird of montane forests between 800 m and 2500 m.
Habits: Keeps to dense forest or forest edge where it joins with mixed-species flocks. Hunts for insects in the lower layers of the forest.

650. YELLOW-BREASTED WARBLER *Seicercus montis* PLATE 73
(AG: Flycatcher-Warbler)

Description: Small (10 cm) olive warbler with rufous head, black supercilium, and white eye-ring. Underparts, rump, and two wing bars yellow.
Iris—brown; bill—black above, pale below; feet—pinkish.
Voice: Irritating high-pitched song of disjointed notes, starting at dawn.
Range: Palawan, Malay Peninsula, Sumatra, Borneo, Flores, and Timor.
Distribution and status: On Sumatra and Borneo a common bird on main mountains, between 1000 and 2200 m, often lower on Borneo.
Habits: Active bird of lower storeys in montane forests, mixing with other species in flock.

651. YELLOW-BELLIED WARBLER *Abroscopus superciliaris* PLATE 73
(Bamboo/White-throated Warbler)

Description: Small (11 cm) yellow-bellied warbler with conspicuous white supercilium. Forehead and crown grey; back of head and back greenish olive; chin, throat, and upper breast white; rest of underparts yellow.
Iris—brown; bill—black with whitish base; feet—pink.
Voice: Sweet little song, descending and rising scale.
Range: E Himalayas, S China, SE Asia, Malay Peninsula, and Greater Sundas.
Distribution and status: A fairly common bird of lowlands and hills, up to about 1500 m, throughout Sumatra, Borneo, and Java. Not recorded on Bali.
Habits: Frequents secondary forests, particularly in bamboo areas. Generally in small flocks in low bushes and bamboo thickets.

652. INORNATE WARBLER *Phylloscopus inornatus* PLATE 73
(Yellow-browed Warbler)

Description: Small (11 cm), bright olive-green warbler with usually two whitish wing bars visible, clear white or cream supercilium, and no visible crown stripes. Underparts vary from white to yellowish green. Distinguished from Arctic Warbler by brighter upperparts, bolder wing bars, and white tips of tertiaries. Distinguished from Eastern Crowned and Mountain Leaf-Warblers by lack of distinct yellow-green median crown stripe, and by white undertail coverts and bolder wing bars.
Iris—brown; bill—dark above with yellow tip, darker below; feet—pinkish brown.
Voice: Noisy; loud *we-est* rising on second note frequently uttered.
Range: Breeds in Himalayas, N Asia, and China; migrant south in winter to India, SE Asia, and Malay Peninsula.

Distribution and status: Recorded once in Bukittinggi, W Sumatra, but probably often overlooked.

Habits: Forms active flocks, often mixing with other small insect feeders, working through foliage in middle and upper canopy.

653. ARCTIC WARBLER *Phylloscopus borealis* PLATE 73
(Willow Warbler)

Description: Small (12 cm) greyish-olive warbler with conspicuous, long yellowish-white supercilium. Upperparts dark olive with a faint pale wing bar; underparts whitish with brownish olive flanks; lores and eye-stripe blackish. Distinguished from Inornate Warbler by longer, slightly upturned bill, duller colour, less prominent wing bars, and lack of white tips on tertials.
Iris—dark brown; bill—dark brown above, yellow below; feet—brown.
Voice: Rattling series of *chweet* notes with last note on higher pitch, and characteristic *zit* occasionally given by wintering birds.
Range: Breeds N Europe, N Asia, and Alaska; migrant south in winter to China, SE Asia, Philippines, and Indonesia.
Distribution and status: A winter visitor in small numbers to primary and secondary forests, up to 2500 m, throughout the Greater Sundas (including islands).
Habits: Frequents open wooded areas, mangroves, secondary forests, and forest edge. Joins mixed-species flocks, working through the foliage of trees, searching for food.

654. EASTERN CROWNED-WARBLER *Phylloscopus coronatus*
(Temminck's Crowned-Warbler, AG: Crowned Leaf-Warbler) PLATE 73

Description: Small (12 cm) yellowish-olive warbler with whitish supercilium and median crown stripe. Upperparts greenish olive with yellow edges to flight feathers including two yellowish wing bars; underparts whitish with contrasting yellow vent; lores and eye-stripe blackish. Distinguished from Arctic and Inornate Warblers by obvious crown stripes, and from Mountain Leaf-Warbler by wing bars and whitish underparts.
Iris—dark brown; bill—brown above, yellow below; feet—grey.
Voice: Dissyllabic penetrating *swee-ett*.
Range: Breeds in NE Asia; migrant south in winter to China, SE Asia, Sumatra, and Java.
Distribution and status: Few records from Sumatra and W Java but may be overlooked.
Habits: Frequents mangroves, wooded areas, and forest edge from sea-level to the

highest peaks. Joins mixed-species flocks and is generally seen in the canopy of larger trees.

655. MOUNTAIN LEAF-WARBLER *Phylloscopus trivirgatus* PLATE 73
(Island Leaf-Warbler/Green Flycatcher-Warbler)

Description: Smallish (11 cm) green and yellow warbler with conspicuous yellowish median crown stripe and supercilium. Upperparts greenish with no wing bars; yellowish underparts diagnostic. A greyer, less yellow race is confined to Mt. Kinabalu.
Iris—nearly black; bill—black above, reddish below; feet—greyish.

Voice: A scolding alarm call and an unmusical, quiet, high-pitched song *tsee-chee-chee-weet*, and variations.

Range: Palawan, Malay Peninsula, and Greater Sundas.

Distribution and status: On Sumatra, Java, and Bali confined to montane forests between 800 and 3000 m but locally abundant. On Borneo found on Mt. Kinabalu south to Tama Abo; also in Penrissen and Poi ranges.

Habits: Frequents the tops of taller trees in the sub-montane and montane forests, and forest edge, up to the alpine zone. Single or in flocks, generally mixing with other species. Feeds mostly in the canopy or in epiphytic ferns and orchids.

Grass-Warblers and Bush-Warblers

656. CLAMOROUS REED-WARBLER *Acrocephalus stentoreus* PLATE 73
(Southern Great Reed-Warbler > Heinroth's/Large-billed Reed-Warbler)

Description: Largish (18 cm) brown warbler with elongated tail and whitish supercilium. Upperparts uniform olive-brown; underparts whitish with buffy flanks and undertail coverts.
Iris—brown; bill—greyish brown; feet—greyish brown.

Voice: Harsh alarm note *chack*, sweet song, interspersed with higher notes. Generally calls at night.

Range: N Africa to S China, Philippines, SE Asia (except Malay Peninsula), Borneo, Java, Moluccas, Lesser Sundas, New Guinea, and Australia.

Distribution and status: Recorded in SE Borneo where known from Rantau lakes and other wetlands. In W Java this is an uncommon bird of wetlands. These populations are sedentary; not recorded in Sumatra or Bali.

Habits: Inhabits swampy reed beds, paddy fields near reed beds, and mangroves. Clings sideways to reed stems when perched and puffs out throat feathers when singing. Generally single or in pairs in reeds or other vegetation close to the ground.

OLD WORLD WARBLERS 335

657. EASTERN REED-WARBLER *Acrocephalus orientalis* PLATE 73
(Oriental Reed-Warbler, AG: Great Reed-Warbler)

Description: Largish (18 cm) brown warbler with conspicuous buff supercilium. Distinguishable in field from Clamorous Reed-Warbler only with difficulty by shorter, thicker bill and slight streaking on side of breast, or in the hand by the outer (ninth) primary longer than sixth, and gape is pinkish not yellow.
Iris—brown; bill—brown above, pinkish below; feet—grey.
Voice: On winter grounds it sometimes sings, usually only a single, harsh, grating *chack* at intervals.
Range: Breeds in E Asia; migrating south in winter to SE Asia, Philippines, Indonesia, and rarely as far as Australia and New Guinea.
Distribution and status: An occasional to common visitor to the Greater Sundas (including islands), recorded from throughout the region.
Habits: Favours reed beds, rice fields, marshes, and secondary scrub in lowlands.

658. BLACK-BROWED REED-WARBLER *Acrocephalus bistrigiceps*
(Von Schrenck's Reed-Warbler) PLATE 73

Description: Medium-sized (13 cm) brown warbler with buffy white supercilium bordered above and below with distinctive black stripes. Underparts whitish.
Iris—brown; bill—dark above, pale below; feet—pinkish.
Voice: Harsh *churr* in alarm.
Range: Breeds in NE Asia; migrates in winter to India, S China, SE Asia, and Malay Peninsula.
Distribution and status: Recorded once in N Sumatra but possibly overlooked.
Habits: Typical reed warbler living among tall reeds and grasses close to water.

659. PALLAS'S WARBLER *Locustella certhiola* PLATE 74
(Grey-naped Warbler, AG: Grasshopper-Warbler)

Description: Medium-sized (15 cm) brown streaked warbler with a buff eyestripe and white tips to rufous tail. Upperparts brown streaked with grey and black; wings and tail reddish brown, tail with blackish subterminal bar; underparts whitish with buff breast and flanks, in juveniles washed yellow and with triangular black spots on breast.
Iris—brown; bill—brown above; yellowish below; feet—pinkish.
Voice: Prolonged, harsh trill *chir-chirrrr*; also thin alarm note *tik tik tik*.
Range: Breeds in N and C Asia; migrates south in winter to China, SE Asia, Palawan, Sulawesi, and the Greater Sundas.
Distribution and status: Probably not uncommon winter visitor throughout the

336 OLD WORLD WARBLERS

Greater Sundas (including islands), though there are few records from Sumatra, Java, and Bali. Common visitor to N Borneo and recorded throughout the island.
Habits: Inhabits reed beds, swamps, paddy fields, and grassy thickets and bracken near water; also forest edge. Skulks in dense vegetation and when flushed, flies only a few metres before diving into cover again.
Note: A sight record of Pleske's Warbler *L. pleskei* on Bali may warrant addition of a new species to the region's list if confirmed, but some authors treat Pleske's Warbler as a race of Pallas's.

660. MIDDENDORFF'S WARBLER *Locustella ochotensis*   PLATE 74
(AG: Grasshopper-Warbler)

Description: Largish (16 cm) olive-brown warbler with buffy brown flanks and whitish belly. Immature birds are streaked on breast and flanks. Distinguished from Pallas's Warbler by unstreaked upperparts.
Iris—brown; bill—dark above, pale below; feet—pinkish.
Voice: *Chi-chirr.*
Range: Breeds in NE Asia; in winter south to S China, Philippines, Sulawesi, and Borneo.
Distribution and status: Recorded N Borneo where this is a rare winter visitor.
Habits: Prefers patches of grass or reeds.

661. LANCEOLATED WARBLER *Locustella lanceolata*   PLATE 74
(Streaked Warbler, AG: Grasshopper-Warbler)

Description: Small (12 cm) brown-streaked warbler. Upperparts olive-brown streaked with black; underparts white, washed with ochre and streaked with black on breast and flanks; supercilium buff; tail lacks white tip.
Iris—dark brown; bill—brown above, yellowish below; feet—pinkish.
Voice: A prolonged, rapid, high-pitched trill.
Range: Breeds in NE Asia; migrates south in winter to Philippines, SE Asia, Greater Sundas, and N Moluccas.
Distribution and status: A scarce winter visitor recorded throughout most of the Greater Sundas; not recorded on Bali.
Habits: Frequents wet rice fields, swampy scrub, fallow fields, and bracken near water.

662. STRIATED GRASSBIRD *Megalurus palustris*   PLATE 74
(AG: Warbler/Canegrass-Warbler)

Description: Large (26 cm) brown warbler with back boldly streaked with black, buffish supercilium, and very elongated, pointed tail. Upperparts bright reddish

brown with black streaks on back and wing coverts; underparts whitish with narrow blackish streaks on breast and flanks, and a rufous wash on flanks and undertail coverts.
Iris—brown; bill—black above, pinkish below; feet—pink.
Voice: Pretty song uttered from perch and in flight, and a sharp clicking call.
Range: India, China, Philippines, SE Asia (except Malay Peninsula), Java, and Bali.
Distribution and status: On Java and Bali this is a fairly common bird up to about 2000 m.
Habits: Inhabits open grassy fields, especially *Saccharum* beds, bamboo clumps, and secondary scrub. Lives partly on the ground where it runs under thick cover. Often perches conspicuously in the open where it sometimes sings a short song, but more often sings in flight like a lark.

Tailorbirds

663. COMMON TAILORBIRD *Orthotomus sutorius* PLATE 75
(Long-tailed Tailorbird)

Description: Small (10 cm), rufous-crowned, white-bellied warbler with a long, often cocked up, tail. Forehead and crown rufous; buffy eyebrow; lores and side of head whitish; nape greyish; back, wings, and tail olive-green; underparts white with grey flanks. In breeding plumage the central tail feathers of the male become elongated further.
Iris—pale buff; bill—black above, pinkish below; feet—pinkish grey.
Voice: Very loud, repetitive, monotonous call *te-chee-te-chee-te-chee*, or single *twee*. (See Appendix 6.)
Range: India to China, SE Asia, Malay Peninsula, and Java.
Distribution and status: On Java it is a widespread species up to 1500 m, but is more erratic in occurrence and less common than the Olive-backed Tailorbird.
Habits: Frequents light forest, secondary forest and gardens. A lively bird, always on the move or cockily giving its penetrating call. Keeps to the understorey and generally stays in thick cover.

664. DARK-NECKED TAILORBIRD *Orthotomus atrogularis* PLATE 75
(Black-necked/Dark-cheeked Tailorbird)

Description: Small (10 cm), rufous-crowned, white-bellied warbler with a long, often cocked-up tail, yellow vent, and diagnostic blackish throat (lacking in immatures). Upperparts olive-green; sides of head grey. Female is duller with less red on head and less black on throat. Distinguished from Common Tailorbird by rufous nape, greener back, and yellow undertail coverts and thighs.

Iris—brown; bill—black above, pinkish below; feet—pinkish grey.
Voice: Sweet, blubbering, clear *kri-ri-ri* unlike other tailorbirds. (See Appendix 6.)
Range: N India to SW China, Philippines, SE Asia, Malay Peninsula, Sumatra, and Borneo.
Distribution and status: On Sumatra (including islands) and Borneo (including Natunas and Anambas) a common lowland bird up to 1200 m.
Habits: Frequents light forest, secondary forest, river banks, and gardens.

665. ASHY TAILORBIRD *Orthotomus ruficeps* PLATE 75
(Red-headed Tailorbird)

Description: Small (11 cm) grey, rufous-headed warbler. Male: crown, chin, and upper throat and cheeks rufous; otherwise plumage ashy grey with white belly. Female is less rufous on head, and chin and upper throat are white.
Iris—reddish brown; bill—brown; feet—flesh.
Voice: Dissyllabic, trilled *trrree-yip* and plaintive *choe-choee*. (See Appendix 6.)
Range: Palawan, Malay Peninsula, and Greater Sundas.
Distribution and status: On Sumatra (including islands) and Borneo this is a common bird, up to 950 m. In Java restricted to coastal mangroves and wetlands of N Java and replaced inland by the Olive-backed Tailorbird. Not recorded on Bali.
Habits: Frequents open forest, forest edge, mangroves, coastal scrub, gardens, secondary vegetation, and bamboo thickets. An active bird of the understorey and tree crowns.

666. OLIVE-BACKED TAILORBIRD *Orthotomus sepium* PLATE 75
(Javan/Bali Tailorbird)

Description: Small (11 cm), rufous-headed, grey warbler. Male: crown, throat, and cheeks rufous; otherwise plumage greenish grey with white belly washed yellow. Female is less rufous on head; chin and upper throat white. Differs from Ashy Tailorbird in more olive back and yellower less grey flanks.
Iris—reddish brown; bill—brown; feet—flesh.
Voice: Notoriously variable in voice, including monotonous repeated calls such as *chew-chew-chew*, *turr-turr tsee-weet . . . tsee-weet*.
Range: Endemic to Java and Bali.
Distribution and status: On Java and Bali this is a common bird up to 1500 m.
Habits: Frequents open forest, forest edge, secondary vegetation, and bamboo thickets. An active bird of the understorey and also crowns of trees.
Note: Treated by some authors as a subspecies of Ashy Tailorbird.

OLD WORLD WARBLERS 339

667. RUFOUS-TAILED TAILORBIRD *Orthotomus sericeus*   PLATE 75
(Rufous-backed Tailorbird)

Description: Small (11 cm) warbler with rufous crown and nape, and diagnostic rufous tail. Buffy white cheeks and red tail distinguish this from all other tailorbirds.
Iris—brown; bill—dark above, pale below; feet—pinkish.
Voice: Persistent shrill *dogjeh* with equal stress but first note higher; slower *too-wee-to*; also scolding churrs. (See Appendix 6.)
Range: Palawan, Malay Peninsula, Sumatra, and Borneo.
Distribution and status: On Sumatra (including islands) and Borneo (including Natunas) this is a rather common resident of low-lying areas, below 500 m.
Habits: Typical tailorbird flicking its erect tail. Lives in pairs or family flocks, keeping to secondary scrub and mangroves. Less active and noisy than Ashy Tailorbird.

668. MOUNTAIN TAILORBIRD *Orthotomus cuculatus*   PLATE 75
(Golden-headed Tailorbird)

Description: Small (12 cm), orange-capped, yellow-bellied forest warbler with a pronounced white supercilium. Upperparts olive-green; chin, throat, and upper breast greyish white; lower breast and belly bright yellow.
Iris—brown; bill—black above, pale below; feet—pink.
Voice: A variable, sweet, tinkling call of 2 or 3 repeated notes followed by a trill *pee-pee-cherrrree*, quite different from other tailorbirds. (See Appendix 6.)
Range: N India to S China, Philippines, SE Asia, Malay Peninsula, and Indonesia.
Distribution and status: On Sumatra, Java, and Bali not uncommon on the higher mountains, between 1000 and 2500 m. In Borneo restricted to the mountains of N Borneo from Mt. Kinabalu south to Tama Abo; also Mts Mulu and Niut.
Habits: Inhabits montane forests, open montane scrub, and bamboo thickets. A gregarious bird, often found in small parties, but generally skulking in thick cover and difficult to see. Easily recognized by its song. Does not make a leaf-purse nest.

Wren-Warblers and Fantail-Warblers

669. HILL PRINIA *Prinia atrogularis*   PLATE 75
(Black-throated/White-browed Prinia, AG: Wren-Warbler)

Description: Largish (16 cm), long-tailed, brown warbler with diagnostic black-streaked breast. Upperparts brown, flanks buffy rufous, and belly buffy white. Grey cheeks, white supercilium, and very long tail distinctive.

Iris—pale brown; bill—dark above, pale below; feet—pinkish.
Voice: Loud piercing *cho-ee*, cho-ee, *cho-ee*, like Common Tailorbird but slower (Smythies).
Range: Himalayas, S China, SE Asia, Malay Peninsula, and Sumatra.
Distribution and status: On Sumatra a common bird of hills and mountains, between 600 and 2500 m.
Habits: Lives in noisy, active family flocks in grass and low vegetation of submontane and montane forests, including dwarf moss forest and sub-alpine scrub.

670. PLAIN PRINIA *Prinia inornata* PLATE 75
(Tawny/Plain-coloured Prinia, AG: Wren-Warbler)

Description: Largish (15 cm), long-tailed, brownish warbler with a whitish supercilium. Upperparts dull greyish brown; underparts buff to rufous. Back paler and more uniform than Brown Prinia.
Iris—light brown; bill—brown above, pale pinkish below; feet—yellowish.
Voice: Shrill *chee-cheerrrrrr-rooweet* or *cheerrrrlet*.
Range: Africa, India, China, SE Asia (except Malay Peninsula), and Java.
Distribution and status: In Java common up to about 1500 m.
Habits: Inhabits areas of long grass, reed beds, marshes, maize and paddy fields. A cocky, active bird often in small parties, calling regularly from trees, grass stems, or in flight.
Note: Sometimes treated as a race of Tawny-flanked Prinia *P. subflava*.

671. YELLOW-BELLIED PRINIA *Prinia flaviventris* PLATE 75

Description: Largish (13 cm), long-tailed, olive-green warbler with white breast and diagnostic yellow breast and belly. Head grey with sometimes faint whitish supercilium; upperparts olive-green; orange yellow eye-ring; chin, throat, and upper breast white.
Iris—brown; bill—black to brown above; pale below; feet—orange.
Voice: Weak, harsh *schink-schink-schink*, and soft mewing like young cat. Hurried bubbling *tidli-idli-u* with stress on last falling syllable. (See Appendix 6.)
Range: Pakistan to S China, SE Asia, Malay Peninsula, and Greater Sundas.
Distribution and status: On Sumatra (including Nias) and Borneo this is a frequent bird, up to 900 m. In W Java it is a rather uncommon bird in suitable habitat at all altitudes. Not recorded on Bali.
Habits: Inhabits reedy swamps, tall grasslands, and scrub. A fairly shy bird, keeping out of sight in long grass or reeds except when singing, perched on a tall stem.

OLD WORLD WARBLERS 341

672. BAR-WINGED PRINIA *Prinia familiaris*   PLATE 75

Description: Largish (13 cm), long-tailed, olive warbler with two diagnostic white wing bars and tail feathers with black and white tips. Upperparts olive brown; throat and middle of breast white; sides of breast and flanks grey; belly and vent pale yellow.
Iris—brown; bill—black above, yellowish below; feet—pink.
Voice: Loud high-pitched *chweet-chweet-chweet*.
Range: Endemic to Sumatra, Java, and Bali.
Distribution and status: In Sumatra it is not uncommon up to 900 m, though apparently absent from extreme N Sumatra. On Java and Bali a very common bird, up to 1500 m.
Habits: Inhabits mangroves and open secondary habitats, especially gardens and parks. It is a noisy, gregarious bird. Hunts on ground and up in treetops.

673. BROWN PRINIA *Prinia polychroa*   PLATE 75
(Javan Brown Prinia)

Description: Largish (15 cm), long-tailed, brown-streaked prinia. Upperparts brown, slightly streaked or mottled; tail brown with small white tips; indistinct whitish supercilium; underparts yellowish buff, whiter on throat, grey on breast, and tawny on flanks and thighs. Back is darker and more streaked than Plain Prinia.
Iris—reddish brown; bill—brown above, pale below; feet—whitish.
Voice: Loud *twee-ee-ee-ee-eet* or *chook-chook*.
Range: SW China, SE Asia (except Malay Peninsula), and Java.
Distribution and status: On Java this is a widespread but uncommon bird, up to 1500 m.
Habits: Inhabits *Imperata* grasslands and low scrub. A shy elusive bird, keeping to thick cover. Lives in pairs or family parties but less noisy and conspicuous than Bar-winged Prinia.

674. ZITTING CISTICOLA *Cisticola juncidis*   PLATE 74
(Fan-tailed/Straw-headed/Common/Streaked Cisticola, AG: Fantail-Warbler)

Description: Small (10 cm) brown-streaked warbler with buffy rufous rump and distinctive white-tipped tail. Distinguished from non-breeding Golden-headed Cisticola by white supercilium, noticeably paler than sides of neck and nape.
Iris—brown; bill—brown; feet—pink to reddish.
Voice: Series of clicking *zit* notes given in undulating display flight.
Range: Africa, S Europe, India, China, Japan, Philippines, SE Asia, Malay Peninsula, Sumatra, Java, Sulawesi, Lesser Sundas, and N Australia.

Distribution and status: On Sumatra (including islands), Java (including Kangean), and Bali common up to 1200 m.

Habits: Lives in open grassland, paddy fields, and sugarcane beds, in generally wetter areas than Golden-headed Cisticola. In courtship flight the male hovers and circles high over his mate, calling. In non-breeding season this is a shy inconspicuous bird.

675. GOLDEN-HEADED CISTICOLA *Cisticola exilis* PLATE 74
(Bright-capped/Golden-capped Cisticola, AG: Fantail-Warbler)

Description: Small (11 cm) brown-streaked warbler with a bright golden crown and brown rump in breeding male. Female and non-breeding male have crown heavily streaked with black but distinguished from Zitting Cisticola by buffish supercilium the same colour as side of neck and nape. Underparts buff, whitish on throat; tail dark brown with buff tip.

Iris—brown; bill—black above, pink below; feet—light brown.

Voice: Breeding male gives a scratching *buzz* followed by a loud, liquid *plook* from perch or in flight; also harsh, high-pitched scolding.

Range: India, China, Philippines, SE Asia (except Malay Peninsula), Sulawesi, Greater Sundas, Moluccas, Lesser Sundas to New Guinea and Australia.

Distribution and status: Now locally common on Sumatra though probably a recent arrival following deforestation. On Borneo recorded in Pontianak. On Java and Bali this is a common bird in suitable habitat, up to 1500 m.

Habits: Inhabits *Imperata* grassland, reeds, and rice fields. A secretive bird of long grass, sometimes seen perched on a tall grass stem or bush. Fluttering flight.

Ground-Warblers

676. JAVAN TESIA *Tesia superciliaris* PLATE 74
(Eyebrowed/Malaysian Tesia, AG: Ground-Warbler]

Description: Tiny (7 cm), greenish-grey, wren-like bird with very short tail and pronounced pale supercilium. Head blackish with pale grey supercilium; upperparts greyish olive; underparts whitish grey.

Iris—brown; bill—brown above, yellow below; feet—brown.

Voice: A loud, explosive, rather fast repetitive song of about 15 notes.

Range: Endemic to W Java.

Distribution and status: A locally common bird of mountain forests, from 1000 to 3000 m, recorded east to Papandayan and Ciremay.

Habits: Lives on or close to the forest floor, in dense thickets or dense ground cover in forest gaps, often near dead trees.

677. BORNEAN STUBTAIL *Urosphena whiteheadi* PLATE 74
(Short-tailed/Whitehead's Stubtail, AG: Bush-Warbler)

Description: Very small (10 cm), short-tailed, brown warbler with long buffish supercilium and orange-buff face. Underparts white with faint dark grey mottling on breast and flanks.
Iris—brown with yellow eye-ring; bill—dark; feet—flesh.
Voice: Very high-pitched, repeated, mechanical *tzi-tzi-tzeee*.
Range: Endemic to Borneo.
Distribution and status: Confined to the highest mountains in N Borneo where found from Mt. Kinabalu south to Liang Kubung and Mueller range. A common bird above 2000 m.
Habits: Active, ground-living warbler of montane forest vegetation. Skulks in thick undergrowth but rather tame and sometimes inquisitive.
Note: Sometimes treated under genus *Cettia*.

678. SUNDA BUSH-WARBLER *Cettia vulcania* PLATE 74
(Mueller's/Mountain/Malaysian/Javan/Indonesian Bush-Warbler)

Description: Small (13 cm), nondescript, brown warbler with longish tail and pale whitish supercilium. Upperparts dark brown and brownish olive; underparts buffish white, washed with brown on sides and across breast.
Iris—brown; bill—black above, yellow below; feet—brown.
Voice: Various loud, sharp, monotone call notes *chee-heeoow*, *cheeoow-wee-ee-eet*, or *trr-trr*. Prolonged wavering note. (See Appendix 6.)
Range: Greater Sundas, Lombok, and Timor.
Distribution and status: On Sumatra locally common between 2100 and 3400 m on the highest peaks from Mts Leuser to Kerinci, and Dempu. Uncommon on Borneo and restricted to the mountains of N Borneo from Mt. Kinabalu south to Mts Murud and Mulu. On Java and Bali it is locally not uncommon above 1500 m.
Habits: Lives in thick undergrowth of mountain forest, generally in upper zone such as moss forest, open *Vaccinium* heath, and edelweiss meadows. Creeps about in a rodent-like manner.
Note: May be conspecific with Brownish-flanked Bush-Warbler *C. fortipes* of SE Asia.

679. RUSSET BUSH-WARBLER *Bradypterus seebohmi* PLATE 74
(Mountain Bush-Warbler > Javan/Timor Bush-Warbler, AG: Scrub-Warbler)

Description: Medium-sized (15 cm) dark brown warbler with a longish, broad, graduated tail. Upperparts olive-brown with a rufous tinge; tail more olive; chin

and throat white, streaked with black; rest of underparts white, washed grey on sides of neck, and olive brown on sides of breast and belly.
Iris—brown; bill—black above, pinkish below; feet—pinkish.
Voice: Mechanical, endlessly repeated, rasping *zree-ut*.
Range: S China, SE Asia, Taiwan, Philippines, Java, Bali, and Timor.
Distribution and status: Confined to the high mountains of C and E Java where recorded from Mt. Lawu east to Bromo; not uncommon in suitable habitat but apparently long overlooked. Recent records of a scrub-warbler from Bali are almost certainly this species.
Habits: Skulks in dense scrub at forest edge and on open, scrubby hillsides, and *Casuarina* groves.
Note: Some authors split off the Javan form as a separate species *montis* but similarity of song indicates that it is best to keep it in this species.

680. FRIENDLY BUSH-WARBLER *Bradypterus accentor* PLATE 74
(Kinabalu Friendly Warbler/Kinabalu Bush-Warbler, AG: Scrub-Warbler)

Description: Medium-sized (15 cm), reddish-brown, skulking warbler. Throat white with blackish spots; breast and belly washed grey; rufous eyebrow.
Iris—brown; bill—black; feet—black.
Voice: Sharp, hissing alarm call, weak single whistle.
Range: Endemic to Borneo.
Distribution and status: On Borneo confined to Mt. Kinabalu and Trus Madi where this is a familiar bird between 2000 and 3600 m.
Habits: A skulker of the undergrowth of upper montane forests, responds well to 'pishing', noted for its tameness, will approach to within inches of a quiet observer. Has strange wing flicking behaviour when agitated.

OLD WORLD FLYCATCHERS—FAMILY MUSCICAPIDAE

A VERY large and varied Old World family of smaller insectivorous birds. Flycatchers have rounded heads and small, broad-based, pointed bills. The wide gape and fringe of stiff rictal bristles help them to snap up small insects. They have short, slender legs and small feet.

Males of most flycatchers are brightly coloured but most females are drab. They regularly join mixed-species flocks. The nests are neat, cup-shaped structures lined with hair and decorated with moss.

A total of 43 species occur in the Greater Sundas, some of which are winter visitors. Flycatchers can be divided into 3 main groups.

OLD WORLD FLYCATCHERS 345

1. Typical Flycatchers: have an upright posture and tend to hawk after insects from a perch.
2. Fantail Flycatchers: restless, active birds which tend to droop the wings and twitch their tails from side to side, or flick tails open as a raised fan.
3. Monarch Flycatchers: more active searchers, picking insects off branches and trunks of trees. Include the spectacular, long-tailed paradise flycatchers.

On the basis of DNA-hybridization studies, Sibley and Monroe (1990) reassigned the last two groups to drongos within the enlarged crow family Corvidae.

Typical Flycatchers

681. FULVOUS-CHESTED JUNGLE-FLYCATCHER *Rhinomyias olivacea* (Olive-backed Jungle-Flycatcher, AG: Flycatcher)   PLATE 76

Description: Medium-sized (15 cm) brownish flycatcher. Upperparts greyish brown with a rufous wash on rump and tail; chin and throat whitish; broad band across breast brownish buff (not grey); belly and undertail whitish.
Iris—brown; bill—black; feet—pink.
Voice: Drawn-out churrs; constant ticking notes interspersed with hurried song; song phrases of 7 to 9 notes at different pitch, each phrase lasting about 1.5 seconds; typical flycatcher song (D.A.H.).
Range: Malay Peninsula and Greater Sundas.
Distribution and status: On Sumatra (including Belitung), Java, and Bali this is a rather uncommon lowland bird, up to 1200 m. In Borneo, found on Natunas, Banggi, and Balambangan Islands and very local in N Borneo.
Habits: Frequents forest edge, secondary forests, and plantations. Keeps to the lower canopy. Hunts singly through foliage and flies after insects.

682. BROWN-CHESTED JUNGLE-FLYCATCHER *Rhinomyias brunneata* (AG: Flycatcher)   PLATE 76

Description: Medium-sized (15 cm) brownish flycatcher. Similar to Fulvous-chested Jungle-Flycatcher but with pale brown breast band and usually slight dark scaling on whitish throat, pale lower mandible. Immature has scaly buff upperparts and black tip to lower mandible.
Iris—brown; bill—blackish above, whitish base below; feet—pink.
Voice: Harsh churrs.
Range: SE Asia, migrating south in winter as far as Malay Peninsula.

Distribution and status: This is a rare migrant recorded once for Brunei.
Habits: Keeps to lower canopy of forest edge, secondary forests, and plantations.

683. GREY-CHESTED FLYCATCHER *Rhinomyias umbratilis*
(White-throated Jungle-Flycatcher, AG: Flycatcher)   PLATE 76

Description: Medium-sized (15 cm) brown flycatcher with white throat contrasting with brownish grey breast band. Immature has rufous buff wing bar and scaling on wing. Distinguished from similarly coloured babblers by behaviour and lack of supercilium, eye-ring, or moustache. Distinguished from other *Rhinomyias* flycatchers by grey rather than brown breast band.
Iris—dark brown; bill—black; feet—pink-grey.
Voice: A thin, tinkling song of 4–6 notes on descending scale *tii, ti-ti-tu-ti-tu* (M. & W.). Three notes on same pitch followed by a trill.
Range: Malay Peninsula, Sumatra, and Borneo.
Distribution and status: On Sumatra (including islands) and Borneo a local resident of lowland primary and secondary forests, peat swamp forest, and plantations, up to 1000 m.
Habits: Typical solitary flycatcher habits. Remains in undergrowth of forest, chasing insects but not coming to the ground.

684. RUFOUS-TAILED JUNGLE-FLYCATCHER
Rhinomyias ruficauda   PLATE 76
(Grey-faced/Chestnut-tailed Jungle-Flycatcher, AG: Flycatcher)

Description: Medium-sized (15 cm) rufous brown flycatcher with bright chestnut tail. Underparts whitish with grey breast band and fulvous vent. The very rufous tail distinguishes it from Grey-chested Flycatcher.
Iris—dull brown; bill—black; feet—pink-bluish.
Voice: Prolonged *chirr* call. (See Appendix 6.)
Range: Philippines and Borneo.
Distribution and status: On Borneo it is confined to higher mountain ranges, between 1000 and 2000 m. Rarely recorded at sea-level.
Habits: Hunts insects from low perch in flight or in vegetation. Keeps to forest edge and openings.

OLD WORLD FLYCATCHERS

685. EYEBROWED JUNGLE-FLYCATCHER *Rhinomyias gularis*
(Kinabalu Jungle-Flycatcher, AG: Flycatcher)   PLATE 76

Description: Medium-sized (15 cm) reddish-brown flycatcher. Upperparts rufous brown with reddish face and diagnostic buff lores and supercilium; underparts grey with sharply contrasting white throat; lower belly almost white. Iris—brown; bill—black; feet—grey.
Voice: Loud churring call when nesting. Otherwise rather silent.
Range: Philippines and Borneo.
Distribution and status: On Borneo this is a not uncommon resident of higher mountains, between 1000 and 2200 m, recorded from Mt. Kinabalu south to Tama Abo, and Mt. Mulu.
Habits: Inquisitive and robin-like. It keeps close to the ground, singly or in small parties.
Note: Some authors prefer to split off the Philippine forms *goodfellowi*, *albigularis*, and *insignis* as two separate species leaving *R. gularis* as a Bornean endemic.

686. DARK-SIDED FLYCATCHER *Muscicapa sibirica*   PLATE 76
(Siberian/Sooty Flycatcher)

Description: Small (13 cm) sooty brown flycatcher with dark flanks. Upperparts sooty brown; faint buff wing bar; underparts white with sooty grey mottled flanks and mottled grey band across upper breast; conspicuous white eye-ring and half collar; malar area streaked with black. Immature has white spots on face and back. Iris—dark brown; bill—black; feet—black.
Voice: Lively *chi-up, chi-up, chi-up*.
Range: Breeds in NE Asia and Himalayas; migrates in winter to S China, Palawan, SE Asia, and Greater Sundas.
Distribution and status: In N Sumatra and N Borneo (including Natunas and Anambas) a regular but uncommon visitor to the hills, mostly to 1000 m. In W Java a very rare visitor to montane forests up to 1500 m. Not recorded on Bali.
Habits: Inhabits undergrowth and middle storeys of montane or sub-montane forests. Sits rather upright on a low bare branch, making dashes to catch passing insects.

687. GREY-STREAKED FLYCATCHER *Muscicapa griseisticta*   PLATE 76
(Grey-Spotted/Spot-breasted Flycatcher)

Description: Medium-sized (15 cm) grey-brown flycatcher with white eye-ring and white underparts prominently streaked with grey. Narrow white band across forehead (barely visible in field) and faint pale wing bar.

Iris—brown; bill—black; feet—black.
Voice: Not reported from Borneo.
Range: Breeds in NE Asia; migrates in winter to S China, Philippines, and Sulawesi to New Guinea.
Distribution and status: Rare winter visitor recorded in Sabah, N Borneo.
Habits: A shy bird found near streams in dense forest, open forest, and forest edge.

688. ASIAN BROWN FLYCATCHER *Muscicapa dauurica* PLATE 76
(Grey-breasted/Brown Flycatcher > Brown-streaked/Chocolate/ Williamson's Flycatcher)

Description: Small (12 cm) greyish brown flycatcher. Migrant *latirostris*: upperparts grey-brown; underparts whitish with brownish grey on sides of breast and flanks; eye-ring white. Bornean resident race *umbrosa* is smaller and darker, especially on head. Another migratory form *williamsoni*: browner with rusty wash on upperparts, buffy streaking on flanks, and buff eye-ring.
Iris—brown; bill—black with yellow base to lower mandible; feet—black.
Voice: Soft vibrant *churr* and quiet thin song, but generally silent.
Range: Breeds in NE Asia and Himalayas; migrant south in winter to India, SE Asia, Philippines, Sulawesi, and Greater Sundas. Also resident and/or migratory populations in Philippines, Malay Peninsula, Sumatra, Borneo, and Sumba.
Distribution and status: N Asian population visits fairly regularly in winter, up to 1500 m throughout the Greater Sundas (including islands). Subtropical and tropical races are known from Sumatra (resident or migrants from Malay Peninsula) and N Borneo.
Habits: Prefers forest edge or hilly or sub-montane forests, but also found occasionally in open forest and gardens. Partial to small offshore islands. Generally solitary or joining mixed-species flocks. Catches insects from perch and shivers tail in a characteristic manner on returning to perch.
Note: The resident tropical forms are sometimes treated as *M. williamsoni* of Malay Peninsula and Borneo, *M. randi* of the Philippines, and *M. segregata* of Sumba. Apparently intermediate specimens indicate interbreeding with Asian Brown Flycatcher.

689. FERRUGINOUS FLYCATCHER *Muscicapa ferruginea* PLATE 76

Description: Small (12 cm) reddish brown flycatcher with a buffy eye-ring and white throat patch. Head slaty; back brown; rump rufous; underparts white with rufous flanks and undertail coverts.
Iris—brown; bill—black; feet—black.
Voice: Generally silent in winter.

Range: Breeds in Himalayas and S China; migrates south in winter as far as the Greater Sundas.
Distribution and status: Recorded in Sumatra and N Borneo where it is an uncommon visitor to lower mountain slopes, generally from 500 to 1500 m. In W Java it is an uncommon visitor at all altitudes. Not recorded on Bali.
Habits: A shy bird of glades and stream sides in thick forest. Hawks after insects from low perch.

690. VERDITER FLYCATCHER *Eumyias thalassina*   PLATE 76
(Indian Verditer Flycatcher)

Description: Largish (16 cm), uniform greenish blue flycatcher. Male has black lores; female duller with dusky lores. Both sexes have whitish scaling on undertail coverts. Immature grey-brown with greenish wash, scaled and spotted buff and blackish. Distinguished from blue-phase forms of Rufous-winged Philentoma by lack of red eyes; from male Pale Blue-Flycatcher and Indigo Flycatcher by greener colour and whitish scaling on blue-grey vent.
Iris—brown; bill—black; feet—blackish.
Voice: Melodious warble, less husky than Pale Blue-Flycatcher. Distinctive loud rattly note (T.H.).
Range: India to S China, SE Asia, Malay Peninsula, Sumatra, and Borneo.
Distribution and status: On Sumatra and Borneo it is an occasional bird of lowlands and hills, up to 1400 m but generally lower.
Habits: Hawks after flying insects from exposed perch in canopy of open forest or at edge of forest clearings.

691. INDIGO FLYCATCHER *Eumyias indigo*   PLATE 76
(Sunda Island Blue-Flycatcher)

Description: Medium-sized (14 cm) dark indigo blue flycatcher. General colour indigo blue; darkest, almost black, around base of bill; whitish forehead extending as a whitish brow over the eye; lower breast greyish, grading into whitish belly; vent buff. Immature has breast and throat blotched with pink.
Iris—red-brown; bill—black; feet—black.
Voice: Squeaky, ringing *fee-foo-fu-fee-fee-fee*, and a hard *turrrr-tur*.
Range: Endemic to the Greater Sundas.
Distribution and status: On Sumatra, Borneo (recorded on Mt. Kinabalu south to Murud and Mulu), and Java a fairly common sub-montane and montane resident, from 900 to 3000 m. Not recorded on Bali.
Habits: Keeps to dark montane forest but is quite tame and approachable. Generally perches low, near the ground; joins mixed flocks.

350 OLD WORLD FLYCATCHERS

692. YELLOW-RUMPED FLYCATCHER *Ficedula zanthopygia*
(Tricoloured/Korean Flycatcher) PLATE 77

Description: Small (13 cm), yellow, white, and black (male) or brown (female) flycatcher. Male has yellow rump, throat, breast, and upper belly; lower belly and undertail coverts white; otherwise black except for white supercilium and wing bar. Female is dull brown above, paler below, and has dull yellow rump. White supercilium and blacker back of male and yellow rump of female distinguish from respective sexes of Narcissus Flycatcher.
Iris—brown; bill—black; feet—black.
Voice: Thin call *pirip*, *pirip*, *pirip*, and short warble recorded for passage birds.
Range: Breeds in NE Asia; migrates in winter south to S China, SE Asia, and Greater Sundas.
Distribution and status: On Sumatra this is an uncommon winter visitor, up to 900 m. In Borneo recorded only from Anambas. Rare south to Java and Bali.
Habits: Frequents scrub and wooded areas.
Note: Formerly treated by some authors as a subspecies of Narcissus Flycatcher.

693. NARCISSUS FLYCATCHER *Ficedula narcissina* PLATE 77

Description: Small (13 cm) black and yellow flycatcher. Male has upperparts black with yellow rump, white wing patch and diagnostic yellow supercilium; underparts mainly yellow. Female: olive-grey upperparts with rufous tail; pale brown underparts with yellowish wash. Distinguished from female Yellow-rumped by olive rump.
Iris—dark brown; bill—bluish black; feet—lead blue.
Voice: Generally silent in winter.
Range: Breeds in NE Asia; migrates in winter to SE Asia, Philippines, and Borneo.
Distribution and status: On Borneo a very uncommon winter visitor to open woodlands and forest edge, up to 1400 m in Kelabit highlands.
Habits: Typical flycatcher, hawking insects from canopy perches and middle storey.

694. MUGIMAKI FLYCATCHER *Ficedula mugimaki* PLATE 77

Description: Small (13 cm), orange, black, and white (male) or brown and orange (female) flycatcher. Male: upperparts blackish grey with narrow white supercilium behind eye; white patch on wing and white edge to base of tail; throat, breast, and sides of belly orange; centre of belly and undertail coverts white. Female: upperparts brown, underparts as male but paler. Immature: plain brown above with buff underparts and white belly.

Iris—dark brown; bill—dark grey; feet—brownish.

Voice: Soft *turrrr*.

Range: Breeds in N Asia; migrates south in winter to SE Asia, Philippines, Sulawesi, and Greater Sundas.

Distribution and status: An uncommon winter visitor to lowland forest and mountains, up to 1500 m, throughout Sumatra and Borneo, less common south to Java and Bali.

Habits: Inhabits hill forests where it frequents the canopy of forest edge, clearings, and deep forest. It often sits quietly on a dead trunk or branch, making rapid sallying flights to catch flying insects.

695. RED-BREASTED FLYCATCHER *Ficedula parva* PLATE 77
(Red-throated Flycatcher)

Description: Small (13 cm) brown flycatcher with conspicuous white lateral flashes at base of dark tail. Breeding male has red breast but is rarely seen in SE Asia, and never recorded in Borneo. Female and non-breeding male are dull grey-brown with whitish throat.

Iris—dark brown; bill—black; feet—black.

Voice: Harsh, sharp *tzick* in alarm.

Range: Breeds in Palaearctic; migrates in winter to China, Philippines, SE Asia, and Borneo.

Distribution and status: Rare vagrant to N Borneo.

Habits: Keeps to smaller trees at forest edge and along rivers. Dashes to cover when alarmed. Flicks dark tail to reveal white base patches, and gives harsh clicking notes.

696. RUFOUS-BROWED FLYCATCHER *Ficedula solitaris* PLATE 77
(Malaysian Flycatcher/White-throated/Solitary Flycatcher]

Description: Small (12 cm) brownish flycatcher with prominent triangular white throat patch, white belly, and brown flanks and breast bar. White throat patch is sometimes outlined with black, especially in N Sumatran form. Crown and sides of head rufous with buffy eye-ring and lores. Immature olive-brown with whitish throat and underparts streaked with rusty brown.

Iris—dark brown; bill—black; feet—brown.

Voice: Thin, sibilant whistle of 3 descending notes *three blind mice*, and harsh churring.

Range: SE Asia, Malay Peninsula, and Sumatra.

Distribution and status: On Sumatra a locally common bird of undergrowth in dense forests, from 900 to 2400 m.
Habits: Active and noisy flycatcher of forest undergrowth, staying close to forest floor.

697. SNOWY-BROWED FLYCATCHER *Ficedula hyperythra*   PLATE 77
(White-fronted/Snow-browed/Thicket/Dull Flycatcher)

Description: Small (11 cm) grey-blue and rufous flycatcher. Male: upperparts slaty blue with prominent but short, white eyebrow; underparts orange-buff on throat, breast, and flanks. Female: brown above, buffy below with buff eyebrow. Immature: mottled brown.
Iris—dark brown; bill—black; feet—grey to brown.
Voice: Quiet song of 3 or 4 shrill notes *cheet-chee-chee-chaw*; single shout *chee*.
Range: N India to S China, Philippines, SE Asia, Malay Peninsula, Greater Sundas, Sulawesi, Moluccas, and Lesser Sundas.
Distribution and status: On Sumatra, Borneo (found from Mt. Kinabalu south to Tama Abo and Mulu; also Niut and Poi ranges), Java, and Bali this is a common bird of montane forests, from 900 to 3100 m.
Habits: Unobtrusive. Sits quietly on low perch or log, then flits down to catch ground insects. Spends a lot of time on the ground where it hops like a robin. Generally solitary and rather tame. Diet includes some berries.

698. RUFOUS-CHESTED FLYCATCHER *Ficedula dumetoria*
(Orange-breasted/Short-tailed Flycatcher)   PLATE 77

Description: Small (11 cm), orange, black, and white flycatcher (male). Male: upperparts black with white eyebrow, white wing bar, and white edge to tail base; chin pinkish; breast and sides of belly orange; belly and vent white. Female: upperparts brown; underparts as male but paler; buffy lores. Borneo race has less prominent eyebrow. Distinguished from Mugimaki Flycatcher by pale chin, blacker back, and longer bill.
Iris—brown; bill—brownish; feet—grey.
Voice: High-ringing call *tsst-tsst*.
Range: Malay Peninsula and Sundas.
Distribution and status: On Sumatra a local resident of hill forest, from 600 to 1500 m, not recorded south of Kaba. On Borneo an uncommon sub-montane resident recorded throughout the hills and locally to sea-level. In Java recorded mainly in W Java; not recorded on Bali but may occur there.
Habits: Inhabits primary forest, feeding close to the ground. Rather quiet and generally solitary.

OLD WORLD FLYCATCHERS 353

699. LITTLE PIED FLYCATCHER *Ficedula westermanni* PLATE 77

Description: Small (11 cm), black and white (male) or brown and white (female) flycatcher. Male: upperparts black with white eyebrow, white wing bar, and white edge to base of tail; underparts white. Female: greyish brown upperparts, whitish underparts, and rufous tail. Immature is brown, mottled with tawny.
Iris—brown; bill—black; feet—black.
Voice: Regularly emits a high-pitched, thin, descending *pi-pi-pi-pi* followed by a low rattle *churr-r-r-r*.
Range: India to S China, Philippines, SE Asia, Malay Peninsula, Sundas, Sulawesi, and Moluccas
Distribution and status: Locally common in montane forest from 1000 to 2600 m throughout the Greater Sundas.
Habits: Frequents sub-montane and mossy forests, and montane Casuarina forests. Feeds at all levels in the canopy and sometimes joins mixed-species flocks.

700. BLUE-AND-WHITE FLYCATCHER *Cyanoptila cyanomelana*
(Japanese Blue-Flycatcher) PLATE 78

Description: Large (17 cm) blue, black, and white (male) or brown and white (female) flycatcher. Male: face, throat and upper breast black; upperparts glossy blue with white patches at base of tail; lower breast, belly, and undertail coverts white. Female: grey-brown upperparts with brown wings and tail, white centre of throat and belly.
Iris—brown; bill—black; feet—black.
Voice: Generally silent on winter grounds.
Range: Breeds in NE Asia; migrates south in winter to China, SE Asia, Philippines, and Greater Sundas.
Distribution and status: A regular winter visitor up to 1400 m to N Borneo and less commonly throughout Borneo. On Sumatra and W Java it is a rare winter visitor to sub-montane forests, up to 1200 m. Not recorded on Bali.
Habits: Frequents wooded areas in primary and secondary forests, feeding quite high in the canopy. Includes some fruit in diet.

701. LARGE NILTAVA *Niltava grandis* PLATE 78
(Great Niltava)

Description: Large (22 cm) dark flycatcher. Male: blue upperparts with shining blue crown, stripe on side of neck, shoulder patch, and rump; black underparts. Female: rufous olive-brown with blue-grey crown, pale blue neck patch, and

whitish throat. **Immature:** brown with white speckling on head, rusty spots on back, and black scaling on underparts.
Iris—dark brown; bill—black; feet—grey.
Voice: Clear, rising whistle of 3 notes introduced by a grace note *k'tu-tu-ti*; also scolding rattle. (See Appendix 6.)
Range: Nepal to SW China, SE Asia, Malay Peninsula, and Sumatra.
Distribution and status: On Sumatra not uncommon in hill and mountain forest, between 900 and 1500, locally up to 2500 m.
Habits: Solitary flycatcher, keeping to dark undergrowth in sub-montane and montane forests.

702. RUFOUS-VENTED NILTAVA *Niltava sumatrana* PLATE 78
(Malaysian/Rufous-bellied Niltava)

Description: Medium-sized (15 cm) dark flycatcher. **Male:** dark blue upperparts with shining blue crown, patch on side of neck, shoulder patch, rump, and tail coverts; throat and sides of head black; underparts orange. **Female:** brown with narrow white throat band and shining blue shoulder patch. Immature is brown with rusty spots on upperparts and black scaling on underparts.
Iris—dark brown; bill—black; feet—bluish grey.
Voice: Hard *tchik*.
Range: Malay Peninsula and Sumatra.
Distribution and status: In N Sumatra it is a locally common bird of montane forests, from 1000 m to the treeline, recorded south to Mt. Kerinci.
Habits: Solitary flycatcher of dark forest undergrowth and middle storey on higher mountains. Very common on Kerinci peak.

703. WHITE-TAILED BLUE-FLYCATCHER *Cyornis concretus*
(Short-tailed Blue-Flycatcher) PLATE 78

Description: Largish (19 cm) dark flycatcher. Sumatran race has conspicuous white patches in spread tail. **Male:** dark blue upperparts with black sides to head and black flight feathers; sooty breast grading to white vent. **Female:** brown with broad white throat band, belly, and undertail coverts. **Immature:** brown with rusty spots on upperparts and black scaling on underparts.
Iris—dark brown; bill—black; feet—dark grey.
Voice: Variable sibilant whistles and harsh *scree* in alarm. (See Appendix 6.)
Range: Assam, SE Asia, Malay Peninsula, Sumatra, and Borneo.
Distribution and status: On Sumatra and Borneo this is an occasional bird of hill forests, from 300 to 1300 m, locally down to sea-level on Borneo.
Habits: Solitary flycatcher of undergrowth in hill and sub-montane forests.

OLD WORLD FLYCATCHERS 355

704. RUECK'S BLUE-FLYCATCHER *Cyornis ruckii* PLATE 78

Description: Largish (17 cm) blue flycatcher. Male has head, throat, and breast blue, and shining blue rump and uppertail coverts. Distinguished from Pale Blue-Flycatcher by darker colour and shining blue rump. Female has rufous-brown upperparts with rufous rump and tail; rusty breast grading to whitish belly. Distinguished from female Pale Blue-Flycatcher by rusty breast. Immature: upperparts brown spotted with buff; rufous forehead, eye-ring, throat, and breast. Underparts scaled with black grading to whitish centre of belly.
Iris—brown; bill—black; feet—black.
Voice: Unknown.
Range: Endemic to Sumatra.
Distribution and status: Known by 4 specimens, 2 collected in secondary lowland forest in the Medan area of N Sumatra; the 2 skins from the Malay Peninsula are of doubtful origin. This is probably a Sumatran endemic.
Habits: Found in logged forest.

705. PALE BLUE-FLYCATCHER *Cyornis unicolor* PLATE 78

Description: Largish (16 cm) pale blue (male) or brownish (female) flycatcher. Male: upperparts bright turquoise blue with black lores; throat and breast paler blue; belly greyish white and undertail coverts white. Female: upperparts grey-brown, tail more rufous-brown; underparts greyish brown; eye-ring and lores fulvous. Immature: brown, mottled with black and fulvous buff.
Iris—brown; bill—brown; feet—brown.
Voice: Loud, sweet song running down scale then up again on last 3 notes; also occasional husky notes.
Range: Himalayas to S China, SE Asia, Malay Peninsula, and Greater Sundas.
Distribution and status: On Sumatra found in the mountains south to Mt. Kerinci and on Borneo and Java, uncommon and local in hill forest between 500 and 1400 m, locally down to 200 m on Borneo. Not recorded on Bali.
Habits: Keeps to primary forests where it stays in the canopy. Rather shy.

706. HILL BLUE-FLYCATCHER *Cyornis banyumas* PLATE 78

Description: Medium-sized (15 cm) blue, orange, and white (male) or brownish (female) flycatcher. Male: upperparts dark blue; forehead and short supercilium pale blue; lores, around eye, forecheeks, and chin spot black; throat, breast, and flanks orange; belly white. Orange throat, black chin spot, and lack of shining rump distinguishes it from all other orange-breasted blue-flycatchers. Female: upperparts brown; eye-ring buff; underparts as male but paler. Juvenile: brown, mottled with buffy-orange spots on upperparts.

Iris—brown; bill—black; feet—brown.
Voice: Loud, melodious, warbling song with several husky notes. In alarm a harsh *chek-chek*. (See Appendix 6.)
Range: Nepal to SW China, Palawan, SE Asia, Malay Peninsula, Borneo, and Java.
Distribution and status: On Borneo and Java this is one of the commoner flycatchers, at moderate to high elevations, between 400 and 2000 m.
Habits: Frequents shady open parts of the undergrowth in primary and secondary forests at all altitudes. Where Mangrove Blue-Flycatcher is absent this species also inhabits coastal forests. Sits quietly, hunting from low perches.

707. LARGE-BILLED BLUE-FLYCATCHER *Cyornis caerulatus*
(Sunda Blue-Flycatcher)   PLATE 78

Description: Smallish (14 cm) blue and orange flycatcher. Male: blue upperparts with shining blue forehead, lower back, and rump; dark orange-rufous breast grading to buff belly; throat paler than breast. Races vary: Sarawak males have black chin, W Bornean have rufous wash on forehead, Sumatran birds have whiter vent. Distinguished from Hill Blue-Flycatcher by shining blue rump. Female: brown upperparts with blue rump and tail, and bluish wash on mantle; underparts as male, lacks black chinspot.
Iris—dark brown; bill—black; feet—dark grey.
Voice: No description.
Range: Endemic to Sumatra and Borneo
Distribution and status: On Sumatra it is a rare lowland bird known by few records. On Borneo this is an occasional bird of inland forests at moderate altitudes.
Habits: Chases insects from low exposed perches at edge of forest clearings. Away from the coasts and riversides this species replaces Mangrove and Malaysian Blue-Flycatchers.

708. BORNEAN BLUE-FLYCATCHER *Cyornis superbus*   PLATE 78

Description: Medium-sized (15 cm) blue and orange flycatcher. Male: upperparts blue with shining blue forehead, eyebrows, nape, and lower back; breast orange, paler on throat, and grading to white vent. Female: brown with distinctive rufous forehead, rump, and tail.
Iris—dark brown; bill—black; feet—blue-grey.
Voice: No information.
Range: Endemic to Borneo.

OLD WORLD FLYCATCHERS 357

Distribution and status: On Borneo an occasional bird of sub-montane forests, between 600 and 1600 m, locally found in lowlands.
Habits: Keeps to dark forest near streams and hawks insects from low perch.

709. MALAYSIAN BLUE-FLYCATCHER *Cyornis turcosus* PLATE 78

Description: Small (13 cm) dark blue flycatcher. Male: blue upperparts with bright blue throat, blackish lores and flight feathers, and shining blue rump. Chest orange-rufous, belly white. Races vary in darkness. Female: like male but chin and throat white. Immature: brown above, spotted with buff; blue wings and tail; breast buffish grading to dirty white belly with black scaling on breast.
Iris—brown; bill—black; feet—blackish.
Voice: Grating alarm call *chrrk* and weak song *diddle diddle dee diddle dee*.
Range: Malay Peninsula, Sumatra, and Borneo.
Distribution and status: On Sumatra not uncommon but on Borneo rather scarce; resident of lowland forests, up to 800 m, but generally below 100 m.
Habits: Prefers lowland and swampy forests, generally close to streams and rivers.

710. TICKELL'S BLUE-FLYCATCHER *Cyornis tickelliae* PLATE 78

Description: Medium-sized (15 cm) dark blue flycatcher. Male like Hill Blue-Flycatcher but brighter and with a sharper division between orange-red breast and white belly. Female: like male but upperparts much greyer. Immature is indistinguishable from immature Hill Blue-Flycatcher.
Iris—brown; bill—black; feet—black.
Voice: 5- to 7-note, metallic trill on a slow descending scale and harsh *trrrt* alarm call. (See Appendix 6.)
Range: India, SE Asia, Malay Peninsula, and Sumatra.
Distribution and status: On Sumatra presumably a rare bird of coastal scrub, known only by 1 specimen from the Tasik river, N Sumatra.
Habits: Hunts for insects from low perches in forest and on ground.

711. MANGROVE BLUE-FLYCATCHER *Cyornis rufigastra* PLATE 78

Description: Medium-sized (15 cm), blue, orange, and white flycatcher. Very similar to male Hill Blue-Flycatcher but lacks pale blue forehead, has blacker chin, and rufous on underparts extends further down belly. Female: like male but paler and with whitish lores (forming V shape above bill) and buffish-white chin. Different races vary slightly.
Iris—brown; bill—black; feet—bluish flesh.
Voice: Clear melodious song, similar to Hill Blue-Flycatcher.
Range: Philippines, Sulawesi, Malay Peninsula, and Greater Sundas.

Distribution and status: On Sumatra (including islands) mainly coastal forest and mangroves in E lowlands and islands. On Borneo (including islands) not uncommon in coastal districts. On Java (including Karimunjawa) now a rather rare bird of coastal forests and small islands, mostly in W Java though recorded on Bali.
Habits: Inhabits small islands where it is usually the common flycatcher in the absence of Hill Blue-Flycatcher. On the main islands it is confined to beach and mangrove forests, and coastal plantations. Often found in pairs which makes it easy to identify due to the blue female. Hunts close to the ground; fond of nipa thickets.

712. PYGMY BLUE-FLYCATCHER *Muscicapella hodgsoni* PLATE 77

Description: Very small (10 cm) narrow-billed flycatcher. Male: blue upperparts with shining blue crown and rump, and black mask; yellowish-rufous underparts with white centre of belly and vent. Female: brown upperparts with rufous rump and tail; underparts whitish with buff wash on breast.
Voice: No information.
Range: Himalayas, SE Asia, Sumatra, and Borneo.
Distribution and status: Found in mountains of N Sumatra south to Mt. Kerinci, from 1100 to 2400 m, but there are few records. On Borneo known only from Mts Kinabalu, Mulu, and Dulit, and in Mueller range in Kalimantan.
Habits: Prefers understorey of primary forests, sometimes visiting ground, and rarely in middle storey. Has habit of flicking wings open and cocking tail (P.R.).

713. GREY-HEADED FLYCATCHER *Culicicapa ceylonensis* PLATE 77
(AG: Canary Flycatcher)

Description: Small (12 cm) distinctive flycatcher with greyish head and breast, and slight crest; olive upperparts and yellow underparts.
Iris—brown; bill—black above, grey below; feet—yellowish brown.
Voice: Clear, sweet whistle *chiree-chilee* with stress on first syllable of each pair and with upward inflection on last note; also rattly *churrru* call.
Range: India to S China, SE Asia, Malay Peninsula, and Sundas.
Distribution and status: On Sumatra (including islands), Borneo, Java, and Bali this is a widespread and not uncommon bird of forests, most common in submontane forests between 600 and 1600 m, but recorded from the lowlands to 2200 m.
Habits: A noisy, active bird, flitting from branch to branch, hunting keenly and chasing flying insects. It regularly flicks its tail open. Generally in lower or middle storeys. Often joins mixed flocks.
Note: DNA studies show that this genus belongs in the Australian Robin family Eopsaltriidae.

Fantail Flycatchers

714. RUFOUS-TAILED FANTAIL *Rhipidura phoenicura*   PLATE 79
(Red-tailed Fantail, AG: Fantail-Flycatcher)

Description: Medium-sized (17 cm) red-tailed fantail. Head and back grey with a fine white eyebrow; chin and throat white; upper breast grey grading into orange-chestnut belly and vent; wings dark chestnut; diagnostic rump and tail bright orange-chestnut.
Iris—brown; bill—black; feet—black.
Voice: Similar to the Pied Fantail but harder *he-tee-tee-tee-oh-weet*, and variations.
Range: Endemic to Java.
Distribution and status: Confined to montane forest of Java, between 1000 and 2500 m where it is locally common.
Habits: Similar to other fantails. Regularly joins mixed flocks; often in dense thickets and bushes quite close to the ground. Typical fantail fanning and swaying habits.

715. WHITE-BELLIED FANTAIL *Rhipidura euryura*   PLATE 79
(AG: Fantail-Flycatcher)

Description: Medium-sized (18 cm) grey and white fantail. Upperparts uniform dark blue-grey with broad white eyebrow; tail slaty grey with broad white tips to outer feathers, conspicuous when fanned; throat and breast dull grey; belly and vent white.
Iris—brown; bill—black; feet—black.
Voice: Excited squeaks *cheet cheet*.
Range: Endemic to Java.
Distribution and status: Confined to montane forest of Java where it is a locally not uncommon resident.
Habits: Similar to other fantails. An active forest bird, single or in pairs, often joining mixed flocks travelling through the middle canopy of the forest. Performs typical fanning and swaying displays.

716. WHITE-THROATED FANTAIL *Rhipidura albicollis*   PLATE 79
(Spot-breasted/White-spotted Fantail, AG: Fantail-Flycatcher)

Description: Medium-sized (18 cm) dark fantail. Almost totally dark grey (looks black in field) with white chin, throat, eyebrow, and tips of tail. Dark grey underparts distinguishes it from other fantails but some individuals are paler underneath.
Iris—brown; bill—black; feet—black.

Voice: A song of high-pitched thin notes; 3 evenly spaced *tut* notes followed by 3 or more descending notes; also sharp *cheet*.
Range: Himalayas, S China, SE Asia, Malay Peninsula, Sumatra, and Borneo.
Distribution and status: On Sumatra a common bird of open forest and secondary forest on hills and mountains, between 900 and 2400 m. On Borneo restricted to the ranges from Mt. Kinabalu south to Kelabit uplands, the Mueller range, and Niut and Poi ranges.
Habits: As other fantails.

717. SPOTTED FANTAIL *Rhipidura perlata* PLATE 79
(Pearlated/Perlated/Pearl-spotted Fantail, AG: Fantail-Flycatcher)

Description: Medium-sized (18 cm) fantail with white eyebrow, white wing bar and diagnostic white spots on grey breast. Outer tail feathers with broad white tips.
Iris—brown; bill—black; feet—black.
Voice: Harsh, high-pitched calls and rising 2-phrase song *chilip*, *pechilip-chi* (M. & W.). (See Appendix 6.)
Range: Malay Peninsula, Sumatra, and Borneo.
Distribution and status: On Sumatra and Borneo this is a common bird up to 1200 m, locally up to 1700 m, in lowland and hill primary and secondary forest. A few specimens have been collected in W Java where it is very rare and is replaced ecologically by the endemic White-bellied Fantail.
Habits: An active bird of middle and lower storeys of primary and old secondary forests. Typical fantail habits.

718. PIED FANTAIL *Rhipidura javanica* PLATE 79
(Malaysian Fantail, AG: Fantail-Flycatcher)

Description: Medium-sized (19 cm) black and white fantail. Adult: upperparts sooty grey with white eyebrow; chin and throat white; black breast band diagnostic; rest of underparts white; tail feathers with broad white tips. Immature: rump and uppertail coverts reddish and breast band less distinct.
Iris—brown; bill—black; feet—black.
Voice: High-pitched squeaky call *chee-chee-wee-weet*. (See Appendix 6.)
Range: Philippines, Malay Peninsula, Greater Sundas, and Lombok.
Distribution and status: A common resident throughout the Greater Sundas (including islands), up to 1500 m.
Habits: Typical active fantail of open wooded areas, including secondary forests, gardens, and mangroves. Sometimes single, in pairs, or family parties; sometimes follows domestic animals or monkeys catching disturbed insects. Joins mixed flocks.

Monarch Flycatchers

719. BLACK-NAPED MONARCH *Hypothymis azurea* PLATE 79
(> Pacific/Small Monarch)

Description: Medium-sized (16 cm) greyish blue flycatcher. Male: head, breast, back, and tail blue; greyer on wings; whitish belly; short black crest; small patch above bill and narrow throat band black. Female: head blue-grey; greyer on breast; back, wing, and tail brownish grey; lacks black crest and throat bar of male. Iris—dark brown; eye-ring—bright blue; bill—bluish black with black tip; feet—bluish black.

Voice: Song is ringing *pwee-pwee-pwee-pwee*, contact call is harsh, chirping *chee*, *chweet*. (See Appendix 6.)

Range: India to China, SE Asia, Philippines, Sulawesi, Malay Peninsula, and Sundas.

Distribution and status: A common resident throughout the Greater Sundas (including islands), up to 900 m, locally to 1500 m.

Habits: A lively, inquisitive bird of the lowland forests and secondary forests. Readily attracted to an imitation of its contact call. Often joins in mixed-species flocks. Usually in the lower parts of the forest.

720. MAROON-BREASTED PHILENTOMA *Philentoma velatum*
(Maroon-breasted Flycatcher/Maroon-breasted Monarch) PLATE 79

Description: Largish (20 cm) indigo blue flycatcher. Male has distinctive black face mask and dark maroon breast (often looks black in field). Female is duller indigo blue all over with greyish wash on belly and blackish face and throat. Iris—red; bill—black; feet—bluish black.

Voice: Noisy, grating, sharp metallic churrs given by both sexes; also a long descending series of spaced, bell-like notes (M. & W.). (See Appendix 6.)

Range: Malay Peninsula and Greater Sundas.

Distribution and status: On Sumatra and Borneo this is a locally common bird of lowlands, up to 1400 m. On Java this is an uncommon bird of lowland forest, up to 800 m. Not recorded on Bali.

Habits: Inhabits dark swampy forest where it catches insects in flight among the lower branches and vines. Generally found near water.

Note: This genus was reassigned by Sibley and Monroe (1990) to the vangines within the enlarged Crow family Corvidae.

721. RUFOUS-WINGED PHILENTOMA *Philentoma pyrhopterum*
(Chestnut-winged Flycatcher, AG: Monarch)   PLATE 79

Description: Medium-sized (16 cm) flycatcher. Male has two colour forms. In common phase, head, breast, and mantle dull blue with rufous wings and tail, and buff belly. In rarer blue phase, upper plumage entirely blue except for whitish-streaked belly. Distinguished from all other blue-flycatchers except female Maroon-breasted by red eyes. Female has rufous wings and tail, grey-brown head and back, and buffy underparts. Distinguished from other brown females by red eyes.
Iris—red; bill—black; feet—brown.
Voice: Soft whistle *tu-huuuu* with emphasis on lower second note; also rising *tew-ii* (P.R.) and harsh metallic scolds. (See Appendix 6.)
Range: Malay Peninsula, Sumatra, and Borneo.
Distribution and status: On Sumatra (including islands) and Borneo (including Natunas) a common flycatcher of forests up to 1000 m, locally up to 1600 m on Borneo.
Habits: An active flycatcher of lower and middle storeys of primary and secondary forests, peat swamp forests, and heath forests.
Note: See taxonomic note of previous species.

722. JAPANESE PARADISE-FLYCATCHER *Terpsiphone atrocaudata*
(Black Paradise-Flycatcher)   PLATE 79

Description: Medium-sized (20 cm, plus 20 cm long tail of male) black-crested flycatcher. Male distinguished from Asian Paradise-Flycatcher by black wings and tail and purplish back. Female similar to female Asian Paradise-Flycatcher, but duller crown lacks metallic gloss.
Iris—dark brown; bare skin around eye—blue; bill—black; feet—bluish.
Voice: As Asian Paradise-Flycatcher.
Range: Breeds in Japan and Korea; winters in SE Asia.
Distribution and status: Occasional birds recorded in N and E Sumatra.
Habits: As Asian Paradise-Flycatcher.

723. ASIAN PARADISE-FLYCATCHER *Terpsiphone paradisi*
(Paradise/Asiatic Paradise Flycatcher)   PLATE 79

Description: Medium-sized (22 cm, plus 20 cm tail on male) sexually dimorphic, glossy black-headed flycatcher with prominent crest. Male is notable with greatly elongated central pair of tail feathers, up to 25 cm beyond rest of tail. Male has two colour phases, both quite different from Japanese Paradise Flycatcher. In one phase, upperparts white streaked with black shafts and underparts pure white; wings black. In other phase, upperparts rufous, underparts greyish. Female is rufous brown with glossy black head.

Iris—brown; bare skin around eye—blue; bill—blue with black tip; feet—blue.

Voice: Ringing whistled song and very loud *chee-tew* contact call, similar to but stronger than calls of Black-naped Monarch. (See Appendix 6.)

Range: Turkestan, India, China, SE Asia, and Sundas.

Distribution and status: On Sumatra (including islands) and Borneo this is a fairly common bird of lowland forest, locally up to 1200 m. On Java restricted to lowland forest below 800 m, more common in the southern half of the island. Not recorded on Bali.

Habits: The white male is conspicuous in flight. Generally hunts from a perch in the lower half of the canopy, often within mixed-species flocks.

WHISTLERS—FAMILY PACHYCEPHALIDAE

WHISTLERS are robust birds with thick, round heads and short, thick, shrike-like bills. They are allied to flycatchers. They live in all levels of the forest canopy. They eat insects gleaned among foliage. Most species have fine, loud whistling songs with a characteristic whipcrack cut-off at the end.

Four species occur in the Greater Sundas.

724. BORNEAN WHISTLER *Pachycephala hypoxantha* PLATE 80
(Bornean Mountain Whistler, AG: Thickhead)

Description: Medium-sized (16 cm) olive-green whistler with black lores and distinctive yellow ear coverts and underparts. Female has more olive throat and breast.

Iris—brown; bill—black; feet—dark brown.

Voice: Intermittent loud songs with whiplash end.

Range: Endemic to Borneo.

Distribution and status: Mountains of N Borneo from Mt. Kinabalu south to the Mueller range; also Niut and Poi ranges. A common bird of montane forests on Mt. Kinabalu, from 900 to 2600 m, and on other mountains up to 2000 m.

Habits: An active and conspicuous bird of the crowns of smaller trees, often with mixed flocks and parties of laughingthrushes.

725. MANGROVE WHISTLER *Pachycephala grisola* PLATE 80
(White-bellied Whistler/Grey Thickhead > Palawan Whistler)

Description: Medium-sized (14 cm), nondescript, greyish-brown whistler. Crown and nape grey; back, wings, and tail greyish brown; chin, throat, breast, and flanks pale grey; belly whitish.
Iris—brown; bill—dark grey; feet—bluish grey.
Voice: Melodious loud whistle of 3 to 6 repeated staccato notes, ending in a whiplash last note, sometimes a double whiplash. (See Appendix 6.)
Range: SE India, SE Asia, Philippines, Malay Peninsula, Greater Sundas, and Lombok.
Distribution and status: On Sumatra (including islands) and Borneo (including islands) a common coastal bird, locally occurring inland and occasionally up to 900 m. On Java (including islands) and Bali this is an uncommon bird of coastal and lowland forest, up to 800 m.
Habits: Frequents mangroves, casuarinas, coastal scrub, rubber plantations, secondary forests, and clumps of bamboos or palms, generally near the sea. A quiet, elusive, and inconspicuous bird found singly or in pairs, generally in the treetops. Heard more often than seen.
Note: Formerly included White-vented Whistler *P. homeyeri* of Philippines (see below). *P. grisola* is synonymous with *P. cinerea*.

726. WHITE-VENTED WHISTLER *Pachycephala homeyeri* PLATE 80

Description: Medium-sized (14 cm), nondescript, rufous whistler with whitish underparts. Upperparts, wing, and tail rufous brown; sides of head rufous; underparts white with throat and breast streaked with cinnamon.
Iris—brown; bill—dark grey; feet—bluish grey.
Voice: Similar to Mangrove Whistler.
Range: Philippines, Sula Islands.
Distribution and status: Occurs on Siamil Island off coast of Sabah.
Habits: As Mangrove Whistler.
Note: Formerly included in Mangrove Whistler *P. grisola*.

727. GOLDEN WHISTLER *Pachycephala pectoralis* PLATE 80
(Common Golden Whistler, AG: Thickhead)

Description: Medium-sized (17 cm), black, white, and yellow (male) whistler. Male: crown, sides of head, nape, and throat band black; chin and throat white; upperparts olive-green; tail blackish; underparts golden yellow. Female: dull olive-brown upperparts, greyish buff underparts, vent washed with yellow.
Iris—red; bill—brown; feet—grey.

Voice: Three or 4 melodious repeated notes ending with a whipcrack cut-off note of lower pitch *dee-dee-dee-awit*.
Range: Java, Bali, Lesser Sundas, Moluccas, New Guinea, and Australia.
Distribution and status: Restricted to E Java and Bali where it is an occasional bird of hill and montane forests.
Habits: Frequents forest and dense woodland or secondary growth. Generally single or in pairs. Forages in middle and upper canopy. During courtship males perform see-saw posturing displays. Mixes with other species in flocks.
Note: This species is an assemblage of over 60 subspecies whose precise relationships remain far from clear.

Pipits and wagtails—FAMILY MOTACILLIDAE

A MODERATELY large, worldwide family of slender, terrestrial birds which walk with a deliberate gait. Many species 'wag' their tails, giving the family its English name. They have slender bills and long, thin legs. All species are insectivorous but also eat other small invertebrates. Many are migratory. Most pipits superficially resemble larks but the longer legs and finer bills are diagnostic. Reassigned by Sibley and Monroe (1990) as a tribe under the Sparrow family Passeridae.
Eight species occur in the Greater Sundas but only 1 is resident.

728. PIED WAGTAIL *Motacilla alba* PLATE 81
(White/Common Pied Wagtail > Masked Wagtail)

Description: Medium-sized (20 cm) black, grey, and white wagtail. General plumage grey above, white below. Wings and tail marked black and white. Hind crown, nape, and breast marked black but less extensively than in breeding season. Extent of black varies considerably with race. Immatures grey where adults are black.
Iris—brown; bill—black; feet—black.
Voice: Clear, hard *chissick*.
Range: Africa, Europe, and Asia. Birds breeding in E Asia winter south to SE Asia and Philippines.
Distribution and status: Migrant to N Borneo in winter where it is common at moderate altitudes, up to 1500 m.
Habits: A bird of open spaces, paddy fields, along stream edges, and on roads. Flies with low dipping flight giving alarm call when disturbed.

729. GREY WAGTAIL *Motacilla cinerea* PLATE 81

Description: Medium-sized (19 cm), long-tailed, greyish wagtail with yellow-green rump and yellow underparts. Distinguished from Yellow Wagtail by grey mantle, white wing bar and yellowish rump in flight, and longer tail. Underparts yellow in adult or whitish in young.
Iris—brown; bill—brownish black; feet—pinkish grey.
Voice: Shrill *tzit-zee* or single hard *tzit*, uttered in flight.
Range: Breeds Europe to Siberia and Alaska; migrates south to Africa, India, SE Asia, Philippines, Indonesia to New Guinea and Australia.
Distribution and status: A regular and locally common visitor at all altitudes throughout the Greater Sundas (including islands).
Habits: Frequents rocky streams where it searches for food in damp gravel or sand; also on alpine meadows of highest mountains.

730. YELLOW WAGTAIL *Motacilla flava* PLATE 81
(> Siberian/Siberian Yellow/Grey-headed/Dark-headed Wagtail)

Description: Medium-sized (18 cm) brownish or olive wagtail. Similar to Grey Wagtail but olive-green or olive-brown, not grey, on back and shorter tail, no white wing bar or yellow on rump visible in flight. Races vary: male of commoner *simillima* race has grey crown, white supercilium, and yellow throat; *taivana* has crown same olive colour as back, and yellow supercilium and throat; rare *tchutchensis* has grey crown, white supercilium, and white throat; very rare *macronyx* has grey head, no supercilium, white chin, and yellow throat. Non-breeding plumage is browner and duller than breeding but by March and April birds are assuming full coloration. Female and immature lack yellow vent. Immature has white belly.
Iris—brown; bill—brown; feet—brown.
Voice: A shrill, musical *tsweep* given in flock flight.
Range: Breeds in Europe to Siberia and Alaska; migrates south to India, China, SE Asia, Philippines, Indonesia to New Guinea and Australia.
Distribution and status: A common winter visitor and passage migrant to the lowlands, particularly coastal areas, of the Greater Sundas (including islands).
Habits: Frequents rice fields, marsh edges, and pastures. Often in very large flocks, feeding around cattle and buffalo.

731. FOREST WAGTAIL *Dendronanthus indicus* PLATE 81
(Tree Wagtail)

Description: Medium-sized (17 cm) brown and pied forest wagtail. Upperparts olive grey with white eyebrow; wings boldly patterned black and white; underparts

white with two black bars across the breast, the lower one sometimes incomplete.
Iris—grey; bill—black; feet—pinkish.
Voice: A loud *chirrup* frequently uttered. Short *tsep* call in flight.
Range: Breeds in E Asia; migrates south in winter to India, SE China, SE Asia, Philippines, and Greater Sundas.
Distribution and status: A common winter visitor to Sumatra, up to 900 m, less common south to Java where it is known only from W and C Java, up to 1500 m. A rather rare visitor to N Borneo. Not recorded on Bali.
Habits: Walks about singly or in pairs on open parts of the forest floor. Gentle lateral swaying motion of tail, unlike the vertical tail wagging of other wagtails. Rather tame; when disturbed flies in low undulating flight to alight again some metres ahead. Also perches in trees.

732. OLIVE-BACKED PIPIT *Anthus hodgsoni* PLATE 81
(Spotted Pipit/Olive/Indian/Oriental Tree-Pipit)

Description: Smallish (16 cm) olive pipit with bold white supercilium. Distinguished from other pipits by less streaked upperparts, buffy throat and flanks, and heavier black streaking on breast and flanks.
Iris—brown; bill—pinkish below, greyish above; feet—pink.
Voice: Thin, hoarse *tseez* call in flight, and a single phrase *tsi..tsi..* repeated at rest from tree perch or on ground.
Range: Breeds in Himalayas and E Asia; migrates in winter to India, SE Asia, Philippines, and Borneo.
Distribution and status: A regular winter visitor to N Borneo where it is quite common at moderate altitudes in open wooded areas.
Habits: Prefers more wooded habitat than other pipits, alights in trees when disturbed.

733. COMMON PIPIT *Anthus novaeseelandiae* PLATE 81
(> Richard's/Paddyfield Pipit)

Description: Medium-sized (18 cm), long-legged, brown, streaked pipit of open grassland. Upperparts streaked brown with buff eyebrow; underparts buff with dark streaks on breast.
Iris—brown; bill—brown above, yellowish below; feet—pink.
Voice: Harsh, high-pitched *shree-ep* in flight and when flushed.
Range: Africa, Asia, India, China, and Siberia through SE Asia, Philippines, Malay Peninsula, Sundas, and Sulawesi to New Guinea, Australia.
Distribution and status: A widespread and common resident on Sumatra

(including islands), Java, and Bali, up to 1500 m. On Borneo a very local resident recorded N and SE Borneo and a rare migrant *richardi* to N Borneo.

Habits: Favours open coastal or montane grassy meadows, burnt *Imperata* grassland, and dry paddy fields. Occurs singly or in small flocks. Stays on the ground where it stands, very upright. Flies with undulating flight, calling with each dip.

Note: Sibley and Monroe (1990) place resident races in Paddyfield Pipit *A. rufulus* and migrant race in Richard's Pipit, but it is safer to leave all under superspecies *novaeseelandiae* until status of several regional races is clarified.

734. PETCHORA PIPIT *Anthus gustavi* PLATE 81
(Siberian Pipit > Menzbier's Pipit)

Description: Smallish (16 cm), brown pipit. Similar to Olive-backed Pipit but white streaks on back form a double V-shape and distinguished by browner coloration. Black moustache prominent. Distinguished from Red-throated Pipit by white bars on back and wing, whiter belly, and lack of white edge to tail.
Iris—brown; bill—horn; feet—pink.

Voice: Hard *pwit* call.

Range: Breeds in NE Asia and China; migrates in winter to SE Asia, Philippines, Sulawesi, and Borneo.

Distribution and status: Winter visitor to N Borneo where it is uncommon along the north coast, recorded south to Kuching.

Habits: Prefers open, wet, grassy areas and coastal forests. Sometimes alights in trees.

735. RED-THROATED PIPIT *Anthus cervinus* PLATE 81

Description: Medium-sized (16 cm) brown pipit. Distinguished from Olive-backed Pipit by browner upperparts and rump more heavily streaked and blotched with black, less bold black streaking on breast, and pinker colouring of throat. Distinguished from Petchora Pipit by pinkish buff rather than white belly, lack of white bars on back and wing and by call.
Iris—brown; bill—horn with yellow base; feet—flesh.

Voice: Thin, high-pitched *pseeoo* call in flight, more musical than other pipits.

Range: Breeds in N Palaearctic; migrates to Africa, India, SE Asia, reaching Malay Peninsula, Philippines, Sulawesi, and Borneo.

Distribution and status: A not uncommon visitor to N Borneo.

Habits: Prefers cultivated wet areas including rice fields.

Wood-swallows—FAMILY ARTAMIDAE

A SMALL, mostly Australasian, family of medium-sized insectivorous birds with short tails, long triangular wings, and powerful beaks. Wood-swallows catch insects on the wing in circular gliding flights, reminiscent of true swallows although they are not related. Wood-swallows tend to be gregarious, congregating on bare high perches where they huddle close together. The nest is a simple cup in a tree fork.
Wood-Swallows were reassigned by Sibley and Monroe (1990) as a tribe of the enlarged Crow family Corvidae. Only one species occurs in the Greater Sundas.

736. WHITE-BREASTED WOOD-SWALLOW *Artamus leucorhynchus*
(White-rumped/Lesser Wood-Swallow) PLATE 80

Description: Medium-sized (18 cm), grey and white, swallow-like bird. Heavy bill bluish grey; head, chin, throat, back, wings, and tail slaty grey; rump and rest of underparts pure white. Distinguishable from true swallows in flight by the broad triangular wings, square tail, and much heavier bill.
Iris—brown; bill—bluish grey; feet—grey.
Voice: Unmusical chattering notes *tee-tee*, *chew-chew-chew*.
Range: Philippines and Indonesia to New Guinea and Australia.
Distribution and status: A common bird of open spaces from sea-level to 1500 m throughout the Greater Sundas (including islands).
Habits: Perches on bare tree, *Casuarina*, telegraph wire, post, or other perch, and makes circular hawking flights catching insects on the wing; sometimes over water. Flight is swallow-like with effortless glides. Birds sit close together, preen each other, and wag their tails. Boldly mobs hawks and crows.

Shrikes—FAMILY LANIIDAE

A MODERATELY large family found throughout the Old World and N America. Shrikes are medium-sized, powerfully built, predatory birds. They have large heads and powerful, deeply notched bills with a strong hooked tooth at the tip. Shrikes perch on low bushes, telegraph wires, or poles and pounce on their prey, usually large insects and small vertebrates; some species impale their prey on thorns. The nests are open, cup-shaped structures built in branch forks.
In the Greater Sundas there are 1 resident shrike, 2 visitors, and 1 vagrant.

737. BROWN SHRIKE *Lanius cristatus* PLATE 80
(Red-tailed Shrike)

Description: Medium-sized (20 cm) brown shrike. Adult: forehead and supercilium white; broad eye-stripe black; crown and upperparts brown; underparts buffy white. Immature: similar but back and sides barred with wavy dark brown lines. Black eye-stripe distinguishes it from immature Tiger Shrike.
Iris—brown; bill—black; feet—blackish grey.
Voice: Generally silent but continuous, harsh chattering *cheh-cheh-cheh* calls and song are given prior to return migration in spring.
Range: Breeds in E Asia; migrates south in winter to India, SE Asia, Philippines, Sundas, Sulawesi, Moluccas, and New Guinea.
Distribution and status: A common winter visitor up to 1500 m to Sumatra and N Borneo, less common south to S Borneo, Java, and Bali.
Habits: Frequents open cultivated and secondary habitats including gardens and plantations. Single birds perch on bushes, wires, and small trees, chase flying insects or pounce on insects or small animals on the ground.

738. TIGER SHRIKE *Lanius tigrinus* PLATE 80
(Thick-billed Shrike)

Description: Medium-sized (19 cm) rufous-backed shrike. Noticeably thicker bill, shorter tail, and larger eye than Brown Shrike. Male: crown and nape grey; back, wings, and tail rich chestnut with fine black bars; broad black eye-stripe; underparts white, faintly barred with brown on flanks. Female: similar but with white lores and eyebrow line. Immature is duller brown with indistinct barred black eye-stripe; pale eyebrow line; buffy underparts and stronger barring on belly and flanks than Brown Shrike.
Iris—brown; bill—blue with black tip; feet—grey.
Voice: Harsh, grating chatter, similar to Brown Shrike.
Range: Breeds in E Asia, China, and Japan; migrates south in winter to Malay Peninsula and the Greater Sundas.
Distribution and status: Irregular winter visitor up to 900 m throughout the Greater Sundas (including islands), common in Sumatra and N Borneo but recorded south to Java and Bali.
Habits: Typical shrike behaviour of hawking for insects from a prominent perch in wooded areas, generally on forest edge. Less conspicuous than Brown Shrike, keeping more to forest.

739. LONG-TAILED SHRIKE *Lanius schach* PLATE 80
(Schach/Rufous-headed/Black-capped/Black-headed Shrike)

Description: Largish (25 cm), long-tailed, brown, black, and white shrike. Adult: forehead, mask, wings, and tail black with a white wing spot; crown and nape grey or grey and black; back, rump, and sides reddish brown; chin, throat, breast, and centre of belly white. Extent of black on head and back varies with race. Immature: duller with barred flanks and back, and greyer head and nape.
Iris—brown; bill—black; feet—black.
Voice: Harsh screeches *terrr* and a warbled song when it sometimes mimics other bird calls.
Range: Iran to China, India, SE Asia, Malay Peninsula, Philippines, and Sundas to New Guinea.
Distribution and status: On Sumatra, Java, and Bali this is a common resident, up to 1600 m. On Borneo a vagrant to N Borneo and a local resident in SE Borneo.
Habits: Frequents open spaces, grassland, scrub, tea and clove plantations, and other open areas. Sits on a low perch and makes darting sallies after flying insects, or more commonly pounces on grasshoppers and beetles on the ground.

740. NORTHERN SHRIKE *Lanius excubitor*
(Great Grey Shrike)

Description: Large (24 cm), grey, black, and white shrike. Male: crown, nape, back, and rump grey, bold black eye-stripe topped by white eyebrow; wings black with white bar; tail black with white edge; underparts whitish. Female and immature: duller with buff scaly underparts.
Iris—brown; bill—black; feet—blackish.
Voice: Shrill cries and rattles.
Range: Northern Eurasia.
Distribution and status: One vagrant recorded for N Borneo.
Habits: Hunts from prominent tree perch or wire in open and wooded country. Sometimes hovers. Often impales prey on tree thorns.

Northern Shrike, *Lanius excubitor* (740)

Starlings—FAMILY STURNIDAE

A LARGE Old World family of robust birds with powerful, sharp, straight beaks and long legs. They are mostly gregarious and most species feed on the ground where they walk in a characteristic jaunty manner. They feed on fruits and invertebrates. Most species nest in tree holes. They are noisy, garrulous birds with harsh calls and a capacity to imitate other birds' calls.

Twelve species, of which nine are resident, occur in the Greater Sundas but two of these are probably introduced. The status of another species, the Coleto, recorded from one small island off Borneo, remains unclear.

741. SHORT-TAILED STARLING *Aplonis minor* PLATE 82
(Lesser Glossy Starling)

Description: Smallish (18 cm) glossy black starling. Very similar to Asian Glossy Starling and distinguished in the field with difficulty only by smaller size and purplish (rather than green) iridescence of head. Immature birds are streaked with black and white.
Iris—red; bill—black; feet—black.
Voice: Clear metallic shriek given mostly in flight.
Range: Philippines, Sulawesi, Java, Bali, and Lesser Sundas.
Distribution and status: On Java and Bali a frequent bird of cultivated land close to forest, up to 1500 m, particularly in E Java and Bali.
Habits: As Asian Glossy Starling.

742. ASIAN GLOSSY STARLING *Aplonis panayensis* PLATE 82
(Indian/Great/Philippine Glossy Starling/Philippine Starling)

Description: Medium-sized (20 cm) glossy black starling. Similar to Lesser Glossy Starling but larger and iridescence of head is green rather than purplish. Immature birds are buffy, streaked with black beneath and streaked with brown and black above.
Iris—red; bill—black; feet—black.
Voice: Single, ringing, metallic *chiew* given mostly in flight.
Range: E. India, SE Asia, Philippines, Malay Peninsula, Greater Sundas, and Sulawesi.
Distribution and status: A common bird of the lowlands, locally up to 1200 m, throughout the Greater Sundas (including islands).
Habits: Lives in gregarious flocks which roost, feed, and nest together. Feeds in

trees and bushes on fruit and insects. Frequents open areas near forest, including villages and cities, and especially coconut plantations.

743. WHITE-SHOULDERED STARLING *Sturnus sinensis* PLATE 82
(Chinese/Grey-backed Starling)

Description: Smallish (18 cm) grey starling. Male: distinguished from other starlings by entirely white upperwing coverts and scapulars. General plumage grey; crown and belly whitish and flight feathers black with white tips to outer tail feathers. Female has less white on wing coverts. Immature birds are browner.
Iris—bluish white; bill—grey; feet—grey.
Voice: Harsh squawks and squeals.
Range: Breeds NE India, S China, and N Vietnam; migrates in winter to SE Asia, Philippines, and Borneo.
Distribution and status: A rare straggler to N Borneo.
Habits: Lives in noisy flocks which congregate to feed in figs and other fruiting and flowering trees, in open country and gardens.

744. CHESTNUT-CHEEKED STARLING *Sturnus philippensis*
(Red-cheeked/Violet-backed Starling) PLATE 82

Description: Smallish (18 cm) dark-backed starling. Male has pale grey or buffy head, whitish underparts, glossy dark violet back, black wings with white shoulder bar and black tail. Distinguished from Purple-backed Starling by chestnut ear coverts and cheeks. The female greyish brown above and whitish below with black wings and tail.
Iris—brown; bill—black with pink base; feet—dark green.
Voice: Loud squealing calls and shrieks.
Range: Breeds in Japan; migrates in winter to Philippines and Borneo.
Distribution and status: An uncommon winter visitor at lower altitudes to N Borneo, rare south to Kalimantan.
Habits: Lives in small flocks, preferring open country. Feeds in trees.

745. PURPLE-BACKED STARLING *Sturnus sturninus* PLATE 82
(Daurian Starling/Daurian Starlet)

Description: Smallish (18 cm) dark-backed starling. Adult male: back glossy, iridescent purple; wings iridescent green-black with pronounced white wing bars; head and breast grey with black patch on nape; belly white. Distinguished from Chestnut-cheeked Starling by black nape patch and lack of chestnut on

cheeks. Female: ashy grey above; brown nape spot; black wings and tail.
Immature: pale brown; underparts mottled with brown.
Iris—brown; bill—blackish; feet—green.
Voice: Typical harsh starling whistling and whickering notes.
Range: Breeds in Himalayas and China, migrates in winter to SE Asia and Greater Sundas.
Distribution and status: On Sumatra (including islands) and W Java an uncommon visitor primarily to coastal areas, but recorded up to 1100 m in Sumatra. Vagrant to N Borneo. Not recorded on Bali.
Habits: Feeds on the ground in open coastal areas.

746. ASIAN PIED STARLING *Sturnus contra* PLATE 82
(Pied/Asiatic Pied Starling)

Description: Medium-sized (24 cm), black and white starling. White forehead, cheeks, wing bar, rump, and belly; throat, breast, and upperparts black (brown in immature).
Iris—grey; bare skin around eye—orange; bill—red with white tip; feet—yellow.
Voice: Noisy, discordant, jaunty cries.
Range: India, SW China, SE Asia (except Malay Peninsula), Sumatra, Java, and Bali.
Distribution and status: In S Sumatra, Java, and Bali a common bird of the cultivated lowlands. Records from Borneo are assumed to be escaped birds.
Habits: Lives in small parties, inhabiting open land. Feeds mostly on the ground, probing for earthworms and other small animals. Congregates in communal roosts at night.

747. BLACK-WINGED STARLING *Sturnus melanopterus* PLATE 82

Description: Medium-sized (23 cm), black and white starling. Plumage entirely white except for black wings and tail. Back colour becomes paler as one progresses westward. Adults have a short crest. Bare skin around eye pinkish yellow.
Iris—dark brown; bill—yellowish; feet—yellow.
Voice: Loud harsh whistles.
Range: Endemic to Java, Bali, and Lombok. Introduced on St. John's Island, Singapore.
Distribution and status: Quite common in open lowlands including towns and gardens, especially in E Java and Bali.
Habits: Lives in pairs or small flocks, feeds on open ground such as grass lawns. Roosts in trees and even on houses in cities.

STARLINGS 375

748. BALI MYNA *Leucopsar rothschildi* PLATE 82
(Rothschild's/White Myna/Bali Starling)

Description: Medium-sized (25 cm) white starling with a long crest. Plumage entirely snowy white except for black wing tips and tip of tail; bare skin around eye bright blue. Crest very long, especially in male. Distinguished from Black-winged Starling by much less extensive black on wings and blue skin around eye.
Iris—grey; bill—grey and yellow; feet—blue grey.
Voice: Loud harsh whistles.
Range: Endemic to Bali.
Distribution and status: Confined to NW Bali where only about 30 birds survive.
Habits: A bird of drier lowlands of W Bali. It roosts communally but travels in pairs to feed. In display the long crest of the male is raised and lowered during song.

749. COMMON MYNA *Acridotheres tristis* PLATE 82
(Indian Myna)

Description: Medium-sized (24 cm), jaunty, brownish myna with dark head. Distinguished from other mynas by lack of crest and bare yellow skin around eye. In flight white wing flash is conspicuous. Immature birds are duller.
Iris—reddish; bill—yellow; feet—yellow.
Voice: Liquid gurgles, sharp screeches, and a musical whistle, including mimicked notes.
Range: Afghanistan to SW China and SE Asia, and Malay Peninsula. Introduced into several towns.
Distribution and status: Occasional records from many localities in the Greater Sundas are escaped cage birds. Feral populations may establish locally but so far no extensive populations have become established. Not recorded on Bali.
Habits: Generally in flocks on the ground. Prefers towns, fields, and gardens.

750. JAVAN MYNA *Acridotheres javanicus* PLATE 82

Description: Medium-sized (25 cm) dark grey starling. Plumage dark grey, almost black, except for white patch on primaries conspicuous in flight, white vent, and white tip of tail. Has a slight crest. Distinguished from Crested Myna by broader white tip of tail, all yellow bill, and white vent. Immature browner.
Iris—orange; bill—yellow; feet—yellow.
Voice: Garrulous, harsh, creaky notes; whistles and rattles. Sometimes mimics other birds.

STARLINGS

Range: E Asia, SE Asia (except Malay Peninsula), Sulawesi, Sumatra (introduced), Java, and Bali.
Distribution and status: Locally common in Sumatra, probably established from escaped cage birds released in Medan area, but now found throughout. In Java and Bali this is the commonest starling of cultivated areas and cities, up to 1500 m.
Habits: Lives in small to large flocks, feeding largely on the ground in open grassy areas and paddy fields. Frequently settles on or around domestic animals, catching insects disturbed by their movements and flies attracted to them.
Note: Sometimes treated as a race of Jungle Myna *A. fuscans* but more likely to be conspecific with White-vented Myna *A. grandis*. Name *javanicus* takes precedence.

751. CRESTED MYNA *Acridotheres cristatellus* PLATE 82
(Chinese Jungle Myna)

Description: Large (26 cm) black myna with prominent crest. Distinguished from Javan Myna by longer crest, red or pink base to bill, narrow white tail tip, and undertail coverts barred with black and white.
Iris—orange; bill—yellow with red base; feet—yellow.
Voice: Like White-vented Myna. Captive birds may talk.
Range: China and Indochina. Introduced into Philippines and Borneo.
Distribution and status: Feral populations established around Kota Kinabalu and perhaps elsewhere, but not yet widespread.
Habits: Lives in small flocks, generally seen strutting on ground in open fields or in towns and gardens.

752. HILL MYNA *Gracula religiosa* PLATE 82
(Talking Myna/Common Grackle/Grackle > Eastern Hill Myna)

Description: Large (30 cm) glossy black starling with conspicuous white wing patches and diagnostic orange wattles and lappets on sides of head.
Iris—dark brown; bill—orange; feet—yellow.
Voice: Loud, clear, piercing *tiong* and an enormous range of clear whistles and calls mimicking other birds and even gibbons. (See Appendix 6.)
Range: India to China, SE Asia, Palawan, Malay Peninsula, and Greater Sundas.
Distribution and status: Locally common throughout the lowlands of Sumatra (including islands) and Borneo (including islands), up to 1000 m. On Java and Bali it was formerly common at the lowland forest edge, but is now rather rare due to trapping and habitat loss; more common in S Java.
Habits: Keeps to tall trees and lives in pairs, sometimes gathering in flocks.

753. COLETO *Sarcops calvus*

Description: Medium-sized (22 cm), black-winged, dark starling with white shoulder patch and diagnostic yellow and red facial wattles. Back and rump buffish grey scaled boldly with black. Plumage otherwise black.
Iris—red, bill—black; feet—black.
Voice: Loud harsh screeches.
Range: Philippines (except Palawan) and Sulu Islands.
Distribution and status: A few specimens collected on Banggi Island off Sabah. Status uncertain.
Habits: Similar to Hill Myna. Occupies forest and woodlands, up to 1500 m.

Coleto, *Sarcops calvus* (753)

Sunbirds and spiderhunters—FAMILY NECTARINIIDAE

Sunbirds and spiderhunters are an Old World tropical family of small, mostly very colourful birds with long curved beaks. Their metallic plumage and ability to hover in front of flowers resemble those of American hummingbirds. Most species are nectar feeders but take some insects and pollen. The long-beaked spiderhunters have become partly insect feeders. All are active, restless birds constantly on the move, looking for food. Many tropical flowers are adapted to attract these birds as pollinating agents by having small trumpet flowers and red or orange colour. The nests of sunbirds are beautiful hanging structures made of fine grass-heads and other soft materials. Spiderhunter nests are sewn on to the underside of large leaves with spider web threads piercing the leaf as plugs.

Twenty species occur in the Greater Sundas.

754. PLAIN SUNBIRD *Anthreptes simplex* PLATE 83
(Plain-coloured Sunbird)

Description: Medium-sized (12 cm) dull coloured sunbird. Upperparts olive green; grey throat and yellowish-green belly. Male has dark iridescent patch (purple in Borneo, green in Sumatra) on forehead (olive-green in female). Distinguished from most sunbirds by greyish underparts and lack of white tips of tail. Distinguished from female Crimson and Brown-throated Sunbirds by greenish tail.
Iris—reddish brown; bill—black; feet—brown or greenish.
Voice: Typical metallic trills and chirps.
Range: Malay Peninsula, Sumatra, and Borneo.
Distribution and status: On Sumatra (including Nias) and Borneo (including Natunas) a locally common but inconspicuous bird of the lowlands, up to 1200 m. Common on offshore islands.
Habits: Prefers open forests and scrub.

755. PLAIN-THROATED SUNBIRD *Anthreptes malacensis* PLATE 83
(Brown-throated > Grey-throated Sunbird)

Description: Medium-sized (13 cm) colourful sunbird. Male: crown and back iridescent green; rump, wing coverts, tail, and malar stripe iridescent purple; cheeks, chin, and throat dull dark brown; rest of underparts yellow. Female: olive-green upperparts, pale yellow underparts.
Iris—red; bill—black; feet—grey black.
Voice: High-pitched chirps *kelichap*, *tweet-tweet-tweet*, or simple melodies *wee-chuuw, wee-chuuw*.
Range: SE Asia, Philippines, Malay Peninsula, Sundas, and Sulawesi.
Distribution and status: A widespread and common lowland bird, up to 1200 m, throughout the Greater Sundas (including islands).
Habits: As common as the Olive-backed Sunbird and a familiar resident of open gardens, coconut plantations, coastal scrub, and mangroves. Aggressively territorial, chasing other sunbirds from favoured food trees such as *Loranthus*, *Musa*, and *Hybiscus*.

756. RED-THROATED SUNBIRD *Anthreptes rhodolaema* PLATE 83
(Shelley's/Rufous-throated/Red-shouldered Sunbird)

Description: Medium-sized (12 cm) colourful sunbird. Male has glossy metallic green crown and mantle, and violet back and rump with yellow-olive underparts. Distinguished from Plain-throated Sunbird by dark maroon-red cheeks and

upperwing coverts, and light red throat. Female: like female Plain-throated Sunbird but duller and more olive, with only small yellowish eye-ring.
Iris—red; bill—black; feet—olive.
Voice: Typical chirps and metallic trill.
Range: Palawan, Malay Peninsula, Sumatra, and Borneo.
Distribution and status: An uncommon bird of lowland forests, up to 500 m, on Sumatra and Borneo.
Habits: Lives among smaller trees and bushes of primary and secondary forests. Visits orchid clusters.

757. RUBY-CHEEKED SUNBIRD *Anthreptes singalensis*   PLATE 83
(Rubycheek)

Description: Small (10 cm) colourful sunbird. Male: crown and upperparts dark iridescent green; cheeks deep red; belly yellow; throat and breast orange-brown. Female: upperparts greenish olive; underparts like male but paler.
Iris—red brown; bill—black; feet—greenish black.
Voice: A shrill chirp *seet-seet*, also shrill, rising trill ending in a brief double note, followed immediately by descending trill ending with 2 separated notes (M. & W.).
Range: Nepal to SW China, SE Asia, Malay Peninsula, and Greater Sundas.
Distribution and status: On Sumatra (including islands), Borneo (including Natunas and N Bornean islands), and Java it is a widespread but uncommon bird of lowlands, up to 1000 m. Not recorded on Bali.
Habits: Lives singly or in pairs, sometimes mixes with other species. Prefers forest edge, light undergrowth, coconut plantations, and *Casuarina* groves where it feeds on pollen.

758. PURPLE-NAPED SUNBIRD *Hypogramma hypogrammicum*
(Blue-naped Sunbird)   PLATE 83

Description: Large (15 cm) sunbird with diagnostic heavily streaked yellow underparts. Male has metallic purple nape, rump, and tail coverts.
Iris—red or brown; bill—black; feet—brown or olive.
Voice: Strident, single *schewp* call.
Range: SW China, SE Asia, Malay Peninsula, Sumatra, and Borneo.
Distribution and status: On Sumatra and Borneo (including Natunas) a locally common resident of lowland forest, up to 1000 m.
Habits: Prefers smaller trees and undergrowth of forest, swamp forest, and secondary scrub. Fans and flicks tail.

759. PURPLE-THROATED SUNBIRD *Nectarinia sperata* PLATE 83
(Van Hasselt's Sunbird > Philippine Sunbird)

Description: Small (10 cm) colourful dark sunbird. Male: upperparts iridescent dark bluish with iridescent green cap; throat iridescent purple; breast dull red. Distinguished from Copper-throated Sunbird by red breast and small size. Female: olive above with yellow below, easily confused with other female sunbirds but duller than other species.
Iris—brown; bill—black; feet—black.
Voice: Sharp metallic chirp *si-si-si*, occasional *wheep* or double-noted whistle with first note rising and second falling.
Range: NE India, SE Asia, Philippines, Malay Peninsula, Sundas, and Sulawesi.
Distribution and status: On Sumatra (including islands), Borneo (including islands), and W Java this is an occasional bird of lowland forest, beach forests, and mangroves, up to 200 m; locally higher on Borneo. Not recorded on Bali.
Habits: Prefers forest edge, clearings, and other marginal habitats including rubber plantations. Usually alone or in pairs.

760. COPPER-THROATED SUNBIRD *Nectarinia calcostetha* PLATE 83
(Macklot's Sunbird)

Description: Medium-sized (13 cm) blackish sunbird. Male: upperparts black with iridescent green gloss; malar stripe and breast iridescent purple; throat and upper breast iridescent dark copper. Distinguished from Purple-throated Sunbird by larger size, lack of red breast, and yellow flanks. Female: grey head, olive back, whitish throat, and dirty yellow belly with white undertail coverts.
Iris—brown; bill—black; feet—black.
Voice: A deep trill.
Range: SE Asia, Palawan, Malay Peninsula, and Greater Sundas.
Distribution and status: On Sumatra (including islands) and Borneo (including islands) a not uncommon resident of mangrove and coastal plantations and woodland. On Java (including Karimunjawa) it is an uncommon bird in beach forest and mangroves. Not recorded on Bali.
Habits: A bird of mangrove forests, coastal coconut groves, and *Casuarina* stands. Has characteristic dipping flight.

761. OLIVE-BACKED SUNBIRD *Nectarinia jugularis* PLATE 83
(Yellow-breasted/Yellow-bellied Sunbird > Black-throated Sunbird)

Description: Small (10 cm) bright yellow-bellied sunbird. Male has black, metallic purple chin and breast, and olive-green back. In eclipse plumage,

metallic purple reduced to a narrow stripe down centre of throat. Female lacks black and is olive green above, yellow below; usually shows pale yellow supercilium.
Iris—dark brown; bill—black; feet—black.
Voice: Musical chirps *cheep*, *cheep*, *chee weet*, and a short melody ending in a clear trill.
Range: China, SE Asia, Philippines, Malay Peninsula, and Indonesia to New Guinea and Australia.
Distribution and status: Throughout the Greater Sundas (including islands) this is the commonest sunbird of open lowland areas, occasionally up to 1700 m.
Habits: A noisy bird which flits from one flowering tree or bush to another in small parties. Males sometimes chase back and forth aggressively. Frequents gardens, coastal scrub, and mangroves, visiting *Loranthus* flowers, *Morinda*, papaya, and many others.

762. WHITE-FLANKED SUNBIRD *Aethopyga eximia* PLATE 83
(Kuhl's Sunbird)

Description: Medium-sized (13 cm including male's long tail) colourful sunbird. Adult male has iridescent purple-blue crown and narrow throat band, red throat and upper breast, olive back and wings, yellow rump, long bluish-green tail, and white tuft of feathers on flanks. Female is dull olive above, dark olive-green below with white flanks and shorter tail.
Iris—brown; bill—black; feet—black.
Voice: Clear precise *tee-tee-tee-leet* and minor variations.
Range: Endemic to Java.
Distribution and status: Common in the mountains of Java, in forest and alpine scrub above 1200 m.
Habits: Travels singly, in pairs, or in small groups, generally not far off the forest floor, visiting flowering trees and vines in thick forest, clearings, and forest edge.

763. CRIMSON SUNBIRD *Aethopyga siparaja* PLATE 83
(Yellow-backed/Scarlet-throated Sunbird)

Description: Medium-sized (13 cm including long tail) bright red (male only) sunbird. Very similar to Scarlet Sunbird but with purple, not red, forehead, shorter tail, and darker grey belly. Female is dull dark olive-green without red wash on wings or tail.
Iris—dark; bill—blackish; feet—bluish.
Voice: A soft *seeseep-seeseep*.
Range: India, S China, SE Asia, Philippines, Sulawesi, Malay Peninsula, and Greater Sundas.

382 SUNBIRDS

Distribution and status: On Sumatra (including islands) and Borneo (including islands) a common resident of the lowlands, found to 900 m on Sumatra and locally to 1300 m on Borneo. In Java it is a rare and local bird of lowland, W Java recorded to 800 m. Not recorded on Bali.

Habits: Seen singly or in pairs, visiting *Erythrina* bushes and similar flowering trees on estates and at the forest edge.

764. JAVAN SUNBIRD *Aethopyga mystacalis*   PLATE 83

Description: Small (12 cm including long tail) bright red (male only) sunbird. Male: crown, malar stripe, and long tail dark iridescent purple; head, breast, and back crimson; rump pale yellow; wings olive; belly pale grey. Distinguished from Crimson Sunbird by red forehead, longer tail, and whiter belly. Female: very small; dull olive grey. Distinguished by red wash on wings and tail.
Iris—dark brown; bill—brown; feet—brown.

Voice: Soft, ringing *tzeep-tzeep*, *cheet-cheet*.

Range: Endemic to Java.

Distribution and status: On Java it is not uncommon in forest and forest edge, up to 1600 m. Absent from Bali.

Habits: Lives in pairs, generally rather noisy in its habits. Keeps to the upper canopy where it is a regular visitor of mistletoe flowers.

Note: Formerly considered to include Temminck's Sunbird.

765. TEMMINCK'S SUNBIRD *Aethopyga temminckii*   PLATE 83
(Scarlet)

Description: Small (male 13 cm, female 10 cm) sharp-tailed sunbird. Male: crimson with yellow rump, scarlet tail, and metallic violet eyebrows, side of crown, nape, and uppertail-coverts; underparts greyish white. Female: very small; olive brown above with grey head and distinguished by red tinge on wings and tail; underparts yellowish olive.
Iris—brown; bill—black; feet—black.

Voice: Soft *cheet cheet*.

Range: Malay Peninsula, Sumatra, and Borneo.

Distribution and status: On Sumatra and Borneo this is an occasional bird of hill and mountain forests, between 800 and 1800 m, locally in the lowlands and to 2000 m on Borneo.

Habits: Frequents secondary and open forests. Actively works through *Loranthus* mistletoe clumps.

Note: This species was formerly treated as 2 subspecies of the Scarlet Sunbird *A. mystacalis*, which is now considered endemic to Java.

SUNBIRDS 383

766. LITTLE SPIDERHUNTER *Arachnothera longirostra* PLATE 84

Description: Smallish (15 cm) olive and yellow spiderhunter. Upperparts olive-green; underparts bright yellow. Whitish grey throat diagnostic. Iris—brown; bill—black above, grey below; feet—bluish indigo.

Voice: Sharp *weechoo* or *cheek-cheek-cheek* in flight, or a simple high-pitched song *tik-ti-ti-ti*, the first note higher and stressed, endlessly repeated, about 3 notes per second.

Range: India, China, SE Asia, Philippines, Malay Peninsula, and Greater Sundas.

Distribution and status: A fairly common bird in lowland and hill forests throughout the Greater Sundas (including islands); also in the mountains of Sumatra, Java, and Bali where found to 2000 m.

Habits: Secretive; keeping to dark thickets such as wild bananas and tall gingers. Most frequently seen flying speedily across jungle trails, giving its characteristic flight call. Also found in secondary forests, plantations, and gardens. Sips nectar from banana and ginger flowers.

767. THICK-BILLED SPIDERHUNTER *Arachnothera crassirostris*
PLATE 84

Description: Smallish (16 cm) spiderhunter with olive-green upperparts and greenish-grey throat grading into yellow underparts. Distinguished from Little Spiderhunter by shorter, thicker bill, grey throat, more distinct pale eyebrow, and darker eye-stripe. Immature birds are duller and greyer.
Iris—brown; bill—blackish; feet—blackish.

Voice: Hard, nasal *chit chit*, also wheezy rattle.

Range: Malay Peninsula, Sumatra, and Borneo.

Distribution and status: An uncommon and local resident of lowland forest on Sumatra and Borneo, recorded up to 1300 m on Sumatra.

Habits: Similar to other spiderhunters. Prefers thickets of wild bananas and gingers in forest and secondary growth.

768. LONG-BILLED SPIDERHUNTER *Arachnothera robusta*
PLATE 84

Description: Largish (21 cm), long-billed, olive and yellow spiderhunter. Olive upperparts and yellow underparts; throat and breast streaked with green. Large size, long bill, lack of cheek patch, eye-ring, and white chin diagnostic. White tips of dark tail feathers characteristic. Bornean birds are more heavily streaked.
Iris—brown; bill—black; feet—blackish olive.

Voice: High-pitched *chit-chit, chit-chit* in flight, or a monotonous harsh *chuu-luut-chuu-luut* from a high tree perch.

Range: Malay Peninsula and Greater Sundas.

Distribution and status: On Sumatra and Java it is an occasional bird of hill forests, mostly from 400 to 1400 m, though locally at sea-level on Java. On Borneo a scarce bird of lowland forest. Not recorded on Bali.

Habits: Similar to other spiderhunters. An aggressive solitary bird chasing other spiderhunters out of its territory. Sits on bare high branches to sing.

769. SPECTACLED SPIDERHUNTER *Arachnothera flavigaster*
(Greater Yellow-eared Spiderhunter) PLATE 84

Description: Large (21 cm), olive spiderhunter with yellow ear patch and broad yellow eye-ring. Distinguished from Yellow-eared Spiderhunter by larger size, shorter bill, smaller ear-patch, broader eye-ring and lack of streaking on throat and breast.
Iris—brown; bill—black; feet—yellowish brown.

Voice: High-pitched *chit-chit*.

Range: Malay Peninsula, Sumatra, and Borneo.

Distribution and status: On Sumatra and Borneo not uncommon in open forest, coconut groves, villages, and secondary scrub, up to 1300 m, locally up to 1800 m on Borneo.

Habits: Generally found in secondary forests. Highly aggressive in defending its feeding territory.

770. YELLOW-EARED SPIDERHUNTER *Arachnothera chrysogenys*
(Lesser Yellow-eared Spiderhunter) PLATE 84

Description: Medium-sized (17 cm) olive and yellow spiderhunter. Upperparts olive-green; underparts yellow with dark streaking on breast; pectoral tufts grey; large yellow cheek patch and narrower eye-ring distinguishes it from Spectacled Spiderhunter.
Iris—brown; bill—blackish; feet—pale brown.

Voice: High *twit-twit-twit-twee-ee* in flight with last note prolonged.

Range: Malay Peninsula and Greater Sundas.

Distribution and status: On Sumatra (including islands) and Borneo it is local and uncommon in lowland forest, recorded to 1400 m on Sumatra. There are few records from the lowlands of W Java though formerly a breeding species. Not recorded on Bali.

Habits: Inhabits forests, secondary scrub, and gardens.

771. GREY-BREASTED SPIDERHUNTER *Arachnothera affinis*
PLATE 84

Description: Medium-sized (17 cm), green and grey spiderhunter. Upperparts olive-green; underparts grey with fine black streaks on throat and breast, distinguishing it from all except Bornean Spiderhunter. Sumatran birds are less distinctly streaked than Javan. Bornean birds are fairly heavily streaked. Immature lacks streaking.
Iris—brown (blue in one Sarawak bird); bill—black above, paler below; feet—pinkish brown.
Voice: Various raucous, piercing ringing calls *chee-wee-dee-weet . . . tee-ree, chee chee-chur,* and others.
Range: Malay Peninsula and Greater Sundas.
Distribution and status: On Sumatra (including Mentawai), Java, and Bali the commonest spiderhunter in dry coastal woodland, also found inland up to 900 m. On Borneo it is quite common in lowland and hill forests and secondary growth, up to 1000 m.
Habits: Usually single. Frequents thickets of wild bananas and flowering trees such as *Erythrina*, often quite high off the ground. Flies fast and low through the forest like Little Spiderhunter.
Note: Some authors include the montane Bornean Spiderhunter *A. everetti* in this species.

772. BORNEAN SPIDERHUNTER *Arachnothera everetti*   PLATE 84
(Everett's/Kinabalu Spiderhunter)

Description: Large (21 cm) green and grey spiderhunter. Similar to Sumatran race of Grey-breasted Spiderhunter but significantly larger and living at higher altitude.
Iris—brown; bill—blackish; feet—pinkish.
Voice: Similar to Grey-breasted.
Range: Endemic to N Borneo.
Distribution and status: Confined to environs of Mt. Kinabalu, between 900 and 1600 m.
Habits: Typical of genus.
Note: Many authors include it with Grey-breasted Spiderhunter *A. affinis.*

773. WHITEHEAD'S SPIDERHUNTER *Arachnothera juliae*   PLATE 84

Description: Medium-sized (18 cm) brown spiderhunter diagnostically streaked with white. Tail and wings blackish; rump and undertail coverts yellow.
Iris—brown; bill—black; feet—black.

Voice: Repetitive, loud shrieking calls in flight or from tree perch; long twittering call and high-pitched squeaking song.
Range: Endemic to Borneo.
Distribution and status: Confined to higher mountains of N Borneo, recorded from Mt. Kinabalu south to Usun Apau and Mt. Dulit; also on Mt. Mulu. Locally common in montane forests, from 1000 to 1500 m and above on Mt. Kinabalu and other high peaks.
Habits: Frequents treetops more than other spiderhunters. Feeds among orchid clusters in mossy forest.

HONEYEATERS—FAMILY MELIPHAGIDAE

HONEYEATERS belong to a large Australo-Papuan family which is well represented in eastern Indonesia but barely reaches the Sundaic region. The family is diverse with some large species such as friarbirds occupying a hornbill-niche while smaller forms occupy niches similar to the oriental sunbirds and spiderhunters. Most are rather drab in appearance, have slender, sharp, decurved bill, and feed on nectar, fruit, and insects. They build cup-like nests.

The only representative of the family to occur in the Greater Sundas, the Indonesian Honeyeater, is typical of the smaller members of the family.

774. INDONESIAN HONEYEATER *Lichmera limbata*   PLATE 84

Description: Smallish (14 cm), plain, dull brownish to olive-grey honeyeater with longish, decurved bill. Upperparts olive-brown or grey; underparts grey with white belly. Diagnostic yellow triangular patch behind the eye. Chin and throat greyish in male, yellowish in female. Immature browner and lacks bare ear patch. Iris—dark brown; bill—black; feet—greyish.
Voice: Sweet, strong song like a warbler, and a harsh, grating alarm note.
Range: Bali and Lesser Sundas.
Distribution and status: On Bali it is not uncommon at all altitudes.
Habits: Frequents gardens, coconut groves, parks, mangroves, forest edge, scrub, and woodlands. Generally single or in small flocks. Acrobatic, active birds visiting flowers like large sunbirds.
Note: Sometimes included in Brown Honeyeater *L. indistincta*.

Flowerpeckers—FAMILY DICAEIDAE

FLOWERPECKERS are very small, active birds found mostly in the Oriental and Australian regions. Several species are brightly coloured with red and orange plumage, hence the general Indonesian name 'cabe' (chilli pepper). Bill shape is variable, from sharp and pointed to thick. Flowerpeckers live in the treetops, eating tiny insects and small fruits, but have a particular association with mistletoes *Loranthus*. They are the main dispersal agents of *Loranthus* seeds and are most abundant where there are many mistletoes such as gardens, mangroves, coastal scrub, etc. Some species are more forest-loving than others. Their beautiful purse nests are suspended from leafy twigs, made of leaves and grass fibres felted together with spider webs.

There are fifteen species in the Greater Sundas. The females are often difficult to identify but flowerpeckers usually travel in pairs or little flocks so identification can usually be made from the males.

775. SCARLET-BREASTED FLOWERPECKER *Prionochilus thoracicus*
PLATE 85

Description: Small (10 cm), colourful (male) flowerpecker with thick bill. Male: black tail, wings, and head, with scarlet crown and breast patch, yellow rump and belly quite distinctive. Female much duller with grey head, white throat, and dirty yellow belly and rump. Distinguished from female Crimson-breasted Flowerpecker by white throat, yellowish rump, and black tail. Generally indistinguishable from female Yellow-rumped Flowerpecker but larger and brighter.
Iris—brown; bill—black; feet—black.
Voice: Typical metallic clicking twitters.
Range: Malay Peninsula, Sumatra, and Borneo
Distribution and status: Recorded N and C Sumatra (including Lingga Archipelago and Belitung) but there are very few records. On Borneo it is a common lowland bird of forest.
Habits: Like other flowerpeckers, mostly seen visiting mistletoe clusters in tree crowns. Favours primary and secondary forests, *Casuarina* groves and heath forest.

776. YELLOW-BREASTED FLOWERPECKER *Prionochilus maculatus*
(Yellow-throated Flowerpecker) PLATE 85

Description: Small (10 cm) olive flowerpecker with thick bill and yellow underparts heavily streaked with dark olive-green. Sexes alike. The orange-red patch on crown is difficult to see in the field; malar stripe white; tail with black tip. Immature like Crimson-breasted and Scarlet-breasted Flowerpeckers but underparts yellower.
Iris—red; bill—black; feet—grey.
Voice: Rasping disyllabic *tsweet, tsweet*.
Range: Malay Peninsula, Sumatra, and Borneo.
Distribution and status: A not uncommon bird of lowland forest on Sumatra (including Nias and Belitung) and Borneo (including Natunas).
Habits: As other flowerpeckers, favouring mistletoe clumps. Found in primary, logged forests, secondary forests, and scrub.

777. YELLOW-RUMPED FLOWERPECKER *Prionochilus xanthopygius*
(Borneo Yellow-rumped Flowerpecker) PLATE 85

Description: Small (9 cm), brightly coloured (male), thick-billed flowerpecker. Male has slaty-blue head, back, wings, and tail with red crown and small orange patch on upper breast; belly and rump yellow. Distinguished from Crimson-breasted Flowerpecker by yellow rump and smaller breast patch. Female: like female Scarlet-breasted Flowerpecker but smaller and duller.
Iris—brown; bill—black; feet—black.
Voice: Harsh, metallic *chip chip* calls.
Range: Endemic to Borneo and Natunas.
Distribution and status: A common bird of lowland forest throughout Borneo and the Natunas; also in Kelabit highlands.
Habits: As other flowerpeckers, but favouring forest edge and secondary scrub.

778. CRIMSON-BREASTED FLOWERPECKER *Prionochilus percussus*
PLATE 85

Description: Small (10 cm) gaudily coloured flowerpecker. Adult male: blue upperparts with black forehead and crimson crown, black primaries, and blue tail; underparts yellow with crimson breast patch; pronounced white moustache underlined with black. Distinguished from Yellow-rumped Flowerpecker by lack of yellow rump. Female: olive-green throat, yellow breast, and faint white moustache.

Iris—brown; bill—black; feet—black.
Voice: Similar to other flowerpeckers.
Range: Malay Peninsula and Greater Sundas.
Distribution and status: On Sumatra (including islands) and W Java a scarce and local bird of lowlands, up to 1000 m. On Borneo a rare bird of lowland forest, recorded mainly in S Borneo and Natunas. Not recorded on Bali.
Habits: Occupies lower and middle storeys of primary forests and upper canopy of secondary forests, plantations, and swamp forests.

779. THICK-BILLED FLOWERPECKER *Dicaeum agile*   PLATE 85
(> Striped/Streaky-breasted Flowerpecker)

Description: Small (9 cm), nondescript, brownish flowerpecker. Crown olive-brown; cheeks grey; back olive; throat and underparts greyish white; breast faintly streaked with grey. Bill is noticeably thick. Underside of tail with white tip.
Iris—red; bill—black; feet—grey.
Voice: Distinct thin *pseeow*, often in flight.
Range: India, SE Asia (except Malay Peninsula), Sumatra, and Java to Lesser Sundas.
Distribution and status: Rare or overlooked in Sumatra where known by 1 specimen collected in Aceh and a few sight records from the mountains of N Sumatra and peat swamp forest in S Sumatra. Recorded throughout Java where there are few records but it is probably largely overlooked. Not recorded for Bali but could occur.
Habits: Similar to other flowerpeckers. A bird of lowland forest and swamp forest; wags tail from side to side when perched.

780. BROWN-BACKED FLOWERPECKER *Dicaeum everetti*
(Everett's Flowerpecker)   PLATE 85

Description: Small (10 cm) greyish-brown flowerpecker with plain brown upperparts, grey streaky breast grading to white vent; thick, finch-like bill. Distinguished from Thick-billed Flowerpecker by browner colour and lack of white edge to tail, but the two species do not overlap in the Greater Sundas.
Iris—pale yellow to orange; bill—grey; feet—grey.
Voice: Typical sharp, metallic *chip-chip* calls like other flowerpeckers.
Range: Malay Peninsula, Riau Archipelago, Natunas, and Borneo.
Distribution and status: In Sumatra known by 2 specimens collected on Bintang, in Riau Archipelago. In Borneo recorded in Sarawak and Natunas where a locally common lowland resident in secondary forests, gardens, and coffee plantations.

Habits: Like other flowerpeckers but prefers open forests.
Note: Treated as a subspecies of Thick-billed Flowerpecker by some authors.

781. YELLOW-VENTED FLOWERPECKER *Dicaeum chrysorrheum*
PLATE 85

Description: Small (9 cm) white-bellied flowerpecker. Adult: olive-green upperparts; bright yellow or orange undertail coverts; rest of underparts white, heavily streaked with diagnostic bold black marks.
Iris—red-orange; bill—black; feet—black.
Voice: Repeated *zit-zit-zit* in flight, and a repeated call *zip-a-zip-treee*.
Range: N India, SW China, SE Asia, Malay Peninsula, and Greater Sundas.
Distribution and status: In Sumatra and Borneo it is uncommon in the lowlands and hills, up to 700 m, on Borneo in Kelabit uplands, to 1400 m. It is rare on Java but recently recorded for Bali.
Habits: A bird of gardens and open forest. Typical busy forager of small fruits and insects, aggressively chasing other birds from its food trees.

782. ORANGE-BELLIED FLOWERPECKER *Dicaeum trigonostigma*
(Orange-breasted Flowerpecker) PLATE 85

Description: Tiny (8 cm) blue and orange flowerpecker. Adult male: bluish head, wings, and tail; diagnostic orange back, rump, and belly; throat grey. Female: olive back, wings, and tail; yellow belly; rump greenish orange. Immature: like female, but without yellow or orange.
Iris—brown; bill—black; feet—dark grey.
Voice: Prolonged, buzzing *brrr-brrr*. A repeated *zit-zit-zit* by male birds even during flight.
Range: E India, Burma, Malay Peninsula, Greater Sundas, and Philippines.
Distribution and status: A common bird from sea-level to 1000 m, locally to 1800 m, throughout the Greater Sundas (including islands).
Habits: A bird of forests, scrub, mangroves, and gardens. Flashes about the crowns of small trees in small parties, feeding on berries. Flies jerkily from tree to tree with a fast wing-beat.

783. PLAIN FLOWERPECKER *Dicaeum concolor*
(Plain-coloured/Nilgiri Flowerpecker) PLATE 85

Description: Tiny (8 cm) nondescript flowerpecker. Olive-green above, pale greyish below with a creamy centre of belly and fine white tufts at bend of wing. Distinguished from Thick-billed and Brown-backed Flowerpeckers by fine bill.

Iris—brown; bill—black; feet—dark blue-grey.
Voice: Metallic *tsit-si-si-si-sew*.
Range: India, S China, SE Asia, Malay Peninsula, and Greater Sundas.
Distribution and status: On Sumatra, Java, and Bali a rarely recorded species in hill forest, from 500 to 1500 m; probably overlooked through confusion with females of other flowerpeckers. On Borneo common in Kelabit uplands, otherwise known at few localities in N Borneo and Natunas.
Habits: Typical flowerpecker inhabiting hill forests, secondary growth, and cultivated areas; visits mistletoes frequently.

784. SCARLET-BACKED FLOWERPECKER *Dicaeum cruentatum*
PLATE 85

Description: Small (9 cm) black and red flowerpecker. Male: scarlet crown, back and rump; black wings, sides of head, and tail; white underparts with grey flanks. Bornean form has black throat. Female: brown with scarlet rump and uppertail coverts. Female differs from Scarlet-headed Flowerpecker by lack of red wash on crown and mantle. Immature plain grey with orange bill and dull orange tinge on rump.
Iris—brown; bill—blackish green; feet—blackish green.
Voice: Hard metallic *dik* calls and song *tissit-tissit*.
Range: India, S China, SE Asia, Malay Peninsula, Sumatra, and Borneo.
Distribution and status: On Sumatra (including islands) and Borneo (including islands) a frequent resident in secondary forest, gardens, and plantations, up to 1000 m.
Habits: Active and aggressive visitors of mistletoe clusters.

785. RED-CHESTED FLOWERPECKER *Dicaeum maugei* PLATE 85
(Blue-cheeked/Timor Flowerpecker)

Description: Tiny (8 cm) blackish flowerpecker with a red rump and red throat patch, outlined in black. Distinguished from Blood-breasted Flowerpecker by red rump and throat, and whiter underparts.
Iris—dark; bill—black; feet—black.
Voice: High-pitched *tsit* similar to other flowerpeckers.
Range: Islands in Flores Sea and Lesser Sundas.
Distribution and status: Recorded on Nusa Penida Island off Bali.
Habits: As other flowerpeckers, flitting among small trees, visiting mistletoes.

786. BLOOD-BREASTED FLOWERPECKER *Dicaeum sanguinolentum*
(Javan Fire-breasted/Javan Flowerpecker) PLATE 85

Description: Tiny (8 cm) colourful flowerpecker. Adult male: dark blue upperparts; creamy buff belly and throat; scarlet breast underlined with irregular black line. Female: dull olive-brown above with scarlet rump; ochre below streaked with olive-grey.
Iris—blue/brown; bill—black; feet—dark grey.
Voice: As other flowerpeckers, a variety of sharp, high-pitched clicks and buzzing.
Range: Java, Bali, and Lesser Sundas.
Distribution and status: On Java and Bali widespread in the hill and mountain forest, and forest edge, mostly from 800 to 2400 m. Status on Sumatra uncertain; one sight record in lowland S Sumatra perhaps a vagrant from Java.
Habits: As other flowerpeckers, flitting among treetops, especially mistletoes and *Viscum* bushes.
Note: May be conspecific with Mistletoebird *D. hirundinaceum*.

787. BLACK-SIDED FLOWERPECKER *Dicaeum monticolum*
(Bornean Fire-breasted/Bornean/Mountain Flowerpecker) PLATE 85

Description: Tiny (8 cm) dark flowerpecker. Male: glossy bluish-black upperparts, scarlet breast lined with black, and buff vent. Female is dull grey like Plain Flowerpecker but slightly larger. Both sexes have white pectoral tuft.
Iris—brown; bill—black; feet—brown.
Voice: Sharp, piercing, metallic *zit* calls like other flowerpeckers.
Range: Endemic to Borneo.
Distribution and status: On Borneo it is a common montane species on most of the higher mountains from Mt. Kinabalu south to Liang Kubung; also Mts Dulit and Mulu, but absent from Penrissen and Poi ranges.
Habits: As other flowerpeckers a regular visitor of mistletoe clumps in tree crowns.
Note: Some authors include it in the Grey-sided Flowerpecker *D. celebicum* or in the Fire-breasted Flowerpecker *D. ignipectus*, or all three of them in the Mistletoebird *D. hirundinaceum*.

788. FIRE-BREASTED FLOWERPECKER *Dicaeum ignipectus*
(Buff-breasted/Green-backed/Bronze-backed Flowerpecker) PLATE 85

Description: Small (9 cm) dark flowerpecker. Male: upperparts glossy greenish black; underparts buffy with cinnamon breast band and narrow black stripe in centre of belly. Female: buffy ochre below. Immature: like immature Plain Flowerpecker but found at higher altitudes.
Iris—brown; bill—black; feet—black.
Voice: High-pitched, metallic chittering.
Range: Himalayas, S China, SE Asia, Malay Peninsula, and Sumatra.
Distribution and status: On Sumatra, it is an occasional bird of montane forests, from 800 to 2200 m.
Habits: As other flowerpeckers, mostly seen visiting clumps of *Loranthus* mistletoe in tree crowns.

789. SCARLET-HEADED FLOWERPECKER *Dicaeum trochileum*
PLATE 85

Description: Tiny (8 cm) scarlet and black flowerpecker. Adult male: head, back, rump, and breast scarlet. Wings and tip of tail black; belly greyish white; white patch at bend of wing. Female: rump scarlet, otherwise upperparts brown with red wash on head and mantle, underparts dull white. Immature: greenish brown upperparts and orange patch on rump.
Iris—brown; bill—black; feet—black.
Voice: Typical flowerpecker high-pitched *zit-zit-zit*.
Range: Sumatra, Borneo, Java, Bali, and Lombok
Distribution and status: Found in coastal and lowland S Sumatra and S Borneo; on Sumatra probably a recent immigrant from Java. On Java (including islands) and Bali it is a common lowland bird of gardens and open country.
Habits: Frequently seen in gardens and open areas, including towns, coastal areas, and mangroves, where it visits mistletoe bushes to eat the sticky berries.

WHITE-EYES—FAMILY ZOSTEROPIDAE

WHITE-EYES are in a large family occurring in Africa, Asia, and Australia. They derive their name from the ring of silvery feathers around the eye in most species. White-eyes are generally small, tit-like birds with greenish-olive or yellowish plumage, small, slender, slightly curved bills, short wings, and small, strong feet. They are extremely agile, restless birds, often forming mixed flocks which work

394 WHITE-EYES

through the treetops in search of small fruits and insects. They visit flowers to feed on nectar like sunbirds. The calls are shrill twitters and chirps. The nests are neat cup-shaped structures woven into a branch fork.
Ten species occur in the Greater Sundas.

790. ORIENTAL WHITE-EYE *Zosterops palpebrosus* PLATE 86
(Indian/Small White-eye > Yellow-bellied White-eye)

Description: Small (11 cm) yellowish-green white-eye. Montane races in Java and Sumatra are characterized by all yellow underparts and yellow patch above the beak; lowland forms in W Java, Borneo, and Sumatra are very similar to Mountain White-eye but are distinguished by a narrow yellow bar down centre of belly and pale grey thighs. Upperparts olive green; throat and vent yellow, and very little or no yellow above black lores.
Iris—yellow-brown; bill—dark brown; feet—olive-grey.
Voice: Soft, high twittering *dzi-da-da*, or repeated metallic *dza dza*. Flocks twitter continuously.
Range: N India to S China, SE Asia, Malay Peninsula, and Sundas.
Distribution and status: On Sumatra (including islands), Java, and Bali a common bird of the lowlands and hills, up to 1400 m. In Borneo it is a scarce bird of mangroves and coastal areas, recorded from S Natunas and SW Sarawak.
Habits: Frequents primary and secondary vegetation. Forms large flocks which join freely with other birds such as minivets, travelling through the tops of the highest trees.

791. ENGGANO WHITE-EYE *Zosterops salvadorii* PLATE 86
(Engano White-eye)

Description: Small (10 cm) olive-backed white-eye. Like lowland Oriental White-eye with yellow breast and pale grey sides; belly creamy white.
Iris—brown; bill—black; feet—black.
Voice: As Oriental White-eye.
Range: Endemic to Enggano Island off SW Sumatra.
Distribution and status: Common.
Habits: As Oriental White-eye.
Note: An insular form of the Oriental White-eye found only on Enggano Island but treated by Mees (1957) as a separate species.

792. BLACK-CAPPED WHITE-EYE *Zosterops atricapilla* PLATE 85

Description: Small (11 cm) olive-backed white-eye. Similar to lowland Oriental White-eye but forehead and crown blackish, and upperparts and underparts darker.

Iris—brown; bill—black; feet—black.
Voice: Soft twitters like Oriental White-eye.
Range: Malay Peninsula, Sumatra, and Borneo.
Distribution and status: On Sumatra common in hill and mountain forest, particularly high mountain tops between 700 and 3000 m. On Borneo common on Mts Kinabalu, Mulu, and Batu Patap, generally above 1000 m.
Habits: Lives in flocks, actively working through crowns of trees and bushes in sub-montane and montane forests.

793. EVERETT'S WHITE-EYE *Zosterops everetti* PLATE 86

Description: Small (11 cm) olive white-eye. Similar to lowland form of Oriental White-eye but with broader yellow band down centre of breast and darker grey on flanks. Safely distinguished by altitude.
Iris—brown; bill—black; feet—black.
Voice: Soft twitters *tsee tsee* and sweet little trilly song (T.H.).
Range: Philippines, Malay Peninsula, and Borneo.
Distribution and status: On Borneo a sub-montane resident confined to the mountains of N Borneo from Mt. Kinabalu to the Poi range; locally common between 800 and 1700 m.
Habits: As other white-eyes.

794. MOUNTAIN WHITE-EYE *Zosterops montanus* PLATE 86

Description: Small (11 cm), white or grey-bellied white-eye. Upperparts olive green; wings and tail darker; throat and vent yellow; belly whitish with brownish flanks. Three races vary in minor details but are distinguished by lack of yellow on belly and diagnostic white iris.
Iris—white; bill—black above, paler below; feet—black.
Voice: High-pitched twitters.
Range: Philippines, Sumatra, Java, Bali, Sulawesi, Moluccas, and Lesser Sundas.
Distribution and status: On Sumatra recorded from ranges south of Sibayak, and on Java and Bali from main peaks except Mt. Pangrango; a common bird in the mountains between 2200 to 3100 m, recorded down to 900 m in the Kerinci valley, but rare below 1800 m.
Habits: Social flocking bird of treetops, calling incessantly and hunting for small insects.

795. JAVAN WHITE-EYE *Zosterops flavus* PLATE 86
(Yellow White-eye)

Description: Small (10 cm) yellow-bellied white-eye. Upperparts olive-yellow; underparts yellow. Smaller and brighter than Lemon-bellied White-eye, and lacks the black loral spot.
Iris—brown, bill—blackish; feet—blackish.
Voice: High-pitched contact notes between flock members.
Range: Endemic to Borneo and Java.
Distribution and status: On Borneo common in Banjarmasin area of S Borneo, otherwise known by 1 specimen from Kuching. In Java found only in mangrove and coastal scrub of NW coast east to Tanjung Krawang.
Habits: Inhabits mangroves, coastal scrub, and relict coastal forests. Habits presumably typical of genus but few details available. Flocks were described as visiting the flowers of langsat trees *Lansium domesticus* in S Borneo. Javan specimens are reported to visit *Erythrina* flowers.

796. LEMON-BELLIED WHITE-EYE *Zosterops chloris* PLATE 86
(Mangrove/Moluccan/Banda/Yellow-bellied/Pale White-eye)

Description: Small (11 cm) yellow-bellied white-eye. Upperparts olive-yellow, underparts pale lemon yellow. Very similar to Javan White-eye but larger and with dark black lores.
Iris—brown; bill—blackish; feet—blackish.
Voice: Thin, high-pitched notes.
Range: Islands in Java Sea, Sulawesi, Moluccas, and Lesser Sundas to the W Papuan Islands.
Distribution and status: In Borneo recorded on Karimata Islands. In Sumatra recorded on islets off Belitung, and in Java restricted to a few islands in the Java Sea (Pulau Seribu, Karimunjawa, Masalombo Besar, Kalambau) and islands off Bali (Manjangan and Nusa Penida). Also recorded on the coast of NW Java and NW Bali where probably a seasonal visitor.
Habits: Typical restless white-eye in small flocks, working through the trees and bushes at all heights, particularly in coastal scrub forest.

797. JAVAN GREY-THROATED WHITE-EYE *Lophozosterops javanicus*
(Grey-throated/Java White-eye) PLATE 86

Description: Largish (13 cm) dull olive white-eye. Head, throat and breast grey; upperparts olive-green; belly pale yellow. Three races vary in amount of white

markings on head and whiteness of the eye-ring, with W Java forms having the least white. All are distinguished from other white-eyes by grey throat.
Iris—brown; bill—black; feet—black.
Voice: Ringing, high-pitched notes *chee-ee-weet, chee-ee-weeweet*, or buzzing *turrr-turrr*. More rattling than other white-eyes (D.A.H.).
Range: Endemic to Java and Bali.
Distribution and status: On Java and Bali confined to the highest mountains, above 1500 m; 4 subspecies are described and populations are presumably isolated, but it can be locally common in the higher forest.
Habits: Freely mixes with flocks of other birds, especially warblers, travelling through the low canopy of montane forests. Active restless bird like other white-eyes.

798. PYGMY WHITE-EYE *Oculocincta squamifrons* PLATE 86
(Pygmy Grey White-eye)

Description: Very small (9 cm) olive-grey bird with white-spotted forecrown and yellowish grey and white underparts. Narrow white eye-ring is barely visible in the field.
Iris—buff; bill—brown-black; feet—dark green.
Voice: *Chit chit chit* (T.H.).
Range: Endemic to Borneo.
Distribution and status: Confined to mountains of N Borneo where recorded from Mts Kinabalu to Penrissen; also from Mts Dulit, Mulu, and Magdalena. Locally common between 1000 and 2500 m but recorded down to 200 m.
Habits: Inhabits moss forests and scrub. Rather tame and generally low in vegetation.

799. MOUNTAIN BLACK-EYE *Chlorocharis emiliae* PLATE 86
(Olive Black-eye)

Description: Large (14 cm), yellow and green white-eye with black lores and diagnostic black eye-ring.
Iris—brown; bill—black; feet—black.
Voice: Thrush-like melodious song; also stuttering call and two-note flight call.
Range: Endemic to Borneo.
Distribution and status: Confined to the high mountains of N Borneo where recorded from Mts Kinabalu and Trus Madi south to Tama Abo, and Niut and Poi range. It is one of the commonest birds on the peaks of Mts Kinabalu and Trus Madi, found between 1200 and 3600 m.
Habits: Active and conspicuous bird in the crowns of small trees and bushes of high altitude forests.

Weavers—FAMILY PLOCEIDAE

This is a very large family, distributed through Australia, Asia, Africa, and Europe which includes the familiar finch-like groups such as sparrows, munias, and weaverbirds.

Weavers are small, short-tailed birds with short, thick bills for eating seeds. They build covered, ball-shaped nests which reach their most complex and elaborate construction among the weaverbirds. They are gregarious, flock-forming birds and this, combined with their preferred diet of grass seeds, makes them serious pests to human agriculture, stealing rice and other cereals.

In the Greater Sundas there are 17 species of weavers.

800. EURASIAN TREE SPARROW *Passer montanus*   PLATE 88
(Tree/European Tree Sparrow)

Description: Medium-sized (14 cm) brown sparrow. Crown chestnut; chin, throat, cheek patch, and eye-stripe black; underparts greyish buff; upperparts mottled brown with black and white markings. Young birds are paler with less distinct markings.
Iris—brown; bill—grey; feet—brown.
Voice: Boisterous chirps and rapid chattering notes.
Range: Eurasia, India, China, SE Asia, Malay Peninsula, Sumatra, Java, and Bali. Introduced or recent immigrant populations occur throughout Philippines and Indonesia to Australia and Pacific islands.
Distribution and status: On Sumatra (including islands), Java, and Bali this is a very common bird in towns and villages where grain is handled or processed, up to 1500 m. On Borneo the species was first noticed in 1964 and is now well established in several coastal towns.
Habits: Closely associated with man, living in flocks around houses, warehouses, etc., and feeding in gardens and cultivated areas on the ground, pecking at tiny seeds or rice. Flocks raid paddy fields in the harvest season.

801. BAYA WEAVER *Ploceus philippinus*   PLATE 88
(Common Weaver/Baya)

Description: Medium-sized (15 cm) golden-crowned weaver. Breeding male: crown and nape golden yellow; side of face black; underparts buff; upperparts dark grey-brown with pale feather edges. Female: lacks yellow and black head markings; has tawny eyebrow stripe and breast.
Iris—brown; bill—blackish grey to brown; feet—pale brown.
Voice: Constant raucous, chattering calls and high-pitched wheezy notes.

Range: India, China, SE Asia, Malay Peninsula, Sumatra, Java, and Bali.

Distribution and status: On Sumatra (including Nias) formerly common in the lowlands and hills, up to 1000 m; apparently populations are now much reduced. On Java and Bali this is a rather local bird with extant colonies mostly in W Java; may now be scarce or absent in E Java and Bali.

Habits: Lives in large social colonies centred on communal nest trees in open areas. Habits similar to the Streaked Weaver.

Baya Weavers, *Ploceus philippinus*, at nest (801)

802. STREAKED WEAVER *Ploceus manyar*   PLATE 88
(Striated/Manyar Weaver)

Description: Medium-sized (14 cm) golden-capped weaver. Breeding male: crown golden yellow; rest of head, chin, and throat black; underparts white with black streaks on breast; upperparts blackish brown with fulvous edges to feathers. Non-breeding male and female: head brown with black streaks on crown and buff eyebrow, whitish patch on neck.

Iris—brown; bill—blackish grey to brown; feet—pale brown.

Voice: Constant chattering and whistling calls.

Range: Pakistan to SW China, SE Asia (except Malay Peninsula), Java, and Bali.
Distribution and status: On Java (including Bawean) and Bali this is a widespread but local species occurring in large colonies and mobile flocks in the lowlands.
Habits: Lives in large colonies around breeding trees or, at other times of year, in mobile flocks. Males are polygamous, each female making her own elaborated woven nest. Preferred habitats are grassy swamps and reed beds or paddy fields.

803. ASIAN GOLDEN WEAVER *Ploceus hypoxanthus* PLATE 88
(Golden Weaver)

Description: Medium-sized (15 cm) golden yellow weaver with black face. Breeding male; crown, nape, rump, and underparts yellow; side of face, chin, and throat black; mantle black with yellow edges to feathers; wings and tail black with whitish edges to feathers. Female and non-breeding male distinguished by buff eyebrow and yellowish underparts.
Iris—brown; bill—black or brown; feet—brown.
Voice: Chattering and wheezing calls.
Range: SE Asia (except Malay Peninsula), Sumatra, and Java.
Distribution and status: Breeding reported in Lake Toba, N Sumatra, lends credence to old record from W Sumatra. On Java now local and uncommon with colonies restricted to W Java.
Habits: Lives in large flocks and colonies around suitable nesting trees, in open areas where there are plenty of rice fields. Habits similar to other weavers.

804. RED AVADAVAT *Amandava amandava* PLATE 87
(Strawberry Finch/Strawberry Waxbill/Red Munia)

Description: Small (10 cm), red-rumped finch with white spots. Male: crimson with blackish wings and tail; regular spots on flanks; wings and rump with small white spots. Female: underparts grey buff; mantle brown; rump red; wings and tail blackish; a few white spots on wings.
Iris—brown; bill—red; feet—flesh.
Voice: Rather feeble *cheep* notes.
Range: Pakistan to SW China, SE Asia, Java, Bali, and Lesser Sundas. Introduced to Malay Peninsula, Sumatra, Borneo, and Philippines.
Distribution and status: On Sumatra the species was known around Medan in 1912–20 but now appears extinct. First recorded on Borneo near Kota Kinabalu in 1969 and now established there. In W Java now fairly rare as a result of trapping though still locally not uncommon in E Java and Bali, up to 1500 m.
Habits: Social, living in small flocks. Frequents scrub, grassland, cultivated areas,

paddy fields, and reed beds. Fast-flying, restless flocks are conspicuous because of the crimson rump patch.

805. PIN-TAILED PARROTFINCH *Erythrura prasina* PLATE 87
(Long-tailed Munia, AG: Parrot-finch/Parrot-mannikin)

Description: Small (15 cm including male's extended tail) green finch with very long tail. Male: face blue; upperparts green; underparts variable dark buff with red central patch to all chestnut; rump and elongated tail red. Female has greenish head and shorter tail. Immature has brown rump.
Iris—dark; bill—grey; feet—red.
Voice: Fine, high-pitched *zit-zit-zit* calls while feeding in flocks.
Range: Local in Thailand, Laos, Malay Peninsula, and Greater Sundas.
Distribution and status: On Sumatra and Java this formerly common bird is now local in the lowlands and hills, up to 1200 m. On Borneo, in season, it is a dreaded rice pest, living in migratory flocks. Not recorded on Bali.
Habits: Lives in small flocks and feeds in rice fields, mixing with munias. Flocks move seasonally with the harvest and can cause heavy local losses. Mostly a bird of bamboo thickets.

806. TAWNY-BREASTED PARROTFINCH *Erythrura hyperythra*
PLATE 87
(Green-tailed/Bamboo Parrotfinch, AG: Parrot-finch/Parrot- mannikin)

Description: Small (10 cm), parrot-like, green finch. Crown, back, and upperparts green; flight feathers dark brown; blue spot on forecrown; underparts tawny pinkish, redder towards head.
Iris—reddish; bill—grey; feet—red.
Range: Malay Peninsula, Borneo, Java, Lesser Sundas, Sulawesi, and Philippines.
Distribution and status: On Java this is now a rare bird of montane areas, between 800 and 2500 m; most recent records are from W Java. On Borneo it is a rare resident of mountains in N Borneo, recorded from Mts Kinabalu and Mulu.
Habits: Frequents moss forests and montane bamboo thickets where it lives in pairs or small flocks. It has a quick flight and visits open grassy areas and neighbouring rice fields to feed.

807. JAVA SPARROW *Padda oryzivora* PLATE 87

Description: Largish (16 cm), colourful, red-billed finch. Adult: head black with conspicuous white cheek patch; upperparts and breast grey; belly pink; undertail white; tail black. Immature has pinkish head with grey crown; breast pink.
Iris—red; bill—deep pink; feet—red.

Voice: Low churring calls. Quiet *tup* and soft, chattering song ending with whined phrase *ti tui* (M. & W.).
Range: Endemic to Java, Kangean, and Bali. Introduced widely from SE Asia to Australia.
Distribution and status: Formerly one of the common birds of cultivated areas in Java and Bali, up to 1500 m, but is now rather scarce as a result of massive capture for the pet trade. Local colonies exist on Sumatra (including islands) and Borneo, but it is less common than a few decades ago.
Habits: A bird of towns, gardens, and cultivated fields. Congregates in large flocks in cane thickets or tall trees. Raids regularly rice and maize fields as well as grain stores. Birds are highly social, rubbing next to each other at perches. In disputes over nest sites protagonists perform an elaborate body weaving display.

808. WHITE-RUMPED MUNIA *Lonchura striata* PLATE 87
(Striated/White-backed/Sharp-tailed Munia, AG: Mannikin)

Description: Medium-sized (11 cm) munia with dark brown upperparts and distinctive pointed black tail, white rump, and buffy white belly. Back thinly streaked with white and underparts finely scaled and streaked with darker buff. Immature is paler with buffy rump.
Iris—brown; bill—grey; feet—grey.
Voice: Lively chirps and trilled *prrrit* (King *et al.*).
Range: India, S China, SE Asia, Malay Peninsula, and Sumatra.
Distribution and status: On Sumatra (including Bangka) a locally common bird of forest edge, secondary scrub, farmland, and gardens at low altitudes, up to 1600 m.
Habits: Lives in small noisy flocks; habits like other munias.

809. WHITE-BELLIED MUNIA *Lonchura leucogastra* PLATE 87
(White-breasted Munia, AG: Mannikin)

Description: Smallish (11 cm), plump, white-bellied munia. Upperparts brown with pale feather shafts; face and breast almost black; flanks dark brown, tail brown washed with yellow.
Iris—reddish; bill—dark above, grey below; feet blue—grey.
Voice: Soft cheeps *chee-ee-ee*, and shrill *prip* alarm call.
Range: Philippines, Malay Peninsula, and Greater Sundas.
Distribution and status: Status on Sumatra uncertain; presumably resident but there are few satisfactory records. On Borneo recorded throughout the island but rather local and apparently nomadic. A nest with eggs was found in W Java but this species is not believed to be resident.

Habits: Less social than other munias, generally living in pairs and skulking about in paddy fields or even at grain stores, associated with forest edge.

810. JAVAN MUNIA *Lonchura leucogastroides* PLATE 87

Description: Smallish (11 cm), plump, black, brown and white munia. Upperparts brown, unstreaked; face and upper breast black; belly sides and flanks white; undertail dark brown. Distinguished from White-bellied Munia by lack of pale streaks on back, lack of yellowish tinge on tail, clean edge between black breast and white belly, and white, rather than brown, flanks.
Iris—brown; bill—dark above, blue below; feet—greyish.
Voice: Soft cheeps *chee-ee-ee* and shrill *pi-i* flocking call.
Range: Sumatra, Java, Bali, and Lombok. Introduced to Singapore.
Distribution and status: On Sumatra locally common in the extreme south where it may have been introduced or colonized from Java. On Java and Bali a very common and widespread species, up to 1500 m.
Habits: Frequents all kinds of cultivated areas and natural grassy patches. Forms flocks during rice harvest but usually lives in pairs or in small parties. Feeds on the ground or plucks seeds from grass heads; spends much time in noisy chirping and grooming in large trees.
Note: Sometimes treated as a race of White-rumped Munia.

811. DUSKY MUNIA *Lonchura fuscans* PLATE 87
(Bornean Munia)

Description: Medium-sized (11 cm) dark munia. Distinguished from other munias by entirely blackish-brown plumage.
Iris—brown; bill—grey below, black above; feet—black.
Voice: Shrill *pee pee* or *chirrup*, and low *teck teck* in flight.
Range: Endemic to Borneo.
Distribution and status: A common bird of forest edge, secondary scrub, grassland, and cultivated areas from sea-level to 500 m throughout Borneo (including islands).
Habits: As other munias; living in rice fields and along river banks far inland.

812. BLACK-FACED MUNIA *Lonchura molucca* PLATE 87
(Moluccan Munia, AG: Mannikin)

Description: Smallish (11 cm) brown munia with white belly and diagnostic white rump. Nape and back light brown; wings and tail dark brown; rump white; forehead, crown, throat, and breast blackish brown; belly white with fine brown streaks.

Iris—brown; bill—grey; feet—grey.
Voice: Typical munia chirps.
Range: Sulawesi, Moluccas, Kangean, Nusa Penida, and Lesser Sundas.
Distribution and status: Found on Kangean and Nusa Penida Islands.
Habits: Similar to other munias. Small to large flocks live in grass beds and rice fields.

813. SCALY-BREASTED MUNIA *Lonchura punctulata* PLATE 87
(Nutmeg/Spotted/Spice Munia/Ricebird, AG: Mannikin)

Description: Smallish (11 cm) brown munia. Upperparts brown, streaked with white feather shafts; throat reddish brown; underparts white scaled with dark brown on breast and flanks. Immature has underparts rich buff without scales.
Iris—brown; bill—bluish grey; feet—grey-black.
Voice: Dissyllabic chirps *ki-dee*, *ki-dee* or in alarm *tret-tret*.
Range: India, China, Philippines, SE Asia, Malay Peninsula, Sundas, and Sulawesi. Introduced to Australia and elsewhere.
Distribution and status: On Sumatra, Java, and Bali a common and widespread species, up to 1800 m. A presumably feral population recently reported in S Borneo.
Habits: Frequents open grassy patches in cultivated lands, paddy fields, gardens, and secondary scrub. Pairs or small flocks mix readily with other species of munias. Shows typical munia tail-wagging and active, flighty behaviour.

814. BLACK-HEADED MUNIA *Lonchura malacca* PLATE 87

Description: Smallish (11 cm) chestnut munia with black head. Young birds are dirty brown all over.
Iris—red; bill—blue grey; feet—pale blue.
Voice: Shrill, reedy *pwi-pwi*.
Range: India, China to SE Asia, Malay Peninsula, Sumatra, Borneo, Philippines, and Sulawesi.
Distribution and status: On Sumatra mainly E lowlands and islands, less common than Scaly-breasted or White-headed Munias. On Borneo (including Natunas) an abundant munia of the lowlands, recorded up to 1800 m on Mt. Kinabalu and Kelabit uplands.
Habits: Forms large flocks which do not mix with other species. Flocks move through paddy fields with a whirring of wings as they rise and settle. The birds are in constant movement and weave their bodies from side to side like other munias. They roost at night in their own nests.
Note: Some authors include the Chestnut Munia of Java and Bali with this species.

815. CHESTNUT MUNIA *Lonchura ferruginosa* PLATE 87

Description: Smallish (11 cm) chestnut finch with white head and black chin and throat.
Iris—red; bill—blue grey; feet—pale blue.
Voice: As Black-headed Munia.
Range: Endemic to Java and Bali.
Distribution and status: A locally common bird of rice fields, where it can be a serious pest; up to 1800 m.
Habits: As Black-headed Munia.
Note: May be conspecific with Black-headed Munia.

816. WHITE-HEADED MUNIA *Lonchura maja* PLATE 87
(Pale-headed Munia)

Description: Smallish (11 cm), white-headed, brown finch. Similar to Chestnut Munia but paler brown, and entire head and throat white. Young birds have brown upperparts and buff underparts and face.
Iris—brown; bill—bluish grey; feet—pale blue.
Voice: High-pitched, piping *puip* flocking call.
Range: Malay Peninsula, Sumatra, Java, Bali, and Sulawesi.
Distribution and status: On Sumatra (including islands), Java, and Bali this is a fairly common and widespread bird, up to 1500 m.
Habits: Frequents marshes and reed beds. Like other munias forms large flocks during rice harvest but spreads out in pairs during breeding season. General behaviour similar to other munias.

FINCHES AND BUNTINGS—FAMILY FRINGILLIDAE

FINCHES and Buntings are a large, almost worldwide family of small, thick-billed, seed-eating birds. They resemble closely the weavers but have longer, notched tails, slightly less massive bills, and make open, cup-shaped, rather than covered, nests. They are flighty, flocking birds of open meadows and scrubland. Several northern breeding species migrate south into tropical Asia in winter.

In the Greater Sundas there is only one resident species of this family, and two definite and one possible visitor.

817. MOUNTAIN SERIN *Serinus estherae* PLATE 88
(Malaysian/Malay Serin > Indonesian Serin, AG: Goldfinch)

Description: Small (11 cm) yellow and grey finch. Male: forehead and breast band yellow, streaked with black; rump bright yellow; wings black with 3 transverse yellow bars and white edges to secondaries; nape and mantle grey; throat black; belly white, streaked with black. Female: similar but yellow rump duller and breast less spotted.
Iris—brown; bill—brownish; feet—black.
Voice: Short, tinkling song given in flight, and dull metallic chittering.
Range: Philippines, Sulawesi, Sumatra, and Java.
Distribution and status: On Sumatra it is only known from alpine meadows and scrub of Mt. Leuser and adjacent peaks. On Java this is a local bird known from Pangrango, Tengger, and Yang. Perhaps primarily a bird of the ericaceous zone, but on Pangrango recorded in forest and pine plantations, down to 1300 m.
Habits: Single birds or small flocks sit in the tops of small bushes in alpine meadows or moss forest; also visit ground. Shy, fast-flying, with undulating flight.

818. LITTLE BUNTING *Emberiza pusilla* PLATE 88

Description: Smallish (13 cm) streaked bunting with striped head. Wintering male and female both have cheeks and crown stripe dull rufous, and malar stripe and edge of ear coverts greyish black; eyebrow and second malar stripe dull buffy rufous. Upperparts brown with dark streaks. Underparts whitish streaked with black on breast and flanks. Has conspicuous pale eye-ring and single wing bar.
Iris—dark brown; bill—grey; feet—pink.
Voice: High-pitched, quiet *pwick*.
Range: Breeds in Europe and N Asia; migrating in winter to India and SE Asia.
Distribution and status: Stragglers reach N Borneo.
Habits: Mixes with pipits. Hides in thick cover.

819. YELLOW-BREASTED BUNTING *Emberiza aureola* PLATE 88
(White-shouldered Bunting)

Description: Largish (15 cm) bunting with 2 distinctive white wing bars. Wintering male has face and throat blackish, scaled with dark buff; crown and back chestnut, streaked with dark brown and scaled dark buff. Underparts yellowish with narrow chestnut bar across breast. In flight the white wing patches are conspicuous. Female is duller with buffy eyebrow, blackish eye-stripe, and white wing bars less distinct. Iris—dark brown; bill—grey; feet—pink.
Voice: Short, loud *tick*.
Range: Breeds in NE Asia; migrating in winter to S Asia.

Distribution and status: A rare straggler to N Borneo.
Habits: Lives in fields or tall grasses and moist thickets. Mixes with munia flocks.

[820. **BLACK-HEADED BUNTING** *Emberiza melanocephala* PLATE 88

Description: Largish (18 cm) mottled brown bunting with unstreaked underparts. Breeding male has black head but in winter head and back are brownish, streaked with black, sometimes with a rufous wash on rump. Female and immature greyish brown above, streaked with black. Both sexes have 2 whitish wing bars and unstreaked grey underparts with yellowish vent, and sometimes yellow wash on breast.
Iris—dark brown; bill—grey; feet—pinkish.
Voice: Short musical *tweet* notes, sometimes ending in short whistled warble, given from exposed perch.
Range: Breeds in C Eurasia; migrates in winter to India and China. Vagrants have been recorded in Thailand and Japan.
Distribution and status: Two birds sighted in Brunei in 1985 and 1988 were probably this species (Mann 1987, 1989).
Habits: A bird of open country with scattered bushes.]

APPENDICES

Appendix 1
Endemic and threatened and endangered species in main reserves

(Borneo: K = Mt. Kinabalu National Park, M = Mt. Mulu National Park, D = Danum Valley and Ulu Segama, S = Samunsam Wildlife Sanctuary, L = Lanjak Entimau Wildife Sanctuary, K = Kutai National Park, P = Gunung Palung Nature Reserve, B = Barito Ulu Research Area, T = Tanjung Puting National Park; Sumatra: L = Gunung Leuser National Park, K = Gunung Kerinci National Park, T = Taitaibatti Nature Reserve, B = Berbak Game Reserve, S = Barisan Selatan Game Reserve, W = Way Kambas National Park; Java and Bali: U = Ujung Kulon National Park, P = Gunung Pangrango National Park, M = Meru Betiri Game Reserve, B = Baluran National Park, W = West Bali (Bali Barat) National Park.)

	Borneo									Sumatra						Java				
	K	M	D	S	L	K	P	B	T	L	K	T	B	S	W	U	P	M	B	W
Fregata andrewsi				*																
Mycteria cinerea																*	*		*	*
Ciconia stormi	*	*	*									*	*	*						
Leptoptilos javanicus						*			*				*	*		*			*	*
Pseudibis davisoni							*		*											
Cairina scutulata														*						
Spilornis kinabaluensis	*	*																		
Spizaetus bartelsi																	*			
Spizaetus nanus	*	*	*							*	*	*	*	*						
Microhierax latifrons	*	*																		
Falco moluccensis																*	*	*	*	*

ENDANGERED SPECIES IN MAIN RESERVES   411

	Borneo									Sumatra						Java				
	K	M	D	S	L	K	P	B	T	L	K	T	B	S	W	U	P	M	B	W
Arborophila javanica																		★		
Arborophila rubrirostris											★									
Arborophila hyperythra	★	★					★													
Arborophila charltonii	★	★																		
Haematortyx sanguiniceps	★	★	?																	
Lophura ignita	★	★		★	★			★	★	★										
Lophura bulweri	★	★		?				★	★											
Lophura inornata																				
Lophura hoogerwerfi										★										
Gallus varius																★				
Polyplectron chalcurum											★									
Polyplectron schleiermacheri	★								★											
Pavo muticus															★	★				
Heliopais personata										★				★						
Vanellus macropterus (extinct)																				
Charadrius javanicus																		?		
Limnodromus semipalmatus																				
Scolopax saturata										★	★					★		★		
Treron oxyura											★									
Treron griseicauda										★										
Treron capellei	★	★		★	★	★	★	★	★	★	★									
Ptilinopus porphyreus											★					★	★			
Ptilinopus melanospila																	★	★	★	★
Ptilinopus cinctus																			★	★
Ducula lacernulata																★	★		★	★

Appendix 1 (cont.)

	Borneo									Sumatra						Java				
	K	M	D	S	L	K	P	B	T	L	K	T	B	S	W	U	P	M	B	W
Ducula pickeringi (small islands)																				
Columba argentina	★	★								★										
Caloenas nicobarica												★				★				
Loriculus pusillus																		★	?	★
Carpococcyx radiceus				★						★										
Centropus rectunguis		★					★			★										
Centropus nigrorufus																	★			
Otus sagittatus					★					★										
Otus stresemanni											★									
Otus angelinae																		★		
Otus umbra (Simeulue only)										★										
Otus enganensis (Enggano only)													★							
Otus brookii			★							★								★		
Otus mentawi												★								
Glaucidium castanopterum			?															★		
Batrachostomus harterti										★										
Batrachostomus poliolophus										★										
Batrachostomus cornutus									★	★										
Caprimulgus concretus					★															
Caprimulgus pulchellus																★	★			
Hydrochous gigas																	★			
Collocalia vulcanorum																	★			
Collocalia linchi	★									★				★		★		★	★	★
Harpactes reinwardtii										★				★		★		★	★	★

ENDANGERED SPECIES IN MAIN RESERVES

	Borneo									Sumatra						Java				
	K	M	D	S	L	K	P	B	T	L	K	T	B	S	W	U	P	M	B	W
Harpactes whiteheadi	★																		★	★
Alcedo coerulescens																			★	★
Halcyon cyanoventris																		★	★	★
Aceros corrugatus		★	★	★	★	★	★	★	★	★	★	★								
Aceros subruficollis		★	★	★	★	★	★	★		?	?	★								
Buceros vigil	★	★	★	★	★	★	★	★		★	★	★	★	★						
Megalaima corvina																	★	★		
Megalaima javensis					★											★	★			
Megalaima monticola	★							★							★					
Megalaima armillaris	★																	★		
Megalaima pulcherrima	★														★					
Megalaima eximia	★			★											★					
Calyptomena hosii	★						★													
Calyptomena whiteheadi	★																			
Pitta schneideri										★										
Pitta arquata	★					★	★													
Pitta baudii			★			★	★													
Pitta venusta									★											
Pitta cyanea																				
Coracina javensis																?	★			
Pericrocotus miniatus										★	★	★	★							
Chloropsis venusta										★	★	★								
Pycnonotus leucogrammicus										★	★	★	★							
Pycnonotus tympanistrigus										★	★									
Pycnonotus nieuwenhuisii										★	★							★		
Pycnonotus bimaculatus														★						

Appendix 1 (cont.)

	Borneo									Sumatra						Java				
	K	M	D	S	L	K	P	B	T	L	K	T	B	S	W	U	P	M	B	W
Setornis criniger	★	★					★	★	★											
Ixos virescens											★					★				
Oriolus hosii										★	★	★								
Dicrurus sumatranus										★	★	★	★							
Dendrocitta occipitalis										★	★			★						
Dendrocitta cinerascens	★	★																		
Pityriasis gymnocephala		★		★	★	★	★	★	★											
Psaltria exilis																	★	★		
Pellorneum pyrrogenys	★	★					★			★						★	★	★		?
Pellorneum buettikoferi										★		★								
Malacocincla perspicillata	★	★		★																
Malacocincla vanderbilti		★																		
Ptilocichla leucogrammica	★	★				★														
Napothera rufipectus		★								★							★	★		
Napothera atrigularis		★																		
Napothera crassa	★																			
Stachyris grammiceps																?	★	★		
Stachyris thoracica																★	★	★		
Stachyris melanothorax																?	★	★	★	★
Macronous flavicollis																				★
Garrulax palliatus	★						★			★										
Garrulax rufifrons																	★	★		
Alcippe pyrrhoptera																★	★	★		
Crocias albonotatus																	★	★		

ENDANGERED SPECIES IN MAIN RESERVES

	Borneo									Sumatra						Java				
	K	M	D	S	L	K	P	B	T	L	K	T	B	S	W	U	P	M	B	W
Yuhina everetti	★	★																		
Copsychus stricklandi	★	★																		
Cinclidium diana										★	★									
Enicurus velatus										★	★						★			
Chlamydochaera jefferyi			★														★			
Cochoa azurea												★								
Cochoa beccarii										★	★						★			
Myiophoneus melanurus										★	★								★	★
Myiophoneus glaucinus	★	★								★	★						★	★	★	★
Zoothera everetti	★	★																★	★	★
Zoothera andromedae										★	★						★	★	★	★
Seicercus grammiceps										★	★						★	★	★	★
Orthotomus sepium																	★	★	★	★
Prinia familiaris																	★			
Tesia superciliaris																★				
Urosphena whiteheadi	★	★					★													
Cettia vulcania	★	★								★	★						★	★	★	
Bradypterus accentor	★									★	★									
Eumyias indigo	★									★	★						★			
Cyornis ruckii											★									
Cyornis sumatranus			★	★																
Cyornis caerulatus	★	★	★	★																
Cyornis superbus	★	★	★	★																
Rhipidura phoenicura																			★	★
Rhipidura perlata																			★	★

Appendix 1 (cont.)

	Borneo									Sumatra						Java				
	K	M	D	S	L	K	P	B	T	L	K	T	B	S	W	U	P	M	B	W
Pachycephala hypoxantha	★						★													
Sturnus melanopterus																	★	★	★	★
Leucopsar rothschildi																			★	★
Aethopyga eximia																★	?			
Aethopyga mystacalis	?															★	?			
Arachnothera everetti	★	★																		
Arachnothera juliae		★																		
Prionochilus xanthopygius			★																	
Dicaeum everetti				★																
Dicaeum trochileum							★									★	★	★	★	
Dicaeum sanguinolentum	★																	★		
Dicaeum monticolum	★																			
Zosterops salvadorii											★									
Zosterops atricapilla								★			★									
Zosterops flavus									★											
Zosterops chloris																★			★	★
Lophozosterops javanicus																	★			
Oculocincta squamifrons	★						★													
Chlorocharis emiliae	★																			
Padda oryzivora																★	★	★	★	★
Lonchura leucogastroides																★	★	★	★	
Lonchura fuscans	★	★		★	★	★	★	★												
Lonchura ferruginosa																	★	★	★	★
Serinus estherae														★				★		

Appendix 2
Endangered and threatened species by island

Endangered birds of Java

Podiceps ruficollis	Absent from many former haunts.
Nycticorax caledonicus	Few recent records (Brantas).
Mycteria cinerea	Few records from W Java.
Threskiornis melanocephalus	Few records from W Java; breeds P. Rambut.
Nettapus coromandelianus	Few recent records (Pulau Dua).
Cairina scutulata	No recent records, possibly extinct.
Aythya australis	Former resident, possibly extinct
Hieraaetus kienerii	Rare (Meru Betiri, Puncak, Gn. Kawi).
Spizaetus bartelsi	Very rare, few recent records.
Falco peregrinus	Rare; a few sightings.
Falco severus	Recent sightings only in E Java.
Arborophila orientalis	Only recent report Baluran.
Pavo muticus	Now confined to very few localities.
Fulica atra	Few recent records (Kamojang).
Metopidius indicus	Very rare; C Java (Meleman).
Vanellus macropterus	Probably extinct.
Treron bicincta	Few recent records.
Treron capellei	Few recent records.
Treron curvirostra	Only recent records from P. Deli/Tinjil.
Treron griseicauda	Rare; local records only (e.g. Bogor).
Treron oxyura	Gunung Halimun.
Ptilinopus cinctus	Rare; resident on Bali only.
Ducula badia	No recent records.
Chrysococcyx malayanus	No recent records.
Chrysococcyx xanthorhynchus	Rare, only records from Meru Betiri.
Phaenicophaeus tristis	No recent records.
Centropus nigrorufus	Very rare (Segara Anakan).
Tyto alba	Rarely seen.
Phodilus badius	Only two recent records.
Otus angelinae	Only recent records from Cibodas.
Otus brookii	No recent records.
Otus rufescens	No recent records.
Strix leptogrammica	Rare.

Strix seloputo	Rare.
Ninox scutulata	Few recent records.
Batrachostomus javensis	Rare (Gn Pancar, Ujung Kulon).
Caprimulgus pulchellus	Only recent records from Pangrango.
Harpactes reinwardtii	Rare; records only from W Java.
Alcedo euryzona	Rare; only two records.
Lacedo pulchella	Rare; few records (Cikepuh).
Halcyon capensis	Now rare in E Java, very rare in Bali.
Halcyon coromanda	Rare; two records W Java.
Picus mentalis	Rare; few records (Cikepuh).
Picus miniaceus	Recent records from Cikepuh, Tukung Gede.
Meiglyptes tristis	No recent records.
Chrysocolaptes lucidus	Rare; few records from E Java and Bali.
Reinwardtipicus validus	Rare; only records from W Java (Baluran).
Pycnonotus zeylanicus	Very rare; two recent records from W Java.
Pycnonotus squamatus	No recent records. Possibly extinct.
Cochoa azurea	Few records from W Java.
Zoothera andromedae	Rare; few records.
Rhinomyas olivacea	Rare; few records from Java and Bali.
Ficedula dumetoria	Recent records from Salak, Puncak.
Cyornis rufigaster	Recent record from Segara Anakan.
Philentoma velatum	Rare; only records from Meru Betiri, Ujung Kulon.
Terpsiphone paradisi	Rare; few records (Ujung Kulon).
Pachycephala pectoralis	Rare (Bali, Yang highlands).
Prionochilus percussus	Rare; only two records from W Java.
Dicaeum agile	No recent records.
Dicaeum chrysorrheum	Rare (Meru Betiri, Tukung Gede).
Nectarinia calcostetha	Recent record from Segara Anakan.
Nectarinia sperata	Rare; few records from W Java.
Aethopyga siparaja	Rare (Ujung Kulon, Rawa Danau).
Arachnothera chrysogenys	Rare; few records.
Arachnothera robusta	Rare; few records from Segara Anakan, Salak.
Zosterops chloris	Rare; few records.
Zosterops flavus	Few records from Pulau Dua and nearby coast and Pamanukan.
Amandava amandava	Rare; few records (Puncak).
Erythrura hyperythra	Rare; E Java (Puncak and Cibodas).
Ploceus hypoxanthus	Rare; few records from W Java.
Ploceus philippinus	Rare, only records from SW Java.
Leucopsar rothschildi	Very rare; less than 50 in Bali Barat National Park.
Oriolus cruentus	Rare; few records from W Java.

Endangered birds of Sumatra

Phalacrocorax carbo	Formerly resident now extinct.
Rhizothera longirostris	Rare.
Melanoperdix nigra	Rare.
Arborophila charltonii	No recent records.
Caloperdix oculea	Rare.
Lophura all spp.	Rare.
Ptilinopus jambu	Habitat loss.
Carpococcyx radiceus	Very rare.
Centropus rectunguis	Rare or overlooked.
Caprimulgus pulchellus	Very rare.
Caprimulgus concretus	Rare.
Otus sagittatus	Habitat loss.
Otus rufescens	Rare, habitat loss.
Otus stresemanni	Species validity in doubt.
Harpactes erythrocephalus	Rare.
Harpactes orrhophaeus	Rare.
Harpactes oreskios	Rare.
Picus vittatus	Habitat loss.
Pitta granatina	Habitat loss.
Pitta venusta	No recent records.
Pitta caerulea	No recent records.
Pitta schneideri	One recent record.
Setornis criniger	Rare or overlooked.
Criniger finschii	Very rare.
Pycnonotus nieuwenhuisi	No recent records.
Pycnonotus jocosus	Introduced but now extinct.
Napothera marmorata	Rare or overlooked.
Kenopia striata	Rare or overlooked.
Pellorneum buettikoferi	Rare, habitat loss.
Trichastoma perspicillatum	No recent records.
Myiophoneus glaucinus	Rare.
Cochoa beccarii	No recent records.
Cyornis tickelliae	Rare, habitat loss.
Cyornis ruckii	Rare.
Cyornis caerulatus	Rare, habitat loss.
Melanochlora sultanea	No recent records.
Amandava amandava	Probably extinct.
Crypsirina temia	No recent records.

Endangered birds of Borneo

Plegadis falcinellus	Not recorded this century.
Plegadis papillosa	Very local and rare.
Nettapus coromandelianus	Rare; status uncertain.
Lophura erythrophthalma	Restricted range and rare.
Lophura bulweri	Widespread but generally rare.
Polyplectron schleiermacheri	Local and rare.
Gallinula tenebrosa	No recent records.
Gallinula chloropus	Uncommon and limited distribution.
Collocalia maxima	Rare or overlooked.
Chrysocolaptes lucidus	Very restricted range.
Pycnonotus nieuwenhuisi	No recent records.
Trichastoma perspicillatum	No recent records.
Rhinomyias olivacea	Limited distribution.
Padda oryzivora	Introduced but now rare.
Cissa chinensis	Very limited distribution.
Crypsirina temia	No recent records.

Appendix 3
Land birds found on offshore island groups

(S = Simeulue, B = Batu Island, N = Nias, B = Banyak Island, M = N Mentawai (Siberut), P = S Mentawai (Pagi, Sipora), E = Enggano, R = Riau Archipelago, L = Lingga Archipelago, B = Bangka, B = Belitung, S = Sunda Straits Island, N = Natunas, B = Bawean, K = Kangean, K = Karimunjawa, K = Karimata, M = Maratuas, T = Tambelan, B = Banggi and NE Island, A = Anambas.)

	S	B	N	B	M	P	E	R	L	B	B	S	N	B	K	K	M	T	B	A
Ardea sumatrana	*	*	*	*	*	*	*							*				*		
Ardea cinerea													*							
Ardea purpurea			*	*		*		*					*	*						
Butorides striatus	*		*	*		*		*		*	*	*	*	*						
Ardeola bacchus/speciosa	*		*					*					*	*	*					
Bubulcus ibis								*						*				*		
Egretta sacra	*		*	*	*	*	*	*	*	*	*	*	*	*	*	*	*	*	*	*
Egretta eulophotes				*																
Egretta alba				*																
Egretta intermedia			*								*			*						
Egretta garzetta			*											*						
Nycticorax nycticorax						*				*			*	*						
Gorsachius melanolophus			*							*	*					*				
Ixobrychus eurhythmus											*					*				
Ixobrychus sinensis											*	*								
Ixobrychus cinnamomeus						*								*						

Appendix 3 (*cont.*)

	S	B	N	B	M	P	E	R	L	B	B	S	N	B	K	K	K	M	T	B	A
Dupetor flavicollis	★																				
Ciconia episcopus				★	★																
Leptoptilos javanicus									★	★											
Dendrocygna javanica				★					★	★											
Anas gibberifrons																					
Anas superciliosa																					
Nettapus coromandelianus							★														
Pandion haliaetus							★			★	★		★		★	★					
Aviceda leuphotes								★													
Pernis ptilorhynchus									★												
Machieramphus alcinus							★		★	★											
Haliastur indus	★	★					★		★	★	★	★	★		★	★					
Haliaeetus leucogaster	★	★					★	★	★	★	★	★	★		★	★		★			
Ichthyophaga ichthyaetus									★	★											
Ichthyophaga nana																					
Spilornis cheela	★				★		★		★	★	★	★	★	★	★	★			★		★
Accipiter gularis		★							★							★					
Accipiter virgatus	★	★					★	★	★	★	★	★	★		★	★					
Accipiter trivirgatus				★					★		★		★								
Accipiter soloensis	★																				
Accipiter badius		★																			
Butastur indicus			★				★														
Ictinaetus malayensis															★						
Spizaetus cirrhatus	★	★				★			★	★	★							★			
Spizaetus alboniger	★	★				★			★	★	★										★

	S	B	N	M	P	E	R	L	B	B	S	N	B	K	K	K	M	T	B	A
Spizaetus nanus				★										★						
Falco tinnunculus/moluccensis		★												★						
Falco severus														★						
Falco peregrinus	★	★			★									★						
Megapodius cumingii/reinwardt																				★
Coturnix chinensis							★	★												
Rollulus rouloul							★	★	★											
Lophura ignita							★	★												
Gallus varius								★												
Turnix suscitator						★	★	★						★						
Gallirallus striatus	★			★		★	★	★												
Rallina fasciata				★		★	★	★												
Porzana cinerea						★	★	★	★				★							
Amaurornis phoenicurus	★			★	★	★	★	★	★				★	★	★			★		
Gallicrex cinerea						★	★													
Gallinula chloropus					★		★	★	★				★	★						
Porphyrio porphyrio				★												★				
Glareola maldivarum	★				★		★	★			★									
Glareola isabella								★												
Treron curvirostra	★			★		★	★	★		★			★	★			★			
Treron griseicauda							★	★		★			★	★						
Treron fulvicollis	★	★					★	★		★										
Treron olax	★	★					★	★		★	★		★	★				★		
Treron vernans							★	★		★	★	★	★	★	★			★		
Treron capellei	★						★	★		★										
Ptilinopus jambu								★		★			★	★						
Ptilinopus porphyreus									?											

Appendix 3 (cont.)

	S	B	N	M	P	E	R	L	B	B	S	N	B	K	K	K	M	T	B	A
Ptilinopus melanospila	★	★	★	★	★	★	★	★	★	★	★	★	★							★
Ducula aenea	★	★	★	★	★	★	★	★	★	★	★	★	★					★	★	★
Ducula bicolor	★	★	★															★	★	★
Ducula badia				★																
Ducula pickeringi																				
Ducula rosacea																				★
Columba vitiensis	★												★							
Columba argentina	★			★	★			★												
Macropygia emiliana			★	★	★				★	★										
Macropygia ruficeps	★				★															★
Streptopelia chinensis							★	★	★	★	★	★								
Geopelia striata							★	★	★	★	★	★								
Chalcophaps indica	★	★	★			★	★	★	★	★	★	★	★							
Caloenas nicobarica	★	★	★	★	★			★	★	★	★	★	★					★	★	★
Psittacula alexandri	★	★	★					★					★							
Psittacula longicauda							★	★	★		★									
Psitinus cyanurus	★					★					★									
Tanygnathus lucionensis														★						
Loriculus galgulus	★				★			★										★		★
Clamator coromandus				★	★				★	★	★									
Cuculus fugax								★	★	★										
Cuculus micropterus			★	★					★	★							★			
Cuculus canorus										★										
Cuculus saturatus	★					★		★						★						
Cacomantis sonneratii			★					★				★								

LAND BIRDS FOUND ON OFFSHORE ISLAND GROUPS

	S	B	N	M	P	E	R	L	B	B	S	N	B	K	K	K	M	T	B	A
Cacomantis merulinus	★	★																		
Cacomantis sepulcralis	★	★																		
Chrysococcyx xanthorhynchus			★					★												
Chrysococcyx basalis					★															
Chrysococcyx minutillus																	★			
Surniculus lugubris			★	★			★	★	★	★										
Eudynamys scolopacea	★	★	★	★		★	★	★	★	★	★	★	★						★	
Phaenicophaeus sumatranus			★	★		★	★	★	★	★	★	★	★							
Phaenicophaeus tristis			★	★																
Phaenicophaeus chlorophaeus				★				★	★		★	★	★	★	★	★				
Phaenicophaeus curvirostris						★	★	★	★		★	★	★		★					
Centropus sinensis			★	★			★	★	★	★	★	★								
Centropus bengalensis				★							★	★								
Tyto alba												★	★							
Phodilus badius		★								★										
Otus rufescens									★											
Otus mantanensis																			★	
Otus umbra	★																			
Otus enganensis					★															
Otus lempiji							★	★	★	★		★	★		★					
Otus mentawi				★																
Bubo sumatranus									★											
Ketupa ketupu			★	★		★	★	★	★		★									
Ninox scutulata							★	★	★	★	★	★							★	
Strix seloputo													★							
Strix leptogrammica		★								★	★									
Batrachostomus auritus				★					★		★									

Appendix 3 (*cont.*)

	S	B	N	M	P	E	R	L	B	B	S	N	B	K	K	K	M	T	B	A
Batrachostomus stellatus							★	★			★									★
Batrachostomus javensis				★			★													
Batrachostomus cornutus								★	★											
Eurostopodus temmincki			★					★	★											
Eurostopodus macrotis	★																			
Caprimulgus indicus							★					★								★
Caprimulgus macrurus							★													
Caprimulgus affinis		★					★	★												
Caprimulgus concretus							★	★	★											
Collocalia fuciphaga	★	★									★									★
Collocalia maxima	★	★												★						
Collocalia salangana		★																		
Collocalia esculenta	★	★					★		★		★									★
Collocalia linchi										★										
Hirundapus giganteus							★	★				★								
Rhaphidura leucopygialis										★	★	★								
Apus pacificus				★			★	★												
Apus affinis				★			★	★												
Cypsiurus balasiensis						★														
Hemiprocne longipennis	★	★					★	★	★	★	★									
Hemiprocne comata		★					★	★	★	★										
Harpactes diardii					★														★	
Harpactes duvaucelii		★	★				★	★			★									
Harpactes oreskios	★	★									★									
Alcedo atthis	★	★					★	★	★		★									

LAND BIRDS FOUND ON OFFSHORE ISLAND GROUPS 427

	S	B	N	B	M	P	E	R	L	B	B	S	N	B	K	K	K	M	T	B	A
Alcedo meninting	★	★	★	★	★	★	★	★	★	★	★				★						
Alcedo coerulescens	★	★	★	★	★	★	★	★	★	★	★				★						
Ceyx erithacus/rufidorsa	★	★	★	★	★	★		★	★	★	★			★	★						
Pelargopsis capensis			★	★	★	★				★	★										
Lacedo pulchella					★				★	★	★										
Halcyon coromanda	★	★	★	★	★	★		★	★	★	★										
Halcyon smyrnensis					★	★		★	★	★	★										
Halcyon pileata	★	★	★	★	★	★		★	★	★	★										
Todirhamphus chloris	★	★	★	★	★	★	★	★	★	★	★			★	★	★			★		
Todirhamphus sanctus	★	★	★	★				★	★	★	★								★		
Actenoides concretus					★					★	★										
Merops leschenaulti	★																				
Merops philippinus	★	★	★	★	★	★		★	★	★	★	★		★	★			★			
Merops viridis		★							★	★	★						★				
Nyctyornis amictus										★	★										
Eurystomus orientalis	★	★	★	★	★	★	★	★	★	★	★		★	★	★						
Anorrhinus galeritus				★			★	★					★	★							
Aceros corrugatus								★	★												
Aceros undulatus								★	★												
Anthracoceros malayanus									★	★	★										
Anthracoceros albirostris			★	★	★	★				★	★									★	
Megalaima rafflesii										★	★										
Megalaima mystacophanos				★																	
Megalaima australis			★	★						★	?										
Megalaima haemacephala																					
Sasia abnormis			★	★						★	★										
Celeus brachyurus			★	★							★		★								

Appendix 3 (*cont.*)

	S	B	N	B	M	P	E	R	L	B	B	S	N	B	K	K	K	M	T	B	A
Picus vittatus															★						
Picus paniceus		★																			
Picus mentalis																					
Picus miniaceus				★												★					
Dinopium javanense						★															
Dinopium rafflesi							★														
Meiglyptes tristis			★						★												
Meiglyptes tukki	★	★							★	★											
Mulleripicus pulverulentus							★	★	★	★	★										
Dryocopus javensis	★						★	★	★	★	★										
Dendrocopus canicapillus	★	★					★	★													
Dendrocopus moluccensis							★	★	★			★									
Hemicircus concretus					★		★	★													
Reinwardtipicus validus							★	★	★	★			★								
Chrysocolaptes lucidus							★	★	★	★											
Corydon sumatranus										★	★		★								
Cymbirhynchus macrorhynchus	★						★	★	★	★	★										
Eurylaimus javanicus		★					★	★	★	★	★										
Eurylaimus ochromalus							★	★		★	★										
Calyptomena viridis						★			★	★	★										
Pitta moluccensis		★	★				★		★	★	★										
Pitta megarhyncha		★	★					★	★												
Pitta sordida			★						★	★	★										
Hirundo rustica	★	★	★				★	★	★	★	★	★	★	?	★	★	★	★	★	★	★
Hirundo tahitica	★	★	★				★	★	★	★	★	★	★	★	★	★	★	★	★	★	★

LAND BIRDS FOUND ON OFFSHORE ISLAND GROUPS

	S	B	N	B	M	P	E	R	L	B	B	S	N	B	K	K	K	M	T	B	A
Delichon dasypus					★																
Hemipus hirundinaceus	★	★					★	★	★	★	★			★					★	★	
Coracina striata	★	★			★	★													★	★	★
Coracina fimbriata	★					★															
Lalage nigra		★							★	★	★										
Pericrocotus divaricatus														?							
Pericrocotus igneus	★								★	★	★										
Pericrocotus flammeus	★						★		★	★	★							★			
Aegithina viridissima			★						★	★	★	★	?	★	★						
Aegithina tiphia			★				★		★	★	★										
Chloropsis sonnerati			★							★	★										
Chloropsis cochinchinensis				★																	
Pycnonotus zeylanicus										★	★										
Pycnonotus aurigaster																					
Pycnonotus eutilotus										★	★										
Pycnonotus melanoleucos						★	★														
Pycnonotus atriceps	★					★	★		★	★	★	★	★	★		★		★			
Pycnonotus cyaniventris						★	★														
Pycnonotus goiavier		★	★		★	★	★		★	★	★			★	★		★				
Pycnonotus plumosus		★	★		★	★	★		★	★	★		★			★					
Pycnonotus simplex							★		★	★	★										
Pycnonotus brunneus	★	★					★		★	★	★						★				
Pycnonotus erythrophthalmos									★	★											
Alophoixus phaeocephalus							★		★	★								★	★		
Setornis criniger								★	★												
Tricholestes criniger								★	★	★									★	★	
Iole olivacea								★	★	★									★	★	★

Appendix 3 (*cont.*)

	S	B	N	B	M	P	E	R	L	B	B	S	N	B	K	K	M	T	B	A
Ixos malaccensis							★	★												
Dicrurus leucophaeus	★	★	★																	
Dicrurus sumatranus/hottentottus					★	★														
Dicrurus paradiseus	★	★			★	★														
Oriolus xanthonotus					★															
Oriolus chinensis	★		★	★	★	★	★	★												
Oriolus xanthornus							★	★												
Irena puella		★					★	★						★						
Platysmurus leucopterus							★	★												
Corvus enca	★		★		★															
Corvus splendens											★	★			?					
Corvus macrorhynchos					★		★				★	★				★	★			
Sitta frontalis	★																			
Pomatorhinus montanus									★	★	★	★								
Pellorneum capistratum		★							★	★	★							★	★	
Malacocincla malaccensis					★				★	★									★	★
Trichastoma rostratum									★	★										
Trichastoma bicolor										★										
Malacocincla abbotti									★	★			★							
Malacopteron magnirostre	★									★									★	★
Malacopteron affine			★																	
Malacopteron cinereum			★						★	★							★			
Malacopteron albogulare									★											
Stachyris poliocephala									★											
Stachyris maculata			★					★												

Species	S	B	N	B	M	P	E	R	L	B	B	S	N	B	K	K	K	M	T	B	A
Stachyris erythroptera	★																				
Stachyris melanothorax																					★
Macronous flavicollis																		★			★
Macronous gularis		★			★			★							★				?		
Macronous ptilosus		★	★			★															
Alcippe brunneicauda			★																		
Copsychus saularis	★	★	★	★	★	★		★	★	★	★	★			★	★					
Copsychus malabaricus	★	★	★		★	★		★	★	★	★	★	★	★		★		★			
Enicurus leschenaulti			★	★	★																
Monticola solitarius									★		★	★									
Zoothera interpres							★														
Zoothera andromedae							★														
Zoothera sibirica		★																			
Turdus obscurus									★	★		★									
Gerygone sulphurea			★						★	★			★								
Phylloscopus borealis								★				★									
Acrocephalus orientalis	★											★	★								
Locustella certhiola	★																				
Orthotomus atrogularis		★	★		★			★	★	★	★	★		★							
Orthotomus ruficeps								★	★	★		★		★							
Orthotomus sericeus												★									
Prinia flaviventris		★	★					★	★	★	★		★	★							
Cisticola juncidis	★																?				
Rhinomyias olivacea												★		★							
Rhinomyias umbratilis			★									★		?							
Ficedula zanthopygia										★	★									★	
Muscicapa dauurica					★					★	★	★			★					★	

Appendix 3 (*cont.*)

	S	B	N	M	P	E	R	L	B	B	S	N	B	K	K	K	M	T	B	A
Cyanoptila cyanomelana											★			★						
Cyornis rufigaster														★	★	★	★	★	★	★
Culicicapa ceylonensis	★	★	★	★	★		★	★		★										
Rhipidura javanica																		★	★	★
Hypothymis azurea	★	★	★	★	★		★	★	★	★				★	★	★	★	★	★	★
Philentoma pyrhopterum			★						★	★										
Terpsiphone paradisi	★	★	★	★	★		★	★	★	★										
Pachycephala grisola	★	★	★	★	★	★	★		★	★	★	★	★						★	
Motacilla cinerea	★	★	★								★	★								
Motacilla flava	★	★	★	★	★			★			★	★			★		★			
Dendronanthus indicus	★	★	★	★																
Anthus richardi								★	★											
Artamus leucorhynchus									★	★	★	★		★	★			★		
Lanius cristatus												★		★	★					
Lanius tigrinus	★	★	★					★												
Lanius schach													★	★						
Aplonis panayensis	★	★	★	★	★	★	★	★	★	★		★		★	★	★	★	★	★	★
Sturnus sinensis	★	★		★																
Acridotheres javanicus														?						
Gracula religiosa	★	★	★	★	★	★	★	★	★	★	★	★	★	★	★	★	★	★	★	★
Anthreptes simplex		★										★								
Anthreptes malacensis	★	★	★	★	★	★	★	★	★	★	★	★	★	★	★	★	★	★	★	★
Anthreptes singalensis		★	★						★	★									★	★
Hypogramma hypogrammicum												★								
Nectarinia sperata	★	★	★	★	★		★	★			★	★								★

LAND BIRDS FOUND ON OFFSHORE ISLAND GROUPS

Species	S	B	N	M	P	E	R	L	B	B	S	N	B	K	K	M	T	B	A
Nectarinia calcostetha	★	★	★	★	★	★	★	★	★	★	★	★	★	★	★	★	★		
Nectarinia jugularis		★	★	★	★	★	★	★	★	★	★	★	★					★	★
Aethopyga siparaja		★	★	★	★		★	★	?							★	★	★	★
Aethopyga temminckii									?	★									
Arachnothera longirostra				★	★		★		★	★									
Arachnothera chrysogenys				★	★		★												
Arachnothera affinis					★														
Prionochilus thoracicus							★		★	★									
Prionochilus maculatus			★							★									
Prionochilus xanthopygius																	★		
Prionochilus percussus					★			★		★	★	★							
Dicaeum agile/everetti					★			★			★			★					
Dicaeum trigonostigma	★			★	★			★		★	★	★							
Dicaeum cruentatum				★	★	★										★	★		
Dicaeum trochileum	★											★							
Zosterops palpebrosa/salvadorii									★		★				★				
Zosterops chloris										★				★					
Passer montanus								★											
Ploceus philippinus			★											★	?				
Ploceus manyar													★	?					
Padda oryzivora									★										
Lonchura striata												★							
Lonchura fuscans																			
Lonchura molucca							★												
Lonchura punctulata									★				★	★	?	★		★	
Lonchura malacca									★	★									
Lonchura maja					★						★	?							

Appendix 4
Bornean montane birds by mountain group

(K = Kinabalu, T = Trus Madi, M = Murud, M = Mulu, K = Kelabits, B = Brassey Range (including Magdalena, Maliau, and Danum/Segama), U = Usun Apau, B = Batu Song, D = Dulit, H = Hose Mountains, L = Lanjak Entimau, L = Liang Kubung, M = Mueller Range, P = Niut/Penrissen, P = Poi Mountains, S = Schwaner Range, P = Palung, K = Kayan Mentarang.)

	K	T	M	M	K	B	U	B	D	H	L	L	M	P	P	S	P	K
Spilornis kinabaluensis	★	★																
Spizaetus alboniger	★	★			★									★	★			★
Falco peregrinus	★	★			★													
Arborophila hyperythra	★	★					?					★						
Caloperdix oculea				★	★			★	★									
Haematortyx sanguiniceps	★		★	★	★	★	★	★	★	★	★	★						
Lophura bulweri			★	★	★	★	?	★	★	★	★	★	★		★			
Ducula badia	★	★	★	★	★	★	★	★	★	★	★	★	★	★	★			★
Macropygia emiliana	★	★	★	★	★	★		★	★	★	★	★	★	★	★			
Macropygia ruficeps	★	★	★	★	★	★	★	★	★	★	★	★	★	★	★			★
Cuculus sparverioides	★	★	★	★	★	★	★	★	★	★	★	★	★	★	★	★		
Cuculus saturatus	★	★	?						★	★	★	★						
Otus rufescens	★		★								★			★		★	★	★
Otus brookii			★					★										
Glaucidium brodiei	★			★		★					★							

Bornean Montane Birds by Mountain Group

	K	T	M	M	K	B	U	B	D	H	L	L	M	P	P	S	P	K
Batrachostomus harterti	?																	
Batrachostomus poliolophus		?																
Harpactes whiteheadi	★	★			★													
Harpactes orrhophaeus	★	★				★												
Harpactes oreskios	★	★				★												
Aceros undulatus	★	★	★		★	★												
Megalaima monticola	★	★	★	★	★	★	★								★		★	★
Megalaima pulcherrima	★	★	★	★			★											★
Megalaima eximia	★	★	★	★	★	★	★											
Psarisomus dalhousiae	★	★	★												★			
Calyptomena hosii	★	★	★		★		★											
Calyptomena whiteheadi	★	★	★			★	★											
Pitta arquata	★	★	★		★	★	★											
Hemipus picatus	★	★	★		★	★	★	★	★	★	★			★				★
Coracina larvata	★	★	★		★	★	★	★	★	★	★	★		★	★		★	★
Pericrocotus solaris	★	★	★		★	★	★	★	★	★	★	★	★	★	★		★	
Pericrocotus flammeus	★	★	★		★	★	★	★					★					
Chloropsis cochinchinensis	★	★	★		★	★	★	★										
Pycnonotus melanicterus	★	★	★		★	★	★	★	★	★	★	★	★	★	★		★	★
Pycnonotus flavescens	★	★	★		★	★	★	★	★	★	★	★		★	★		★	★
Alophoixus ochraceus	★	★	★		★	★	★	★	★	★	★	★		★	★		★	
Hypsipetes flavala	★	★	★				★											
Dicrurus leucophaeus	★	★							★									
Oriolus hosii															★	★		
Oriolus cruentus	★	★	★		★	★	★		★									
Cissa thalassina	★	★	★		★			★										

Appendix 4 (*cont.*)

	K	T	M	M	K	B	U	B	D	H	L	L	M	P	P	S	P	K
Dendrocitta cinerascens	★	★			★	★	★	★	★	★	★	★	★					
Pellorneum pyrrogenys	★	★			★	★	★	★	★	★	★	★	★					
Napothera crassa	★				★		★											
Napothera epilepidota			★		★						★							
Stachyris leucotis					★													
Garrulax palliatus	★				★	★	★	★			★	★	★					
Garrulax lugubris	★	★			★	★	★	★			★							
Garrulax mitratus	★	★			★	★	★	★	★	★	★	★	★					
Pteruthius flaviscapis	★	★			★	★	★		★	★	★							
Alcippe brunneicauda	★				★	★	★	★	★	★	★	★						
Yuhina everetti																★	★	★
Brachypteryx montana	★	★			★		★				★	★	★					
Enicurus leschenaulti	★																	
Chlamydochaera jefferyi	★						★				★						★	★
Myiophonus glaucinus	★										★	★	★					
Zoothera citrina	★				★						★							
Zoothera everetti	★	★			★		★				★	★	★				★	
Turdus poliocephalus	★	★			★													
Seicercus montis	★				★	★	★				★	★	★			★	★	
Abroscopus superciliaris	★				★	★	★											
Phylloscopus trivirgatus	★	★			★	★	★	★	★	★	★	★	★					
Orthotomus cuculatus	★	★	★		★	★	★										?	★
Urosphena whiteheadi	★				★				★		★	★						
Cettia vulcania	★	★	★															
Bradypterus accentor	★																	

	K	T	M	M	K	B	U	B	D	H	L	L	M	P	P	S	P	K
Rhinomyias ruficauda	★	★	★	★	★	★							★					
Rhinomyias gularis	★	★	★	★	★													
Eumyias indigo	★	★	★	★	★		★											
Ficedula hyperythra	★	★	★	★	★													
Ficedula westermanni	★	★	★	★	★													
Cyornis superbus	★				★													
Muscicapella hodgsoni	★		★	★														★
Rhipidura albicollis	★	★	★	★	★	★												
Pachycephala hypoxantha	★	★	★	★	★	★	★	★	★	★	★	★	★	★	★			
Aethopyga mystacalis	★	★	★	★	★	★	★	★	★	★	★						★	
Arachnothera everetti	★	?	?	?	★	★		★										
Arachnothera juliae	★	★	★	★	★		★		★									
Dicaeum monticolum	★	★	★	★	★	★	★	★	★	★	★	★	?	★	★			
Zosterops atricapilla	★	★	★	★	★	★		★										
Zosterops everetti	★	★	★	★	★	★	★	★	★	★	★	★		★	★		★	
Oculocincta squamifrons	★	★	★		★									★	★			
Chlorocharis emiliae	★					★												
Erythrura hyperythra	★				★													

Appendix 5
Annotated list of birds of the Malay Peninsula not described in the text

Painted Stork   *Mycteria leucocephalus* (vagrant)
Like Milky Stork, but black wing coverts.

Red-headed Vulture   *Sarcogyps calvus* (marginal)
Black vulture, with bare red head.

Long-billed Vulture   *Gyps indicus* (marginal)
Brown vulture, without white rump.

Malay Peacock-Pheasant   *Polyplectron malacense* (endemic)
Like Bornean Peacock-Pheasant.

Crested Argus   *Rheinardia ocellata*
Like Great Argus, but longer crest and white eyebrow.

Sarus Crane   *Grus antigone* (vagrant)
Distinctive large crane.

Wood Snipe   *Gallinago nemoricola* (visitor)
Large snipe without white trailing edge to wing.

Crab Plover   *Dromas ardeola* (vagrant)
Distinctive pied wader with stout black bill.

Yellow-vented Green-Pigeon   *Treron seimundi*
Like Sumatran Green-Pigeon, but with greener breast and white vent.

Red Collared-Dove   *Streptopelia tranquebarica*
Like Island Collared-Dove, but darker and redder.

Blossom-headed Parakeet   *Psittacula roseata* (marginal)
Parakeet with pink or purple head.

Dusky Eagle Owl   *Bubo coromandus*
Like Buffy Fish-Owl, but greyer and with feathered tarsus.

Brown Fish-Owl   *Ketupa zeylonicus*
Like Buffy Fish-Owl, but lacks white patch above base of bill.

Brown-winged Kingfisher   *Pelargopsis amauroptera*
Like Stork-billed Kingfisher, but brown wings and tail.

Golden-throated Barbet   *Megalaima franklinii*
Like Black-browed Barbet, but white instead of blue on face.

Streak-breasted Woodpecker   *Picus viridanus*
Like Laced Woodpecker, but lacing extends over breast.

Bamboo Woodpecker   *Gecinulus viridis*
Green above, uniform buff cheek, red rump, male red cap.

Bay Woodpecker   *Blythipicus pyrrhotis*
Like Maroon Woodpecker, but larger and with barred upperparts.

Gurney's Pitta   *Pitta gurneyi* (endemic to Kra Isthmus area)
Blue crown, orange-barred underparts.

Rosy Minivet   *Pericrocotus roseus* (visitor)
Grey minivet, male pinkish, only wing bar vivid.

Great Iora   *Aegithinia lafresnayei*
Larger than Common Iora and no wing bars.

Orange-bellied Leafbird   *Chloropsis hardwickii*
Leafbird with orange-yellow belly.

Stripe-throated Bulbul   *Pycnonotus finlaysoni*
Green bulbul with yellow streaks on face and throat.

Puff-throated Babbler   *Pellorneum ruficeps*
Rufous cap, white throat, bold dark streaks on breast.

Large Scimitar-Babbler   *Pomatorhinus hypoleucos*
Brown bill, necklace of white streaks from eyebrow to flanks.

Streaked Wren-Babbler   *Napothera brevicaudata*
   Brown wren-babbler with white streaks on throat.

Chestnut-crowned Laughingthrush   *Garrulax erythrocephalus* (marginal)
   Rufous cap and golden-olive wings and tail.

Cutia   *Cutia nipalensis*
   Grey crown, black mask, white unders, black-scaled flanks.

Black-eared Shrike-Babbler   *Pteruthius melanotis*
   Like Chestnut-fronted Shrike-Babbler, with black patch on ear coverts.

Rufous-winged Fulvetta   *Alcippe castaneceps*
   Bold pied pattern on side of head, streaky rufous crown.

Grey-cheeked Fulvetta   *Alcippe morrisonia*
   Fulvetta with white eye-ring and black supercilium.

Brown-cheeked Fulvetta   *Alcippe poioicephala*
   Like Brown Fulvetta with buff unders.

Chestnut-tailed Minla   *Minla strigula*
   Black-scaled white throat and red/black on wings and tail.

Blue-winged Minla   *Minla cyanouroptera*
   White eyebrow, bluish on crown, primaries, and tail.

White-hooded Babbler   *Gampsorhynchus rufulus*
   Striking brown babbler with white head.

Rufous-headed Robin   *Luscinia ruficeps* (vagrant)
   Rufous head, white throat, black band on breast.

White-tailed Robin   *Cinclidium leucurum*
   Blue-black male with two white flashes in tail.

Slaty-backed Forktail   *Enicurus schistaceus*
   Like White-crowned Forktail, but grey cap and nape, and dark crown.

White-throated Rock-Thrush   *Monticola gularis* (visitor)
   Male has red unders with white throat patch.

Malayan Whistling-Thrush *Myiophonus robinsoni* (endemic)
Like Blue Whistling-Thrush, but shorter tail, spangles only on breast.

Rufescent Prinia *Prinia rufescens*
Like Yellow-bellied Prinia, but duller and more rufous on back.

Golden-crested Myna *Ampeliceps coronatus*
Black myna with golden crest.

Black-throated Sunbird *Aethopyga saturata*
Male with long tail, black throat, white rump, maroon back.

Streaked Spiderhunter *Arachnothera magna*
Large, streaky spiderhunter with orange legs.

Plain-backed Sparrow *Passer flaveolus*
Like Eurasian Tree Sparrow, but cheeks buff and cinnamon.

Brown Bullfinch *Pyrrhula nipalensis*
Brownish bullfinch with white rump.

Appendix 6
Sonosketches of characteristic bird calls (not to scale)

Pheasants and Pigeons
125. *Rhizothera longirostris*
144. *Polyplectron schleiermacheri*
143. *Polyplectron chalcurum*
141. *Gallus gallus*
145. *Argusianus argus* (male)
145. *Argusianus argus* (both)
264. *Ducula aenea*
266. *Ducula badia*
274. *Macropygia emiliana*
275. *Macropygia ruficeps*
286. *Streptopelia bitorquata*
277. *Streptopelia chinensis*
278. *Geopelia striata*
279. *Chalcophaps indica*

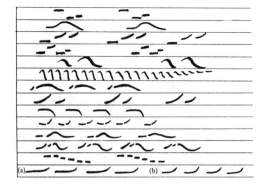

Cuckoos
291. *Cuculus sparverioides*
293. *Cuculus fugax*
294. *Cuculus micropterus*
295. *Cuculus canorus*
296. *Cuculus saturatus*
297. *Cacomantis sonneratii*
298. *Cacomantis merulinus*
299. *Cacomantis sepulcralis*
305. *Surniculus lugubris*
306. *Eudynamys scolopacea*
310. *Phaenicophaeus chlorophaeus*
315. *Centropus sinensis*

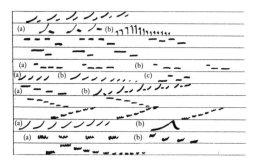

SONOSKETCHES OF CHARACTERISTIC BIRD CALLS 443

Owls, Trogons, and others
319. *Phodilus badius*
322. *Otus spilocephalus*
321. *Otus rufescens*
325. *Otus sunia*
330. *Otus lempiji*
334. *Glaucidium brodiei*
335. *Glaucidium castanopterum*
337. *Strix seloputo*
336. *Ninox scutulata*
344. *Batrachostomus javensis*
346. *Eurostopodus temminckii*
349. *Caprimulgus macrurus*
370. *Harpactes kasumba*
371. *Harpactes diardii*
374. *Harpactes duvaucelii*
375. *Harpactes oreskios*
384. *Lacedo pulchella*
398. *Upupa epops*

Barbets
409. *Psilopogon pyrolophus*
410. *Megalaima lineata*
411. *Megalaima corvina*
412. *Megalaima chrysopogon*
413. *Megalaima rafflesii*
414. *Megalaima mystacophanos*
415. *Megalaima javensis*
416. *Megalaima oorti*
417. *Megalaima monticola*
418. *Megalaima henricii*
419. *Megalaima armillaris*
420. *Megalaima pulcherrima*
421. *Megalaima australis*
422. *Megalaima eximia*
423. *Megalaima haemacephala*

Woodpeckers and Honeyguide
425. *Indicator archipelagicus*
426. *Picumnus innominatus*
428. *Celeus brachyurus*
432. *Picus puniceus*
433. *Picus chlorolophus*
435. *Picus miniaceus*
439. *Meiglyptes tukki*
440. *Mulleripicus pulverulentus*
441. *Dryocopus javensis*

444 APPENDIX 6

Broadbills
449. *Corydon sumatranus*
451. *Eurylaimus javanicus*
452. *Eurylaimus ochromalus*
455. *Calyptomena viridis*

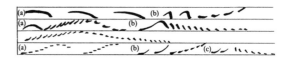

Pittas
466. *Pitta granatina*
462. *Pitta moluccensis*
464. *Pitta megarhyncha*
468. *Pitta sordida*
469. *Pitta guajana*

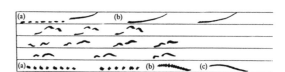

Orioles and others
480. *Tephrodornis gularis*
494. *Aegithina tiphia*
496. *Chloropsis sonnerati*
498. *Chloropsis cochinchinensis*
537. *Oriolus xanthonotus*
538. *Oriolus chinensis*
549. *Platysmurus leucopterus*
752. *Gracula religiosa*
542. *Irena puella*

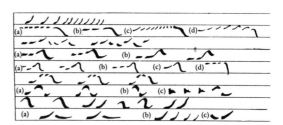

Babblers
559. *Pellorneum capistratum*
564. *Malacocincla malaccensis*
562. *Trichastoma rostratum*
563. *Trichastoma bicolor*
565. *Malacocincla sepiarium*
566. *Malacocincla abbotti*
569. *Malacopteron magnirostre*
570. *Malacopteron affine*
571. *Malacopteron cinereum*
572. *Malacopteron magnum*
574. *Pomatorhinus montanus*
580. *Napothera macrodactyla*
583. *Napothera epilepidota*
584. *Pnoepyga pusilla*
585. *Stachyris rufifrons*
586. *Stachyris chrysaea*
589. *Stachyris poliocephala*
591. *Stachyris maculata*
595. *Stachyris erythroptera*
598. *Macronous gularis*
605. *Garrulax mitratus*
606. *Leiothrix argentauris*
607. *Pteruthius flaviscapis*
609. *Alcippe brunneicauda*
615. *Eupetes macrocerus*

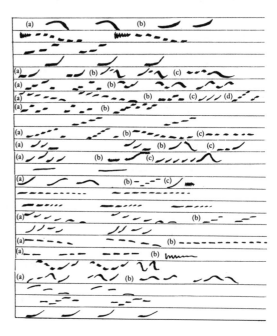

SONOSKETCHES OF CHARACTERISTIC BIRD CALLS 445

Shortwings and Warblers
616. *Brachypteryx leucophrys*
617. *Brachypteryx montana*
647. *Gerygone sulphurea*
663. *Orthotomus sutorius*
664. *Orthotomus atrogularis*
665. *Orthotomus ruficeps*
668. *Orthotomus cuculatus*
671. *Prinia flaviventris*
678. *Cettia vulcania*

Flycatchers and Whistler
684. *Rhinomyias ruficauda*
701. *Niltava grandis*
703. *Cyornis concretus*
706. *Cyornis banyumas*
710. *Cyornis tickelliae*
717. *Rhipidura perlata*
718. *Rhipidura javanica*
719. *Hypothymis azurea*
720. *Philentoma velatum*
721. *Philentoma pyrhopterum*
723. *Terpsiphone paradisi*
725. *Pachycephala grisola*

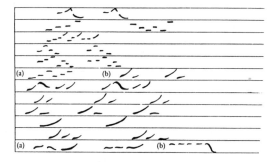

Appendix 7
Regional ornithological clubs, journals, and museums

The following organizations are all concerned with the study and conservation of the birds of the Greater Sundas.

The International Council for Bird Preservation (ICBP, now BirdLife International) studies and monitors bird populations worldwide, with particular interest in conservation. ICBP publishes many useful books and lists including the *Birds to watch* world checklist of threatened birds. ICBP acts as the Bird Specialist group of IUCN (World Conservation Union). Address: Wellbrook Court, Girton Road, Cambridge CB3 0NA, UK.

The Asian Wetland Bureau (AWB) is an independent international body based in Kuala Lumpur, engaged in surveys, monitoring, and consultancy work relating to the promotion of conservation and sustained use of wetlands in SE Asia. They have a branch office in Bogor, Indonesia. AWB publishes a Regional newsletter, *Asian Wetland News*. Address: c/o Institute of Advanced Studies, University of Malaya, 59100 Kuala Lumpur, Malaysia.

The Oriental Bird Club promotes interest in and conservation of the birds of the Oriental region. They sometimes organize special tours to parts of the region. The club produces an annual journal *Forktail* and a biennial bulletin. Address: c/o The Lodge, Sandy, Bedfordshire SG19 2DL, UK.

The World Wide Fund for Nature (WWF) is an international non-government organization working in Malaysia and Indonesia to assist the governments in conservation and sustainable use of living resources. A national organization is established in Malaysia. Address: 3rd Floor Wisma IJM Annexe, Jalan Yong Shook Lin, Locked bag No. 911, 46990 Petaling Jaya, Malaysia. There is a Representative's Office in Indonesia. Address: Jl. Tampak Siring 1b, Cipete, Jakarta Selatan, Indonesia.

The Malayan Nature Society is concerned with wildlife in Malaysia, including Sarawak and Sabah, and the broader region. The society holds regular meetings and publishes a quarterly journal *The Malayan Nature Journal*, as well as bird checklists and an annual Bird Report which relates the year's ringing data and other Malaysian bird findings. Address: PO Box 10750, 50724 Kuala Lumpur, Malaysia.

The Indonesian Ornithological Society promotes interest in and study of birds in Indonesia. It has become the main records screening authority in the country. Its regular journal *Kukila* publishes most of the new ornithological findings. The society has recently commissioned the publication of a new Indonesian checklist. Address: PO Box 4087, Jakarta 4087, Indonesia.

The Bali Bird Society is a small group of enthusiasts concentrating on birdwatching on Bali. Address: PO Box 400, Denpasar, Bali, Indonesia.

In addition to the latter two purely ornithological societies, Indonesia has a very large number of more general nature-lover groups and several bird-fancier clubs.

Museums and collections

The main ornithological collection for Indonesia is in the **Museum Bogoriense** situated in the Botanical Gardens of Bogor. The museum is managed by the Indonesian Institute for Sciences (LIPI) and has an excellent collection of bird specimens from all parts of the archipelago. The staff are currently engaged on the major task of preparing a list of vernacular names for all Indonesian species.

The **Sarawak Museum** in Kuching has a fine collection of Borneo bird specimens and publishes the famous *Sarawak Museum Journal*. The museum has for many years sponsored expeditions and studies over all of Northern Borneo.

The **Sabah Museum** is relatively new but has already established itself as a respected and important centre and has amassed a good representative collection of bird skins for the state.

The **National Museum** in Singapore has an excellent bird collection from the whole region and publishes a journal and special publications.

Important overseas collections of skins from the region are contained in the Natural History Museum, London and Tring, UK; The Smithsonian Institution, Washington, DC, USA; The Rijksmuseum van Natuurlijke Historie, Leiden, Netherlands, and the Musée d'Histoire Naturelle, Paris, France.

Bibliography

Allport, G. and Milton, G. R. (1988). A note on the recent sighting of *Zosterops flava* Javan White-eye. *Kukila*, **3**, 142–9.
Anderson, J. A. R., Jermy, A. C., and the Earl of Cranbrook (1982). *Gunung Mulu National Park, Sarawak—A management and development plan*. Appendix VI (Birds). Royal Geographical Society, London.
Andrew, P. (1985). A Sabine's Gull *Larus sabini* off the coast of Sumatra. *Kukila*, **2**, 9.
Andrew, P. (1985). An annotated checklist of the birds of the Cibodas-Gunung Gede Nature Reserve. *Kukila*, **2**, 10–27.
Andrew, P. (1990). The status of the Sunda Coucal *Centropus nigrorufus* Cuvier. *Kukila*, **5**, 56–61.
Andrew, P. and Milton, G. R. (1988). A note on the Javan Scops-Owl *Otus angelinae* Finsch. *Kukila*, **3**, 79–81.
Ash, J. S. (1984). Bird observations on Bali. *Bull. Brit. Orn. Club*, **104**, 24–35.
Balen, S. van (1984). Sight records of the Black baza *Aviceda leuphotes* on Java. *Ardea*, **72**, 234.
Balen, S. van (1986). Short note on the occurrence of grey phase Black-headed Bulbuls *Pycnonotus atriceps* especially on Java. *Kukila*, **2**, 86–7.
Balen, S. van (1987). Some biological notes on the Tawny-breasted Parrotfinch *Erythrura hyperythra*. *Kukila*, **3**, 54–7.
Balen, S. van (1991). Jaegers in Indonesian waters. *Kukila*, **5**, 117–24.
Balen, S. van (1991). Faunistic notes from Bali, with some new records. *Kukila*, **5**, 125–32.
Balen, S. van and Compost, A. R. (1989). Overlooked evidence of the Short-toed Eagle *Circaetus gallicus* on Java. *Kukila*, **4**, 44–6.
Balen, S. van and Lewis, A. (1991). Blue-crowned Hanging-Parrot on Java. *Kukila*, **5**, 140–1.
Balen, S. van and Marhadi, A. (1989). A breeding record of the Pink-necked Fruit Dove *Ptilinopus porphyreus* in Java. *Kukila*, **4**, 144–5.
Balen, S. van and Noske, R. (1991). Note on two sight records of the Yellow-rumped Flycatcher on Bali. *Kukila*, **5**, 142.
Balen, S. van, Margawati, E. T., and Sudaryanti (1988). A checklist of the birds of the Botanical Gardens of Bogor, West Java. *Kukila*, **3**, 82–92.
Banks, E. (1983). Mountain birds and plate tectonics. *Brunei Museum J.*, **5**, 105–7.
Becking, J. H. (1975). New evidence of the specific affinity of *Cuculus lepidus* Muller. *Ibis*, **117**, 275–84.
Beebe, W. (1918–22). A monograph of the Pheasants. 4 vols. London.
Bemmel, A. C. van and Hoogerwerf, A. (1940). The Birds of Goenoeng Api. *Treubia*, **17**, 421.

Bishop, K. D. (1985). Scarlet-headed Flowerpecker *Dicaeum trochileum*: a new bird for Sumatra. *Kukila*, **2**, 7-8.

Bishop, K. D. (1988). The Long-legged Pratincole *Stiltia isabella* in Bali. *Kukila*, **3**, 141.

Bishop, K. D. and Cook, S. (1985). Orange-flanked Bluetails *Tarsiger cyanurus* wintering in south-east Sumatra. *Kukila*, **2**, 7-8.

Bohap bin Jalan and Galdikas, B. M. F. (1987). Birds of Tanjung Puting National Park, Kalimantan Tengah. A preliminary list. *Kukila*, **3**, 33-7.

Bucknill, Sir J. A. S. and Chasen, F. N. (1990). *Birds of Singapore and South-east Asia*. Graham Brash, Singapore.

Burton, K. M. (1987). Unseasonal occurrence of Blue Rock-Thrush *Monticola solitarius* in Sumatra. *Kukila*, **3**, 61.

Chasen, F. N. (1935). A Handlist of Malaysian Birds. *Bull. Raffles Mus.*, **11**, 1-389.

Chasen, F. N. (1937). The Birds of Billiton Island. *Treubia*, **16**, 205-38.

Chasen, F. N. and Hoogerwerf, A. (1941). The birds of the Netherlands-Indian Mt. Leuser Expedition 1937 to North Sumatra. With a general survey, an itinerary, and field notes by A. Hoogerwerf. *Treubia*, **18**, Suppl. 1-125.

Chasen, F. N. and Kloss, C. B. (1926). Spolia Mentawiensis—Birds. *Ibis* 269-306.

Collar, N. J. and Andrew, P. (1988). *Birds to watch*, ICBP Technical Publication No. 8. Cambridge.

Compost, A. R. and Milton, G. R. (1986). An early arrival of the Malayan Night Heron *Gorsachius melanolophus* in Java. *Kukila*, **2**, 88-9.

Dammerman, K. W. (1926). The fauna of Durian and Rhio-Lingga Archipelago. *Treubia*, **8**, 281-326.

Danielsen, F., Purwoko, A., Silvius, M. J., Skov, H., and Verheugt, W. (1991). Breeding colonies of Milky Stork in South Sumatra. *Kukila*, **5**, 133-5.

Danielsen, F., Skov, H., and Suwarman, U. (1991). Breeding colonies of waterbirds along the coast of Jambi Province, Sumatra, August 1989. *Kukila*, **5**, 135-7.

Davies, G. and Payne, J. (1982). *A faunal survey of Sabah*. WWF Malaysia, Kuala Lumpur.

Davison, G. W. H. (1979). Alleged occurrence of *Rheinartia ocellata* in Sumatra. *Bull. Brit. Orn. Club*, **99**, 80-1.

Davison, G. W. H. (1979). The status of pheasants in Malaysia and Indonesia. Unofficial Report, pp. 2-7.

Davison, G. W. H. (1982). Systematics within the genus *Arborophila* Hodgson. *Federation Museum Journal*, **27**, 125-33.

De Wulf, R., Supomo, D., and Rauf, K. (1981). Kerinci-Seblat Proposed National Park—Preliminary Management Plan 1982-1987. FAO, FO/INS/78/061 Field Report 14, Bogor.

De Wulf, R., Supomo, D., and Rauf, K. (1981). Barisan Selatan Game Reserve—Management Plan 1982-1987. FAO, FO/INS/78/061 Field Report 21, Bogor.

Delacour, J. (1947). *Birds of Malaysia*. Macmillan, New York.

Delacour, J. (1951). *The pheasants of the world*. Country Life Ltd, London.
Diamond, J. (1991). World of the living dead. *Natural History*, **91**, 30–7.
Dickinson, E. C. (1989). A review of larger Philippine swiftlets of the genus *Collocalia*. *Forktail*, **4**, 19–53.
Erftemeijer, P. (1989). The occurrence of nests of Nankeen Night Heron *Nycticorax caledonicus* in East Java. *Kukila*, **4**, 146–7.
Fogden, M. P. L. (1972). The seasonality and population dynamics of equatorial forest birds in Sarawak. *Ibis*, **114**, 307–43.
Forshaw, J. M. and Cooper, W. (1973). *Parrots of the world*. Doubleday, New York.
Francis, C. M. (1987). The management of edible birds nest caves in Sabah. Wildlife Section, Sabah Forest Department.
Freeman, J. D. (1960). Iban augury. In Smythies' *Birds of Borneo*. Oliver and Boyd, London, pp. 73–97.
Galdikas, M. F. and King, B. (1989). Lesser Adjutant nests in SW Kalimantan. *Kukila*, **4**, 151.
Gibson-Hill, C. A. (1952). A revised list of the birds known from the Rhio-Lingga Archipelago. *Bull. Raffles Mus.*, **24**, 344–79.
Giersen, W. (1992). *Preliminary management plan of Danau Sentarum, W. Kalimantan*. WWF/PHPA, Bogor.
Glenister, A. G. (1955). *The Birds of the Malay Peninsula, Singapore, Penang*. Oxford University Press, Oxford.
Goodwin, D. (1967). *Pigeons and doves of the world*. British Museum, London.
Gore, M. E. J. (1968). A check-list of the birds of Sabah, Borneo. *Ibis*, **110**, 165–96.
Green, C. A. (1991). First sight record of Yellow-vented Flowerpecker for Bali. *Kukila*, **5**, 143–4.
Harrison, C. J. O. (1985). A re-assessment of the affinities of some small Oriental babblers (Timaliidae). *Forktail*, **1**, 81–3.
Harrisson, T. (1960). Birds and Men in Borneo. In Smythies' *Birds of Borneo*. Oliver and Boyd, London, pp. 21–60.
Harvey, W. G. and Holmes, D. A. (1976). Additions to the avifaunas of Sumatra and Kalimantan, Indonesia. *Bull. Brit. Orn. Club*, **96**, 90–2.
Heinroth, O. (1938). The pairing display of Bulwer's Pheasant. *J. Ornith.*, **86**, 1–4.
Hellebrekers, W. Ph. J. and Hoogerwerf, A. (1967). A further contribution to our zoological knowledge of the Island of Java (Indonesia). *Zool. Verhandelingen*, **88**, 164.
Helvoort, B. van (1981). *Bird populations in the rural ecosystems of West Java*. Nature Conservation Dept., Univ. of Wageningen.
Helvoort, B. E. van (1985). A breeding record of Great Thick-knee *Esacus magnirostris* on Bali. *Kukila*, **2**, 68.
Helvoort, B. E. van and M. N. Soetawijaya (1987). First sighting of Black-crested Bulbul *Pycnonotus melanicterus* on Bali. *Kukila*, **3**, 52–3.

Holmes, D. A. (1969). Bird notes from Brunei: December, 1967–September 1968. *Sarawak Mus. J.*, **17**, 399–402.
Holmes, D. A. (1977). Faunistic notes and further additions to the Sumatran avifauna. *Bull. Brit. Orn. Club*, **97**, 68–71.
Holmes, D. A. (1989). Status report on Indonesian Galliformes. *Kukila*, **4**, 133–43.
Holmes, D. A. (1990). Note on the occurrence of the White-winged Wood Duck *Cairina scutulata* on the west coast of North Sumatra. *Kukila*, **5**, 69–72.
Holmes, D. A. (1991). Note on the status of the White-shouldered Ibis in Kalimantan. *Kukila*, **5**, 145–7.
Holmes, D. A. and Burton, K. (1987). Recent note on the avifauna of Kalimantan. *Kukila*, **3**, 2–32.
Holmes, D. A. and Nash, S. (1989). *The Birds of Java and Bali*. Oxford University Press, Singapore.
Holmes, D. A. and Nash, S. (1990). *The birds of Sumatra and Kalimantan*. Oxford University Press, Singapore.
Holmes, D. A. and Wall, D. (1989). Letter: *Setornis criniger*, *Malacopteron albogulare* and conservation in Indonesia. *Forktail*, **4**, 123–5.
Hoogerwerf, A. (1942). *Report of a trip to the Nature Reserve Poeloe Doewa, Bantam*. 17th Official Report.
Hoogerwerf, A. (1948). Distribution of Birds in Java. *Treubia*, **19**, 116–27.
Hoogerwerf, A. (1949a). *De Avifauna van Tjibodas en Omgeving*. Koninklijke Plantentuin van Indonesie, Buitenzorg.
Hoogerwerf, A. (1949b). *De Avifauna van de Plantentuin te Buitenzorg*. Koninklijke Plantentuin van Indonesie, Buitenzorg.
Hoogerwerf, A. (1949c). *Een Bijdrege tot de Oologie van het Eiland Java*. Plantentuin van Indonesie, Buitenzorg.
Hoogerwerf, A. (1962a). Some ornithological notes on the smaller islands around Java. *Ardea*, **50**, 180.
Hoogerwerf, A. (1962b). Ornithological notes on the Sunda Strait area and the Karimundjawa, Bawean and Kangean Islands. *Bull. Brit. Orn. Club*, **82**, 147.
Hoogerwerf, A. (1962c). On *Aegithina tiphia* (Linn.) the Common Iora, from Udjung Kulon, Western Java. *Bull. Brit. Orn. Club*, **82**, 15.
Hoogerwerf, A. (1962d). Notes on Indonesian birds with special reference to the avifauna of Java and the surrounding small islands (I, II, and III). *Treubia*, **26**, 11–291.
Hoogerwerf, A. (1966/67). Notes on the island of Bawean (Java Sea) with special reference to the birds. *Nat. Hist. Bull. Siam Soc.*, **21**, 313–40 and **22**, 15–103.
Hoogerwerf, A. (1969). On the ornithology of the Rhino Sanctuary Udjung Kulon in West Java (Indonesia). Reprinted from the *Nat. Hist. Bull. Siam Soc.*, **23**.
Hoogerwerf, A. (1970). *Udjung Kulon: the land of the last Javan Rhinoceros*. E. J. Brill, Leiden.
Hooferwerf, A. and Rengers Hora Siccama, G. F. H. W. (1938). *De Avifauna van Batavia en Omstreken*. E. J. Brill, Leiden.

Hornskov, J. (1987). More birds from Berbak Game Reserve, Sumatra. *Kukila*, **3**, 58–9.

Hurrell, P. (1989). On the rediscovery of Schneider's Pitta in Sumatra. *Kukila*, **4**, 53–6.

Indrawan, M. (1991). A winter flock of Pheasant-tailed Jacanas at Ciamis, West Java. *Kukila*, **5**, 138–9.

Iskandar, J. (1985). Survey on waders, herons, storks, and terns in south-east coast of Sumatra, Indonesia. Unpublished paper presented at Third East-Asian Bird Protection Conference in Japan, 29–31 May, 1985.

Jenkins, D. V. and de Silva, G. S. (1978). An annotated check-list of the Birds of the Mount Kinabalu National Park, Sabah, Malaysia. In Dingley, E. R. (ed.), *Kinabalu. Summit of Borneo*. Sabah Soc., Monogr., pp. 347–402.

Johns, A. D. (1989). Rainforest birds and logging. *Forktail*, **4**, 103–5.

Kemp, A. C. (1978). A review of the hornbills: biology and radiation. *Living Bird*, **17**, 105–36.

Kennerley, P. R. (1989). A record of a *Bradypterus* warbler probably breeding in Bali. *Kukila*, **4**, 155–7.

King, B. F., Woodcock, M. W., and Dickinson, E C. (1975). *A Field Guide to the Birds of South-East Asia*. Collins, London.

Klapste, J. (1984). Occurrence of the Long-billed Dowitcher on Bali, Indonesia, and other observations. *Austral. Bird Watcher*, **10**, 186–95.

Korte, J. de (1984). Status and conservation of seabird colonies in Indonesia. *ICBP Technical Publ.*, **2**, 527–45.

Kuroda, N. (1933). *Non-Passeres: Birds of the Island of Java I*. Kuroda, Tokyo.

Kuroda, N. (1936). *Passeres: Birds of the Island of Java II*. Kuroda, Tokyo.

Lambert, F. and Erftemeijer, P. (1989). The waterbirds of Pulau Rambut, Java. *Kukila*, **4**, 109–18.

Lambert, F. R. and Howes, J. R. (1989). A recent sighting of Salvadori's Pheasant. *Kukila*, **4**, 56–8.

Lekagul, B. and Round, P. D. (1991). *A guide to the birds of Thailand*. Saha karn Bhaet Co. Ltd., Bangkok.

MacKinnon, J. (ed.) (1983). *Tanjung Puting National Park. Management plan for development*. WWF/PPA, Bogor.

MacKinnon, J. (1990). *Field guide to the birds of Java and Bali*. Gadjah Mada University Press, Jogjakarta.

MacKinnon, J. and Artha, M. B. (1981/2). *National conservation plan for Indonesia*. Vol. II (Sumatra), Vol. III (Java and Bali), and Vol. V (Kalimantan). FAO, FO/INS/78/061 Field Reports 39, 36, and 17, Bogor.

MacKinnon, J. and Setiono, D. (1983). *Recommendations for the development of an elephant reserve Padang-Sugihan, Sumatra Selatan Province*. WWF Report, Bogor.

MacKinnon, J. and Warsito (1982). *Gunung Palung Reserve, Kalimantan Barat. Preliminary management plan*. FAO/UNDP Report, Bogor.

MacKinnon, J. and Wind, J. (1980). *Birds of Indonesia*. FAO, FO/INS/78/061 Special Report, Bogor. 67 pp.

Mahfudz, and Marshall, J. T. (1979). *Report of gibbon study and birds in Kalimantan*. PPA, WWF, Laboratorium Fak. Biologi Univ. National.

Mann, C. F. (1987). Notable bird observations from Brunei, Borneo. *Forktail*, 3, 51–6.

Mann, C. F. (1989). More notable bird observations from Brunei, Borneo. *Forktail*, 5, 17–22.

Marle, J. G. van and Voous, K. H. (1988). *The Birds of Sumatra: an annotated check-list*. Brit. Ornithol. Union, Check-list No. 10. British Ornithologists' Union, London.

Marsh, C. W. (1989). *Final Report—Expedition to Maliau Basin, Sabah, April–May 1988*. Yayasan Sabah/WWF No. MY5 126/88.

Marshall, J. T. (1978). Systematics of smaller Asian night birds based on voice (with supplementary gramophone record). *Orn. Monogr. AOU* No. 25.

Mason, V. (1988). The Oriental Pratincole *Glareola maldivarum* in Bali. *Kukila*, 3, 41.

Mason, V. (1990). Note on the Large Hawk-Cuckoo *Cuculus sparverioides* in Bali. *Kukila*, 5, 73.

Mason, V. and Jarvis, F. (1989). *Birds of Bali*. Periplus, Singapore.

Mayr, E. (1938). Notes on a collection of birds from South Borneo. *Bull. Raffles Mus.*, 14, 5–46.

Medway, Lord (1962). The Swiftlets (Collocalia) of Niah Cave, Sarawak. Part 1, Breeding biology. *Ibis*, 104, 45–66.

Medway, Lord (1963). *Pocket checklist of the birds of Malaya and Singapore*. Malay Nature Society, Kuala Lumpur.

Medway, Lord (1966). Field characters as a guide to the specific relations of swiftlets. *Pro. Linn. Soc. Lond.*, 177, 151–72.

Medway, Lord and Wells, D. R. (1976). *The birds of the Malay Peninsula*, Vol. 5. H. F. and G. Witherby, London.

Mees, G. F. (1957). A systematic review of the Indo-Australian Zosteropidae (part I). *Zool. Verhand. Rijksmus. Nat. Hist. Leiden*, 35, 1–204.

Mees, G. F. (1961). A systematic review of the Indo-Australian Zosteropidae (part II). *Zool. Verhand. Rijksmus. Nat. Hist. Leiden*, 50, 1–168.

Mees, G. F. (1971). Systematics and faunistic remarks on birds from Borneo and Java with new records. *Zool. Medel.*, 45, 225–44.

Mees, G. F. (1986). A list of the birds recorded from Bangka Island, Indonesia. *Zool. Verhand.* 232. 1–176.

Mikhola, H. (1986). Barn Owl *Tyto alba* in Bali. *Kukila*, 2, 95.

Milton, G. R. (1985). Notes on the distribution of the Masked Finfoot *Heliopais personata* in Indonesia. *Kukila*, 2, 41–3.

Milton, G. R. and Marhadi, A. (1985). The bird life of the nature reserve Pulau Dua. *Kukila*, 2, 32–41.

Milton, G. R. and Marhadi, A. (1989). An investigation into the market-netting of birds in West Java, Indonesia. WWF Report, Bogor.

Mitchell, A. H. (1981). *Report on a survey of Palau Simeulue, Aceh, with a proposal for a Suaka Margasatwa*. WWF Report. Bogar.
Nash, A. D. and Nash, S. V. (1985a). Breeding notes on some Padang-Sugihan birds. *Kukila*, 2, 59-63.
Nash, A. D. and Nash, S. V. (1985b). An incidence of duetting between an Abbott's Babbler and a Magpie Robin. *Kukila*, 2, 66.
Nash, A. D. and Nash, S. V. (1985c). Large Frogmouth *Batrachostomus auritus* mobbed by a Greater Racket-tailed Drongo *Dicrurus paradiseus*. *Kukila*, 2, 67.
Nash, A. D. and Stephen, V. (1985). An extreme example of aggression displayed by the Greater Racket-tailed Drongo. *Kukila*, 2, 7.
Nash, S. V. and Nash, A. D. (1985). A checklist of the forest and forest edge birds of the Padang-Sugihan Wildlife Reserve, South Sumatra. *Kukila*, 2, 51-9.
Nash, S. V. and Nash, A. D. (1985). The song of Abbott's Babbler *Trichastoma abbotti* in South Sumatra. *Kukila*, 2, 64-5.
Nash, S. V. and Nash, A. D. (1986). Records of the White-winged Duck *Cairina scutulata* in Sumatran peatswamp forest. *Bull. Oriental Bird Club*, 3, 17.
Nash, S. V. and Nash, A. D. (1986). *Burung-Burung Taman Nasional Tanjung Puting*. WWF Report, Bogor.
Nash, S. V. and Nash, A. D. (1986). *The ecology and natural history of birds in the Tanjung Puting National Park, Central Kalimantan, Indonesia*. WWF Report, Bogor.
Nash, S. V. and Nash, A. D. (1987). An atypical spontaneous song by Abbott's Babbler *Trichastoma abbotti*. *Kukila*, 3, 62-4.
Nash, S. V. and Nash, A. D. (1987). Song variations in the White-chested Babbler *Trichastoma rostratum*. *Kukila*, 3, 65-8.
Nash, S. V. and Nash, A. D. (1987). Synchronized antiphonal duetting by Short-tailed Babblers *Trichastoma malaccense*. *Kukila*, 3, 68-71.
Nash, S. V. and Nash, A. D. (1988). An annotated checklist of the birds of Tanjung Puting National Park, Central Kalimantan. *Kukila*, 3, 93-116.
Nisbet, I. C. T. and Medway, Lord (1972). Dispersion, population ecology, and migration of Eastern Great Reed Warblers *Acrocephalus orientalis* wintering in Malaysia. *Ibis*, 114, 451-94.
Oberholser, H. C. (1917). The birds of Bawean Island, Java Sea. *Proc. US Nat. Mus.*, 52, 183-98.
Oberholser, H. C. (1919). The birds of the Tambelan Islands, South China Sea. *Proc. U.S. Nat. Mus.*, 55, 129-43.
Oberholser, H. C. (1919). Notes on Birds Collected by Dr W. L. Abbott on Pulo Taya, Berhala Strait, Southeastern Sumatra. *Proc. U.S. Nat. Mus.*, 55, 267-74.
Ollington, R. F. and Parish, D. (1989). Lesser Yellowlegs *Tringa flavipes* in Sumatra: new to SE Asia. *Kukila*, 4, 58-61.
Parker, S. A. (1981). Prolegomenon to further studies in the *Chrysococcyx* 'malayanus' group. *Zool. Verhand. Rijksmus. Nat. Hist. Leiden*, 187, 1-56.
Peters, J. L. (1931-51). *Check-list of birds of the world*. 1-7. Continued 8-15; (1960-

1986). Harvard University Press and Museum of Comparative Zoology, Cambridge.
Petersen, S. (1991). A record of White-shouldered Ibis in East Kalimantan. *Kukila*, **5**, 144–5.
Prieme, A. and Heegaard, M. (1988). A visit to Gunung Niut in West Kalimantan. *Kukila*, **3**, 138–40.
Rice, C. G. (1989). A further range extension of the Black-breasted Thrush *Chlamydochaera jefferyi* in Kalimantan. *Kukila*, **4**, 47–8.
Richmond, C. W. (1903). Birds collected by Dr W. L. Abbott on the coast and islands of Northwest Sumatra. *Proc. US Nat. Mus.*, **26**, 485–524.
Riley, J. H. (1930). Birds from the small islands off the northeast coast of Dutch Borneo. *Proc. US Nat. Mus.*, **77**, 1–23.
Ripley, S. D. (1944). The bird fauna of the West Sumatra Islands. *Bull. Mus. Comp. Zool. Harvard Coll.*, **44**, 307–430.
Ripley, S. D. (1977). *Rails of the world*. Fehely, Toronto.
Rozendaal, F. G. (1989). Notes on Asian-Pacific Birds. *Dutch Birding*, **11**, 164–7.
Salomonsen, F. (1983). Revision of the Melanesian Swiftlets and their conspecific forms in the Indo-Australian and Polynesian Region. *K. Dansk Videnskab. Selsk. Biol. Ser.*, **23**, 1–112.
Sheldon, F. H. (1987). Habitat preferences of the Hook-billed Bulbul *Setornis criniger* and the White-throated Babbler *Malacopteron albogulare* in Borneo. *Forktail*, **3**, 17–25.
Shelford, R. W. C. (1900). The birds of Mt. Penrissen and neighbouring districts. *J. Straits Branch Royal. As. Soc.*, **33**, 10–21.
Short, L. L. (1978). Sympatry in woodpeckers of lowland Malayan forest. *Biotropica*, **10**, 122–33.
Sibley, C. G. and Monroe, B. L., Jr. (1990). *Distribution and taxonomy of birds of the world*. Yale University Press, New Haven and London.
Siebers, H. C. (1927). Midden-Oost-Borneo Expeditie 1925. In *Batavia*, pp. 313–87. Indisch Comite Voor Wetenschappeliyke Onderzoekingen. Batavia.
Silvius, M. (1987). Notes on new wader records for Berbak Game Reserve, Sumatra. *Kukila*, **3**, 59–60.
Silvius, M. J. (1988). On the importance of Sumatra's east coast for waterbirds with notes on the Asian Dowitcher *Limnodromus semipalmatus*. *Kukila*, **3**, 117–37.
Silvius, M. J. and de Jongh, H. (1989). White-winged Wood Duck, a new site for Jambi Province. *Kukila*, **49**, 150–51.
Silvius, M. J. and Erftemeijer, P. L. A. (1989). A further revision of the main wintering range of the Asian Dowitcher *Limnodromus semipalmatus*. *Kukila*, **4**, 49–50.
Silvius, M. S. L. and Verheught, W. J. M. (1986). The birds of Barbak Game Reserve, Jambi Province, Sumatra. *Kukila*, **2**, 76–84.
Silvius, M. J. and Verheugt, W. J. M. (1989). The status of storks, ibises, and spoonbills in Indonesia. *Kukila*, **4**, 119–32.

Simons, H. (1987). *Gunung Niut Nature Reserve, proposed management plan*. WWF/PHPA, Bogor.

Simpson, K. and Day, N. (1989). *Field guide to the birds of Australia*. Viking O'Neil Victoria.

Sims, R. W. (1959). The *Ceyx erithacus* and *rufidorsum* problem. *J. Linn. Soc. (Zoology)* XLIV, 296, 212–21.

Slater, P., Slater, P., and Slater, R. (1990). *The Slater field guide to Australian birds*. Weldon, Sydney.

Smith, A. P. (1977). Observations of birds in Brunei. *Sarawak Mus. J.*, **25**, 235–69.

Smythies, B. E. (1960). *The birds of Borneo*. Oliver and Boyd, London.

Smythies, B. E. (1981). *The birds of Borneo* (3rd edn). The Sabah Society and the Malayan Nature Society. Kuala Lumpur.

Sody, H. J. V. (1927). Lijst van Buitenzorg Vogels en Zoogdieren. *Natuurk. Tydschr. Ned. Ind.*, **87**, 181–204.

Somadikarta, S. (1973). The White-breasted Kingfisher *Halcyon smyrnensis* in Java. *Ardea*, **61**, 186–7.

Somadikarta, S. (1982). The White-bellied Swiftlet *Collocalia esculenta* from Java. *Bull. Brit. Orn. Club*, **102**, 18–20.

Somadikarta, S. (1986). *Collocalia linchi*; Horsfield & Moore—a revision. *Bull. Brit. Orn. Club*, **106**, 32–40.

Somadikarta, S. and Holmes, D. A. (1979). An influx of Australian pelicans *Pelecanus conspicillatus* in Indonesia. *Bull. Brit. Orn. Club*, **99**, 154.

Steffee, N. D. (1981). *Field checklist of the birds of Java and Bali*. Russ Mason's Natural History Tours, Kissimmee, Florida.

Strien, N. J. van (1977). *Sumatran birds*. Nature Conservation Dept., Agricultural University, Wageningen, Comm. 157.

Strien, N. J. van (1981). *Birds of Pulau Dua and Pulau Rambut*. School of Environmental Conservation Management, Ciawi, Java.

Tuck, G. and Heinzel, H. (1979). *A field guide to the seabirds of Southern Africa and the world*. Collins, London and Johannesburg.

Vaurie, C. (1949). A revision of the bird family Dicruridae. *Bull. AMNH*, **93**, 199–342.

Vaurie, C. (1951). *Annex. Mus. Novitates*, **1529**, 1–47.

Verheugt, W. (1985). Endangered birds thriving in Sumatra. *Bull. IUCN*, **16**, 12.

Voous, K. H. (1950). The breeding seasons of birds in Indonesia. *Ibis*, **92**, 279–87.

Voous, K. H. (1968). D. J. van Balen—'Birds of Sumatra'. *Ardea*, **56**, 289.

Wells, D. R. (1989). Eyebrowed Thrushes *Turdus obscurus* in Bali. *Kukila*, **4**, 149.

Wells, D. R. and Becking, J. H. (1975). Vocalizations and status of Little and Himalayan Cuckoos, *Cuculus poliocephalus* and *C. saturatus* in Southeast Asia. *Ibis*, **117**, 366–71.

Wells, D. R. and Francis, C. M. (1984). Further evidence of a resident Brown Flycatcher *Muscicapa latirostris*. *Bull. Brit. Orn. Club*, **104**, 124–7.

Wells, D. R. and Medway, Lord (1976). Taxonomic and faunistic notes on birds of the Malay Peninsula. *Bull. Brit. Orn. Club*, **96**, 20–34.

Wells, D. R., Hails, C. J., and Hails, A. J. (1978). *A study of the birds of the Gunung Mulu National Park, Sarawak, with special emphasis on those of lowland forests.* Cyclostyled Report of Royal Geographic Society.

White, C. M. N. and Bruce, M. D. (1986). *The birds of Wallacea.* Brit. Ornithol. Union, Check-list No. 7. British Ornithologist's Union, London.

White, T. (1984). *A field guide to the bird songs of South-east Asia.* National Sound Archive (tape and notes).

Whitmore, T. C. (ed.) (1981). *Biogeographic evolution of the Malay Archipelago.* Clarendon Press, Oxford.

Whitten, A. (1980). *Saving Siberut, a conservation master plan. Appendix I. Birds of Siberut.* WWF, Bogor, Indonesia.

Whitten, A. J. (1989). Pelicans at Cengkareng. *Kukila*, **4**, 64.

Wiegant, W. and Helvoort, B. van (1987). First sighting of *Tachybaptus novaehollandiae* on Bali. *Kukila*, **3**, 50–1.

Wilkinson, R., Dutson, G., Sheldon, B., Darjono, and Noor, Y. R. (1991). The avifauna of the Barito Ulu region, Central Kalimantan. *Kukila*, **5**, 99–116.

Added in proof:

Balen, S. van (1987*b*). Forest fragmentation and the survival of forest birds in Java: A preliminary report. In *Proceedings, Seminar Alumni Daad*, Bogor, 115–65.

Index

Upright numerals refer to species number; italics to plate number; and a number in parentheses to an alternative scientific or common name.

Abroscopus superciliaris 651; *73*
Accipiter
  badius 104; *10, 13*
  gularis 99; *10, 13*
  nisus 101; *10, 13*
  soloensis 103; *10, 13*
  trivirgatus 102; *10, 13*
  virgatus 100; *10, 13*
Aceros
  comatus 400; *47*
  corrugatus 401; *47*
  subruficollis 403; *47*
  undulatus 402; *47*
Acridotheres
  cristatellus 751; *82*
  fuscus (750; *82*)
  grandis (750; *82*)
  javanicus 750; *82*
  tristis 749; *82*
Acrocephalus
  arundinaceus (657; *73*)
  bistrigiceps 658; *73*
  orientalis 657; *73*
  stentoreus 656; *73*
Actenoides concretus 391; *45*
Actitis hypoleucos (199; *23*)
Adjutant
  Greater 59; *7*
  Lesser 60; *7*
Aegithina
  tiphia 494; *57*
  viridissima 493; *57*
Aerodramus
  brevirostris (357; *42*)
  fuciphagus (354; *42*)
  germani (354; *42*)
  maximus (355; *42*)
  salangana (356; *42*)
  vanikorensis (356; *42*)
Aethopyga
  eximia 762; *83*
  mystacalis 764; *83*

  siparaja 763; *83*
  temminckii 765; *83*
Alauda
  arvensis (471; *54*)
  gulgula 471; *54*
Alcedo
  atthis 377; *44*
  coerulescens 380; *44*
  euryzona 379; *44*
  meninting 378; *44*
Alcippe
  brunneicauda 609; *68*
  pyrrhoptera 610; *68*
Alophoixus
  bres 521; *60*
  ochraceus 520; *60*
  phaeocephalus 522; *60*
Amandava amandava 804; *87*
Amaurornis
  fusca (154; *18*)
  phoenicurus 157; *18*
Anas
  acuta 68; *8*
  clypeata 75; *8*
  crecca 69; *8*
  gibberifrons 70; *8*
  penelope 72; *8*
  platyrhynchos 71; *8*
  querquedula 74; *8*
  superciliosa 73; *8*
Anastomus oscitans 55; *7*
Anhinga melanogaster 28; *3*
Anorrhinus galeritus 399; *47*
Anous
  minutus 249; *30*
  stolidus 248; *30*
  tenuirostris (249; *30*)
Anthracoceros
  albirostris 405; *47*
  convexus (405; *47*)
  malayanus 404; *47*

Anthreptes
  malacensis 755; *83*
  rhodolaema 756; *83*
  simplex 754; *83*
  singalensis 757; *83*
Anthus
  cervinus 735; *81*
  gustavi 734; *81*
  hodgsoni 732; *81*
  novaeseelandiae 733; *81*
  richardi (733; *81*)
Aplonis
  minor 741; *82*
  panayensis 742; *82*
Apus
  affinis 365; *42*
  pacificus 364; *42*
Arachnothera
  affinis 771; *84*
  chrysogenys 770; *84*
  crassirostris 767; *84*
  everetti 772; *84*
  flavigaster 769; *84*
  juliae 773; *84*
  longirostra 766; *84*
  robusta 768; *84*
Arborophila
  brunneopectus (128; *14*)
  charltonii 132; *14*
  chloropus (132; *14*)
  hyperythra 131; *14*
  javanica 129; *14*
  orientalis 128; *14*
  rubrirostris 130; *14*
Ardea
  cinerea 33; *5*
  novaehollandiae 35; *5*
  purpurea 34; *5*
  sumatrana 32; *5*
Ardeola
  bacchus 37; *6*
  speciosa 38; *6*
Arenaria interpres 201; *22*
Argus, Great 145; *15*
Argusianus argus 145; *15*
Artamus leucorhynchus 736; *80*
Asio flammeus 339; *39*
Avadavat, Red 804; *87*
Aviceda
  jerdoni 81; *9*
  leuphotes 82; *9*

Aythya australis 77; *8*
  fuligula 76; *8*

Babbler
  Abbott's 566; *65*
  Black-browed 567; *65*
  Black-browed (568; *65*)
  Black-capped 559; *65*
  Black-necked (593; *67*)
  Black-throated (588; *67*)
  Black-throated 593; *67*
  Blyth's (563; *65*)
  Brown-headed (569; *65*)
  Buettikofer's 561; *65*
  Buff-breasted (560; *65*)
  Chestnut-capped 600; *68*
  Chestnut-rumped 591; *67*
  Chestnut-winged 595; *67*
  Common Brown (566; *65*)
  Crescent-chested 596; *67*
  Ferruginous 563; *65*
  Golden 586; *67*
  Golden-headed (586; *67*)
  Greater Red-headed (572; *65*)
  Grey (588; *67*)
  Grey-breasted 573; *65*
  Grey-headed 589; *67*
  Grey-throated 588; *67*
  Horsfield's 565; *65*
  Hume's (565; *65*)
  Javan (587; *67*)
  Koengke (568; *65*)
  Lesser Red-headed (571; *65*)
  Mangrove Brown (563; *65*)
  Moustached 569; *65*
  Pearl-chested (596; *67*)
  Pearly-cheeked (596; *67*)
  Plain (570; *65*)
  Red-capped (600; *68*)
  Red-fronted (585; *67*)
  Red-headed (572; *65*)
  Red-rumped (591; *67*)
  Red-winged (595; *67*)
  Rufous-crowned 574; *65*
  Rufous-fronted 585; *67*
  Rusty Brown (565; *65*)
  Scaly-crowned 571; *65*
  Short-tailed 562; *65*
  Sooty-capped 570; *65*
  Sooty-headed (570; *65*)
  Spot-necked 590; *67*
  Spotted (590; *67*)

Temminck's 560; *65*
Tickell's (560; *65*)
Vanderbilt's 568; *65*
White-bibbed 594; *67*
White-breasted 587; *67*
White-chested 562; *65*
White-collared (594; *67*)
White-eared (592; *67*)
White-necked 592; *67*
White-throated (573; *65*)
Barbet
  Black-banded 415; *49*
  Black-browed 416; *48*
  Black-throated (422; *48*)
  Blue-crowned (419; *49*)
  Blue-eared 421; *49*
  Bornean 422; *48*
  Brown 424; *49*
  Brown-throated 411; *49*
  Coppersmith 423; *49*
  Crimson-breasted (423; *49*)
  Crimson-crowned (422; *48*)
  Fire-tufted 409; *48*
  Flame-fronted (419; *49*)
  Gaudy (414; *48*)
  Gold-whiskered 412; *48*
  Golden-naped 420; *49*
  Javan (415; *49*)
  Javan Brown-throated (411: 49)
  Kinabalu (420; *49*)
  Lineated 410; *48*
  Many-coloured (413; *48*)
  Mountain 417; *49*
  Mueller's (416; *48*)
  Orange-fronted 419; *49*
  Red-crowned 413; *48*
  Red-throated 414; *48*
  Yellow-crowned 418; *48*
Barred-Owlet
  Chestnut-winged (335; *39*)
  Javan (335; *39*)
  Spadiced (335; *39*)
Batrachostomus
  auritus 340; *41*
  cornutus 345; *41*
  harterti 341; *41*
  javensis 344; *41*
  poliolophus 343; *41*
  stellatus 342; *41*
Bay Owl
  Asian (319; *39*)
  Oriental 319; *39*

Baya (801; *88*)
Baza
  Asian (81; *9*)
  Black 82; *9*
  Black-crested (82; *9*)
  Jerdon's 81; *9*
Bee-Eater
  Bay-headed (392; *46*)
  Blue-tailed 393; *46*
  Blue-throated 394; *46*
  Chestnut-headed 392; *46*
  Rainbow 395; *46*
  Red-bearded 396; *46*
Berenicornis comatus (400; *47*)
Besra 100; *10, 13*
Bittern
  Black 52; *6*
  Chestnut (51; *6*)
  Chinese Little (50; *6*)
  Cinnamon 51; *6*
  Common (53)
  Great 53
  Japanese (48; *6*)
  Little Yellow (50; *6*)
  Mangrove (52; *6*)
  Schrenck's 49; *6*
  Tiger (47; *6*)
  Von Schrenck's (49; *6*)
  Yellow 50; *6*
Black-eye
  Mountain 799; *86*
  Olive (799; *86*)
Blue-Flycatcher
  Bornean 708; *78*
  Hill 706; *78*
  Japanese (700; *78*)
  Large-billed 707; *78*
  Malaysian 709; *78*
  Mangrove 711; *78*
  Pale 705; *78*
  Pygmy 712; *77*
  Rueck's 704; *78*
  Short-tailed (703; *78*)
  Sunda (707; *78*)
  Sunda Island (691; *76*)
  Tickell's 710; *78*
  White-tailed 703; *78*
Bluebird, Asian (542; *62*)
Bluechat (619; *64*)
Bluetail
  Orange-flanked (620; *64*)
  Red-flanked (620; *64*)

Bluetail (*cont.*)
  Siberian (620; *64*)
Blythipicus rubiginosus 446; *50*
Booby
  Abbott's 22; *2*
  Blue-faced (21; *2*)
  Brown 23; *2*
  Masked 21; *2*
  Red-Footed 20; *2*
  White (21; *2*)
  White-faced (21; *2*)
Botaurus stellaris 53;
Brachypteryx
  leucophrys 616; *64*
  montana 617; *64*
Bradypterus
  accentor 680; *74*
  montis (679; *74*)
  seebohmi 679; *74*
Bristlehead, Bornean 553; *63*
Broadbill
  Banded 451; *52*
  Black-and-Red 450; *52*
  Black-and-Yellow 452; *52*
  Black-throated (457; *52*)
  Black-throated Green (457; *52*)
  Dusky 449; *52*
  Gould's (453; *52*)
  Green 455; *52*
  Hodgson's (453; *52*)
  Hose's 456; *52*
  Lesser Green (455; *52*)
  Long-tailed 454; *52*
  Magnificent (456; *52*)
  Magnificent Green (456; *52*)
  Silver-breasted 453; *52*
  Whitehead's 457; *52*
Bronze-Cuckoo
  Australian (303; *37*)
  Gould's 304; *37*
  Horsfield's 302; *37*
  Little 303; *37*
  Malay (303; *37*)
  Malayan (303; *37*)
  Malaysian (303; *37*)
  Narrow-billed (302; *37*)
  Rufous-tailed (302; *37*)
Bubo sumatranus 332; *39*
Bubulcus ibis 39; *5*
Buceros
  bicornis 407; *47*
  rhinoceros 406; *47*

vigil 408; *47*
Bulbul
  Ashy 528; *60*
  Black-and-White 503; *58*
  Black-capped (509; *58*)
  Black-crested 505; *58*
  Black-crested Yellow (505; *58*)
  Black-headed 504; *58*
  Black-headed Yellow (505; *58*)
  Blue-wattled 511; *59*
  Blyth's (513; *59*)
  Brown (517; *59*)
  Brown-eared (528; *60*)
  Brown White-throated (520; *60*)
  Buff-vented 525; *60*
  Cream-striped 501; *58*
  Cream-vented 516; *59*
  Crestless White-throated (522; *60*)
  Dull Brown (525; *60*)
  Finsch's 519; *60*
  Flavescent 513; *59*
  Golden-vented (509; *58*)
  Green-backed (527; *60*)
  Green-winged (526; *60*)
  Grey-bellied 507; *58*
  Grey-cheeked 521; *60*
  Grey-headed (522; *60*)
  Grey Mountain (526; *60*)
  Hairy-backed 524; *60*
  Hook-billed 523; *60*
  Large Olive (515; *59*)
  Lesser Brown (518; *59*)
  Long-billed (523; *60*)
  Malaysian Wattled (511; *59*)
  Ochraceous 520; *60*
  Olive-brown (515; *59*)
  Olive-crowned (502; *59*)
  Olive-winged 515; *59*
  Olive White-throated (521; *60*)
  Orange-spotted 512; *59*
  Pale-faced (513; *59*)
  Puff-backed 510; *59*
  Red-eyed 517; *59*
  Red-eyed Brown (517; *59*)
  Red-whiskered 508; *58*
  Scaly-breasted 506; *58*
  Scrub (521; *60*)
  Sooty-Headed 509; *58*
  Spectacled 518; *59*
  Spot-necked 502; *59*
  Straw-crowned (500; *58*)
  Straw-headed 500; *58*

Streaked 527; *60*
Striated (501; *58*)
Sumatran (501; *58*)
Sunda 526; *60*
White-eyed (516; *59*)
White-eyed Brown (516; *59*)
Yellow-bellied 522; *60*
Yellow-crowned (500; *58*)
Yellow-vented 514; *59*
Bulweria
  bulwerii 6; *1*
  fallax 7; *1*
Bunting
  Black-headed 820; *88*
  Little 818; *88*
  Yellow-breasted 819; *88*
  White-shouldered (819; *88*)
Burhinus giganteus 225; *26*
Bushchat, Pied 633; *71*
Bushlark (470; *54*)
Bush-Robin
  Orange-flanked 620; *64*
  Red-flanked (620; *64*)
  Siberian (620; *64*)
Bush-Warbler
  Bornean (677; *74*)
  Friendly 680; *74*
  Javan (679; *74*)/(678; *74*)
  Indonesian (678; *74*)
  Kinabalu (680; *74*)
  Malaysian (678; *74*)
  Mountain (679; *74*)
  Mueller's (678; *74*)
  Russet 679; *74*
  Short-tailed (677; *74*)
  Sunda 678; *74*
  Timor (679; *74*)
  Whitehead's (677; *74*)
Butastur
  indicus 106; *11*
  liventer 105; *11*
Buteo buteo 107; *11*
Butorides striatus 36; *6*
Buttonquail
  Barred 148; *15*
  Common (147; *15*)
  Lesser Sunda (148; *15*)
  Little (147; *15*)
  Small 147; *15*
  Sunda (148; *15*)
Buzzard
  Common 107; *11*

Grey-Faced 106; *11*
Rufous-Winged 105; *11*
Steppe (107; *11*)

Cacatua sulphurea 284; *35*
Cacomantis
  merulinus 298; *37*
  sepulcralis 299; *37*
  sonneratii 297; *37*
  variolosus (299; *37*)
Cairina scutulata 79; *8*
Calidris
  acuminata 214; *25*
  alba 217; *25*
  alpina 215; *25*
  canutus 209; *24*
  ferruginea 216; *25*
  minutilla (213; *24*)
  ruficollis 211; *25*
  subminuta 213; *25*
  temminckii 212; *25*
  tenuirostris 210; *25*
  testacea (216; *24*)
Caloenas nicobarica 280; *34*
Calonectris leucomelas 8; *1*
Caloperdix oculea 133; *15*
Calorhamphus fuliginosus 424; *49*
Calyptomena
  hosei (456; *52*)
  hosii 456; *52*
  viridis 455; *52*
  whiteheadi 457; *52*
Canary-Flycatcher, Grey-Headed (713; *77*)
Canegrass-Warbler, Striated (662; *74*)
Capella, see Gallinago
Caprimulgus
  affinis 350; *41*
  concretus 351; *41*
  indicus 348; *41*
  macrurus 349; *41*
  pulchellus 352; *41*
Carpococcyx radiceus 313; *38*
Casmerodius albus (42; *5*)
Cataracta maccormicki 231; *27*
Celeus brachyurus 428; *50*
Centropus
  bengalensis 316; *38*
  nigrorufus 317; *38*
  rectunguis 314; *38*
  sinensis 315; *38*
Cettia
  fortipes (678; *74*)

Cettia (cont.)
  vulcania 678; 74
  whiteheadi (677; 74)
Ceyx
  erithacus 381; 44
  rufidorsa 382; 44
Chalcophaps indica 279; 34
Charadrius
  alexandrinus 176; 21
  asiaticus (183; 21)
  dubius 175; 21
  hiaticula 174; 21
  javanicus 177; 21
  leschenaultii 182; 21
  mongolus 181; 21
  peronii 179; 21
  placidus 180; 21
  ruficapillus 178; 21
  veredus 183; 21
Chat, Pied (633; 71)
Chlamydochaera jefferyi 629; 55
Chlidonias
  hybridus 235; 28
  leucopterus 236; 28
Chlorocharis emiliae 799; 86
Chloropsis
  aurifrons 497; 57
  cochinchinensis 498; 57
  cyanopogon 495; 57
  sonnerati 496; 57
  venusta 499; 57
Chrysococcyx
  basalis 302; 37
  maculatus 300; 37
  minutillus 304; 37
  russatus 304; 37
  xanthorhynchus 301; 37
Chrysocolaptes
  lucidus 448; 51
  validus (447; 51)
Ciconia
  episcopus 56; 7
  stormi 57; 7
Cinclidium diana 625; 70
Circaetus gallicus 92; 10
Circus
  aeruginosus 96; 10, 13
  cyaneus 97; 10, 13
  melanoleucos 98; 10, 13
  spilonotus 95; 10, 13
Cissa
  chinensis 545; 63

thalassina 544; 63
Cissa, Hunting (545; 63)
Cisticola
  exilis 675; 74
  juncidis 674; 74
Cisticola
  Bright-capped (675; 74)
  Common (674; 74)
  Fan-tailed (674; 74)
  Gold-capped (675; 74)
  Golden-headed 675; 74
  Streak-headed (674; 74)
  Streaked (674; 74)
  Yellow-headed (675; 74)
  Zitting 674; 74
Clamator coromandus 290; 36
Cochoa
  azurea 630; 71
  beccarii 631; 71
Cochoa
  Javan 630; 71
  Sumatran 631; 71
Cockatoo
  Lesser Sulphur-Crested (284; 35)
  Yellow-Crested 284; 35
Coffinbird (349; 41)
Coleto 753
Collared-Dove
  Island 276; 34
  Javan (276; 34)
  Javanese (276; 34)
  Philippine (276; 34)
Collocalia
  brevirostris (357; 42)
  esculenta 358; 42
  fuciphaga 354; 42
  germani (354; 42)
  gigas (353; 42)
  linchi 359
  lowii (355; 42)
  maxima/lowii 355; 42
  salangana (356; 42)
  vanikorensis 356; 42
  vulcanorum 357; 42
Columba
  argentina 271; 33
  livia 272; 33
  vitiensis 270; 33
Coot
  Black (162; 19)
  Common 162; 19

Eurasian (162; *19*)
European (162; *19*)
Purple (161; *19*)
Copsychus
  malabaricus 622; *70*
  pyrrhopygus (624; *70*)
  saularis 621; *70*
  stricklandi 623; *70*
Coracina
  fimbriata 484; *55*
  javensis 481; *55*
  larvata 482; *55*
  novaehollandiae (481; *55*)
  striata 483; *55*
Cormorant
  Big Black (25; *3*)
  Black (24; *3*)
  Common (25; *3*)
  Great 25; *3*
  Javan (27; *3*)
  Large Black (25; *3*)
  Little 27; *3*
  Little Black 24; *3*
  Little Pied 26; *3*
  White-throated (26; *3*)
Corvus
  enca 550; *63*
  levaillanti (552; *63*)
  macrorhynchos 552; *63*
  splendens 551; *63*
Corydon sumatranus 449; *52*
Coturnix chinensis 127; *15*
Coucal
  Greater 315; *38*
  Lesser 316; *38*
  Short-toed 314; *38*
  Sunda 317; *38*
Courser, Australian (227; *26*)
Crake
  Ashy (156; *18*)
  Baillon's 153; *18*
  Band-bellied 155; *18*
  Banded (152; *18*)
  Chinese Banded (155; *18*)
  Grey-bellied (156; *18*)
  Malay (151; *18*)
  Malay Banded (151; *18*)
  Marsh (153; *18*)
  Philippine (152; *18*)
  Red-legged 151; *18*
  Ruddy (154; *18*)
  Ruddy-breasted 154; *18*
  Ryukyu (152; *18*)
  Siberian Ruddy (155; *18*)
  Slaty-legged 152; *18*
  Slaty-legged Banded (152; *18*)
  Tiny (153; *18*)
  White-browed 156; *18*
Crested-Cuckoo
  Chestnut-winged (290; *36*)
  Red-winged (290; *36*)
Crested-Tern
  Chinese 247; *30*
  Great 245; *30*
  Lesser 246; *30*
Criniger
  bres (521; *60*)
  finschii 519; *60*
  ochraceus (520; *60*)
  phaeocephalus (522; *60*)
Crocethia alba (217; *24*)
Crocias
  albonotatus 611; *68*
  guttatus (611; *68*)
Crocias, Spotted 611; *68*
Crow
  Bald-headed (553; *63*)
  Colombo (551; *63*)
  Grey-headed (551; *63*)
  House 551; *63*
  Indian (551; *63*)
  Large-billed 552; *63*
  Little (550; *63*)
  Slender-billed 550; *63*
  Thick-billed (552; *63*)
Crowned Leaf-Warbler, Eastern (654; *73*)
Crowned-Warbler
  Eastern (654; *73*)
  Temminck's (654; *73*)
Crow-Pheasant (315; *38*)
Crypsirina temia 548; *63*
Cuckoo
  Asian Emerald 300; *37*
  Banded (297; *37*)
  Banded Bay 297; *37*
  Chestnut-winged 290; *36*
  Common 295; *36*
  Drongo 305; *37*
  Eurasian (295; *36*)
  Grey (295; *36*)
  Grey-headed (299; *37*)
  Himalayan (296; *36*)
  Indian 294; *36*
  Lesser (295)

Cuckoo (cont.)
  Oriental 296; 36
  Plaintive 298; 37
  Red-winged (290; 36)
  Rusty-breasted 299; 37
  Short-winged (294; 36)
  Violet 301; 37
Cuckoo-Dove
  Barred 273; 34
  Indonesian (274; 34)
  Large (273; 34)
  Little 275; 34
  Red (274; 34)
  Ruddy 274; 34
  Sunda (274; 34)
Cuckoo-Shrike
  Bar-bellied 483; 55
  Barred (483; 55)
  Black-faced (482; 55)
  Lesser 484; 55
  Malaysian 481; 55
  Sunda 482; 55
Cuculus
  canorus 295; 36
  fugax 293; 36
  micropterus 294; 36
  poliocephalus (296; 36)
  saturatus 296; 36
  sparverioides 291; 36
  vagans 292; 36
Culicicapa ceylonensis 713; 77
Curlew
  Australian (187; 22)
  Common (184; 22)
  Eastern (187; 22)
  Eurasian 184; 22
  European (184; 22)
  Far-Eastern 187; 22
  Hudsonian (185; 22)
  Little 186; 22
  Western (184; 22)
Cyanoptila cyanomelana 700; 78
Cymbirhynchus macrorhynchus 450; 52
Cyornis
  banyumas 706; 78
  caerulatus 707; 78
  concreta (703; 78)
  concretus 703; 78
  ruckii 704; 78
  ruecki (704; 78)
  rufigaster 711; 78
  rufigastra (711; 78)
  superbus 708; 78
  tickelliae 710; 78
  turcosa (709; 78)
  turcosus 709; 78
  unicolor 705; 78
Cypsiurus
  balasiensis 366; 42
  batasiensis (366; 42)

Dabchick (1; 26)
Dafila (68; 8)
Darter, Oriental 28; 3
Delichon dasypus 477; 54
Demigretta (40; 5)
Dendrocitta
  cinerascens 547; 63
  occipitalis 546; 63
Dendrocopus
  canicapillus 443; 50
  macei 442; 50
  moluccensis (444; 50)
Dendrocygna
  arcuata 67; 8
  javanica 66; 8
Dendronanthus indicus 731; 81
Dicaeum
  agile 779; 85
  celebicum (787; 85)
  chrysorrheum 781; 85
  concolor 783; 85
  cruentatum 784; 85
  everetti 780; 85
  ignipectus 788; 85
  maugei 785; 85
  monticolum 787; 85
  sanguinolentum 786; 85
  trigonostigma 782; 85
  trochileum 789; 85
Dicrurus
  aeneus 532; 61
  annectans 531; 61
  bracteatus 534; 61
  hottentottus 534; 61
  leucophaeus 530; 61
  macrocercus 529; 61
  paradiseus 536; 61
  remifer 533; 61
  sumatranus 535; 61
Dinopium
  javanense 436; 50
  rafflesii 437; 50

Dollarbird 397; *46*
Dotterel
  Caspian (183; *21*)
  Malaysian (181; *21*)
  Oriental (183; *21*)
  Red-capped (178; *21*)
Dove
  Barred (278; *34*)
  Burmese Spotted (277; *34*)
  Emerald 279; *34*
  Green-winged 279; *34*
  Peaceful (278; *34*)
  Spot-necked (277; *34*)
  Spotted 277; *34*
  Zebra 278; *34*
Dowitcher
  Asian 202; *22*
  Asiatic (202; *22*)
  Long-billed 203; *23*
  Snipe-billed (202; *22*)
Drongo
  Ashy 530; *61*
  Black 529; *61*
  Bronzed 532; *61*
  Crow-billed 531; *61*
  Greater Racket-tailed 536; *61*
  Hair-crested 534; *61*
  Lesser Racket-tailed 533; *61*
  Spangled (534; *61*)
  Sumatran 535; *61*
Dryocopus javensis 441; *50*
Duck
  Pacific Black 73; *8*
  Tufted 76; *8*
  White-winged 79; *8*
Ducula
  aenea 264; *33*
  badia 266; *33*
  bicolor 265; *33*
  lacernulata 267; *33*
  pickeringi 268; *33*
  rosacea 269; *33*
Dunlin 215; *25*
Dupetor flavicollis 52; *6*

Eagle
  Asian Black (108; *11*)
  Black 108; *11*
  Booted 109; *11*
  Chestnut-bellied (110; *11*)
  Indian Black (108; *11*)
  Rufous-bellied 110; *11*
  Short-toed 92; *10*
Eagle-Owl
  Barred 332; *39*
  Malay (332; *39*)
  Malaysian (332; *39*)
Eared Nightjar
  Giant (347; *41*)
  Great 347; *41*
  Malaysian 346; *41*
Egret
  Cattle 39; *5*
  Chinese 41; *5*
  Great 42; *5*
  Great White (42; *5*)
  Intermediate 43; *5*
  Large (42; *5*)
  Lesser (43; *5*)
  Little 44; *5*
  Plumed (43; *5*)
  Puff-backed (39; *5*)
  Smaller (43; *5*)
  Swinhoe's (41; *5*)
  Yellow-billed (43; *5*)
Egretta
  alba 42; *5*
  eulophotes 41; *5*
  garzetta 44; *5*
  intermedia 43; *5*
  novaehollandiae (35; *5*)
  sacra 40; *5*
Elanus caeruleus 85; *9*
Emberiza
  aureola 819; *88*
  melanocephala 820; *88*
  pusilla 818; *88*
Enicurus
  leschenaulti 628; *70*
  ruficapillus 627; *70*
  velatus 626; *70*
Ephippiorhynchus asiaticus 58
Erithacus cyane 619; *64*
Erpornis (614; *68*)
Erythrura
  hyperythra 806; *87*
  prasina 805; *87*
Esacus magnirostris (225; *26*)
Eudynamys scolopacea 306; *36*
Eumyias
  indigo 691; *76*
  thalassina 690; *76*
Eupetes (615; *68*)

Eupetes macrocerus 615; *68*
Eurostopodus
  macrotis 347; *41*
  temminckii 346; *41*
Eurylaimus
  javanicus 451; *52*
  ochromalus 452; *52*
Eurynorhynchus pygmaeus 219; *24*
Eurystomus orientalis 397; *46*

Fairy-Bluebird
  Asian 542; *62*
  Blue-backed (542; *62*)
  Blue-mantled (542; *62*)
Falco
  cenchroides 119; *12, 13*
  moluccensis 118; *12*
  peregrinus 122; *12, 13*
  severus 121; *12, 13*
  subbuteo 120; *12, 13*
  tinnunculus 117; *12, 13*
Falcon, Peregrine 122; *12, 13*
Falconet
  Bornean (116; *12*)
  Black-legged (115; *12*)
  Black-sided (115; *12*)
  Black-tailed (115; *12*)
  Black-thighed 115; *12*
  White-browed (116; *12*)
  White-fronted 116; *12*
Fantail
  Malaysian (718; *79*)
  Pearlated (717; *79*)
  Pearl-spotted (717; *79*)
  Perlated (717; *79*)
  Pied 718; *79*
  Red-tailed (714; *79*)
  Rufous-tailed 714; *79*
  Spot-breasted (716; *79*)
  Spotted 717; *79*
  White-bellied 715; *79*
  White-spotted (716; *79*)
  White-throated 716; *79*
Fantail-Flycatcher, *see* Fantail
Fantail-Warbler
  Bright-capped (675; *74*)
  Common (674; *74*)
  Gold-capped (675; *74*)
  Golden-headed (675; *74*)
  Streaked (674; *74*)
  Streak-headed (674; *74*)
  Yellow-headed (675; *74*)
  Zitting (674; *74*)
Ficedula
  dumetoria 698; *77*
  hyperythra 697; *77*
  mugimaki 694; *77*
  narcissina 693; *77*
  parva 695; *77*
  solitaria (696; *77*)
  solitaris 696; *77*
  westermanni 699; *77*
  zanthopygia 692; *77*
Finch, Strawberry (804; *87*)
Finfoot
  Masked 163; *19*
  Asian (163; *19*)
Fireback
  Crested 137; *16*
  Crestless 136; *16*
  Hoogerwerf's (140; *17*)
  Malaysian (137; *16*)
  Rufous-tailed (136; *16*)
  Salvadori's (139; *16*)
  Sumatran (140; *17*)
  Veillot's (137; *16*)
Fireback Pheasant (*Lophura*), *see* Fireback
Fish-Eagle
  Greater (90; *9*)
  Grey-headed 90; *9*
  Lesser 89; *9*
  White-bellied 88; *9*
Fishing-Eagle, *see* Fish-Eagle
Fish-Owl
  Buffy 333; *39*
  Malay (333; *39*)
  Malaysian (333; *39*)
Flameback
  Common (436; *50*)
  Greater 448; *51*
Flowerpecker
  Black-sided 787; *85*
  Bornean (787; *85*)
  Bornean Fire-breasted (787; *85*)
  Borneo Yellow-rumped (777; *85*)
  Blood-breasted 786; *85*
  Blue-cheeked (785; *85*)
  Bronze-backed (788; *85*)
  Brown-backed 780; *85*
  Buff-backed (788; *85*)
  Crimson-breasted 778; *85*
  Everett's (780; *85*)
  Fire-breasted 788; *85*

INDEX 469

Green-backed (788; *85*)
Javan (786; *85*)
Javan Fire-breasted (786; *85*)
Lesser Sunda (785; *85*)
Mountain (787; *85*)
Nilgiri (783; *85*)
Orange-bellied 782; *85*
Orange-breasted (782; *85*)
Plain 783; *85*
Plain-coloured (783; *85*)
Red-chested 785; *85*
Scarlet-backed 784; *85*
Scarlet-breasted 775; *85*
Scarlet-headed 789; *85*
Streaky-breasted (779; *85*)
Striped (779; *85*)
Thick-billed 779; *85*
Timor (785; *85*)
Yellow-breasted 776; *85*
Yellow-rumped 777; *85*
Yellow-throated (776; *85*)
Yellow-vented 781; *85*
Flycatcher
Asian Brown 688; *76*
Blue-and-White 700; *78*
Brown-chested (682; *76*)
Brown-streaked (688; *76*)
Canary (713; *77*)
Chestnut-tailed (684; *76*)
Chestnut-winged (721; *79*)
Chocolate (688; *76*)
Dark-sided 686; *76*
Dull (697; *77*)
Eyebrowed (685; *76*)
Ferruginous 689; *76*
Fulvous-chested (681; *76*)
Grey-breasted (688; *76*)
Grey-chested (683; *76*)
Grey-headed 713; *77*
Grey-spotted (687; *76*)
Grey-streaked 687; *76*
Indian Verditer (690; *76*)
Indigo 691; *76*
Korean (692; *77*)
Little Pied 699; *77*
Malaysian (696; *77*)
Mangrove Blue 711; *78*
Maroon-breasted (720; *79*)
Mugimaki 694; *77*
Narcissus 693; *77*
Olive-backed (681; *76*)

Orange-breasted (698; *77*)
Paradise (723; *79*)
Red-breasted 695; *77*
Red-throated (695; *77*)
Rufous-browed 696; *77*
Rufous-chested 698; *77*
Rufous-tailed (684; *76*)
Rufous-winged (721; *79*)
Short-tailed (698; *77*)
Siberian (686; *76*)
Snowy-browed 697; *77*
Solitary (696; *77*)
Sooty (686; *76*)
Spot-breasted (686; *76*)
Thicket (697; *77*)
Tricoloured (692; *77*)
Verditer 690; *76*
White-browed (685; *76*)
White-fronted (685; *76*)
White-throated (696; *77*)
Williamson's (688; *76*)
Yellow-rumped 692; *77*
Flycatcher-Shrike
Bar-winged 478; *55*
Black-winged 479; *55*
Pied (478; *55*)
Flycatcher-Warbler
Chestnut (648; *73*)
Chestnut-crowned (648; *73*)
Chestnut-headed (648; *73*)
Green (655; *73*)
Sunda (649; *73*)
Yellow-breasted (650; *73*)
Flyeater (647; *73*)
Forktail
Chestnut-naped 627; *70*
Lesser 626; *70*
White-crowned 628; *70*
Fregata
andrewsi 29; *4*
ariel 31; *4*
minor 30; *4*
Frigatebird
Christmas 29; *4*
Christmas Island (29; *4*)
Great 30; *4*
Greater (30; *4*)
Least (31; *4*)
Lesser 31; *4*
Frog-Hawk (103; *10*)
Frogmouth
Blyth's (344; *41*)

Frogmouth (*cont.*)
  Dulit 341; *41*
  Gould's 342; *41*
  Horned (345; *41*)
  Javan 344; *41*
  Large 340; *41*
  Long-tailed 345; *41*
  Pale-headed (343; *41*)
  Short-tailed 343; *41*
  Sunda (345; *41*)
Fruit-Dove
  Banded (262; *32*)
  Black-backed 262; *32*
  Black-naped (262; *32*)
  Jambu 260; *32*
  Pink-headed 261; *32*
  Pink-necked (261; *32*)
  White-headed 263; *32*
Fruit-Hunter
  Black-breasted 629; *55*
Fulica atra 162; *19*
Fulvetta
  Brown 609; *68*
  Javan 610; *68*

Gallicrex cinerea 158; *19*
Gallinago
  gallinago 206; *22*
  megala 205; *22*
  stenura 204; *22*
Gallinula
  chloropus 159; *19*
  tenebrosa 160; *19*
Gallinule, Purple (161; *19*)
Gallirallus striatus 150; *18*
Gallus
  gallus 141; *17*
  varius 142; *17*
Gannet, *see* Booby
Garganey 74; *8*
Garrulax
  leucolophus 603; *69*
  lugubris 604; *69*
  mitratus 605; *69*
  palliatus 601; *69*
  rufifrons 602; *69*
Gelochelidon nilotica (237; *28*)
Geopelia striata 278; *34*
Gerygone sulphurea 647; *73*
Gerygone
  Golden-bellied 647; *73*
  Yellow-breasted (647; *73*)

Glareola
  isabella 227; *26*
  maldivarum 226; *26*
Glaucidium
  brodiei 334; *39*
  castanopterum 335; *39*
Glossy Starling, *see* Starling
Godwit
  Bar-tailed 189; *22*
  Black-tailed 188; *22*
Goldenback
  Common 436; *50*
  Greater 448; *51*
Golden-Plover
  Asian (173; *20*)
  Asiatic (173; *20*)
  Eastern (173; *20*)
  Pacific 173; *20*
Goldfinch
  Indonesian (817; *88*)
  Malay (817; *88*)
  Malaysian (817; *88*)
  Mountain (817; *88*)
Gorsachius
  goisagi 48; *6*
  melanolophus 47; *6*
Goshawk
  Chinese 103; *10, 13*
  Crested 102; *10, 13*
  Grey (103; *10, 13*)
  Horsfield's (103; *10, 13*)
  Little Banded (104; *10, 13*)
Gracula religiosa 752; *82*
Grackle, Common (752; *82*)
Grassbird, Striated (662; *74*)
Grasshopper-Warbler
  Grey-naped (659; *74*)
  Lanceolated (661; *74*)
  Middendorff's (660; *74*)
  Pallas's (659; *74*)
  Streaked (661; *74*)
Great Reed-Warbler
  Eastern (657; *73*)
  Southern (656; *73*)
Grebe
  Australasian 2; *26*
  Australian (2; *26*)
  Black-throated (2; *26*)
  Little 1; *26*
  Red-throated (1; *26*)
Green Magpie
  Bornean (544; *63*)

Short-tailed (544; *63*)
Green-Pigeon
　Cinnamon-headed 255; *31*
　Grey-cheeked 254; *31*
　Korthal's (252; *31*)
　Large 259; *32*
　Little 256; *31*
　Orange-breasted 258; *32*
　Pink-necked 257; *32*
　Sumatran 251; *31*
　Thick-billed 253; *31*
　Wedge-tailed 252; *31*
Greenshank
　Common 193; *23*
　Eurasian (193; *23*)
　Greater (193; *23*)
　Nordmann's 194; *23*
　Spotted (194; *23*)
Ground-Babbler, Bornean (576; *66*)
Ground-Cuckoo
　Green-billed (313; *38*)
　Malayan (313; *38*)
　Sunda 313; *38*
Ground-Dove
　Barred (278; *34*)
　Peaceful (278; *34*)
　Zebra (278; *34*)
Ground-Thrush
　Andromeda (642; *72*)
　Everett's (641; *72*)
　Sunda (642; *72*)
Ground-Warbler
　Eyebrowed (676; *74*)
　Javan (676; *74*)
　Malaysian (676; *74*)
Gull
　Brown-headed 234; *27*
　Common Black-headed 232; *27*
　Sabine's 233; *27*
Gygis alba 250; *28*
Gyps bengalensis 91

Haematortyx sanguiniceps 134; *14*
Halcyon
　capensis (383; *44*)
　chloris (389; *45*)
　concreta (391; *45*)
　coromanda 385; *45*
　cyanoventris 387; *45*
　pileata 388; *45*
　sancta (390; *45*)

smyrnensis 386; *45*
Haliaeetus leucogaster 88; *9*
Haliastur indus 87; *9*
Hanging-Parrot
　Blue-crowned 288; *35*
　Malaysian (288; *35*)
　Yellow-throated 289; *35*
Harpactes
　diardii 371; *43*
　duvaucelii 374; *43*
　erythrocephalus 376; *43*
　kasumba 370; *43*
　oreskios 375; *43*
　orrhophaeus 373; *43*
　reinwardtii 369; *43*
　whiteheadi 372; *43*
Harrier
　Eastern Marsh 95; *10, 13*
　Hen (97; *10, 13*)
　Northern 97; *10, 13*
　Pied 98; *10, 13*
　Spot-backed (95; *10, 13*)
　Spot-rumped (95; *10, 13*)
　Western Marsh 96; *10, 13*
　White-rumped (97; *10, 13*)
Hawk
　Bat 84; *9*
　Fish (80; *9*)
　Lizard (81; *9*)
Hawk-Cuckoo
　Fugitive (293; *36*)
　Hodgson's 293; *36*
　Large 291; *36*
　Lesser (292; *36*)
　Moustached 292; *36*
Hawk-Eagle
　Blyth's 113; *11*
　Changeable 111; *11*
　Crested (111; *11*)
　Javan 112; *11*
　Marsh (111; *11*)
　Sunda (111; *11*)
　Wallace's 114; *11*
Hawk-Owl
　Brown 336; *39*
　Oriental (336; *39*)
　Philippine (336; *39*)
Heliopais personata 163; *19*
Hemicircus concretus 445; *50*
Hemiprocne
　comata 368; *42*
　longipennis 367; *42*

Hemipus
  hirundinaceus 479; *55*
  picatus 478; *55*
Heron
  Dusky Grey (32; *5*)
  Giant (32; *5*)
  Great-billed 32; *5*
  Great White (42; *5*)
  Green-backed (36; *6*)
  Grey 33; *5*
  Little (36; *6*)
  Mangrove (36; *6*)
  Night (*see* Night-Heron)
  Pond (*see* Pond-Heron)
  Purple 34; *5*
  Reef (40; *5*)
  Striated 36; *6*
  The (33; *5*)
  White-faced 35; *5*
Heterophasia picaoides 612; *68*
Heteroscelus
  brevipes 200; *22*
  incana (200; *22*)
Hieraaetus
  kienerii 110; *11*
  pennatus 109; *11*
Hill-Partridge
  Bar-backed (128; *14*)
  Chestnut-bellied (129; *14*)
  Grey-bellied (128; *14*)
  Grey-breasted (128; *14*)
  Javan (129; *14*)
  Sumatran (128; *14*)
Himantopus
  himantopus 222; *26*
  leucocephalus 221; *26*
Hirundapus
  caudacutus 360; *42*
  cochinchinensis 361; *42*
  giganteus 362; *42*
Hirundo
  daurica 475; *54*
  rustica 473; *54*
  striolata 476; *54*
  tahitica 474; *54*
Hobby
  Eurasian 120; *12, 13*
  European (120; *12, 13*)
  Northern (120; *12, 13*)
  Oriental 121; *12, 13*
Honey-Buzzard
  Asian (83; *9*)
  Crested (83; *9*)
  Eastern (83; *9*)
  Oriental 83; *9*
Honeyeater
  Brown (774; *84*)
  Indonesian 774; *84*
Honeyguide
  Malayan (425; *49*)
  Malaysian 425; *49*
Hoopoe
  Common (398; *46*)
  Eurasian 398; *46*
Hornbill
  Asian Black 404; *47*
  Black (404; *47*)
  Bushy-crested 399; *47*
  Great 407; *47*
  Great Indian (407; *47*)
  Helmeted 408; *47*
  Long-crested (399; *47*)
  Malaysian Black (404; *47*)
  Malaysian Pied (405; *47*)
  Oriental Pied 405; *47*
  Plain-pouched 403; *47*
  Rhinoceros 406; *47*
  Southern Pied (405; *47*)
  White-crested (399; *47*)
  White-crowned 400; *47*
  Wreathed 402; *47*
  Wrinkled 401; *47*
House-Martin
  Asian 477; *54*
  Asiatic (477; *54*)
Hydrochous gigas 353; *42*
Hydrophasianus chirurgus 165; *19*
Hydroprogne caspia (238; *28*)
Hypogramma hypogrammicum 758; *83*
Hypothymis azurea 719; *79*
Hypsipetes
  charlottae (525; *60*)
  criniger (524; *60*)
  flavala 528; *60*
  malaccensis (527; *60*)
  virescens (526; *60*)

Ibis
  Black-headed 61; *7*
  Glossy 63; *7*
  Indian Black-necked (61; *7*)
  Oriental Black-necked (61; *7*)
  White-Shouldered 62; *7*

Ichthyophaga
  humilis 89; *9*
  ichthyaetus 90; *9*
  nana (89; *9*)
Ictinaetus malayensis 108; *11*
Imperial-Pigeon
  Black-backed (267; *33*)
  Dark-backed 267; *33*
  Enggano (264; *33*)
  Green 264; *33*
  Grey 268; *33*
  Javan (269; *33*)
  Mountain 266; *33*
  Nutmeg (265; *33*)
  Pied 265; *33*
  Pink-headed 269; *33*
Indicator archipelagicus 425; *49*
Iole
  olivacea 525; *60*
  viresceus 526; *60*
Iora
  Black-winged (494; *57*)
  Common 494; *57*
  Green 493; *57*
  Small (494; *57*)
Irediparra gallinacea 164; *19*
Irena puella 542; *62*
Ixobrychus
  cinnamomeus 51; *6*
  eurhythmus 49; *6*
  flavicollis (52; *6*)
  sinensis 50; *6*
Ixos
  malaccensis 527; *60*
  virescens 526; *60*

Jacana
  Bronze-winged 166; *19*
  Comb-crested 164; *19*
  Pheasant-tailed 165; *19*
Jaeger
  Arctic (229; *27*)
  Long-tailed 230; *27*
  Pamatorhine (228; *27*)
  Parasitic 229; *27*
  Pomarine 228; *27*
Jay
  Crested 543; *63*
  Crested Malay (543; *63*)
  Malay (543; *63*)

Jewel-Babbler
  Malay (615; *68*)
  Malaysian (615; *68*)
Jungle-Babbler
  Abbott's (566; *65*)
  Buettikofer's (561; *65*)
  Black-browed (567; *65*)
  Black-browed (568; *65*)
  Blyth's (563; *65*)
  Buff-breasted (560; *65*)
  Common Brown (566; *65*)
  Ferruginous (564; *65*)
  Horsfield's (565; *65*)
  Koengke (568; *65*)
  Mangrove Brown (563; *65*)
  Rusty Brown (565; *65*)
  Short-tailed (562; *65*)
  Temminck's (560; *65*)
  Tickell's (560; *65*)
  Vanderbilt's (568; *65*)
  White-chested (563; *65*)
Jungle-Flycatcher
  Brown-chested 682; *76*
  Chestnut-tailed (684; *76*)
  Eyebrowed 685; *76*
  Fulvous-chested 681; *76*
  Grey-chested 683; *76*
  Grey-faced (684; *76*)
  Kinabalu (685; *76*)
  Olive-backed (681; *76*)
  Rufous-tailed 684; *76*
  White-browed (685; *76*)
  White-throated (683; *76*)
Junglefowl
  Green 142; *17*
  Red 141; *17*
  Wild (141; *17*)

Kenopia striata 577; *66*
Kestrel
  Australian 119; *12, 13*
  Common 117; *12, 13*
  Eurasian (117; *12, 13*)
  Moluccan (119; *12*)
  Nankeen (119; *12, 13*)
  Old World (117; *12, 13*)
  Rock (117; *12, 13*)
  Spotted 118; *12*
Ketupa ketupu 333; *39*
Kingfisher
  Banded 384; *44*

Kingfisher (cont.)
  Black-backed 381; *44*
  Black-capped 388; *45*
  Blue-banded 379; *44*
  Blue-eared 378; *44*
  Chestnut-collared (391; *45*)
  Collared 389; *45*
  Common 377; *44*
  Deep Blue (378; *44*)
  European (377; *44*)
  Javan 387; *45*
  Malay (381; *44*)
  Mangrove (389; *45*)
  Oriental (381; *44*)
  Red-backed (382; *44*)
  River (377; *44*)
  Ruddy 385; *45*
  Rufous-backed 382; *44*
  Rufous-collared 391; *45*
  Sacred 390; *45*
  Small Blue 380; *44*
  Smyrna (386; *45*)
  Stork-billed 383; *44*
  Three-toed (381; *44*)
  White-breasted (386; *45*)
  White-collared (389; *45*)
  White-throated 386; *45*
Kite
  Bat (84; *9*)
  Black 86; *9*
  Black-shouldered (85; *9*)
  Black-winged 85; *9*
  Brahminy 87; *9*
  Common (85; *9*)
  Indonesian (85; *9*)
  Pariah (86; *9*)
  Red-backed (87; *9*)
  White-headed (87; *9*)
  Yellow-billed (86; *9*)
Knot
  Great 210; *24*
  Red 209; *24*
Koel
  Asian (306; *36*)
  Common 306; *36*
  Indian (306; *36*)

Lacedo pulchella 384; *44*
Lalage
  fimbriata (484; *55*)
  nigra 485; *55*
  sueurii 486; *55*
Lanius
  cristatus 737; *80*
  excubitor 740
  schach 739; *80*
  tigrinus 738; *80*
Lapwing
  Common (168; *20*)
  Grey-headed 169; *20*
  Javan (170; *20*)
  Javanese 170; *20*
  Javan Wattled (170; *20*)
  Northern 168; *20*
  Red-wattled 171; *20*
Lark
  Australian 470; *54*
  Eastern (470; *54*)
  Horsfields (470; *54*)
  Javan (470; *54*)
  Singing (470; *54*)
Larus
  brunnicephalus 234; *27*
  ridibundus 232; *27*
  sabini (233; *27*)
Laughingthrush
  Black 604; *69*
  Catbird (601; *69*)
  Chestnut-capped 605; *69*
  Grey and Brown (601; *69*)
  Red-fronted (602; *69*)
  Rufous-fronted 602; *69*
  Sunda 601; *69*
  White-crested 603; *69*
Leaf-Warbler
  Inornate (652; *73*)
  Island (655; *73*)
  Mountain 655; *73*
  Yellow-browed (652; *73*)
Leafbird
  Blue-masked 499; *57*
  Blue-winged 498; *57*
  Golden-fronted 497; *57*
  Golden-headed (498; *57*)
  Golden-hooded (497; *57*)
  Greater (496; *57*)
  Greater Green 496; *57*
  Great Green (496; *57*)
  Lesser (495; *57*)
  Lesser Green 495; *57*
  Masked (499; *57*)
  Yellow-headed (497; *57*)
Leiothrix argentauris 606; *69*

Leothrix, Silver-eared (606; *69*)
Leptoptilos
   dubius 59; *7*
   javanicus 60; *7*
Leucopsar rothschildi 748; *82*
Lichmera
   indistincta (774; *84*)
   limbata 774; *84*
Limicola falcinellus 218; *24*
Limnodromus
   scolopaceus 203; *23*
   semipalmatus 202; *22*
Limosa
   lapponica 189; *22*
   limosa 188; *22*
Locustella
   certhiola 659; *74*
   lanceolata 661; *74*
   ochotensis 660; *74*
Lonchura
   ferruginosa 815; *87*
   fuscans 811; *87*
   leucogastra 809; *87*
   leucogastroides 810; *87*
   maja 816; *87*
   malacca 814; *87*
   molucca 812; *87*
   punctulata 813; *87*
   striata 808; *87*
Lophozosterops javanicus 797; *86*
Lophura
   bulweri 138; *16*
   erythrophthalma 136; *16*
   hoogerwerfi 140; *17*
   ignita 137; *16*
   inornata 139; *16*
Loriculus
   galgulus 288; *35*
   pusillus 289; *35*
   vernalis (289; *35*)
Lorikeet
   Blue-crowned (288; *35*)
   Coconut (283; *35*)
   Malaysian (288; *35*)
   Rainbow 283; *35*
   Yellow-throated (289; *35*)
Luscinia calliope 618; *64*

Machaeramphus alcinus 84; *9*
Macronous
   flavicollis 597; *68*
   gularis 598; *68*
   kelleyi (597; *68*)
   ptilosus 599; *68*
Macropygia
   emiliana 274; *34*
   phasianella (274; *34*)
   ruficeps 275; *34*
   unchall 273; *34*
Magpie
   Black 549; *63*
   Black-crested (549; *63*)
   Bornean (544; *63*)
   Chinese Green (545; *63*)
   Green 545; *63*
   Short-tailed 544; *63*
   Short-tailed Green (544; *63*)
   White-winged (549; *63*)
Malacocincla
   abbotti 566; *65*
   malaccensis 564; *65*
   perspicillata 567; *65*
   sepiarium 565; *65*
   vanderbilti 568; *65*
Malacopteron
   affine 570; *65*
   albogulare 573; *65*
   cinereum 571; *65*
   magnirostre 569; *65*
   magnum 572; *65*
Malkoha
   Black-bellied 307; *38*
   Chestnut-bellied 308; *38*
   Chestnut-breasted 312; *38*
   Green-billed 309; *38*
   Raffles's 310; *38*
   Red-billed 311; *38*
   Rufous-bellied (308; *38*)
Mallard 71; *8*
Mannikin, see Munia
Megalaima
   armillaris 419; *49*
   australis 421; *49*
   chrysopogon 412; *48*
   corvina 411; *49*
   eximia 422; *48*
   haemacephala 423; *49*
   henricii 418; *48*
   javensis 415; *49*
   lineata 410; *48*
   monticola 417; *49*
   mystacophanos 414; *48*

Megalaima (cont.)
  oorti 416; *48*
  pulcherrima 420; *49*
  rafflesii 413; *48*
Megalurus palustris 662; *74*
Megapode
  Common (124; *14*)
  Orange-footed (123; *14*)
  Philippine (124; *14*)
  Reinwardt's (123; *14*)
  Tabon (124; *14*)
Megapodius
  cumingii 124; *14*
  freycinet (124; *14*)
  reinwardt 123; *14*
Meiglyptes
  tristis 438; *50*
  tukki 439; *50*
Melanochlora sultanea 556; *64*
Melanoperdix nigra 126; *14*
Merops
  leschenaulti 392; *46*
  ornatus 395; *46*
  philippinus 393; *46*
  superciliosus (393; *46*)
  viridis 394; *46*
Mesia, Silver-eared 606; *69*
Mesophoyx intermedia (43; *5*)
Metopidius indicus 166; *19*
Microhierax
  fringillarius 115; *12*
  latifrons 116; *12*
Micropternus brachyurus (428; *50*)
Milvus migrans 86; *9*
Minivet
  Ashy 487; *56*
  Fiery 489; *56*
  Flame (492; *56*)
  Grey-chinned 490; *56*
  Grey-throated (490; *56*)
  Indian (492; *56*)
  Lesser (488; *56*)
  Little (488; *56*)
  Mountain (490; *56*)
  Scarlet 492; *56*
  Small 488; *56*
  Sunda 491; *56*
  Yellow-throated (490; *56*)
Minla, *see* Crocias
Mirafra javanica 470; *54*
Monarch
  Black-naped 719; *79*
  Chestnut-winged (721; *79*)
  Maroon-breasted (720; *79*)
  Pacific (719; *79*)
  Small (719; *79*)
Monarch-Flycatcher
  Black-naped (719; *79*)
  Pacific (719; *79*)
  Small (719; *79*)
Monticola solitarius 635; *71*
Moorhen
  Black (160; *19*)
  Common 159; *19*
  Dusky 160; *19*
  Purple (161; *19*)
Motacilla
  alba 728; *81*
  cinerea 729; *81*
  flava 730; *81*
Mulleripicus pulverulentus 440; *51*
Munia
  Bornean (811; *87*)
  Black-faced 812; *87*
  Black-headed 814; *87*
  Chestnut 815; *87*
  Dusky 811; *87*
  Javan 810; *87*
  Moluccan (812; *87*)
  Nutmeg (813; *87*)
  Pale-headed (816; *87*)
  Red (804; *87*)
  Scaly-breasted 813; *87*
  Sharp-tailed (808; *87*)
  Spice (813; *87*)
  Spotted (813; *87*)
  Striated (808; *87*)
  White-backed (808; *87*)
  White-bellied 809; *87*
  White-breasted (809; *87*)
  White-headed 816; *87*
  White-rumped 808; *87*
Muscicapa
  dauurica 688; *76*
  ferruginea 689; *76*
  griseisticta 687; *76*
  indigo (691; *76*)
  latirostris (688; *76*)
  sibirica 686; *76*
  thalassina (690; *76*)
  williamsoni (688; *76*)
Muscicapella hodgsoni 712; *77*
Mycteria cinerea 54; *7*

Myiophoneus
  borneensis (637; *71*)
  caeruleus 638; *71*
  castaneus (637; *71*)
  flavirostris (638; *71*)
  glaucinus 637; *71*
  melanurus 636; *71*
Myiophonus, see Myiophoneus
Myna
  Bali 748; *82*
  Chinese Jungle (751; *82*)
  Common 749; *82*
  Crested 751; *82*
  Eastern Hill (752; *82*)
  Hill 752; *82*
  Indian (749; *82*)
  Javan 750; *82*
  Rothschild's (748; *82*)
  Talking 752; *82*
  White (748; *82*)
Myophonus, see Myiophoneus

Napothera
  atrigularis 579; *66*
  brevicauda (582; *66*)
  crassa 582; *66*
  epilepidota 583; *66*
  macrodactyla 580; *66*
  marmorata 581; *66*
  rufipectus 578; *66*
Nectarinia
  calcostetha 760; *83*
  hypogrammica (758; *83*)
  jugularis 761; *83*
  sperata 759; *83*
Needletail
  Brown (362; *42*)
  Brown-backed 362; *42*
  Giant (362; *42*)
  Grey-throated (361; *42*)
  Northern (360; *42*)
  Silver-backed 361; *42*
  White-backed (361; *42*)
  White-throated 360; *42*
  White-vented (361; *42*)
Nettapus coromandelianus 78; *8*
Night-Heron
  Black-crowned 45; *6*
  Common (45; *6*)
  Japanese 48; *6*

  Malay (47; *6*)
  Malayan 47; *6*
  Nankeen (46; *6*)
  Rufous 46; *6*
Nightjar
  Allied (350; *41*)
  Bonaparte's 351; *41*
  Grey 348; *41*
  Indian Jungle (348; *41*)
  Japanese (348; *41*)
  Jungle (348; *41*)
  Large-tailed 349; *41*
  Long-tailed (349; *41*)
  Salvadori's 352; *41*
  Savannah 350; *41*
  White-tailed (349; *41*)
Niltava
  caerulata (707; *78*)
  grandis 701; *78*
  sumatrana 702; *78*
  superba (708; *78*)
Niltava
  Great (701; *78*)
  Large 701; *78*
  Malaysian (702; *78*)
  Rufous-bellied (702; *78*)
  Rufous-vented 702; *78*
Ninox scutulata 336; *39*
Noddy
  Black 249; *30*
  Brown 248; *30*
  Common (248; *30*)
  White (250; *30*)
Numenius
  arquata 184; *22*
  madagascariensis 187; *22*
  minutus 186; *22*
  phaeopus 185; *22*
Nun Babbler, *see* Fulvetta
Nuthatch
  Blue 558; *64*
  Velvet-fronted 557; *64*
Nycticorax
  caledonicus 46; *6*
  nycticorax 45; *6*
Nyctyornis amictus 396; *46*

Oceanites oceanicus 11; *1*
Oceanodroma
  leucorhoa (12; *1*)
  matsudairae 13; *1*

Oceanodroma (cont.)
  melania (13; *1*)
  monorhis 12; *1*
Oculocincta squamifrons 798; *86*
Oenanthe oenanthe 634; *71*
Openbill
  Asian 55; *7*
  Asiatic (55; *7*)
  Oriental (55; *7*)
Oriole
  Asian Black-headed (539; *62*)
  Black 540; *62*
  Black-and-Crimson 541; *62*
  Black-hooded 539; *62*
  Black-naped 538; *62*
  Black-throated (537; *62*)
  Crimson-breasted (541; *62*)
  Dark-throated 537; *62*
  Indian Black-headed (539; *62*)
  Malaysian (537; *62*)
  Oriental Black-headed (539; *62*)
Oriolus
  chinensis 538; *62*
  cruentus 541; *62*
  hosii 540; *62*
  xanthonotus 537; *62*
  xanthornus 539; *62*
Orthotomus
  atrogularis 664; *75*
  cuculatus 668; *75*
  cucullatus (668; *75*)
  ruficeps 665; *75*
  sepium 666; *75*
  sericeus 667; *75*
  sutorius 663; *75*
Osprey 80; *9*
Otus
  angelinae 324; *40*
  bakkamoena (330; *40*)
  brookii 329; *40*
  enganensis 328; *40*
  lempiji 330; *40*
  mantanensis 326; *40*
  mentawi 331; *40*
  rufescens 321; *40*
  sagittatus 320; *40*
  scops (325; *40*)
  spilocephalus 322; *40*
  stresemanni 323; *40*
  sunia 325; *40*
  umbra 327; *40*

Owl
  Barn 318; *39*
  Bay (319; *39*)
  Short-eared 339; *39*
Owlet
  Barred (319; *39*)
  Chestnut-winged (335; *39*)
  Collared 334; *39*
  Javan 335; *39*
  Spadiced (335; *39*)

Pachycephala
  cinerea (725; *80*)
  grisola 725; *80*
  homeyeri 726; *80*
  hypoxantha 724; *80*
  pectoralis 727; *80*
Pachyptila
  belcheri 4; *1*
  desolata 5; *1*
  vittata (5; *1*)
Padda oryzivora 807; *87*
Painted Snipe, Greater 167; *20*
Pandion haliaetus 80; *9*
Papasula abbotti 22; *2*
Paradise-Flycatcher
  Asian 723; *79*
  Asiatic (723; *79*)
  Black (722; *79*)
  Japanese 722; *79*
Parakeet
  Long-tailed 282; *35*
  Moustached (281; *35*)
  Red-breasted 281; *35*
Parrot
  Blue-naped 287; *35*
  Blue-rumped 285; *35*
  Great-billed 286; *35*
  Island (286; *35*)
  Moluccan (286; *35*)
Parrotfinch
  Bamboo (806; *87*)
  Green-tailed (806; *87*)
  Long-tailed (806; *87*)
  Pin-tailed 805; *87*
  Tawny-breasted 806; *87*
Parrot-Finch, *see* Parrotfinch
Parrot-Munia, *see* Parrotfinch
Partridge
  Bar-backed (128; *14*)
  Black 126; *14*

INDEX 479

Bornean (131; *14*)
Chestnut-bellied 129; *14*
Chestnut-breasted (132; *14*)
Chestnut-necklaced 132; *14*
Crested 135; *15*
Crimson-headed 134; *14*
Ferruginous 133; *15*
Grey-bellied (128; *14*)
Grey-breasted 128; *14*
Javan (129; *14*)
Long-billed 125; *14*
Red-billed 130; *14*
Red-breasted 131; *14*
Roulroul (135; *15*)
Scaly-breasted (132; *14*)
Sumatran (128; *14*)
Parus major 555; *64*
Passer montanus 800; *88*
Pavo muticus 146; *15*
Peacock
  Green (146; *15*)
  Green-necked (146; *15*)
Peacock-Pheasant
  Bornean 144; *17*
  Bronze-tailed (143; *17*)
  Lesson's (143; *17*)
  Malay (144)
  Sumatran 143; *17*
Peafowl
  Green 146; *15*
  Green-necked (146; *15*)
Peewit (168; *20*)
Pelagodroma marina 14; *1*
Pelargopsis capensis 383; *44*
Pelecanus
  conspicillatus 19; *3*
  onocrotalus 17; *3*
  philippensis 18; *3*
Pelican
  Australian 19; *3*
  Eastern (17; *3*)
  European (17; *3*)
  Great White 17; *3*
  Grey (18; *3*)
  Philippine (18; *3*)
  Rosy (17; *3*)
  Spectacled (19; *3*)
  Spot-Billed 18; *3*
Pellorneum
  buettikoferi 561; *65*
  capistratum 559; *65*
  pyrrogenys 560; *65*

Peregrine (122; *12*)
Pericrocotus
  cinnamomeus 488; *56*
  divaricatus 487; *56*
  flammeus 492; *56*
  igneus 489; *56*
  miniatus 491; *56*
  solaris 490; *56*
Pernis
  apivorus (83; *9*)
  ptilorhynchus 83; *9*
Petrel
  Barau's 3; *1*
  Bulwer's 6; *1*
  Frigate (14; *1*)
  Jouanin's 7; *1*
  Wilson's (11; *1*)
Phaenicophaeus
  chlorophaeus 310; *38*
  curvirostris 312; *38*
  diardi 307; *38*
  javanicus 311; *38*
  sumatranus 308; *38*
  tristis 309; *38*
Phaethon
  lepturus 16; *2*
  rubricauda 15; *2*
Phalacrocorax
  carbo 25; *3*
  melanoleucus 26; *3*
  niger 27; *3*
  pygmaeus (27; *3*)
  sulcirostris 24; *3*
Phalarope
  Grey (223; *26*)
  Northern (224; *26*)
  Red (223; *26*)
  Red-necked 224; *26*
Phalaropus
  fulicaria 223; *26*
  lobatus 224; *26*
Pheasant
  Argus (145; *15*)
  Bulwer's 138; *16*
  Great Argus (145; *15*)
  Hoogerwerf's 140; *17*
  Salvadori's 139; *16*
  Sumatran (140; *17*)
  Wattled (138; *16*)
  White-tailed Wattled (138; *16*)
Philentoma
  pyrhopterum 721; *79*

Philentoma (cont.)
　velatum 720; *79*
Philentoma
　Chestnut-winged (721; *79*)
　Maroon-breasted 720; *79*
　Rufous-winged 721; *79*
Philomachus pugnax 220; *24*
Phodilus badius 319; *39*
Phylloscopus
　borealis 653; *73*
　coronatus 654; *73*
　inornatus 652; *73*
　trivirgatus 655; *73*
Picoides
　canicapillus (443; *50*)
　macei (442; *50*)
　moluccensis 444; *50*
Piculet
　Rufous 427; *50*
　Speckled 426; *50*
Picumnus innominatus 426; *50*
Picus
　canus 430; *51*
　chlorolophus 433; *51*
　flavinucha 431; *51*
　mentalis 434; *51*
　miniaceus 435; *51*
　puniceus 432; *51*
　vittatus 429; *51*
Pigeon
　Cinnamon-headed (255; *31*)
　Feral (272; *33*)
　Green Spectacled (251; *31*)
　Green-spectacled (251; *31*)
　Grey-cheeked (254; *31*)
　Hackled (280; *34*)
　Korthal's (252; *31*)
　Large (2599; *32*)
　Little (256; *31*)
　Metallic 270; *33*
　Nicobar 280; *34*
　Orange-breasted (258; *32*)
　Pink-necked (257; *32*)
　Rock 272; *33*
　Thick-billed (253; *31*)
　Wedge-tailed (252; *31*)
　White-throated (270; *33*)
　Yellow-bellied (251; *31*)
Pintail (68; *8*)
　Northern 68; *8*
Pintail-Pigeon, Sumatran (251; *31*)

Pipit
　Common 733; *81*
　Indian (732; *81*)
　Mynzbier's (734; *81*)
　Olive (732; *81*)
　Olive-backed 732; *81*
　Oriental (732; *81*)
　Petchora 734; *81*
　Red-throated 735; *81*
　Richard's (733; *81*)
　Siberian (734; *81*)
　Spotted (732; *81*)
Pitta
　arcuata (460; *53*)
　arquata 460; *53*
　baudii 461; *53*
　brachyura (462; *53*)
　caerulea 459; *53*
　elegans 465; *53*
　granatina 466; *53*
　guajana 469; *53*
　megarhyncha 464; *53*
　moluccensis 462; *53*
　nympha 463; *53*
　schneideri 458; *53*
　sordida 468; *53*
　venusta 467; *53*
Pitta
　Banded 469; *53*
　Black-and-Scarlet (467; *53*)
　Black-crowned 467; *53*
　Black-headed (468; *53*)
　Blue-banded 460; *53*
　Blue-headed 461; *53*
　Blue-tailed (469; *53*)
　Blue-winged 462; *53*
　Elegant 465; *53*
　Fairy 463; *53*
　Garnet 466; *53*
　Giant 459; *53*
　Graceful (467; *53*)
　Hooded 468; *53*
　Long-billed (464; *53*)
　Mangrove 464; *53*
　Moluccan (462; *53*)
　Schneider's 458; *53*
Pityriasis gymnocephala 553; *63*
Platalea regia 65; *7*
Platelea minor 64; *7*
Platylophus galericulatus 543; *63*
Platysmurus leucopterus 549; *63*
Plegadis falcinellus 63; *7*

Ploceus
  hypoxanthus 803; *88*
  manyar 802; *88*
  philippinus 801; *88*
Plover
  Black-bellied (172; *20*)
  Caspian (183; *21*)
  Common Ringed 174; *21*
  Geoffrey's (182; *21*)
  Golden (173; *20*)
  Green (168; *20*)
  Grey 172; *20*
  Javan 177; *21*
  Kentish 176; *21*
  Lesser (181; *21*)
  Little Ringed 175; *21*
  Long-billed 180; *21*
  Malay (179; *21*)
  Malaysian 179; *21*
  Mongolian 181; *21*
  Oriental 183; *21*
  Red-capped 178; *21*
  Ringed (174; *21*)
Pluvialis
  dominica (173; *20*)
  fulva 173; *20*
  squatarola 172; *20*
Pnoepyga pusilla 584; *66*
Pochard
  Australian White-eyed 77; *8*
Podiceps
  novaehollandiae (2; *26*)
  ruficollis (1; *26*)
Polyplectron
  chalcurum 143; *17*
  malacense (144)
  schleiermacheri 144; *17*
Pomatorhinus montanus 574; *66*
Pond-Heron
  Chinese 37; *6*
  Javan 38; *6*
  Javanese (38; *6*)
Porphyrio porphyrio 161; *19*
Porzana
  cinerea 156; *18*
  fusca 154; *18*
  paykullii 155; *18*
  pusilla 153; *18*
Pratincole
  Australian 227; *26*
  Eastern Collared (226; *26*)
  Isabelline (227; *26*)
  Large Indian (226; *26*)
  Long-legged (227; *26*)
  Oriental 226; *26*
Prinia
  atrogularis 669; *75*
  familiaris 672; *75*
  flaviventris 671; *75*
  inornata 670; *75*
  polychroa 673; *75*
  subflava (670; *75*)
Prinia
  Bar-winged 672; *75*
  Black-throated (669; *75*)
  Brown 673; *75*
  Hill 669; *75*
  Javan Brown (673; *75*)
  Plain 670; *75*
  Plain-coloured (670; *75*)
  Tawny (670; *75*)
  Tawny-flanked (670; *75*)
  White-browed (669; *75*)
  Yellow-bellied 671; *75*
Prion
  Antarctic 5; *1*
  Narrow-billed (4; *1*)
  Slender-billed 4; *1*
  Thin-billed (4; *1*)
Prionochilus
  maculatus 776; *85*
  percussus 778; *85*
  thoracicus 775; *85*
  xanthopygius 777; *85*
Psaltria exilis 554; *64*
Psarisomus dalhousiae 454; *52*
Pseudibis
  davisoni 62; *7*
  papillosa (62; *7*)
Psilopogon pyrolophus 409; *48*
Psittacula
  alexandri 281; *35*
  longicauda 282; *35*
Psittinus cyanurus 285; *35*
Pterodroma baraui 3; *1*
Pteruthius
  aenobarbus 608; *69*
  flaviscapis 607; *69*
Ptilinopus
  cinctus 263; *32*
  jambu 260; *32*
  melanospila 262; *32*
  porphyreus 261; *32*
Ptilocichla leucogrammica 576; *66*

Pueo (339; *39*)
Puffinus
  carneipes 9; *1*
  pacificus 10; *1*
Pycnonotus
  atriceps 504; *58*
  aurigaster 509; *58*
  bimaculatus 512; *59*
  brunneus 517; *59*
  cyaniventris 507; *58*
  erythrophthalmos 518; *59*
  eutilotus 510; *59*
  flavescens 513; *59*
  goiavier 514; *59*
  jocosus 508; *58*
  leucogrammicus 501; *58*
  melanicterus 505; *58*
  melanoleucos 503; *58*
  nieuwenhuisi 511; *59*
  plumosus 515; *59*
  simplex 516; *59*
  squamatus 506; *58*
  striatus (501; *58*)
  tympanistragus 502; *59*
  zeylanicus 500; *58*
Pygmy-Goose
  Cotton 78; *8*
  White (78; *8*)
Pygmy-Kingfisher, *see* Kingfisher
Pygmy-Owl
  Collared (334; *39*)
  Javan (335; *39*)
Pygmy-Triller
  Bar-winged (478; *55*)
  Black-winged (479; *55*)
  Pied (478; *55*)
Pygmy-Woodpecker
  Brown-capped (444; *50*)
  Fulvous-breast (442; *50*)
  Grey-capped (443; *50*)
  Grey-headed (443; *50*)
  Malaysian (444; *50*)
  Streak-bellied (442; *50*)
  Sunda (444; *50*)

Quail
  Asian Blue (127; *15*)
  Blue-breasted 127; *15*
  Painted (127; *15*)

Rail
  Banded (150; *18*)
  Blue-breasted (150; *18*)
  Slaty-breasted 150; *18*
  Water 149; *18*
Rail-Babbler
  Malay (615; *68*)
  Malaysian 615; *68*
Rainbow-bird (395; *46*)
Rallina
  eurizonoides 152; *18*
  fasciata 151; *18*
  paykulli (155; *18*)
Rallus
  aquaticus 149; *18*
  striatus (150; *18*)
Redshank
  Common 191; *23*
  Spotted 190; *23*
Reed-Warbler
  Black-browed 658; *73*
  Clamorous 656; *73*
  Eastern 656; *73*
  Heinroth's (656; *73*)
  Large-billed (656; *73*)
  Oriental (657; *73*)
  Von Schrenck's (658; *73*)
Reef-Egret (40; *5*)
  Eastern (40; *5*)
  Pacific 40; *5*
Reeve (220; *24*)
Reinwardtipicus validus 447; *51*
Raphidura leucopygialis 363; *42*
Rhinomyias
  brunneata 682; *76*
  gularis 685; *76*
  olivacea 681; *76*
  ruficauda 684; *76*
  umbratilis 683; *76*
Rhinoplax vigil (408; *47*)
Rhipidura
  albicollis 716; *79*
  euryura 715; *79*
  javanica 718; *79*
  perlata 717; *79*
  phoenicura 714; *79*
Rhizothera longirostris 125; *14*
Rhyticeros
  corrugatus (401; *47*)
  subruficollis (403; *47*)
  undulatus (402; *47*)
Ricebird (813; *87*)
Rimator malacoptilus 575; *66*
Riparia riparia 472; *54*

Robin
  Blue-tailed (620; *64*)
  Magpie 621; *70*
  Siberian Blue 619; *64*
  Sunda Blue 625; *70*
Rock-Thrush
  Blue 635; *71*
  Red-bellied (635; *71*)
Roller, Broad-billed (397; *46*)
Rollulus rouloul 135; *15*
Rostratula benghalensis 167; *24*
Roulroul (135; *15*)
Rubycheek (757; *83*)
Rubythroat, Siberian 618; *64*
Ruff 220; *24*

Sand-Martin 472; *54*
  Common (472; *54*)
  Gorgetted (472; *54*)
Sand-Plover
  Great (182; *21*)
  Greater 182; *21*
  Javan (177; *21*)
  Kentish (176; *21*)
  Large (182; *21*)
  Large-billed (182; *21*)
  Lesser (181; *21*)
  Malay (179; *21*)
  Malaysian (179; *21*)
  Mongolian (181; *21*)
  Red-capped (178; *21*)
Sanderling 217; *25*
Sandpiper
  Broad-billed 218; *25*
  Common 199; *23*
  Curlew 216; *25*
  Green 196; *23*
  Marsh 192; *23*
  Red-backed (215; *25*)
  Red-necked (211; *24*)
  Rufous-necked (211; *24*)
  Sharp-tailed 214; *25*
  Siberian Pectoral (214; *25*)
  Spoonbill (219; *25*)
  Spoon-billed 219; *25*
  Terek 198; *23*
  Wood 197; *23*
Sarcops calvus 753
Sasia abnormis 427; *50*
Saxicola
  caprata 633; *71*
  torquata 632; *71*

Scimitar-Babbler
  Chestnut-backed 574; *66*
  Yellow-billed (574; *66*)
Scolopax
  rusticola 207; *23*
  saturata 208; *22*
Scops-Owl
  Angeline's (324; *40*)
  Asian (325; *40*)
  Collared 330; *40*
  Enggano 328; *40*
  Japanese (330; *40*)
  Javan 324; *40*
  Malayan (320; *40*)
  Mantanani 326; *40*
  Mentaur (327; *40*)
  Mentawai 331; *40*
  Mountain 322; *40*
  Oriental 325; *40*
  Rajah 329; *40*
  Rajah's (329; *40*)
  Reddish 321; *40*
  Simeulue 327; *40*
  Simular (327; *40*)
  Simulu (327; *40*)
  Stresemann's 323; *40*
  White-fronted 320; *40*
Scrub-Robin, Malaysian (615; *68*)
Scrub-Warbler
  Friendly (680; *74*)
  Javan (679; *74*)
  Kinabalu (680; *74*)
  Mountain (679; *74*)
  Russet (679; *74*)
  Timor (679; *74*)
Scrubfowl
  Common (124; *14*)
  Orange-footed 123; *14*
  Philippine (124; *14*)
  Reinwardt's (123; *14*)
  Tabon 124; *14*
Sea-Eagle, White-bellied (88; *9*)
Sedge-Warbler
  Black-browed (658; *73*)
  Von Schrenck's (658; *73*)
Seicercus
  castaniceps 648; *73*
  grammiceps 649; *73*
  montis 650; *73*
Serilophus lunatus 453; *52*
Serin
  Indonesian (817; *88*)

Serin (*cont.*)
  Malay (817; *88*)
  Malaysian (817; *88*)
  Mountain 817; *88*
Serinus estherae 817; *88*
Serpent-Eagle
  Crested 93; *11*
  Kinabalu (94; *11*)
  Mountain 94; *11*
Setornis criniger 523; *60*
Shag, *see* Cormorant
Shama
  Rufous-tailed 624; *70*
  White-browed 623; *70*
  White-rumped 622; *70*
Shearwater
  Flesh-footed 9; *1*
  Pale-footed (9; *1*)
  Streaked 8; *1*
  Streak-headed (8; *1*)
  Wedge-tailed 10; *1*
  White-faced (8; *1*)
  White-fronted (8; *1*)
Shikra 104; *10, 13*
Shortwing
  Blue (617; *64*)
  Lesser 616; *64*
  Himalayan Blue (617; *64*)
  Indigo Blue (617; *64*)
  Mrs. La Touche's (616; *64*)
  White-browed 617; *64*
Shoveler (75; *8*)
  Common (75; *8*)
  Northern 75; *8*
Shrike
  Black-cappped (739; *80*)
  Black-headed (739; *80*)
  Brown 737; *80*
  Bristled (553; *63*)
  Great Grey (740)
  Long-tailed 739; *80*
  Northern 740
  Red-tailed (737; *80*)
  Rufous-headed (739; *80*)
  Schach (739; *80*)
  Thick-billed (739; *80*)
  Tiger 738; *80*
Shrike-Babbler
  Black-crowned (607; *69*)
  Chestnut-fronted 608; *69*
  Greater (739; *80*)
  Red-winged (607; *69*)
  White-browed 607; *69*
Sibia
  Long-tailed 612; *68*
  Spotted (611; *68*)
Sitta
  azurea 558; *64*
  frontalis 557; *64*
Skua
  Arctic (229; *27*)
  Long-tailed (230; *27*)
  Pamatorhine (228; *27*)
  Parasitic (229; *27*)
  Pomarine (228; *27*)
  South Polar 231; *27*
Skylark
  Eastern (471; *54*)
  Indian (471; *54*)
  Little (471; *54*)
  Oriental 471; *54*
  Small (471; *54*)
Snakebird, *see* Anhinga
Snake-Eagle, Short-toed (92; *10*)
Snipe
  Chinese (205; *24*)
  Common 206; *24*
  Fantail (205; *24*)
  Greater Painted 167; 24
  Marsh (205; *24*)
  Painted (167; *24*)
  Pintail 204; *24*
  Swinhoe's 205; *24*
Sparrow
  Eurasian Tree 800; *88*
  European Tree (800; *88*)
  Java 807; *87*
  Tree (800; *88*)
Sparrowhawk
  Asiatic (99; *10, 13*)
  Common (101; *10, 13*)
  Eurasian 101; *10, 13*
  Japanese 99; *10, 13*
  Japanese Lesser (99; *10, 13*)
  Northern (101; *10, 13*)
Spiderhunter
  Bornean 772; *84*
  Everett's (772; *84*)
  Greater Yellow-eared (769; *84*)
  Grey-breasted 771; *84*
  Kinabalu (772; *84*)
  Lesser Yellow-eared (770; *84*)
  Little 766; *84*

Long-billed 768; *84*
Spectacled 769; *84*
Thick-billed 767; *84*
Whitehead's 773; *84*
Yellow-eared 770; *84*
Spilornis
  cheela 93; *11*
  kinabaluensis 94; *11*
Spine-tailed Swift
  Brown (362; *42*)
  Brown-backed (362; *42*)
  Giant (362; *42*)
  Grey-throated (361; *42*)
  Silver-backed (361; *42*)
  White-backed (361; *42*)
Spinetail, Silver-rumped (363; *42*)
Spitzaetus nipalensis (113; *11*)
Spizaetus
  alboniger 113; *11*
  bartelsi 112; *11*
  cirrhatus 111; *11*
  nanus 114; *11*
Spoonbill
  Black-faced 64; *7*
  Lesser (64; *7*)
  Royal 65; *7*
Stachyris
  chrysaea 586; *67*
  erythroptera 595; *67*
  grammiceps 587; *67*
  leucotis 592; *67*
  maculata 591; *67*
  melanothorax 596; *67*
  nigriceps 588; *67*
  nigricollis 593; *67*
  poliocephala 589; *67*
  rufifrons 585; *67*
  striolata 590; *67*
  thoracica 594; *67*
Starlet, Daurian (745; *82*)
Starling
  Asian Glossy 742; *82*
  Asian Pied 746; *82*
  Asiatic Pied (746; *82*)
  Bali (748; *82*)
  Black-winged 747; *82*
  Chestnut-cheeked 744; *82*
  Chinese (743; *82*)
  Daurian (745; *82*)
  Grey-backed (743; *82*)
  Lesser Glossy (741; *82*)
  Philippine (742; *82*)
  Pied (746; *82*)
  Purple-backed 745; *82*
  Red-cheeked (744; *82*)
  Rothschild's (748; *82*)
  Short-tailed 741; *82*
  Violet-backed (744; *82*)
  White (748; *82*)
  White-shouldered 743; *82*
Stercorarius
  longicaudus 230; *27*
  maccormicki (231; *27*)
  parasiticus 229; *27*
  pomarinus 228; *27*
Sterna
  albifrons 244; *30*
  anaethetus 242; *29*
  bengalensis 246; *30*
  bergii 245; *30*
  bernsteini 247; *30*
  caspia 238; *28*
  dougallii 240; *29*
  fuscata 243; *29*
  hirundo 239; *29*
  nilotica 237; *28*
  sumatrana 241; *29*
  zimmermanni (247; *30*)
Stilt
  Black-winged 222; *26*
  Common (222; *26*)
  Pied (222; *26*)
  White-headed 221; *26*
Stiltia isabella 227; *26*
Stint
  Long-toed 213; *25*
  Red-necked (211; *24*)
  Rufous-necked 211; *25*
  Temminck's 212; *25*
Stonechat 632; *71*
Stork
  Black-necked 58
  Greater Adjutant (59; *7*)
  Green-necked (58)
  Lesser Adjutant (60; *7*)
  Milky 54; *7*
  Milky Wood (54; *7*)
  Open-billed (55; *7*)
  Storm's 57; *7*
  White-necked (56; *7*)
  Woolly-Necked 56; *7*
Storm-Petrel
  Matsudaira's 13; *1*

Storm-Petrel (*cont.*)
  Swinhoe's 12; *1*
  White-faced 14; *1*
  Wilson's 11; *1*
Streptopelia
  bitorquata 276; *34*
  chinensis 277; *34*
Strix
  leptogrammica 338; *39*
  seloputo 337; *39*
Stubtail
  Bornean 677; *74*
  Short-tailed (677; *74*)
  Whitehead's (677; *74*)
Sturnus
  contra 746; *82*
  melanopterus 747; *82*
  philippensis 744; *82*
  sinensis 743; *82*
  sturninus 745; *82*
Sula
  abbotti (22; *2*)
  dactylatra 21; *2*
  leucogaster 23; *2*
  sula 20; *2*
Sunbird
  Black-throated (761; *83*)
  Blue-naped (758; *83*)
  Brown-throated (755; *83*)
  Copper-throated 760; *83*
  Crimson 763; *83*
  Grey-throated (755; *83*)
  Javan 764; *83*
  Kuhl's (762; *83*)
  Macklot's (760; *83*)
  Olive-backed 761; *83*
  Plain 754; *83*
  Plain-throated 755; *83*
  Purple-naped 758; *83*
  Purple-throated 759; *83*
  Red-throated 756; *83*
  Ruby-cheeked 757; *83*
  Rufous-shouldered (756; *83*)
  Rufous-throated (756; *83*)
  Scarlet 765; *83*
  Shelley's (756; *83*)
  Temminck's 765; *83*
  White-flanked 762; *83*
  Yellow-bellied (761; *83*)
  Yellow-breasted (761; *83*)
Surniculus lugubris 305; *37*
Swallow
  Bank (472; *54*)
  Barn 473; *54*
  Ceylon (475; *54*)
  Common (473; *54*)
  Eastern House (474; *54*)
  Greater Striated (476; *54*)
  House (473; *54*)
  Least (474; *54*)
  Lesser Striated (475; *54*)
  Mosque (476; *54*)
  Pacific 474; *54*
  Red-rumped 475; *54*
  Rustic (473; *54*)
  Small House (474; *54*)
  Striated 476; *54*
Swamphen, Purple 161; *19*
Swift
  Asian Palm 366; *42*
  Fork-tailed 364; *42*
  House (365; *42*)
  Little 365; *42*
  Needle-tailed, *see* Needletail
  Pacific (356; *42*)
  Silver-rumped (363; *42*)
  Spine-tailed (360; *42*)
  White-rumped (364; *42*)
Swiftlet
  Black-nest 355; *42*
  Brown-rumped (354; *42*)
  Cave 359; *42*
  Edible-nest 354; *42*
  German's (354; *42*)
  Giant 353; *42*
  Glossy 358; *42*
  Grey-rumped (354; *42*)
  Himalayan (357; *42*)
  Mossy (356)
  Mossy-nest (356)
  Sunda (356)
  Thunberg's (356)
  Volcano 357; *42*
  White-bellied (358; *42*)
  White-nest (354; *42*)

Tachybaptus
  novaehollandiae 2; *26*
  ruficollis 1; *26*
Tailorbird
  Ashy 665; *75*
  Bali (666; *75*)
  Black-necked (664; *75*)
  Common 663; *75*

Dark-cheeked (664; *75*)
Dark-necked 664; *75*
Golden-headed (668; *75*)
Javan (666; *75*)
Long-tailed (663; *75*)
Mountain 668; *75*
Olive-backed 666; *75*
Red-headed (665; *75*)
Rufous-backed (664; *75*)
Rufous-tailed 667; *75*
Tanygnathus
 lucionensis 287; *35*
 megalorhynchos 286; *35*
Tarsiger cyanurus 620; *64*
Tattler
 Grey-rumped (200; *22*)
 Grey-tailed 200; *22*
 Polynesian (200; *22*)
 Siberian (200; *22*)
Teal
 Common (69; *8*)
 Cotton (78; *8*)
 Garganey (74; *8*)
 Green-winged 69; *8*
 Indonesian Grey (70; *8*)
 Sunda 70; *8*
Tephrodornis
 gularis (480; *55*)
 virgatus 480; *55*
Tern
 Black-naped 241; *29*
 Bridled 242; *29*
 Bronze-winged (242; *29*)
 Caspian 238; *28*
 Common 239; *29*
 Crested (246; *30*)
 Fairy (250; *28*)
 Gull-billed 237; *28*
 Little 244; *30*
 Noddy (248; *30*)
 Roseate 240; *29*
 Sooty 243; *29*
 Swift (245; *30*)
 Whiskered 235; *28*
 White (250; *28*)
 White-winged 236; *28*
 White-winged Black (236; *28*)
Terpsiphone
 atrocaudata 722; *79*
 paradisi 723; *79*
Tesia superciliaris 676; *74*
 Eyebrowed (676; *74*)

Javan 676; *74*
Malaysian (676; *74*)
Thick-knee
 Beach 225; *26*
 Great (225; *26*)
Thickhead
 Bornean (724; *80*)
 Bornean Mountain (724; *80*)
 Common Golden (727; *80*)
 Golden (727; *80*)
 Grey (725; *80*)
 Mangrove (725; *80*)
 Palawan (725; *80*)
 White-bellied (725; *80*)
 White-vented (726; *80*)
Threskiornis melanocephalus 61; *7*
Thrush
 Andromeda (642; *72*)
 Black-breasted (629; *55*)
 Chestnut-capped 639; *72*
 Chestnut-headed (639; *72*)
 Dark (645; *72*)
 Grey-headed (645; *72*)
 Enggano (639; *72*)
 Everett's 641; *72*
 Eyebrowed 645; *72*
 Island 646; *72*
 Kuhl's (639; *72*)
 Orange-headed 640; *72*
 Scaly 644; *72*
 Siberian 643; *72*
 Small-billed (644; *72*)
 Sunda 642; *72*
 Tiger (644; *72*)
 White-browed (645; *72*)
 White-throated (640; *72*)
 White's (644; *72*)
 White's Scaly (644; *72*)
Timalia pileata 600; *68*
Tit
 Cineraceous (555; *64*)
 Great 555; *64*
 Grey (555; *64*)
 Pygmy 554; *64*
 Japanese (555; *64*)
 Sultan 556; *64*
Tit-Babbler
 Fluffy-backed 599; *68*
 Grey-cheeked 597; *68*
 Grey-faced (597; *68*)
 Plume-backed (599; *68*)
 Striated (598; *68*)

Tit-Babbler (*cont.*)
  Striped 598; *68*
  Stripe-throated (598; *68*)
  Yellow-breasted (598; *68*)
  Yellow-throated (597; *68*)
Todirhamphus
  chloris 389; *45*
  sanctus 390; *45*
Tree-Babbler
  Black-necked (593; *67*)
  Black-throated (588; *67*)
  Black-throated (593; *67*)
  Brown-headed (569; *65*)
  Chestnut-rumped (591; *67*)
  Chestnut-winged (595; *67*)
  Crescent-chested (596; *67*)
  Greater Red-headed (572; *65*)
  Grey (588; *67*)
  Grey-breasted (573; *65*)
  Grey-headed (589; *67*)
  Grey-throated (588; *67*)
  Javan (587; *67*)
  Lesser Red-headed (571; *65*)
  Moustached (569; *65*)
  Pearl-chested (596; *67*)
  Pearly-cheeked (596; *67*)
  Plain (570; *65*)
  Red-headed (572; *65*)
  Red-rumped (591; *67*)
  Red-winged (595; *67*)
  Rufous-crowned (572; *65*)
  Scaly-crowned (571; *65*)
  Sooty-capped (570; *65*)
  Sooty-headed (570; *65*)
  Spot-necked (590; *67*)
  Spotted (590; *67*)
  White-bibbed (594; *67*)
  White-breasted (587; *67*)
  White-collared (594; *67*)
  White-eared (592; *67*)
  White-necked (592; *67*)
  White-throated (573; *65*)
Tree-Duck
  Lesser (66; *8*)
  Whistling (67; *8*)
Tree-Partridge
  Bornean (131; *14*)
  Chestnut-breasted (132; *14*)
  Chestnut-necklaced (132; *14*)
  Red-billed (130; *14*)
  Red-breasted (131; *14*)
  Scaly-breasted (132; *14*)

Tree-Pipit
  Indian (732; *81*)
  Olive (732; *81*)
  Olive-backed (732; *81*)
  Oriental (732; *81*)
  Spotted (732; *81*)
Treepie
  Black (548; *63*)
  Black Racket-tailed (548; *63*)
  Bronzed (548; *63*)
  Bornean 547; *63*
  Grey (546; *63*)
  Malaysian (546; *63*)
  Mountain (546; *63*)
  Racket-tailed 548; *63*
  Sumatran 546; *63*
Treeswift
  Grey-rumped 367; *42*
  Lesser (368; *42*)
  Whiskered 368; *42*
Treron
  bicincta 258; *32*
  capellei 259; *32*
  curvirostra 253; *31*
  fulvicollis 255; *31*
  griseicauda 254; *31*
  olax 256; *31*
  oxyura 251; *31*
  sphenura 252; *31*
  vernans 257; *32*
Trichastoma
  abbotti (566; *65*)
  buettikoferi (561; *65*)
  bicolor 563; *65*
  malaccense (564; *65*)
  perspicillatum (567; *65*)
  pyrrogenys (560; *65*)
  rostratum 562; *65*
  sepiarium (565; *65*)
  tickelli (560; *65*)
  vanderbilti (568; *65*)
Trichixos pyrrhopygus 624; *70*
Trichoglossus haematodus 283; *35*
Tricholestes criniger 524; *60*
Triller
  Black-breasted (629; *55*)
  Pied 485; *55*
  Sueur's (486; *55*)
  White-shouldered 486; *55*
  White-winged (486; *55*)
Tringa
  brevipes 200; *22*

cinereus 198; *23*
erythropus 190; *23*
flavipes 195; *23*
glareola 197; *23*
guttifer 194; *23*
hypoleucos 199; *23*
nebularia 193; *23*
ochropus 196; *23*
stagnatilis 192; *23*
totanus 191; *23*
Trogon
  Blue-rumped (369; *43*)
  Blue-tailed 369; *43*
  Cinnamon-rumped 373; *43*
  Diard's 371; *43*
  Orange-breasted 375; *43*
  Red-headed 376; *43*
  Red-naped 370; *43*
  Red-rumped (374; *43*)
  Scarlet-rumped 374; *43*
  Whitehead's 372; *43*
Tropicbird
  Red-tailed 15; *2*
  White-tailed 16; *2*
  Yellow-billed (16; *2*)
Turdus
* obscurus* 645; *72*
* poliocephalus* 646; *72*
Turnix
* suscitator* 148; *15*
* sylvatica* 147; *15*
Turnstone, Ruddy 201; *22*
Tyto alba 318; *39*

Upupa epops 398; *46*
Urosphena whiteheadi 677; *74*

Vanellus
* cinereus* 169; *20*
* indicus* 171; *20*
* macropterus* 170; *20*
* vanellus* 168; *20*
Vulture
  Indian White-backed (91)
  White-backed (91)
  White-rumped 91

Wagtail
  Common Pied (728; *81*)
  Dark-headed (730; *81*)
  Forest 731; *81*
  Grey 729; *81*
  Grey-headed (730; *81*)
  Masked (728; *81*)
  Pied 728; *81*
  Siberian (730; *81*)
  Siberian Yellow (730; *81*)
  Tree (731; *81*)
  White (728; *81*)
  Yellow 730; *81*
Warbler
  Arctic 653; *73*
  Bamboo (651; *73*)
  Chestnut (648; *73*)
  Chestnut-crowned 648; *73*
  Chestnut-headed (648; *73*)
  Grey-naped (659; *74*)
  Inornate 652; *73*
  Kinabalu Friendly (680; *74*)
  Lanceolated 661; *74*
  Middendorf's 660; *74*
  Pallas's 659; *74*
  Pleske's (659)
  Streaked (661; *74*)
  Striated 662; *74*
  Sunda 649; *73*
  White-throated (651; *74*)
  Willow (653; *74*)
  Yellow-bellied 651; *73*
  Yellow-breasted 650; *73*
  Yellow-browed (652; *73*)
Watercock 158; *19*
Waterhen
  Purple (161; *19*)
  White-breasted 157; *18*
Waxbill, Strawberry (804; *87*)
Weaver
  Asian Golden 803; *88*
  Baya 801; *88*
  Common (801; *88*)
  Golden (803; *88*)
  Manyar (802; *88*)
  Streaked 802; *88*
  Striated (802; *88*)
Wheatear 634; *71*
Whimbrel 185; *22*
  Little (186; *22*)
Whistler
  Bornean, 724; *80*
  Bornean Mountain (724; *80*)
  Common Golden (727; *80*)

Whistler (*cont.*)
  Golden 727; *80*
  Grey (725; *80*)
  Mangrove 725; *80*
  Palawan (725; *80*)
  White-bellied (725; *80*)
  White-vented 726; *80*
Whistling-Duck
  Lesser 66; *8*
  Wandering 67; *8*
Whistling-Thrush
  Blue 638; *71*
  Brown-winged (637; *71*)
  Shiny 636; *71*
  Sumatran (637; *71*)
  Sunda 637; *71*
White-eye
  Banda (796; *86*)
  Black-capped 792; *86*
  Engano (791; *86*)
  Enggano 791; *86*
  Everett's 793; *86*
  Grey-throated (797; *86*)
  Indian (790; *86*)
  Java (797; *86*)
  Javan 795; *86*
  Javan Grey-throated 797; *86*
  Lemon-bellied 796; *86*
  Mangrove (796; *86*)
  Moluccan (796; *86*)
  Mountain 799; *86*
  Oriental 790; *86*
  Pale (796; *86*)
  Pygmy 798; *86*
  Pygmy Grey (798; *86*)
  Small (790; *86*)
  Yellow (795; *86*)
  Yellow-bellied (790; *86*)
White Tern, Common 250; *28*
Wigeon (72; *8*)
  Eurasian 72; *8*
Wood-Duck, White-winged (79; *8*)
Wood-Owl
  Brown 338; *39*
  Spotted 337; *39*
Wood-Partridge
  Black (126; *14*)
  Crested (135; *15*)
  Ferruginous (133; *15*)
  Roulroul (135; *15*)
Wood-Pigeon
  Grey (271; *33*)
  Metallic (270; *33*)
  Silver (271; *33*)
  Silvery 271; *33*
  White-throated (270; *33*)
Wood-Shrike
  Bald-headed (553; *63*)
  Brown-tailed (480; *55*)
  Hook-billed (480; *55*)
  Large (480; *55*)
Wood-Stork, Milky (54; *7*)
Wood-Swallow
  Lesser (736; *80*)
  White-breasted 736; *80*
  White-rumped (736; *80*)
Woodcock
  Dusky (208; *22*)
  East Indian (208; *22*)
  Eurasian 207; *23*
  Horsfield's (208; *22*)
  Indonesian (208; *22*)
  Rufous 208; *22*
Woodpecker
  Banded 435; *51*
  Black-naped Green (430; *51*)
  Brown-capped (444; *50*)
  Buff-necked 439; *50*
  Buff-rumped 438; *50*
  Checker-throated 434; *51*
  Chequer-throated (434; *51*)
  Common Golden-backed (436; *50*)
  Crimson-winged 432; *51*
  Fulvous-breasted 442; *50*
  Fulvous-rumped (438; *50*)
  Great Black (441; *50*)
  Great Slaty 440; *51*
  Greater Golden-backed (448; *51*)
  Greater Yellow-naped (431; *51*)
  Grey-and-Buff 445; *50*
  Grey-breasted (445; *50*)
  Grey-capped 443; *50*
  Grey-faced (430; *51*)
  Grey-headed 430; *51*
  Laced 429; *51*
  Laced Green (429; *51*)
  Lesser Yellow-naped (433; *51*)
  Malaysian (444; *50*)
  Maroon 446; *50*
  Olive-backed 437; *50*
  Orange-backed 447; *51*
  Rufous 428; *50*
  Streak-bellied (442; *50*)

Sunda 444; *50*
White-bellied 441; *50*
Woodshrike
  Bald-headed (553; *63*)
  Brown-tailed (480; *55*)
  Hook-billed (480; *55*)
  Large 480; *55*
Wren-Babbler
  Black-throated 579; *66*
  Bornean 576; *66*
  Borneo Streaked (582; *66*)
  Brown (584; *66*)
  Eye-browed 583; *66*
  Large 580; *66*
  Large-footed (580; *66*)
  Long-billed 575; *66*
  Lesser Scaly-breasted (584; *66*)
  Marbled 581; *66*
  Mountain 582; *66*
  Mueller's (581; *66*)
  Plain (670; *75*)
  Plain-coloured (670; *75*)
  Pygmy 584; *66*
  Red-breasted (578; *66*)
  Rufous-breasted (578; *66*)
  Rusty-breasted 578; *66*
  Striped 577; *66*
  Sumatran (575; *66*)
  Tawny (670; *75*)
  Tawny-flanked (670; *75*)
  White-throated (575; *66*)
Wren-Warbler
  Bar-winged (672; *75*)
  Black-throated (669; *75*)
  Brown (673; *75*)
  Hill (669; *75*)
  Javan Brown (673; *75*)

White-browed (669; *75*)
Yellow-bellied (671; *75*)

Xema sabini 233; *27*
Xenorhynchus (58)
Xenus cinereus (198; *23*)

Yellowlegs, Lesser 195; *23*
Yellownape
  Greater 431; *51*
  Lesser 433; *51*
Yuhina
  everetti 613; *68*
  zantholeuca 614; *68*
Yuhina
  Chestnut-crested 613; *68*
  Chestnut-headed (613; *68*)
  White-bellied 614; *68*

Zoothera
  andromedae 642; *72*
  citrina 640; *72*
  dauma 644; *72*
  everetti 641; *72*
  interpres 639; *72*
  sibirica 643; *72*
Zosterops
  atricapilla 792; *86*
  atricapillus (792; *86*)
  chloris 796; *86*
  everetti 793; *86*
  flava (795; *86*)
  flavus 795; *86*
  montanus 794; *86*
  palpebrosa 790; *86*
  salvadorii 791; *86*